Targeting Angiogenesis, Inflammation, and Oxidative Stress in Chronic Diseases

Targeting Angiogenesis, Inflammation, and Oxidative Stress in Chronic Diseases

Edited by

Tapan Behl
Amity School of Pharmaceutical Sciences, Amity University, Mohali, Punjab, India

Sukhbir Singh
Department of Pharmaceutics, MM College of Pharmacy, Maharishi Markandeshwar (Deemed to be University), Mullana, Ambala, Haryana, India

Neelam Sharma
Department of Pharmaceutics, MM College of Pharmacy, Maharishi Markandeshwar (Deemed to be University), Mullana, Ambala, Haryana, India

Series Editor

Domenico Ribatti
Department of Translational Biomedicine and Neuroscience, University of Bari Medical School, Bari, Italy

ACADEMIC PRESS
An imprint of Elsevier

ELSEVIER

Contents

Contributors

Ansab Akhtar School of Health Sciences & Technology, UPES, Dehradun, Uttarakhand, India

Sulaiman Mohammed Alnaseer Department of Pharmacology and Toxicology, Unaizah College of Pharmacy, Qassim University, Qassim, Saudi Arabia

Sumel Ashique Department of Pharmaceutics, Pandaveswar School of Pharmacy, Pandaveswar, West Bengal, India

Tapan Behl Amity School of Pharmaceutical Sciences, Amity University, Mohali, Punjab, India

Manorama Bhandari School of Health Sciences & Technology, University of Petroleum and Energy Studies, Dehradun, Uttarakhand, India

Bhupinder Bhyan Department of Pharmaceutics, Swift School of Pharmacy, Rajpura, Punjab, India

Gunjan Vasant Bonde School of Health Sciences & Technology, University of Petroleum and Energy Studies, Dehradun, Uttarakhand, India

Silpi Chanda Department of Pharmacognosy, Amity Institute of Pharmacy, Lucknow; Amity University, Noida, UP, India

Diksha Chugh Department of Pharmacology, Delhi Pharmaceutical Sciences and Research University (DPSRU), New Delhi, India

Biswajit Dash School of Health and Medical Sciences, Adamas University, Kolkata, West Bengal, India

Pooja Dhami School of Health Sciences and Technology, University of Petroleum and Energy Studies, Dehradun, Uttarakhand, India

Bhavya Dhawan Amity Institute of Pharmacy (AIP), Amity University Uttar Pradesh, Noida, Uttar Pradesh, India

Priya Dhiman Department of Pharmaceutics, MM College of Pharmacy, Maharishi Markandeshwar (Deemed to be University), Mullana-Ambala, Haryana, India

Shubham Dwivedi School of Health Sciences & Technology, UPES, Dehradun, Uttarakhand, India

Arshad Farid Gomal Center of Biochemistry and Biotechnology, Gomal University, Dera Ismail Khan, Khyber Pakhtunkhwa, Pakistan

Ashish Garg Department of Pharmaceutics, Guru Ramdas Khalsa Institute of Science and Technology (Pharmacy), Jabalpur, Madhya Pradesh, India

Ikmeet Kaur Grewal Department of Pharmacy, Government Medical College, Patiala, Punjab, India

Sonam Grewal Department of Pharmaceutics, MM College of Pharmacy, Maharishi Markandeshwar (Deemed to be University), Mullana-Ambala, Haryana, India

Divyanshi Gupta PSIT-Pranveer Singh Institute of Technology (Pharmacy), Kanpur, Uttar Pradesh, India

Sangeetha Gupta Amity Institute of Pharmacy (AIP), Amity University Uttar Pradesh, Noida, Uttar Pradesh, India

Shraddha Manish Gupta Department of Pharmaceutical Sciences, School of Health Sciences and Technology, UPES, Dehradun, Uttarakhand, India

Soumya Gupta School of Health Sciences and Technology, University of Petroleum and Energy Studies, Dehradun, Uttarakhand, India

Sumeet Gupta Department of Pharmacology, MM College of Pharmacy, Maharishi Markandeshwar (Deemed to be University), Mullana-Ambala, Haryana, India

Afzal Hussain Department of Pharmaceutics, College of Pharmacy, King Saud University, Riyadh, Saudi Arabia

Vaibhav Jaiswal IES Institute of Pharmacy, IES University, Bhopal, Madhya Pradesh, India

Mohd Masih Uzzaman Khan Department of Pharmaceutical Chemistry and Pharmacognosy, Unaizah College of Pharmacy, Unaizah, Saudi Arabia

Dhruv Kumar School of Health Sciences and Technology, UPES, Dehradun, Uttarakhand, India

Sanjay Kumar Department of Food Science and Technology, Graphic Era (Deemed to be University), Dehradun, Uttarakhand, India

Shubneesh Kumar Department of Pharmaceutics, Bharat Institute of Technology, School of Pharmacy, Meerut, Uttar Pradesh, India

Vinod Kumar Department of Food Science and Technology, Graphic Era (Deemed to be University), Dehradun, Uttarakhand, India; Peoples' Friendship University of Russia (RUDN University), Moscow, Russian Federation

Kiran Chandrakant Mahajan Sgmspm's Sharadchandra Pawar College of Pharmacy, Pune, Maharashtra, India

Gaurav Malik Department of Pharmaceutics, MM College of Pharmacy, Maharishi Markandeshwar (Deemed to be University), Mullana-Ambala, Haryana, India

Irfan Ahmad Malik Aman Pharmacy College, Jhunjhunu, Rajasthan, India

Shubhrajit Mantry Sgmspm's Sharadchandra Pawar College of Pharmacy, Pune, Maharashtra, India

Arun Kumar Mishra Pharmacy Academy, IFTM University, Moradabad, Uttar Pradesh, India

Neeraj Mishra Amity Institute of Pharmacy, Amity University, Gwalior, Madhya Pradesh, India

Mukesh Nandave Department of Pharmacology, Delhi Pharmaceutical Sciences and Research University (DPSRU), New Delhi, India

Sudeep Pukale Lupin Research Park, Nande, Maharashtra, India

Sandeep Rathor Department of Pharmaceutics, MM College of Pharmacy, Maharishi Markandeshwar (Deemed to be University), Mullana-Ambala, Haryana, India

Shruti Rathore LCIT School of Pharmacy, Bilaspur, Chhattisgarh, India

Indra Rautela Department of Biotechnology, School of Applied and Life Sciences (SALS), Uttaranchal University, Dehradun, Uttarakhand, India

B.S. Rawat Department of Physics, School of Applied and life Sciences, Uttaranchal University, Dehradun, India

Maryam Sarwat Amity Institute of Pharmacy (AIP), Amity University Uttar Pradesh, Noida, Uttar Pradesh, India

Deepika Sharma Department of Pharmaceutical Sciences, School of Health Sciences and Technology, UPES, Dehradun, Uttarakhand, India

Neelam Sharma Department of Pharmaceutics, MM College of Pharmacy, Maharishi Markandeshwar (Deemed to be University), Mullana-Ambala, Haryana, India

Ankit Shokeen Amity Institute of Pharmacy (AIP), Amity University Uttar Pradesh, Noida, Uttar Pradesh, India

Sukhbir Singh Department of Pharmaceutics, MM College of Pharmacy, Maharishi Markandeshwar (Deemed to be University), Mullana-Ambala, Haryana, India

Swati Singh School of Health Sciences and Technology, UPES, Dehradun, Uttarakhand, India

Raj Kumar Tiwari Department of Pharmacognosy, Era College of Pharmacy, Era University, Lucknow, UP, India

Neha Tiwary Department of Pharmaceutics, MM College of Pharmacy, Maharishi Markandeshwar (Deemed to be University), Mullana-Ambala, Haryana, India

Jyoti Upadhyay School of Health Sciences and Technology, University of Petroleum and Energy Studies, Dehradun, Uttarakhand, India

Shuchi Upadhyay Department of Allied Health Sciences, School of Health Science and Technology, University of Petroleum and Energy Studies UPES, Dehradun, Uttarakhand, India

Rohini Verma School of Health Sciences and Technology, University of Petroleum and Energy Studies, Dehradun, Uttarakhand, India

Himangi Vig PSIT-Pranveer Singh Institute of Technology (Pharmacy), Kanpur, Uttar Pradesh, India

Ankita Wal PSIT-Pranveer Singh Institute of Technology (Pharmacy), Kanpur, Uttar Pradesh, India

Pranay Wal PSIT-Pranveer Singh Institute of Technology (Pharmacy), Kanpur, Uttar Pradesh, India

Shahid Nazir Wani Chitkara College of Pharmacy, Chitkara University, Rajpura, Punjab; Aman Pharmacy College, Jhunjhunu, Rajasthan, India

Ishrat Zahoor Department of Pharmaceutics, MM College of Pharmacy, Maharishi Markandeshwar (Deemed to be University), Mullana-Ambala, Haryana, India

About the editors

Tapan Behl
Professor, Amity School of Pharmaceutical Sciences, Amity University, Mohali, Punjab, India

Dr. Tapan Behl has a rich experience in Research and Teaching and is deft in conducting Animal Experimentation, Diabetes and Associated Complications, Neurological Studies, Hepatological Studies, Scientific Literature Extraction, Retina Isolation, Retro-orbital Plexus, and Surgical Procedures. He keeps abreast of recent developments by reading the current literature, liaises with colleagues, and has the ability to manage multitasking projects both as an individual and as part of a team. He completed his PhD in Medical Pharmacology in 2017 from Vallabhbhai Patel Chest Institute, University of Delhi, India, and is also a recipient of the Gold Medal for academic excellence during a Master's tenure. To date, he has over 350 publications in peer-reviewed journals with a cumulative impact factor of over 2000, 5692 citations, an h-index of 36, an i10-index of 167, and 10 patents to his credit. Dr. Behl is efficient in completing projects assigned on time to government and private agencies. He also handled the project from CCRAS, Ministry of AYUSH, Government of India, entitled "Effect of Indian Almond and Sweet Almond in Diabetes induced Nephropathy and Cataract in Rats." He has received an International Travel Grant from the Science and Engineering Research Board, Department of Science and Technology, Government of India, to present research work in the prestigious 18th World Congress of Basic and Clinical Pharmacology in Kyoto, Japan. His many reputed awards include the Young Scientist Award, Tech Research Award, Young Researcher Award, Fellow of Medical Research Council, and Fellowship in Health Promotion and Education. He serves as an Editorial Board Member of reputed journals including Nature. He has guided 3 PhD students and 15 M. Pharmacy students, and 4 research scholars are pursuing their PhD under his guidance. Dr. Behl is guiding students for the award of Masters and PhD degree. He is listed among the top 2% of scientists worldwide per the list released by Stanford University and Elsevier BV.

Sukhbir Singh
Professor, Department of Pharmaceutics, MM College of Pharmacy, Maharishi Markandeshwar (Deemed to be University), Haryana, India.

Dr. Sukhbir Singh, PhD in Pharmaceutical Science, has over 15 years' experience in academics and research. He is a highly enthused, self-determined, and organized pharmaceutical research professional. He has 200 publications in national/international journals of repute, a Scopus h-index of 21, an i10-index of 25, 9 books, and 15 book chapters and filed 19 patents to his credit. He has presented/published 40 review/research abstracts in various national/international conferences. He has been listed in the top 2% of world scientists as per the ranking released by Stanford University, California, USA, and Elsevier. He has completed four consultancy projects on herbal/homeopathic formulation. He and his team received the IEDC, Government of India, a prototyping grant of Rs 2.5 lacs for a project related to management of ARDS in COVID. He has guided 9 PhD students and 30 M. Pharmacy students, and 6 PhD students are pursuing their PhD under his guidance. His areas of interest in research are nanotechnology/liposomes/niosomes/SEDDS-based formulation development and solubility enhancement of BCS class II drugs. He has expertise in the operation of various statistical analysis tools.

Neelam Sharma
Professor, Department of Pharmaceutics, MM College of Pharmacy, Maharishi Markandeshwar (Deemed to be University), Mullana-Ambala 133207, Haryana, India.

Dr. Neelam Sharma, PhD in Pharmaceutical Science, has over 14 years' experience in academics and research. She is a highly self-determined and organized pharmaceutical research professional. She has 170 publications in various national/international journals of repute, a Scopus h-index of 17, an i10-index of 20, 9 books, 15 book chapters, and filed 21 patents to her credit. She has presented/published 35 review/research abstracts in national/international conferences. She has completed four consultancy projects on herbal/homeopathic formulation and received the IEDC, Government of India, a prototyping grant of Rs 2.5 lacs for a project related to management of ARDS in COVID. She has guided 2 PhD students and 20 M. Pharmacy students, and 6 PhD students are pursuing their PhD under her guidance. Her areas of interest are liposomes/niosomes/SEDDS/nanotechnology-based formulation development and solubility enhancement of BCS class II drugs. She has expertise in the operation of various statistical analysis tools.

Understanding the role of angiogenesis, inflammation and oxidative stress in diabetes mellitus: Insights into the past, present and future trends

Sandeep Rathor[a], Sukhbir Singh[a], Neelam Sharma[a], Ishrat Zahoor[a], and Bhupinder Bhyan[b]
[a]Department of Pharmaceutics, MM College of Pharmacy, Maharishi Markandeshwar (Deemed to be University), Mullana-Ambala, Haryana, India, [b]Department of Pharmaceutics, Swift School of Pharmacy, Rajpura, Punjab, India

1 Introduction

Diabetes is a group of metabolic disorders marked over time by an excessive increase in blood glucose levels and becomes a serious health illness with considerable social and economic complications. Diabetes mellitus type 1 and type 2 are dangerous long-term conditions in which the body is unable to make any or enough insulin or fails to utilize the insulin that is generated properly [1,2]. Hyperglycemia, hyperlipidemia, abnormalities in the metabolism of glucose and lipids, and changed liver enzyme levels are all related to diabetes. Diabetes has now been linked to chronic hyperglycemia, which is recognized by persistently and abnormally high postprandial blood glucose levels [3]. With a population of 77 million, the International Diabetes Federation evaluated that the prevalence of diabetes will be 7.7% in 2019 and will rise to 9.5% in 2045 [4,5]. Diabetes primarily affects Western nations, although it is becoming more common in emerging Asian countries such as China and India [6]. It is the seventh leading reason of death in the United States due to its several severe side effects, which include neuropathy, nephropathy, high blood pressure, cardiovascular risk, an unbalanced lipid profile, and retinopathy [7]. Around 90% of all instances of diabetes were in adults with type 2 diabetes mellitus. In addition to reduced pancreatic insulin secretion, type 2 diabetes mellitus is a metabolic condition marked primarily by insulin resistance and limited insulin action in the muscle and liver cells [8,9]. The prevention and treatment of type 2 diabetes mellitus are therefore one of the most important issues of the 21st century to diminish complications, death, and healthcare costs [10]. It also produces polyuria, polydipsia, and polyphagia as symptoms which are characterized by high blood glucose levels.

Targeting Angiogenesis, Inflammation and Oxidative Stress in Chronic Diseases. https://doi.org/10.1016/B978-0-443-13587-3.00012-6

2 Types of diabetes mellitus

2.1 Type 1 diabetes mellitus (T1DM)

The primary cause of type 1 diabetes mellitus is the demolition of pancreatic cells, which leads to an insufficient amount of insulin being produced. Juvenile-onset diabetes was formerly the name given to it since it typically develops in children. The onset often happens throughout childhood, peaking between the ages of 5–7 and during puberty, but it can happen at any age. Commonly, it is an endocrine autoimmune metabolic condition occurring mainly during childhood [11]. The loss of β-cells in the massive majority of patients (70%–90%) is the consequence of type 1 diabetes mellitus autoimmunity nevertheless in a smaller set of the population, no autoantibodies or immune responses are noticed, and the root of β-cell destruction is unknown (idiopathic type 1b diabetes mellitus). It is associated with the presence of autoantibodies many months or years before the onset of symptoms. These autoantibodies act as a biomarker for the development of autoimmunity. The autoantibodies associated with T1DM are those directed against insulin, 65 kDa glutamic acid decarboxylase, insulinoma-associated protein 2, or zinc transporters. The first β-cell autoantibody to appear in early childhood is usually directed against insulin or GAD65 (i.e., antiinsulin or antiGAD65 autoantibodies), however, these autoantibodies can each be present, whereas it is far uncommon to look at islet antigen-2 (IA-2) or zinc transporter (ZNT) autoantibody first. What triggers the appearance of a first-appearing β-cell-targeting autoantibody is doubtful but is beneath scrutiny in several studies of kids who are being followed up since birth [12–14]. The different causes of type 1 diabetes mellitus are autoimmune diseases, endocrine disease, viruses and infection, destruction of the pancreas, environmental factors, drug and chemical toxins, autoimmune destructions of β-cells, etc. [11].

2.2 Type 2 diabetes mellitus (T2DM)

In type 2 diabetes mellitus (T2DM) there is dysregulation of carbohydrate, protein, and fat metabolism leading to decreased insulin secretion, insulin resistance, or a combination of both. Of the three main categories of diabetes, T2DM is more common (in more than 90% of all cases) than type 1 diabetes mellitus or gestational diabetes. Over the past few decades, our understanding of the course and development of T2DM has evolved rapidly. It has become more prevalent between children and adolescents, partly due to an increase in the number of overweight or obese young people. Compared to type 1 diabetes, type 2 diabetes is more strongly associated with family history and heredity. Twin studies have shown that genetics contribute significantly to the occurrence of T2DM [14]. Its chief basis is gradually decreased insulin secretion through pancreatic β-cells, the individual who has a preexisting background of insulin resistance in skeletal muscle, adipose tissue, and liver. Obvious hyperglycemia a high-risk condition preceded by prediabetes that predisposes individuals to T2DM development. Prediabetes is described by one of the following: Impaired fasting glucose

(IFG) levels, impaired glucose tolerance (IGT), or elevated level of glycated hemoglobin A1c (HbA1c) [15,16]. Impaired fasting glucose levels are assessed with fasting plasma glucose levels that are higher than normal, while IGT is characterized by insulin resistance in muscle and impaired late (second-phase) insulin secretion after a meal. Individuals with IFG levels manifest hepatic insulin resistance and impaired early insulin secretion. Annual conversion rates of prediabetes to T2DM range from 3% to 11% per year [17].

Finding persons who have prediabetes and intervening with lifestyle changes (weight reduction and exercise) as well as antidiabetic and antiobesity drugs are necessary for the prevention of diabetes. According to the American Diabetes Association (ADA) Consensus Conference, metformin, pioglitazone, and a combination dosage of metformin and rosiglitazone should be used to treat high-risk patients with IGT and IFG levels (HbA1c $> 6.5\%$; BMI 30 kg per m^2; age 60 years). However, prediabetic people can be anticipated to benefit from a lowered chance of acquiring diabetes, a better lipid profile, and a decreased cardiovascular risk, including a reduced risk of developing hypertension, if they can effectively lose weight and continue physical activity plan. T2DM is a complex chronic disorder that necessitates ongoing medical attention, patient self management for control of abnormal glucose levels, and multifactorial risk reduction strategies to normalize blood glucose, lipid profiles, and blood pressure to prevent or minimize acute and long-term macrovascular complications (such as a heart attack or stroke) and microvascular complications (including retinopathy, nephropathy, and neuropathy) [18,19].

2.3　Gestational diabetes mellitus (GDM)

The term "gestational diabetes" was first introduced by Carrington in 1957. This type of high blood sugar occurs during pregnancy and is usually detected in the later stage of the second trimester or early in the third trimester, this condition is typically resolved after delivery. Therefore, the term "GDM" refers to a wide range of hyperglycemia that includes mild forms like IGT or IFG during late pregnancy to severe forms like those observed in early pregnancy indicative of overt diabetes. While diabetes-related hyperglycemia was initially rare, it has gained prevalence with the global epidemics of diabetes and obesity, as well as the trend of delayed childbearing, especially in regions where early-onset diabetes and obesity are also common [20].

In the current epidemic of hyperglycemia outside of pregnancy, probably, some cases diagnosed as gestational diabetes mellitus (GDM) are undiagnosed prepregnancy hyperglycemia with varying degrees of severity. There is no universal agreement on a single diagnostic protocol or criteria for GDM, which makes international comparisons challenging. GDM poses risks to both the mother, mainly hypertensive disorders of pregnancy, and the fetus, mainly excessive fetal growth and adiposity. A diagnosis of GDM can indicate a higher risk of diabetes, obesity, and premature cardiovascular disease in women and their offspring over the long-term [21].

3 National and international prevalence

More than 37 million adults in the United States have diabetes, which has more than doubled over the past 20 years. Diabetes may lead to major health concerns that can harm the eyes, kidneys, and nerves, among other sections of the body, if it is not managed. Diabetes mellitus type 1 and diabetes mellitus type 2 are the two primary subtypes. When a person has diabetes type 1, their body is unable to produce insulin. The most prevalent kind of diabetes type 2 is characterized by the body's inability to use insulin suitably. According to the International Diabetes Federation, there are 537 million diabetics worldwide. The International Diabetes Federation projects that by 2045, 783 million people worldwide will have diabetes, a 46% rise from the current number of cases [4,6].

India has been the second-highest number of diabetics worldwide. Furthermore, 40 million people in India have impaired glucose tolerance, which places them at a high-risk of emergent diabetes type 2. This is the second-highest total in the entire world. More than half (53.1%) of diabetics in India do not have a diagnosis. When diabetes is not adequately diagnosed or managed, it can have catastrophic and occasionally deadly effects, such as heart attack, stroke, renal failure, blindness, and lower limb amputation. They have the result of reducing living quality and increasing healthcare costs, which heightens the need for treatment availability. Diabetics of type 2 make up around 90% of the population. In 2008, there were 347 (314–382) million cases of diabetes globally, with 90% of those cases being type 2 diabetes, up from 153 (127–182) million in 1980 due to changes in lifestyle and an increase in obesity [22]. According to conservative estimates, 429–552 million people worldwide will develop diabetes by 2030 as a result of being overweight, obese, and having an extended life expectancy [23,24]. Obesity is defined as having a body mass index of more than $30\,kg/m^2$ [25]. According to epidemiology research, those who are obese have a much higher chance of developing diabetes [26]. T2DM has been identified as an obesity-related condition as a result [27].

4 Pathophysiology of diabetes mellitus

The body breaks down the food it consumes and absorbs it into circulation through the digestive system. While some of the food is consumed right once, the majority is preserved for later use. It serves as a fuel, particularly for the brain cells, which are dependent on glucose for their operation. It is stored as glycogen in the liver and muscle cells. The islets of Langerhans of the pancreas create insulin, which is required for the transportation of glucose into the cells for utilization as fuel. A protein hormone, insulin is released into the intracellular environment by the beta cells before entering the circulation to be used. It interacts with the insulin receptor protein, which in turn triggers a series of intracellular events, at the cellular level. In response to this contact, a new protein known as a glucose transporter (GLUT4 in muscle cells) is produced. This protein moves to the cell surface and facilitates the entry of bigger molecular nutrients like protein and glucose [28,29].

Insulin serves a variety of purposes, including facilitating the intake of nutrients and stimulating the production of fat and glycogen while suppressing the breakdown of fat and glycogen-producing enzymes. It is natural for storage compounds to break down, and this process is essential to metabolism. The biological system cannot use the generated metabolites in the absence of insulin, which results in a catabolic condition. The most crucial mechanisms leading to the hyperglycemia of diabetes, whether from insulin insufficiency or insulin resistance, occur in the liver, which is also capable of producing fresh glucose (gluconeogenesis) from protein [30].

Type 1 diabetes mellitus known as an autoimmune condition causes the death of insulin-producing cells in the pancreas by invading macrophages and CD4+ and CD8+ T lymphocytes. Majorities of patients who previously had insulin treatment and those with circulating islet cell antibodies both have these antibodies. A lack of insulin secretion may result from the autoimmune destruction of pancreatic β-cells, which results in metabolic abnormalities linked to T1DM. Additionally, individuals with T1DM secrete an excessive amount of glucagon and have decreased insulin production, which might disrupt pancreatic α-cell function [31]. The metabolic problems brought on by low insulin are made worse by the incorrect rise of glucagon levels. Lack of insulin results in unchecked lipolysis and excessive plasma levels of free fatty acids, which inhibit glucose utilization in peripheral tissues including skeletal muscle. A variety of genes, including glucokinase in the liver and the GLUT 4 class of glucose transporters in adipose tissue, are required for target tissues to respond to insulin correctly, and an insulin shortage may reduce their expression. Impaired protein, glucose, and lipid metabolism are the main metabolic abnormalities caused by insulin insufficiency in T1DM [32,33].

Insulin resistance and decreased insulin production due to pancreatic beta-cell dysfunction are the two primary pathophysiological abnormalities associated with type 2 diabetes [34]. Although "relative" to the degree of insulin resistance, the plasma insulin concentration (both fasting and meal stimulated) is elevated in an absolute sense. Normal glucose homeostasis cannot be sustained by elevated plasma insulin levels. It is essentially difficult to distinguish the contribution of each to the etiopathogenesis of T2DM because of the close association between the production of insulin and the sensitivity of hormone action in the complex management of glucose homeostasis. Impaired glucose tolerance eventually results from insulin resistance and hyperinsulinemia [35].

5 Oxidative stress in diabetes mellitus

Both diabetes mellitus type 1 and diabetes mellitus type 2 have incorrectly managed blood sugar levels that rise to high levels over extended periods. One of diabetes' many consequences, persistent hyperglycemia, is also one of the disease's hallmarks. Even though there are still many unresolved questions about the physiopathology of oxidative stress, it is widely recognized that it plays a crucial role in the development and progression of diabetes. Oxidative stress is a condition when there is an imbalance in the creation and elimination of ROS, which promotes the synthesis of oxidants.

ROS include the hydroxyl radical (OH), superoxide anion (O_2), and peroxynitrite (ONO_2) among other oxygen-free radicals. Because it produces free radicals so easily, hydrogen peroxide (H_2O_2), one of oxygen's nonradical byproducts, is also categorized as ROS. Excessively nicotinamide adenine dinucleotide hydrogen phosphate (NADPH) is generated when cells have an excessively high protein biosynthetic load, which increases the production of superoxide anion radicals (O_2) and causes cells more susceptible to oxidative damage [36].

Organelles like mitochondria are necessary for energetic metabolism because they allow oxidative phosphorylation, which releases energy in the form of adenosine triphosphate (ATP). In this process, NADH and $FADH_2$, which are produced when nutrients are oxidized, by the electron transport chain (ETC), produce ATP, ROS, and mostly O_2. Although mitochondria are the primary generator of intracellular ROS, these organelles also have antioxidant enzymes that, under normal circumstances, control the cellular redox imbalance. The manganese superoxide dismutase (MnSOD), a superoxide dismutase that is specific to mitochondria and suppresses O_2, is shown in Fig. 1 and highlights the crucial function that mitochondria play as a source of ROS in maintaining them under homeostatic processes [37].

The primary nutritional origin of energy for the transport chain is glucose, which is produced as NADH and $FADH_2$. This makes it understandable why ROS are linked to the physiopathology of diabetes. There is proof that people with type 2 diabetes have changed antioxidant enzymes [38], and several investigations have found that people

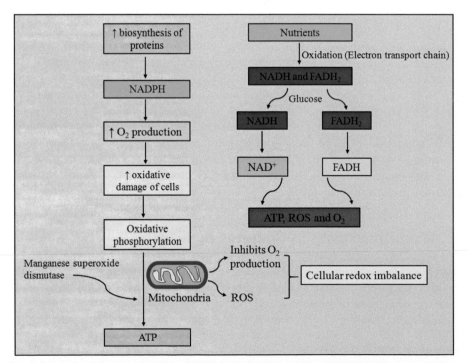

Fig. 1 Mitochondria play a source of ROS and in keeping them under homeostatic regulation.

with diabetes mellitus generally exhibit oxidative stress. Additionally, diabetes has been linked to malfunction of the mitochondrial ETC in terms of mitochondrial disorders [39]. It is known very well that abnormalities in the mitochondrial genome predispose individuals to diabetes. Proteins that are encoded by the mitochondrial genome are required for the synthesis of ATP and are a component of the ETC. For example, recent prospective research conducted on people with mitochondrial dysfunction discovered a greater risk of endocrine illnesses like diabetes mellitus [40].

High-level oxidative stress and mitochondrial disorders have also been linked, according to research [41]. It has been hypothesized that mitochondria are significant in the etiology of diabetes and the subsequent ROS production, which is a significant trigger of the implications of the diseases [42], according to the body of evidence linking T2DM and oxidative stress affecting mitochondria. Advanced glycation metabolites, glucose oxidation, and lipid peroxidation are all increased by hyperglycemia inside cells [43], which causes the production of ROS [44], which reduces insulin release.

6   Inflammation in diabetes mellitus

The microvascular disorders of diabetes (such as renal disease, nerve, and retinal) are heavily influenced by inflammation at the vascular level, while the macrovascular complications of diabetes are strongly influenced by oxidative stress (such as peripheral arterial disease, coronary artery disease, and stroke). Hence, it is clear that obesity has a role in the development of both insulin resistance and diabetes. A rise in blood levels of free fatty acids (FFA) is another effect of poor insulin sensitivity, particularly as a result of the disruption of insulin's antilipolytic effect on adipocytes. In reaction, fat cells release large amounts of FFA into the bloodstream, which leads to more insulin release interference and ectopic lipid deposition, among other systemic lipotoxic effects. In this regard, lipotoxicity is now recognized as a major factor in insulin resistance [45].

6.1   Relationship between oxidative stress and inflammation

A high-calorie diet, exposure to immunological, chronic, radioactive, toxic chemicals, allergies, illnesses, obesity, and pathogenic diseases like cancer are all linked to inflammation, which is the body's natural defensive mechanism against pathogens. Oxidative stress and other protein oxidations are the results of several chronic illnesses associated with the increased generation of ROS [46]. Peroxiredoxin 2 has been identified as an inflammatory signal and protein oxidations cause the production of inflammatory signal molecules. Several researchers have investigated the connection between oxidative stress and inflammation. There is proof that oxidative stress contributes to the pathogenesis of chronic inflammatory disorders. The following are the results of the investigation. By altering synaptic and nonsynaptic transmission between neurons, ROS produced in brain tissues may act as the origin for neuroinflammation and cell death, which in turn leads to neurodegeneration and loss of memory [47,48]. Hyperglycemia causes significant cellular damage to the brain due to the increased production of oxidative stress. Many disorders, including insulin resistance, diabetes mellitus type 2, and cardiovascular diseases, are influenced by chronic inflammation [47,49].

6.2 Important roles for oxidative stress and inflammation in vascular failure

Because of the failure to regulate intracellular glucose content about blood glucose levels and their inability to prevent glucose from entering when blood glucose levels are high, vascular endothelial cells are a common target of hyperglycemic injury [50]. Endothelial cells store a lot of glucose in this circumstance (during hyperglycemia) and may suffer considerable oxidative damage. Both direct advanced glycation end product (AGE) damage caused by glycation and indirect ROS damage caused by hyperglycemia can trigger an inflammatory response in the endothelium. It appears that the immune system alters the integrity of the endothelium by generating ROS through a pulmonary explosion [51]. All of these components cause endothelial tissue to experience oxidative stress and ROS, which encourages inflammation and damages the vascular endothelium. Additionally, ROS contributes to inflammation by increasing inflammatory cytokine levels and upregulating the expression of growth factors and cellular adhesion molecules at the beginning of cardiovascular issues linked to T2DM [52,53]. Fig. 2 explains the mechanisms of hyperglycemia-induced endothelial dysfunction.

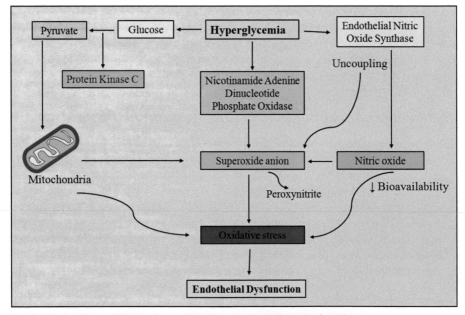

Fig. 2 Mechanisms of hyperglycemia-induced endothelium dysfunction.

7 Role of angiogenesis in diabetes mellitus

The complex process of angiogenesis, which leads to the formation of capillaries, includes the interaction of many vascular endothelial growth factors. The development of new blood vessels from the existing vasculature is known as angiogenesis. It starts in utero and lasts all the way through old age, happening in both health and illness. The distance between a blood capillary, which is created by the angiogenesis process, and any metabolically active tissue in the body is less than a few hundred micrometers. Several molecules are needed for angiogenesis to function effectively, including adhesion receptors, extracellular matrix proteins, angiogenic agents, and proteolytic enzymes [54,55]. The balance of positive and negative angiogenic modulators in the vascular milieu is also necessary for angiogenesis. A few of the linkages between abnormal angiogenesis and the development of chronic diabetic complications are shown in Fig. 3.

Excessive or defective angiogenic processes have been linked to diabetes mellitus, and both of these factors are important contributors to the development of chronic diabetic problems with serious clinical repercussions. The driving factors and key clinical repercussions of diabetes' impact on angiogenesis are diabetic retinopathy is a neurovascular disorder of the retina associated with diabetes that is marked by the development of new blood vessels. Ischemia can result from vascular alterations in the

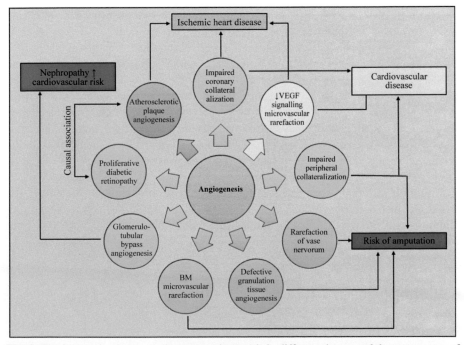

Fig. 3 The interaction between aberrant angiogenesis in different tissues and the emergence of problems from chronic diabetes. *VEGF*, vascular endothelial growth factor.

preproliferative phases, which in turn can cause angiogenesis in the retina and vitreous invasion. Because of the lack of strong intercellular connections, newly created blood vessels are young and brittle, making them vulnerable to ruptures that might result in sight-threatening hemorrhages. The angle of the eye's anterior chamber may neovascularize, which might lead to retinopathy glaucoma [56,57].

Diabetic kidney disease (DKD) is marked by excessive and aberrant angiogenesis, a defective vasculature has been observed to promote glomerular enlargement by forming new blood vessels with blood capillaries. For instance, the generation of vascular endothelial growth factor (VEGF)-A may have proangiogenic effects as well as concurrently cause dysfunction in glomerular endothelial cells. The initial step of angiogenesis that VEGF-A induces is the weakening of endothelial cell connections to enable sprouting; this may enhance the permeability of the ultrafiltration barrier and endorse albuminuria [58,59].

It was commonly seen that newly formed arteries connected to peritubular capillaries bypass the glomerulus. However, it is conceivable to consider the overexpression of VEGF-A in DKD caused by hyperglycemia and shear stress as a compensatory effort to lower intraglomerular pressure. The aberrant control of angiogenesis by VEGF-A in DKD has been linked to angiopoietin-1 and angiopoietin-2 abnormalities, and their suppression may have therapeutic implications [60].

Acute hyperglycemia can cause nerve damage without causing vascular alterations, the clinical data supported the theory that diabetic neuropathy is dependent on an inadequate angiogenic response. Uncertainty exists about the precise processes by which diabetes damages the vasa nervorum, including whether these mechanisms are exclusive to neuropathy or mimic the broad mechanisms by which hyperglycemia damages endothelial cells [61]. It has been hypothesized that excessive glucose stimulates the creation and expression of VEGF by Schwann cells, which may then cause vasa nervorum endothelial cell failure. Reducing or preventing VEGF overexpression can indeed improve diabetic neuropathy-related symptoms [62,63].

The presence of hyperglycemia considerably slows the healing of cutaneous lesions. Diabetic individuals may develop persistent, nonhealing ulcers that are confined to certain pressure areas on their feet, such as the metatarsophalangeal joints, ankles, and heels. The process of healing a wound must include angiogenesis [64]. Insufficient angiogenesis in diabetes individuals manifests as reduced endothelial cell proliferation and diminished cell and growth aspect receptiveness at the site of the lesion. For instance, the synthesis of VEGF by wild-type fibroblasts is enhanced threefold in response to hypoxia, but the production of VEGF by diabetic fibroblasts is not upregulated in hypoxia situations [65].

8   Management of diabetes mellitus

8.1   Nutritional management

Consuming enough dietary fiber, the treatment of cardiovascular risk factors and glycemic control, in particular, fiber-containing natural resources, have been reported to be improved, reducing the risk of cardiovascular death in diabetics [66–68]. It is generally advised that diabetic patients consume fiber and whole grains in amounts that are at least comparable to those advised for the general population; approximately 30 g/d for women and 38 g/d for men, or 14 g per 1000 kcal. This is to account for the modest beneficial effects on cardiovascular risk factors. Consuming carbohydrates containing meals such as fruits, vegetables, legumes, whole grains, and dairy is quite valuable for the diabetic patient [69,70].

According to epidemiological research, fats increase the chance of acquiring obesity and cardiovascular disease [71]. As with the other primary principles, there is no ideal fat proportion. Instead, diabetic patients often follow the guidelines for the overall population (between 20% and 35%), especially if the patients are overweight, in that instance, the percentage must be kept within reasonable limits. According to certain research studies examining the Balanced diet pattern, monounsaturated fatty acids can reduce cardiovascular risk factors and improve glycemic control, particularly if saturated fatty acids are substituted. Eating omega-3-rich food, however, may help to reduce the risk of cardiac disease [72].

8.2 Physical activity

The simplest and most fundamental methods for treating diabetes are physical activity and exercise. General benefits of encouraging exercise within a specific strategy include improvements in glycemic control [73], blood pressure and cholesterol profiles, cardiovascular health advantages, enhanced quality of life, psychological well-being, and treatment of depression only a few of the benefits. Choose an aerobic activity you enjoy doing, such as walking, running, biking, or swimming. People should try to exercise for 150 min a week, or 30 min or more of moderate aerobic activity, on average. Resistance exercise increases your ability for daily chores, balance, and strength. Resistance training includes activities like yoga, dance, and grappling. Those with type 2 diabetes should make an effort to perform two to three resistance exercises per week [74,75].

9   Treatment of diabetes mellitus

9.1 Insulin treatments

Insulin was discovered in 1921, and human clinical trials in 1922 by Banting and Best. It is made up of the peptide chains chain A and chain B. Disulfide bridges bind these two chains together. Connecting-peptide can produce immunogenic reactions. The first diabetic medication was insulin. Different types of insulin based on the onset of time and duration of action are shown in Table 1. The best way to lower hyperglycemia is with insulin treatment, which helps control glucose metabolism. It also raises high-density lipoprotein and lower triglycerides [76–82].

Table 1 Based on the onset and duration of action different types of insulin.

Class	Type	Onset of action (min/h)	Peak effect (h)	Duration of action (h)
Ultra-short acting insulins	Insulin Lispro	0.25 (15 min)	0.5–2	2–6
	Insulin Aspart	0.25 (15 min)	0.5–1	3–6
Short-acting insulins	Regular soluble insulin (crystalline)	0.5–1	2–4	6–8
Intermediate- acting insulins	NPH (isophane)	1–2	6–12	8–24
	Lente	1–2	6–12	8–24
Long-acting insulins	Ultralente	4–6	16–18	24–36
	Protamine zinc	4–6	14–20	24–36
	Insulin	2–5	5–24	22–24

9.2 Noninsulin treatments

9.2.1 Insulin secretagogues

Some medications, primarily sulfonylureas and meglitinides work by interacting with the sulfonylurea receptor present in the pancreatic cells, which increases insulin release from the pancreas [83]. Tolbutamide, chlorpropamide, tolazamide, and acetohexamide are examples of first-generation sulfonylurea whereas glibenclamide, glipizide, and glimepiride are examples of second-generation sulfonylurea [84]. The development of second-generation sulfonylurea was aided by improved efficacy, a faster onset of action, reduced plasma half-lives, and a longer duration of activity. Sulfonylurea's side effects may include symptoms of low sugar levels such as perspiration, disorientation, and agitation [85].

9.2.2 Biguanides

It acts by boosting the body's responsiveness to natural insulin, decreasing gastrointestinal glucose absorption, and lowering hepatic glucose synthesis. By preventing gluconeogenesis and promoting glycolysis, biguanides reduce hepatic glucose production. They increase insulin receptor activation, which enhances insulin signaling [86]. These compounds do not directly affect the production of insulin-like insulin secretagogues. Metformin, phenformin, and buformin are examples of various compounds in this group. Biguanides have antihypertriglyceridemic and vasoprotective characteristics, none of which result in hypoglycemia or weight gain. Nevertheless, biguanides frequently have gastrointestinal side effects, such as diarrhea, vomiting, cramps, nausea, and increased flatulence. Vitamin B_{12} absorption is thought to be diminished with long-term usage [87–89].

9.2.3 Alpha-glucosidase inhibitors

Alpha-glucosidase inhibitors are mostly not suggested for those with inflammatory bowel diseases like Crohn's disease or ulcerative colitis, intestinal blockages, gastrointestinal disorders, or diabetic ketoacidosis, which causes the body to burn fat for energy instead of carbohydrates [90]. If a patient has a big intestinal ulcer, liver cirrhosis, or is pregnant, acarbose is not advised [91].

9.2.4 Incretin mimetics

The incretins or peptides produced by the gut include glucagon-like peptides and insulinotropic polypeptides that are glucose-dependent. A decrease in blood glucose levels is encouraged by a group of naturally occurring metabolic hormones known as incretins [92]. These hormones are released following a meal. The gut's L cells release a peptide of 36 amino acids called glucagon-like peptide-1 after the introduction of a meal. Glucagon-like peptide-1 secretion from pancreatic beta cells is equivalent to insulin secretion [93]. In response to glucagon-like peptide-1, the pancreatic beta cells start to make and secrete insulin. The strategy utilized for treating diabetes type 1 and diabetes type 2 may be the creation of glucagon-like peptide-1 analogs with a longer half-life [94].

9.2.5 Amylin analogs

The hormone amylin is composed of a single chain of 37 amino acids. By way of insulin, pancreatic beta cells release it. By slowing stomach emptying and preventing glucagon release, it keeps blood glucose levels stable throughout fasting and after meals. It regulates how much food is ingested by adjusting the brain's center for appetite [95]. As both diabetes type 1 and diabetes type 2 lack amylin, research, and development of amylin analogs that maintain the homeostasis of glucose. Amylin cannot be used as a medication because it aggregates and is insoluble in solution; thus, chemical analogs that can mimic the effects of amylin were created. The parenteral administration of amylin analogs is utilized to treat both diabetes mellitus type 1 and diabetes mellitus type 2 [96]. The only medication in this family is pramlintide acetate, which is sold under the trade name Symlin and is taken subcutaneously [97].

10 Innovative drugs delivery systems for treatment of diabetes mellitus

Traditional drug delivery methods are subject to several drawbacks, including ineffectiveness brought on by poor or inadequate doses, decreased potency or altered effects brought on by drug metabolism, and a lack of target selectivity [98–100]. Novel drug delivery systems (NDDSs) are emerging fields in recent years due to their advantages in reduced dosing frequency, increased bioavailability, prevention from degradation in acidic gastric environments, targeted therapeutic efficacy with a decrease in associated side effects, and more [101]. Although several NDDSs are being investigated for the treatment of various ailments, only a small number, such as microparticulate and nanoparticulate systems, have been reported for the treatment of diabetes mellitus type 2.

The particulate system is made up of miniature structures that may carry drugs inside of cells, and attaching ligands to them causes them to be recognized by certain receptors. Thus, it is believed that these methods are the most ideal ones for delivering antidiabetic medications [102].

Drugs that are trapped in microparticles can be released selectively at the desired spot. By adjusting the medication's release rate, these systems keep the drug concentration in plasma constant. Due to their smaller size and higher surface-to-volume ratio, microparticles are used to speed up the dissolving of insoluble medicines [103].

Transcellular transport by carrier- or receptor-mediated endocytosis is the method used to move microparticulate systems. Attributable to their size, microparticles cannot overcome the tight connections of the mucosal membrane to enter cells via paracellular transport, whereas nanoparticulate systems exhibit better intracellular uptake than microparticulate systems [104]. Polymeric nanoparticles (NPs), metallic NPs, lipid-based NPs, and biological NPs are the many subcategories of nanoparticles [105]. Via cellular absorption mechanisms such as transcellular and paracellular pathways, nanoparticles transport the medications they have captured [106]. In addition, the NPs exhibit enhanced mucoadhesion because they interact electrostatically with the negatively charged mucus and endothelium layer to remain in the gastrointestinal system.

The insoluble pharmaceuticals of Biopharmaceutical Classification System (BCS) classes II and IV are included in a liquid form of an oil-in-water nanoemulsion with a particle size of 200 nm or less [107]. It increases the solubility of medications and creates a broad interfacial area to speed up the pace at which insoluble pharmaceuticals are absorbed [108]. A different technique of administration of medication than oral and parenteral is the transdermal delivery system. A transdermal delivery system is a low-cost, noninvasive treatment that patients may administer themselves. A transdermal delivery system may solve the issue of medication first-pass metabolism metabolizing very quickly. With the aid of permeability enhancers, the transdermal delivery system may be a viable alternative for the administration of hydrophilic medications, macromolecules, and vaccinations [109] Table 2 summarizes research outcomes of various types of nanoparticles investigated for treatment of diabetes mellitus.

11   Conclusion and future perspectives

Type 2 diabetes is a highly frequent chronic metabolic disease. Although being an important topic of research, there is still much to learn about the physiopathology of this condition because it is unknown what causes it and several factors seem to be involved. The reasons for molecular diabetes development are still being studied. Diabetic cardiovascular issues are the most frequent long-term diabetes complications, which are associated with death as well as disability. The rising costs of its

Table 2 An overview of nanoparticles investigated for treating diabetes mellitus.

Formulation/drugs	Excipients	Preparation technique	Significance	Reference
Nanosuspension/glimepiride	Polyvinyl Pyrrolidone K30, Pluronic F60, and PEG 400	Evaporation followed by sonication technique	Improve solubility, the therapeutic effect of the drug, and exhibited a maximum release of approximately 97.6% in 1h	[110]
Transdermal patches/glimepiride	HPMC M.N 86000, chitosan, acetonitrile, propylene glycol, dimethyl sulfoxide (DMSO), and silicone coated liners	Evaporation followed by sonication technique	Pharmacokinetic parameters and glimepiride skin permeability were found to be enhanced by optimized glimepiride transdermal patches. Increased bioavailability and action time	[111]
Nanoparticles/glipizide	Calcium chloride, sodium alginate, chitosan (MW 60,000, glacial acetic acid (GAA)), sodium hydroxide, and potassium dihydrogen phosphate	Ionotropic controlled gelation method	The goal of the current work was to create glipizide loaded controlled-release nanoparticles that showed high entrapment efficiency and regulated release properties	[112]
Nanoemulsion/repaglinide	Sefsol 218 (propylene glycol-monocaprylic ester), tween-80 (polyoxyethylene sorbitan monooleate)	Solvent injection and ultrasonication methods	Provides a more effective hypoglycemic response than tablet formulation	[113]
Solid lipid nanocarrier/repaglinide	Pluronic F68, soya lecithin, dialysis membrane (12,000–14,000)	Emulsification ultrasonication technique	The highest degree of entrapping efficiency and provide a prolonged in vitro release	[114]
Nanoemulsion/pioglitazone	Capryol 90, labrasol, transcutol-P (diethylene glycol monoethyl ether), and Pluronic F127	Solvent injection and ultrasonication methods	Enhancement of skin barrier function, reduction of lesions, diminished infiltration of inflammatory cells, and expression of pro-inflammatory cytokines	[115]

Continued

Table 2 Continued

Formulation/ drugs	Excipients	Preparation technique	Significance	Reference
Nanoparticles/ pioglitazone	Poloxamer 407 (PL F-127), sodium deoxycholate (SDC)	Media milling technique, planetary ball milling	Increase rate of disintegration, bioavailability, and solubility	[116]
Microspheres/ sitagliptin	Carbopol 934 P	Nano spray drier	Prolong the release and retention time of sitagliptin	[117]
Nanoparticles/ sitagliptin	Chitosan, sodium deoxycholate (purity ≥97%)	Spray dried method	Improve the effectiveness of sitagliptin for oral delivery, enhances bioadhesion quality, and showed sustained drug release	[118]
Nanoparticles/ glibenclamide	Eudragit L100, alloxan	Solvent displacement method	The nanoparticle formulation might increase patient compliance, lower side effects, and reduce dosage frequency	[119]

complications and management make diabetes a huge financial burden as well. The development of diabetes and the course of the illness are both significantly influenced by oxidative stress, as is well documented in this chapter. Many inflammatory mediators implicated in several chronic illnesses are activated by oxidative stress. According to clinical data, oxidative stress and inflammation brought on by excessive ROS generation are likely to play a significant role in the development of a number of illnesses, including chronic disorders connected to inflammation. As a result, oxidative stress not only aids in the progression of the illness but also plays a role in it. To comprehend how ROS are engaged in the illness, we have concentrated on the most significant pathways implicated in ROS formation during the onset and progression of diabetes. As we have demonstrated, inflammaging is linked to vascular function decrease, a significant risk factor for the development of cardiovascular disease. Moreover, oxidative stress, which results from cellular senescence and a lack of adaptive immunological function (immunosenescence), is a hallmark of age-onset illnesses including cardiovascular disease. Even though ocular tissues from diabetes patients were discovered to have considerably higher amounts of VEGF. The principal treatments for type 2 diabetes include insulin secretagogues, biguanides, alpha-glucosidase inhibitors, incretin mimetics, amylin antagonists, and sodium-glucose co-transporter inhibitors. Dual drug regimens are typically advised for patients who are unable to accomplish treatment objectives with first-line oral hypoglycemic medications used as monotherapy. Traditional dosage formulations have variable bioavailability and brief half-lives, which need repeated dosing and amplify adverse effects despite the great therapeutic benefits. Nanotechnology-based approaches are more appealing due to the added benefit of site-specific medicine administration with enhanced bioavailability and a more variable dosing schedule, which is relevant given the pathological complexity of the aforementioned condition.

References

[1] Rathor S, Bhatt DC. Formulation, characterization, and pharmacokinetic evaluation of novel glipizide-phospholipid nano-complexes with improved solubility and bioavailability. Pharm Nanotechnol 2022;10(2):125–36. https://doi.org/10.2174/221173851 0666220328151512.

[2] Arunachalam S, Gunasekaran S. Diabetic research in India and China today: from literature-based mapping to health-care policy. Curr Sci 2002;9(10):1086–97. https://www.jstor.org/stable/24106793.

[3] Nolte MS. Pancreatic hormones and antidiabetic drugs. In: Basic and clinical pharmacology. McGraw Hill Medical; 2001.

[4] Guariguata L, Whiting D, Weil C, Unwin N. The international diabetes federation diabetes atlas methodology for estimating the global and national prevalence of diabetes in adults. Diabetes Res Clin Pract 2011;94(3):322–32. https://doi.org/10.1016/j.diabres.2011.10.040.

[5] Saeedi P, Petersohn I, Salpea P, Malanda B, Karuranga S, Unwin N, Colagiuri S, Guariguata L, Motala AA, Ogurtsova K, Shaw JE. Global and regional diabetes prevalence estimates for 2019 and projections for 2030 and 2045: results from the International

Diabetes Federation Diabetes Atlas. Diabetes Res Clin Pract 2019;157, 107843. https://doi.org/10.1016/j.diabres.2019.107843.

[6] Zheng Y, Ley SH, Hu FB. Global aetiology and epidemiology of type 2 diabetes mellitus and its complications. Nat Rev Endocrinol 2018;14(2):88–98. https://doi.org/10.1038/nrendo.2017.151.

[7] Martín-Timón I, Sevillano-Collantes C, Segura-Galindo A, del Cañizo-Gómez FJ. Type 2 diabetes and cardiovascular disease: have all risk factors the same strength? World J Diabetes 2014;5(4):444. https://doi.org/10.4239/wjd.v5.i4.444.

[8] Galicia-Garcia U, Benito-Vicente A, Jebari S, Larrea-Sebal A, Siddiqi H, Uribe KB, Ostolaza H, Martín C. Pathophysiology of type 2 diabetes mellitus. Int J Mol Sci 2020;21(17):6275. https://doi.org/10.3390/ijms21176275.

[9] Rathor S, Bhatt DC. Novel glibenclamide–phospholipid complex for diabetic treatment: formulation, physicochemical characterization, and in-vivo evaluation. Indian J Pharm Educ Res 2022;56(3):697–705.

[10] De Berardis G, Lucisano G, D'Ettorre A, Pellegrini F, Lepore V, Tognoni G, Nicolucci A. Association of aspirin use with major bleeding in patients with and without diabetes. JAMA 2012;307(21):2286–94. https://doi.org/10.1001/jama.2012.5034.

[11] Katsarou A, Gudbjörnsdottir S, Rawshani A, Dabelea D, Bonifacio E, Anderson BJ, Jacobsen LM, Schatz DA, Lernmark Å. Type 1 diabetes mellitus. Nat Rev Dis Primers 2017;3(1):1–7. https://doi.org/10.1038/nrdp.2017.16.

[12] TEDDY Study Group. The environmental determinants of diabetes in the young (TEDDY) study: study design. Pediatr Diabetes 2007;8(5):286–98. https://doi.org/10.1111/j.1399-5448.2007.00269.x.

[13] Paschou SA, Papadopoulou-Marketou N, Chrousos GP, Kanaka-Gantenbein C. On type 1 diabetes mellitus pathogenesis. Endocr Connect 2018;7(1):R38. https://doi.org/10.1530/EC-17-0347.

[14] Eiselein L, Schwartz HJ, Rutledge JC. The challenge of type 1 diabetes mellitus. ILAR J 2004;45(3):231–6. https://doi.org/10.1093/ilar.45.3.231.

[15] Gardner DS, Tai ES. Clinical features and treatment of maturity onset diabetes of the young (MODY). Diabetes Metab Syndr Obes 2012;5:101–8.

[16] DeFronzo RA, Abdul-Ghani MA. Preservation of β-cell function: the key to diabetes prevention. J Clin Endocrinol Metab 2011;96(8):2354–66. https://doi.org/10.1210/jc.2011-0246.

[17] Ferrannini E, Gastaldelli A, Iozzo P. Pathophysiology of prediabetes. Med Clin 2011;95(2):327–39. https://doi.org/10.1016/j.mcna.2010.11.005.

[18] Garvey WT, Ryan DH, Henry R, Bohannon NJ, Toplak H, Schwiers M, Troupin B, Day WW. Prevention of type 2 diabetes in subjects with prediabetes and metabolic syndrome treated with phentermine and topiramate extended release. Diabetes Care 2014;37(4):912–21. https://doi.org/10.2337/dc13-1518.

[19] Nathan DM, Davidson MB, DeFronzo RA, Heine RJ, Henry RR, Pratley R, Zinman B. Impaired fasting glucose and impaired glucose tolerance: implications for care. Diabetes Care 2007;30(3):753–9. https://doi.org/10.2337/dc07-9920.

[20] McIntyre HD, Catalano P, Zhang C, Desoye G, Mathiesen ER, Damm P. Gestational diabetes mellitus. Nat Rev Dis Primers 2019;5(1):47. https://doi.org/10.1038/s41572-019-0098-8.

[21] Langer O, Yogev Y, Most O, Xenakis EM. Gestational diabetes: the consequences of not treating. Am J Obstet Gynecol 2005;192(4):989–97. https://doi.org/10.1016/j.ajog.2004.11.039.

[22] Danaei G, Finucane MM, Lu Y, et al. National, regional, and global trends in fasting plasma glucose and diabetes prevalence since 1980: systematic analysis of health examination surveys and epidemiological studies with 370 country-years and 2.7 million participants. Lancet 2011;378:31–40. https://doi.org/10.1016/S0140-6736(11)60679-X.

[23] Wild S, Roglic G, Green A, Sicree R, King H. Global prevalence of diabetes: estimates for the year 2000 and projections for 2030. Diabetes Care 2004;27(5):1047–53. https://doi.org/10.2337/diacare.27.5.1047.

[24] Shaw JE, Sicree RA, Zimmet PZ. Global estimates of the prevalence of diabetes for 2010 and 2030. Diabetes Res Clin Pract 2010;87(1):4–14. https://doi.org/10.1016/j.diabres.2009.10.007.

[25] Hedley AA, Ogden CL, Johnson CL, Carroll MD, Curtin LR, Flegal KM. Prevalence of overweight and obesity among US children, adolescents, and adults, 1999-2002. JAMA 2004;291(23):2847–50. https://doi.org/10.1001/jama.291.23.2847.

[26] Mokdad AH, Bowman BA, Ford ES, Vinicor F, Marks JS, Koplan JP. The continuing epidemics of obesity and diabetes in the United States. JAMA 2001;286(10):1195–200. https://doi.org/10.1001/jama.286.10.1195.

[27] Grundy SM. Metabolic complications of obesity. Endocrine 2000;13:155–65. https://doi.org/10.1385/ENDO:13:2:155.

[28] Stumvoll M, Goldstein BJ, Van Haeften TW. Type 2 diabetes: principles of pathogenesis and therapy. Lancet 2005;365(9467):1333–46. https://doi.org/10.1016/S0140-6736(05)61032-X.

[29] Cerf ME. Beta cell dysfunction and insulin resistance. Front Endocrinol 2013;4:37. https://doi.org/10.3389/fendo.2013.00037.

[30] Fu Z, Gilbert ER, Liu D. Regulation of insulin synthesis and secretion and pancreatic beta-cell dysfunction in diabetes. Curr Diabetes Rev 2013;9(1):25–53. https://doi.org/10.2174/157339913804143225.

[31] Boland BB, Rhodes CJ, Grimsby JS. The dynamic plasticity of insulin production in β-cells. Mol Metab 2017;6(9):958–73. https://doi.org/10.1016/j.molmet.2017.04.010.

[32] Rorsman P, Ashcroft FM. Pancreatic β-cell electrical activity and insulin secretion: of mice and men. Physiol Rev 2018;98(1):117–214. https://doi.org/10.1152/physrev.00008.2017.

[33] Dali-Youcef N, Mecili M, Ricci R, Andrès E. Metabolic inflammation: connecting obesity and insulin resistance. Ann Med 2013;45(3):242–53. https://doi.org/10.3109/07853890.2012.705015.

[34] Petersen KF, Befroy D, Dufour S, Dziura J, Ariyan C, Rothman DL, DiPietro L, Cline GW, Shulman GI. Mitochondrial dysfunction in the elderly: possible role in insulin resistance. Science 2003;300(5622):1140–2. https://doi.org/10.3109/07853890.2012.705015.

[35] Stump CS, Short KR, Bigelow ML, Schimke JM, Nair KS. Effect of insulin on human skeletal muscle mitochondrial ATP production, protein synthesis, and mRNA transcripts. Proc Natl Acad Sci 2003;100(13):7996–8001. https://doi.org/10.1073/pnas.1332551100.

[36] Yang H, Jin X, Kei Lam CW, Yan SK. Oxidative stress and diabetes mellitus. Clin Chem Lab Med 2011;49(11):1773–82. https://doi.org/10.1515/cclm.2011.250.

[37] Tiwari BK, Pandey KB, Abidi AB, Rizvi SI. Markers of oxidative stress during diabetes mellitus. J Biomark 2013;2013. https://doi.org/10.1155/2013/378790.

[38] Opara EC. Oxidative stress, micronutrients, diabetes mellitus and its complications. J R Soc Promot Heal 2002;122(1):28–34. https://doi.org/10.1177/146642400212200112.

[39] Maiese K. New insights for oxidative stress and diabetes mellitus. Oxidative Med Cell Longev 2015;2015. https://doi.org/10.1155/2015/875961.

[40] Lipinski B. Pathophysiology of oxidative stress in diabetes mellitus. J Diabetes Complicat 2001;15(4):203–10. https://doi.org/10.1016/S1056-8727(01)00143-X.

[41] Di Meo S, Reed TT, Venditti P, Victor VM. Role of ROS and RNS sources in physiological and pathological conditions. Oxidative Med Cell Longev 2016;2016. https://doi.org/10.1155/2016/1245049.

[42] Di Meo S, Reed TT, Venditti P, Victor VM. Harmful and beneficial role of ROS 2017. Oxidative Med Cell Longev 2018;2018. https://doi.org/10.1155/2016/7909186.

[43] Prentki M, Matschinsky FM, Madiraju SM. Metabolic signaling in fuel-induced insulin secretion. Cell Metab 2013;18(2):162–85. https://doi.org/10.1016/j.cmet.2013.05.018.

[44] Forouhi NG, Misra A, Mohan V, Taylor R, Yancy W. Dietary and nutritional approaches for prevention and management of type 2 diabetes. BMJ 2018;361. https://doi.org/10.1136/bmj.k2234.

[45] Zimmerman RS. Diabetes mellitus: management of microvascular and macrovascular complications. Cleveland Clinic Centers for Continuing Education; 2016.

[46] Berlett BS, Stadtman ER. Protein oxidation in aging, disease, and oxidative stress. J Biol Chem 1997;272(33):20313–6. https://doi.org/10.1074/jbc.272.33.20313.

[47] Salzano S, Checconi P, Hanschmann EM, Lillig CH, Bowler LD, Chan P, Vaudry D, Mengozzi M, Coppo L, Sacre S, Atkuri KR. Linkage of inflammation and oxidative stress via release of glutathionylated peroxiredoxin-2, which acts as a danger signal. Proc Natl Acad Sci 2014;111(33):12157–62. https://doi.org/10.1073/pnas.1401712111.

[48] Popa-Wagner A, Mitran S, Sivanesan S, Chang E, Buga AM. ROS and brain diseases: the good, the bad, and the ugly. Oxidative Med Cell Longev 2013;2013. https://doi.org/10.1155/2013/963520.

[49] Reagan LP, Magariños AM, McEwen BS. Neurological changes induced by stress in streptozotocin diabetic rats. Ann N Y Acad Sci 1999;893:126–37. https://doi.org/10.1111/j.1749-6632.1999.tb07822.x.

[50] Kaiser N, Sasson S, Feener EP, Boukobza-Vardi N, Higashi S, Moller DE, Davidheiser S, Przybylski RJ, King GL. Differential regulation of glucose transport and transporters by glucose in vascular endothelial and smooth muscle cells. Diabetes 1993;42(1):80–9. https://doi.org/10.2337/diab.42.1.80.

[51] Goldin A, Beckman JA, Schmidt AM, Creager MA. Advanced glycation end products: sparking the development of diabetic vascular injury. Circulation 2006;114(6):597–605. https://doi.org/10.1161/CIRCULATIONAHA.106.621854.

[52] Zhu L, He P. fMLP-stimulated release of reactive oxygen species from adherent leukocytes increases microvessel permeability. Am J Phys Heart Circ Phys 2006;290(1):H365–72. https://doi.org/10.1152/ajpheart.00812.2005.

[53] Lavrovsky Y, Chatterjee B, Clark RA, Roy AK. Role of redox-regulated transcription factors in inflammation, aging and age-related diseases. Exp Gerontol 2000;35(5):521–32. https://doi.org/10.1016/S0531-5565(00)00118-2.

[54] Borne RT, O'Donnell C, Turakhia MP, Varosy PD, Jackevicius CA, Marzec LN, Masoudi FA, Hess PL, Maddox TM, Ho PM. Adherence and outcomes to direct oral anticoagulants among patients with atrial fibrillation: findings from the veterans health administration. BMC Cardiovasc Disord 2017;17(1):1–7. https://doi.org/10.1186/s12872-017-0671-6.

[55] Fadini GP, Sartore S, Schiavon M, Albiero M, Baesso I, Cabrelle A, Agostini C, Avogaro A. Diabetes impairs progenitor cell mobilisation after hindlimb ischaemia–reperfusion injury in rats. Diabetologia 2006;49:3075–84. https://doi.org/10.1007/s00125-006-0401-6.

[56] Simo R, Carrasco E, Garcia-Ramirez M, Hernandez C. Angiogenic and antiangiogenic factors in proliferative diabetic retinopathy. Curr Diabetes Rev 2006;2(1):71–98.

[57] Nakagawa T, Kosugi T, Haneda M, Rivard CJ, Long DA. Abnormal angiogenesis in diabetic nephropathy. Diabetes 2009;58(7):1471–8. https://doi.org/10.2337/db09-0119.

[58] Foster RR. The importance of cellular VEGF bioactivity in the development of glomerular disease. Nephron Exp Nephrol 2009;113(1):e8–15. https://doi.org/10.1159/000228078.

[59] Hohenstein B, Hausknecht B, Boehmer K, Riess R, Brekken RA, Hugo CP. Local VEGF activity but not VEGF expression is tightly regulated during diabetic nephropathy in man. Kidney Int 2006;69(9):1654–61. https://doi.org/10.1038/sj.ki.5000294.

[60] Eremina V, Jefferson JA, Kowalewska J, Hochster H, Haas M, Weisstuch J, Richardson C, Kopp JB, Kabir MG, Backx PH, Gerber HP. VEGF inhibition and renal thrombotic microangiopathy. N Engl J Med 2008;358(11):1129–36.

[61] Gnudi L. Angiopoietins and diabetic nephropathy. Diabetologia 2016;59(8):1616–20. https://doi.org/10.1007/s00125-016-3995-3.

[62] Dei Cas A, Gnudi L. VEGF and angiopoietins in diabetic glomerulopathy: how far for a new treatment? Metabolism 2012;61(12):1666–73. https://doi.org/10.1016/j.metabol.2012.04.004.

[63] Samii A, Unger J, Lange W. Vascular endothelial growth factor expression in peripheral nerves and dorsal root ganglia in diabetic neuropathy in rats. Neurosci Lett 1999;262 (3):159–62. https://doi.org/10.1016/S0304-3940(99)00064-6.

[64] Lin TH, Wang CL, Su HM, Hsu PC, Juo SH, Voon WC, Shin SJ, Lai WT, Sheu SH. Functional vascular endothelial growth factor gene polymorphisms and diabetes: effect on coronary collaterals in patients with significant coronary artery disease. Clin Chim Acta 2010;411(21–22):1688–93. https://doi.org/10.1016/j.cca.2010.07.002.

[65] Hochberg I, Hoffman A, Levy AP. Regulation of VEGF in diabetic patients with critical limb ischemia. Ann Vasc Surg 2001;15(3):388–92. https://doi.org/10.1007/s100160 010089.

[66] Wheeler ML, Dunbar SA, Jaacks LM, Karmally W, Mayer-Davis EJ, Wylie-Rosett J, Yancy Jr WS. Macronutrients, food groups, and eating patterns in the management of diabetes: a systematic review of the literature, 2010. Diabetes Care 2012;35(2):434–45. https://doi.org/10.2337/dc11-2216.

[67] Evert AB, Boucher JL, Cypress M, Dunbar SA, Franz MJ, Mayer-Davis EJ, Neumiller JJ, Nwankwo R, Verdi CL, Urbanski P, Yancy Jr WS. Nutrition therapy recommendations for the management of adults with diabetes. Diabetes Care 2014;37(Suppl 1):S120–43. https://doi.org/10.2337/dc14-S120.

[68] Vitolins MZ, Anderson AM, Delahanty L, Raynor H, Miller GD, Mobley C, Reeves R, Yamamoto M, Champagne C, Wing RR, Mayer-Davis E. Action for health in diabetes (look AHEAD) trial: baseline evaluation of selected nutrients and food group intake. J Am Diet Assoc 2009;109(8):1367–75. https://doi.org/10.1016/j.jada.2009.05.016.

[69] Estruch R, Ros E, Salas-Salvadó J, Covas MI, Corella D, Arós F, Gómez-Gracia E, Ruiz-Gutiérrez V, Fiol M, Lapetra J, Lamuela-Raventos RM. Primary prevention of cardiovascular disease with a Mediterranean diet supplemented with extra-virgin olive oil or nuts. N Engl J Med 2018;378(25), e34.

[70] Ferrari R, Censi S, Cimaglia P. The journey of omega-3 fatty acids in cardiovascular medicine. Eur Heart J Suppl 2020;22(Suppl J):49–53. https://doi.org/10.1093/eurheartj/suaa118.

[71] Karlström BE, Järvi AE, Byberg L, Berglund LG, Vessby BO. Fatty fish in the diet of patients with type 2 diabetes: comparison of the metabolic effects of foods rich in n–3 and

n–6 fatty acids. Am J Clin Nutr 2011;94(1):26–33. https://doi.org/10.3945/ajcn. 110.006221.

[72] Post RE, Mainous AG, King DE, Simpson KN. Dietary fiber for the treatment of type 2 diabetes mellitus: a meta-analysis. J Am Board Fam Med 2012;25(1):16–23. https://doi. org/10.3122/jabfm.2012.01.110148.

[73] Phielix E, Meex R, Moonen-Kornips E, Hesselink MK, Schrauwen P. Exercise training increases mitochondrial content and ex vivo mitochondrial function similarly in patients with type 2 diabetes and in control individuals. Diabetologia 2010;53:1714–21. https:// doi.org/10.1007/s00125-010-1764-2.

[74] Chimen M, Kennedy A, Nirantharakumar K, Pang TT, Andrews R, Narendran P. What are the health benefits of physical activity in type 1 diabetes mellitus? A literature review. Diabetologia 2012;55:542–51. https://doi.org/10.1007/s00125-011-2403-2.

[75] Hayes C, Kriska A. Role of physical activity in diabetes management and prevention. J Am Diet Assoc 2008;108(4):S19–23. https://doi.org/10.1016/j.jada.2008.01.016.

[76] Ansari T, Alhamad AR, Aloreyfij A, Alshmas B, Sami W. Asssociation between medication adherence and patients factors with type 2 diabetes mellitus in Majmaah City, Kingdom of Saudi Arabia. Pak Armed Forces Med J 2020;70(3):818–23.

[77] Tamborlane WV, Beck RW, Bode BW, Buckingham B, Chase HP, Clemons R, Fiallo-Scharer R, Fox LA, Gilliam LK, Hirsch IB. Juvenile Diabetes Research Foundation continuous glucose monitoring study group continuous glucose monitoring and intensive treatment of type 1 diabetes. N Engl J Med 2008;359(14):1464–76.

[78] Wong JC, Foster NC, Maahs DM, Raghinaru D, Bergenstal RM, Ahmann AJ, Peters AL, Bode BW, Aleppo G, Hirsch IB, Kleis L. Real-time continuous glucose monitoring among participants in the T1D exchange clinic registry. Diabetes Care 2014;37 (10):2702–9. https://doi.org/10.2337/dc14-0303.

[79] Schwedes U, Siebolds M, Mertes G, SMBG Study Group. Meal-related structured self-monitoring of blood glucose: effect on diabetes control in non-insulin-treated type 2 diabetic patients. Diabetes Care 2002;25(11):1928–32. https://doi.org/10.2337/diacare. 25.11.1928.

[80] Farmer A, Wade A, Goyder E, Yudkin P, French D, Craven A, Holman R, Kinmonth AL, Neil A. Impact of self monitoring of blood glucose in the management of patients with non-insulin treated diabetes: open parallel group randomised trial. BMJ 2007;335 (7611):132. https://doi.org/10.1136/bmj.39247.447431.BE.

[81] Saudek CD, Derr RL, Kalyani RR. Assessing glycemia in diabetes using self-monitoring blood glucose and hemoglobin A1c. JAMA 2006;295(14):1688–97. https://doi.org/ 10.1001/jama.295.14.1688.

[82] Rodbard HW, Blonde L, Braithwaite SS, Brett EM, Cobin RH, Handelsman Y, Hellman R, Jellinger PS, Jovanovic LG, Levy P, Mechanick JI. American Association of Clinical Endocrinologists medical guidelines for clinical practice for the management of diabetes mellitus. Endocr Pract 2007;13:1–68. https://doi.org/10.4158/ep.13.s1.1.

[83] Seino S, Sugawara K, Yokoi N, Takahashi H. β-Cell signalling and insulin secretagogues: a path for improved diabetes therapy. Diabetes Obes Metab 2017;19:22–9. https://doi.org/10.1111/dom.12995.

[84] Kalra S, Bahendeka S, Sahay R, Ghosh S, Md F, Orabi A, Ramaiya K, Al Shammari S, Shrestha D, Shaikh K, Abhayaratna S. Consensus recommendations on sulfonylurea and sulfonylurea combinations in the management of Type 2 diabetes mellitus— International Task Force. Indian J Endocrinol Metab 2018;22(1):132–57. https://doi. org/10.4103/ijem.IJEM_556_17.

[85] Sola D, Rossi L, Schianca GP, Maffioli P, Bigliocca M, Mella R, Corlianò F, Fra GP, Bartoli E, Derosa G. State of the art paper sulfonylureas and their use in clinical practice. Arch Med Sci 2015;11(4):840–8.

[86] Quillen DM, Samraj G, Kuritzky L. Improving management of type 2 diabetes mellitus: 2. Biguanides. Hosp Pract 1999;34(11):41–4. https://doi.org/10.1080/21548331.1999. 11443925.

[87] García Rubiño ME, Carrillo E, Ruiz Alcalá G, Domínguez-Martín A, Marchal J, Boulaiz H. Phenformin as an anticancer agent: challenges and prospects. Int J Mol Sci 2019;20 (13), 3316. https://doi.org/10.3390/ijms20133316.

[88] Bourron O, Daval M, Hainault I, Hajduch E, Servant JM, Gautier JF, Ferre P, Foufelle F. Biguanides and thiazolidinediones inhibit stimulated lipolysis in human adipocytes through activation of AMP-activated protein kinase. Diabetologia 2010;53:768–78. https://doi.org/10.1007/s00125-009-1639-6.

[89] Sanchez-Rangel E, Inzucchi SE. Metformin: clinical use in type 2 diabetes. Diabetologia 2017;60:1586–93. https://doi.org/10.1007/s00125-017-4336-x.

[90] Ghannay S, Snoussi M, Messaoudi S, Kadri A, Aouadi K. Novel enantiopure isoxazolidine and C-alkyl imine oxide derivatives as potential hypoglycemic agents: design, synthesis, dual inhibitors of α-amylase and α-glucosidase, ADMET and molecular docking study. Bioorg Chem 2020;104, 104270. https://doi.org/10.1016/j.bioorg. 2020.104270.

[91] Holt RI, Lambert KD. The use of oral hypoglycaemic agents in pregnancy. Diabet Med 2014;31(3):282–91. https://doi.org/10.1111/dme.12376.

[92] Hansen KB, Vilsbøll T, Knop FK. Incretin mimetics: a novel therapeutic option for patients with type 2 diabetes—a review. Diabetes Metab Syndr Obes 2010;3:155–63.

[93] Sun EW, De Fontgalland D, Rabbitt P, Hollington P, Sposato L, Due SL, Wattchow DA, Rayner CK, Deane AM, Young RL, Keating DJ. Mechanisms controlling glucose-induced GLP-1 secretion in human small intestine. Diabetes 2017;66(8):2144–9. https://doi.org/10.2337/db17-0058.

[94] Fulcher G, Matthews DR, Perkovic V, de Zeeuw D, Mahaffey KW, Mathieu C, Woo V, Wysham C, Capuano G, Desai M, Shaw W. Efficacy and safety of canagliflozin when used in conjunction with incretin-mimetic therapy in patients with type 2 diabetes. Diabetes Obes Metab 2016;18(1):82–91. https://doi.org/10.1111/dom.12589.

[95] Adeghate E, Kalász H. Suppl 2: amylin analogues in the treatment of diabetes mellitus: medicinal chemistry and structural basis of its function. Open Med Chem J 2011;5:78. https://doi.org/10.2174/1874104501105010078.

[96] Schmitz O, Brock B, Rungby J. Amylin agonists: a novel approach in the treatment of diabetes. Diabetes 2004;53(Suppl 3):S233–8. https://doi.org/10.2337/diabetes.53. suppl_3.S233.

[97] Hoogwerf BJ, Doshi KB, Diab D. Pramlintide, the synthetic analogue of amylin: physiology, pathophysiology, and effects on glycemic control, body weight, and selected biomarkers of vascular risk. Vasc Health Risk Manag 2008;4(2):355–62. https://doi.org/ 10.2147/vhrm.s1978.

[98] DiSanto RM, Subramanian V, Gu Z. Recent advances in nanotechnology for diabetes treatment. Wiley Interdiscip Rev Nanomed Nanobiotechnol 2015;7(4):548–64. https:// doi.org/10.1002/wnan.1329.

[99] Dash TK, Konkimalla VB. Poly-ε-caprolactone based formulations for drug delivery and tissue engineering: a review. J Control Release 2012;158(1):15–33. https://doi.org/ 10.1016/j.jconrel.2011.09.064.

[100] Rai VK, Mishra N, Agrawal AK, Jain S, Yadav NP. Novel drug delivery system: an immense hope for diabetics. Drug Deliv 2016;23(7):2371–90. https://doi.org/10.3109/10717544.2014.991001.

[101] Uppal S, Italiya KS, Chitkara D, Mittal A. Nanoparticulate-based drug delivery systems for small molecule anti-diabetic drugs: an emerging paradigm for effective therapy. Acta Biomater 2018;(81):20–42. https://doi.org/10.1016/j.actbio.2018.09.049.

[102] Ozeki T, Kano Y, Takahashi N, Tagami T, Okada H. Improved bioavailability of a water-insoluble drug by inhalation of drug-containing maltosyl-β-cyclodextrin microspheres using a four-fluid nozzle spray drier. AAPS PharmSciTech 2012;13:1130–7. https://doi.org/10.1208/s12249-012-9826-z.

[103] Cao SJ, Xu S, Wang HM, Ling Y, Dong J, Xia RD, Sun XH. Nanoparticles: oral delivery for protein and peptide drugs. AAPS PharmSciTech 2019;20:1. https://doi.org/10.1208/s12249-019-1325-z.

[104] Jeevanandam J, Barhoum A, Chan YS, Dufresne A, Danquah MK. Review on nanoparticles and nanostructured materials: history, sources, toxicity and regulations. Beilstein J Nanotechnol 2018;9(1):1050–74. https://doi.org/10.3762/bjnano.9.98.

[105] Reinholz J, Landfester K, Mailänder V. The challenges of oral drug delivery via nanocarriers. Drug Deliv 2018;25(1):1694–705. https://doi.org/10.1080/10717544.2018.1501119.

[106] Patel G, Shelat P, Lalwani A. Statistical modeling, optimization and characterization of solid self-nanoemulsifying drug delivery system of lopinavir using design of experiment. Drug Deliv 2016;23(8):3027–42. https://doi.org/10.3109/10717544.2016.1141260.

[107] Nasr A, Gardouh A, Ghorab M. Novel solid self-nanoemulsifying drug delivery system (S-SNEDDS) for oral delivery of olmesartanmedoxomil: design, formulation, pharmaco-kinetic and bioavailability evaluation. Pharmaceutics 2016;8(3):20. https://doi.org/10.3390/pharmaceutics8030020.

[108] Izham MN, Hussin Y, Aziz MN, Yeap SK, Rahman HS, Masarudin MJ, Mohamad NE, Abdullah R, Alitheen NB. Preparation and characterization of self nano-emulsifying drug delivery system loaded with citraland its antiproliferative effect on colorectal cells in vitro. Nanomaterials 2019;9(7). https://doi.org/10.3390/nano9071028.

[109] Prausnitz MR, Langer R. Transdermal drug delivery. Nat Biotechnol 2008;26(11):1261–8. https://doi.org/10.1038/nbt.1504.

[110] Shinkar DM, Jadhav SS, Pingale PL, Boraste SS, Vishvnath Amrutkar S. Formulation, evaluation, and optimization of glimepiride nanosuspension by using antisolvent evaporation technique. Pharmacophore 2022;13(4). https://doi.org/10.51847/1yGT4slm1W.

[111] Ahmed OA, Afouna MI, El-Say KM, Abdel-Naim AB, Khedr A, Banjar ZM. Optimization of self-nanoemulsifying systems for the enhancement of in vivo hypoglycemic efficacy of glimepiride transdermal patches. Expert Opin Drug Deliv 2014;11(7):1005–13. https://doi.org/10.1517/17425247.2014.906402.

[112] Emami J, Boushehri MS, Varshosaz J. Preparation, characterization and optimization of glipizide controlled release nanoparticles. Res Pharm Sci 2014;9(5):301. PMC4317998.

[113] Akhtar J, Siddiqui HH, Fareed S, Badruddeen, Khalid M, Aqil M. Nanoemulsion: for improved oral delivery of repaglinide. Drug Deliv 2016;23(6):2026–34. https://doi.org/10.3109/10717544.2015.1077290.

[114] Rawat MK, Jain A, Mishra A, Muthu MS, Singh S. Development of repaglinide loaded solid lipid nanocarrier: selection of fabrication method. Curr Drug Deliv 2010;7(1):44–50. https://doi.org/10.2174/156720110790396472.

[115] Espinoza LC, Vera-García R, Silva-Abreu M, Domènech Ò, Badia J, Rodríguez-Lagunas MJ, Clares B, Calpena AC. Topical pioglitazone nanoformulation for the treatment of

atopic dermatitis: design, characterization and efficacy in hairless mouse model. Pharmaceutics 2020;12(3):255. https://doi.org/10.3390/pharmaceutics12030255.

[116] Alshora DH, Alsaif S, Ibrahim MA, Ezzeldin E, Almeanazel OT, Abou El Ela AE, Ashri LY. Co-stabilization of pioglitazone HCL nanoparticles prepared by planetary ball milling: in-vitro and in-vivo evaluation. Pharm Dev Technol 2020;25(7):845–54. https://doi.org/10.1080/10837450.2020.1744163.

[117] Harsha S, Attimard M, Khan TA, Nair AB, Aldhubiab BE, Sangi S, Shariff A. Design and formulation of mucoadhesive microspheres of sitagliptin. J Microencapsul 2013;30(3): 257–64. https://doi.org/10.3109/02652048.2012.720722.

[118] SreeHarsha N, Ramnarayanan C, Al-Dhubiab BE, Nair AB, Hiremath JG, Venugopala KN, Satish RT, Attimarad M, Shariff A. Mucoadhesive particles: a novel, prolonged-release nanocarrier of sitagliptin for the treatment of diabetics. Biomed Res Int 2019;2019. https://doi.org/10.1155/2019/3950942.

[119] Dora CP, Singh SK, Kumar S, Datusalia AK, Deep A. Development and characterization of nanoparticles of glibenclamide by solvent displacement method. Acta Pol Pharm 2010;67(3):283–90.

Targeting angiogenesis, inflammation, and oxidative stress in obesity

Soumya Gupta and Rohini Verma
School of Health Sciences and Technology, University of Petroleum and Energy Studies, Dehradun, Uttarakhand, India

1 Introduction

Overweight and obesity are persistent public health problems in both developed and developing countries and pose a serious health risk. The health issue has become an epidemic such that as per global estimates [1] more than one billion people worldwide are obese out of which 650 million are adults and approximately 39 million children are affected. Rates of overweight and obesity are continuing to grow in adults and children not only in developed countries with factors like rapid urbanization and inappropriate food environments, The rising trends in the incidence of overweight and obesity are particularly worrisome for many developing countries, since they are now facing a double burden of malnutrition. Low- and middle-income countries (LMICs) have the highest number of people affected by obesity, with a significant increase observed in recent years. The prevalence of obesity has more than doubled across all LMICs and tripled in low-income countries since 2010. According to the National Family Health Survey [2], obesity rates in India have risen from 21% to 24% among women and 19% to 23% among men between the fourth and fifth rounds. In addition, childhood obesity has become a rapidly growing public health concern in LMICs, highlighting the importance of early prevention measures [1].

Obesity is a chronic, relapsing, multifactorial disease [3] as defined by the International Classification of Disease (ICD) [4,5]. It is a complex disease triggered by interlinked biological causes, genetics to dysfunctional food systems play an important role. Despite these complexities, most of the current preventive, and treatment strategies are individualistic and simple, center on the "eat less, move more" mantra, and place the blame solely on individual behavior [4].

Numerous studies have demonstrated that Asian Indians have a greater likelihood of developing insulin resistance and cardiovascular risk factors when compared to white Caucasians of the same age and BMI [6–8]. The disproportionate accumulation of cardiovascular risk factors in Asian Indians may be partly due to differences in their body composition compared to white Caucasians. Asian Indians tend to have higher amounts of total, truncal, intra-abdominal, subcutaneous, and ectopic fat tissues, even at lower or equivalent BMI levels [8–12], and hence are at a higher risk of developing

Targeting Angiogenesis, Inflammation and Oxidative Stress in Chronic Diseases. https://doi.org/10.1016/B978-0-443-13587-3.00007-2

obesity related comorbidities at lesser BMI levels. Effective interventions are urgently required for Asian Indians to tackle the increasing prevalence of obesity and metabolic diseases, which have significant economic implications. Although the prevalence of obesity and overweight in Asian populations is lower than in developed countries like the USA, Asian Indians are at a greater risk of developing comorbidities associated with obesity at a lower BMI. Therefore, prompt action is necessary to address this issue [13].

Overweight and obesity are most commonly assessed and defined using the Body Mass Index (BMI). BMI is measured as the ratio of weight in kilograms divided by the height in meters squared, expressed as kg/m^2. Ideally, BMI should assess adiposity correctly; however, BMI cannot distinguish between muscle mass and fat mass, offering several limitations in its correct assessment as well [14]. WHO/NUT/NCD [15] proposed the following cut-offs for overweight and obesity in adults: a BMI of 25–$29.9 kg/m^2$ defines overweight, while a BMI $\geq 30 kg/m^2$ defines obesity. Keeping ethnicity into consideration, WHO, IASO, and IOTF [16] recommended the following guidelines in the Asia-specific perspective to redefine obesity and its treatment: a BMI of 23–$24.9 kg/m^2$ defines overweight, while a BMI of 25–$29.9 kg/m^2$ defines obesity.

Obesity often starts early in life [4], even before a baby is conceived and hence preventive measures should ideally start during the preconception period. Apparently, along with obesity occur its associated metabolic consequences including hypertension, metabolic syndrome, dyslipidemia, type II diabetes mellitus (T2DM), cardiovascular diseases, cancer, and mental health issues among others, which are hard to dispense with.

However, as the prevalence of overweight and obesity is rising all over the world, our knowledge of its possible metabolic and pathological consequences is still equivocal. The exact mechanism by which obesity is linked to associated medical conditions is not yet completely understood. However, a growing body of evidence points to a potential connection between chronic inflammation and oxidative stress occurring within adipose tissue, leading to an inflammatory state, as a key factor in the development of metabolic problems related to obesity [17].

2   Pathophysiology of obesity

Obesity is caused by a combination of factors including malnutrition, a sedentary lifestyle, and genetic predisposition. Aside from dysfunctional angiogenesis, obesity is characterized by an abnormal inflammatory response, low antioxidant capacity, and decreased insulin sensitivity, all which may eventually lead to inflammation, oxidative stress, and insulin resistance [17]. AT function is important for the human body as it plays a role in energy metabolism, and releases fatty acids and hormones that regulate energy balance in the body. Additionally, adipose tissue functions as an endocrine organ by secreting a variety of adipokines that affect inflammation, insulin sensitivity, and other physiological processes. The defining feature of adipose tissue (AT) is the presence of adipocytes, which have the unique capacity to accumulate large amounts of lipids within specialized droplets, giving adipose depots their characteristic appearance [18].

2.1 Morphology of adipocytes

Adipocytes possess a similar physical structure, yet they constitute a heterogeneous group of cells that exhibit diverse functionalities owing to the presence of varied adipocyte subtypes in each depot. These subtypes include beige/brite, white (WAT), and brown (BAT) adipocytes that exist in varying proportions within individual depots and have been linked with diverse and mixed health outcomes. For example, brown adipocytes, which surround internal organs, are characterized by high mitochondrial density and play a crucial role in adaptive thermogenesis. However, white adipocytes (WAT) are the predominant type of adipocytes, and they can increase in size in response to increased caloric intake [19,20]. However, excessive expansion of WAT can lead to negative health outcomes through its secretion of proinflammatory cytokines. Other depots are more restricted in their ability to expand and have other functions including immunological surveillance [21] and regulation of vessel tone [22]. Adipose tissues are a complex organ that not only consist of adipocytes (40%–50%) but also connective tissue matrix, vascular and neural cells, and non-adipocyte cells called stromal vascular fraction (SVF). SVF includes preadipocytes, immune cells (macrophages, natural killer cells, B-lymphocytes, and T-lymphocytes), endothelial cells (ECs), vascular progenitors, fibroblasts, and mesenchymal stem cells [23].

The white adipose tissues, based on their distribution site, are mainly divided into two types: subcutaneous adipose tissue (SAT) and visceral adipose tissue (VAT). Overnutrition, which is the consumption of excessive amounts of food and nutrients, can lead to surplus energy, weight gain, and obesity. When energy intake exceeds energy expenditure, the surplus energy is stored in the SAT, primarily as triglycerides, leading to AT expansion. Adipose tissue can expand to store increasing amounts of lipids, but eventually, when the capacity of the adipose tissue is exceeded, lipids can accumulate in other organs and tissues [24].

Local subgroups of white adipose tissue (WAT) also have specialized roles, and excessive accumulation of WAT in certain body sites, such as the upper body (android obesity) and lower body (gynoid obesity), can lead to the development of obesity-related diseases. Android obesity is a strong risk factor for inflammatory pathologies [25], while gynoid obesity does not typically result in metabolic complications [25,26].

2.2 Adipose tissue expansion

AT expansion refers to the increase in size and number of adipocytes in the adipose tissue, which can occur in response to overnutrition and a calorie surplus. Adipose tissue can expand through two main mechanisms: hypertrophy and hyperplasia. Hypertrophy is the increase in size of existing adipocytes, which occurs when the cells take up more fat and increase in volume while hyperplasia is the increase in the number of adipocytes, which occurs when new adipocytes are formed from precursor cells called preadipocytes.

The expansion of adipose tissue during obesity involves an increase in both the number and size of adipocytes, leading to significant changes in the structure and function of adipose tissue. This process, known as "adipose tissue remodeling" [24], is associated with dysregulated secretion of adipokines and cytokines, resulting in increased secretion of proinflammatory molecules that promote systemic low-grade inflammation [27]. Such inflammation is known to trigger the infiltration of macrophages and T-cells into adipose tissue, contributing to the pathogenesis of obesity [28]. AT can also become dysfunctional, leading to a state of chronic low-grade inflammation and the release of proinflammatory molecules, which can contribute to the development of metabolic diseases such as type 2 diabetes and cardiovascular disease [29].

2.3   Secretion of adipokines by adipose tissues

Adipocytes produce a variety of signaling molecules that help regulate multiple physiological processes in the body. The cytokines released by adipocytes can either mitigate or exacerbate metabolic complications, depending on their systemic or intracellular levels and the modulation of cellular signaling pathways [30]. These molecules include hormones, cytokines, and growth factors. Some examples of adipokines include leptin, which regulates energy balance and appetite, and adiponectin, which improves insulin sensitivity and glucose metabolism. Adipokines are signaling molecules that are secreted by adipose tissue and play a role in regulating various physiological processes in the body. Adipokines also play a role in the regulation of inflammation, blood pressure, and cardiovascular health. Adipose tissue can secrete both proinflammatory and antiinflammatory adipokines, and the balance of these molecules can have a significant impact on overall health. Therefore, dysregulation of adipokine secretion, characterized by altered levels of antiinflammatory and proinflammatory adipokines, can contribute to the pathogenesis of metabolic disorders like type 2 diabetes and cardiovascular disease [31].

The endocrine factors that modulate insulin signaling, adipogenesis, preadipocyte proliferation and differentiation, and mitochondrial energy dissipation via lipid metabolism regulation are critical mechanisms in the context of obesity. Therefore, targeting these variables at the systemic or intracellular level could be a promising therapeutic strategy for preventing or treating obesity-related complications [17].

2.4   Adipose tissue expansion in an obese state

Adipose tissue expansion in obesity can occur in various depots, such as the subcutaneous (under the skin) and visceral (around organs) depots. Visceral adipose tissue expansion is a significant concern due to its higher metabolic activity, resulting in an augmented secretion of proinflammatory adipokines that may contribute to the pathogenesis of metabolic disorders, including type 2 diabetes, cardiovascular disease, and some cancers [32].

Obesity can lead to an increase in adipocyte size, known as hypertrophic obesity, which is associated with metabolic complications, including insulin resistance and inflammation. Additionally, weight gain can lead to an increase in the number of small adipocytes, known as hyperplastic obesity, which is associated with increased metabolic flexibility and better insulin sensitivity. However, it is important to note that weight loss can lead to a reduction in the size and number of adipocytes but depending on the cause of the obesity and the duration, these mechanisms may or may not return to normal [33].

In individuals with obesity, unregulated expansion of adipose tissue can result in lipid metabolism dysfunction, which can include excessive lipolysis. This, in turn, can cause an increase in the production and release of free fatty acids (FFAs) into the bloodstream [34]. Lipid metabolism refers to the process by which lipids are stored, used, and broken down by the body. The excessive accumulation of lipids within adipocytes leads to dysfunction of the adipose tissue, including decreased insulin sensitivity, increased inflammation, and altered secretion of adipokines. Dysregulation of lipid metabolism manifests in various ways, such as increased hepatic lipogenesis, which is triggered by an influx of glucose and insulin. This results in an accumulation of triglycerides in the liver, also known as nonalcoholic fatty liver disease (NAFLD), which can impair liver function and increase the risk of liver failure [35]. Additionally, excessive accumulation of lipids in muscle tissue, known as intramyocellular lipids (IMCLs), can lead to insulin resistance, reduced muscle function, and a heightened risk of developing type 2 diabetes [36]. The dysfunction of adipose tissue in obesity also results in the overproduction and secretion of free fatty acids (FFAs), which can exacerbate NAFLD and contribute to insulin resistance when transported to muscle tissue.

Adipose tissue expansion is also associated with chronic low-grade inflammation that is mainly characterized by the infiltration of immune cells, especially macrophages, into adipose tissue [37]. The accumulation of adipose tissue, especially in the abdominal area, can trigger the release of proinflammatory cytokines, which are linked to the severity of metabolic dysfunction [38]. Inflammatory cells, such as macrophages, are commonly found in the adipose tissue of obese individuals, and they contribute to the increased production of proinflammatory cytokines, including tumor necrosis factor-alpha (TNF-alpha) and interleukin-6 (IL-6), by adipocytes. This proinflammatory environment can promote the development of metabolic disorders, such as type 2 diabetes and cardiovascular disease [37].

Adipocyte inflammation caused by the accumulation of immune cells can also alter the secretion of adipokines in obesity. This alteration may cause an imbalance between the production of antiinflammatory and proinflammatory adipokines such as adiponectin and leptin, respectively. Consequently, this disturbed adipokine profile can lead to metabolic disorders by affecting insulin sensitivity, glucose metabolism, and inflammation. For example, low adiponectin levels have been linked to insulin resistance and an increased risk of type 2 diabetes, while elevated levels of proinflammatory cytokines such as TNF-alpha and IL-6 have been associated with an increased risk of cardiovascular disease [39]. In addition, the accumulation of immune cells in adipocytes can produce reactive oxygen species (ROS), which can exacerbate adipose tissue dysfunction and metabolic disorders [40].

It is also important to note that the inflammation and altered adipokine profile in obesity can also be influenced by other factors such as genetics, diet, and physical activity. Therefore, a multimodal approach that includes lifestyle modifications such as diet and exercise, as well as pharmacological interventions, may be necessary to improve the inflammation and adipokine profile in obesity.

2.4.1   Molecular mechanisms that lead to adipose tissue expansion

Mechanisms include changes in the transcriptional regulation of genes involved in adipocyte differentiation, hypertrophy, and lipogenesis (the synthesis of fat).

a. One of the key molecular mechanisms that drive the expansion of adipose tissue in obesity is the activation of the peroxisome proliferator-activated receptor-gamma (PPAR-γ) pathway. PPAR-γ is a transcription factor that plays a critical role in the differentiation of pre-adipocytes into mature adipocytes and the regulation of lipid metabolism. In obesity, PPAR-γ is activated, leading to an increase in the number of adipocytes and the accumulation of lipids within these cells [24,41].
b. Another mechanism that contributes to the expansion of adipose tissue in obesity is the activation of the mammalian target of rapamycin (mTOR) pathway. The mTOR pathway regulates cell growth and proliferation, and in obesity, it is activated leading to an increase in the size of adipocytes and the number of adipocytes [42].
c. Insulin resistance is another mechanism that contributes to the expansion of adipose tissue in obesity. Insulin resistance is a common feature of obesity and is characterized by a decrease in the ability of insulin to stimulate glucose uptake in adipose tissue and muscle. This leads to an increase in blood glucose levels and can contribute to the development of type 2 diabetes [43].

These molecular mechanisms are interrelated and a change in one of them can affect the other.

3   Oxidative stress and inflammation in the pathogenesis of obesity

Accumulation of abdominal fat initiates inflammation and a prooxidant state [44]. Oxidative stress (OS) is essentially increased levels of reactive oxygen species (ROS) and reactive nitrogen species (RNS) with decreased antioxidant capacity [45]. Hydrogen peroxide (H_2O_2), superoxide (O_2^-), hydroxyl radicals (OH), hypochlorite (ClO^-), nitric oxide (NO), and peroxynitrite ($ONOO^-$) are few examples of ROS and RNS. They can be generated intracellularly by various cell organelles such as mitochondria, endoplasmic reticulum (ER), lysosomes, peroxisomes as well as a few cytosolic enzymes. At optimal concentrations, they have a wide range of biological impacts, including the immune system and intracellular signaling-mediated defense against harmful microbes. At the cellular level under pathological stress, ROS can lead to the progression of various diseases like obesity, cardiovascular disease, diabetes, cancer, etc. They modulate mitochondrial activity and energy balance in hypothalamic neurons and can cause damage to proteins, lipids, and DNA. They regulate adipocyte maturation, hunger, the level of inflammatory mediators, lipogenesis, and preadipocyte differentiation [46].

Tissues have antioxidant components that collaborate to maintain ROS/RNS at appropriate levels. Glutathione, ubiquinone, thioredoxin, and urate are some examples of naturally occurring antioxidant compounds. Proteins with antioxidant characteristics include ferritin, transferrin, lactoferrin, and caeruloplasmin because they bind to mitigate the action of transition metals that could otherwise start oxidative processes. Other naturally occurring antioxidant enzymes include catalase, thioredoxin reductase, peroxiredoxins (Prx), superoxide dismutase (SOD), glutathione peroxidase (GPx), glutathione reductase, glutathione S-transferase, and ubiquinone oxidoreductase (NQO1) [47].

3.1   Role of ROS

Adipocyte differentiation is a complex process mediated through various pathways and involves the expression of several transcription factors, cell-cycle proteins, hormones, and small compounds. These pathways and transcription factors are important in the development of obesity. Many receptors are involved in the regulation of such pathways such as tyrosine kinase receptor, AMP-activated protein kinase (AMPK), peroxisome proliferator-activated receptor γ (PPARγ), PPAR coactivator 1α (PGC-1α), and CCAAT/enhancer-binding protein β (C/EBPβ). Signal transduction and receptors in these pathways are sensitive to and regulated by ROS [48]. Under oxidative stress, there is an increase in the expression PPARγ, C/EBPβ, and PGC-1α. Further, in preadipocytes, the insulin-like growth factor (IGF) receptor regulates ROS-sensitive downstream signaling pathways like phosphatidylinositol 3- kinase (PI 3-kinase) and the Ras-mitogen-activated protein kinase (MAPK), where; MAPK regulates the growth and differentiation of preadipocyte. MAPK is controlled and modulated by ROS [49].

Renin-angiotensin system (RAS) is associated with Obesity and Insulin resistance. Angiotensin II (Ang II), secreted by preadipocytes, favors the synthesis of prostaglandin I2 which stimulates adipocyte differentiation and accumulation of triglycerides. It also stimulates the release of the satiety hormone, leptin. Nicotine adenine dinucleotide phosphate (NADPH) oxidase (NOX) family of enzymes are key producers of ROS in preadipocytes [48,50]. NOX controls cell growth, differentiation, metabolism, and apoptosis (Fig. 1). Ang II stimulates NOX, which transfers electrons from NADPH to produce superoxide anion (O_2^-) which is further converted to H_2O_2 by superoxide dismutase. In healthy adipocytes, insulin stimulation increases Nox4-mediated H_2O_2 generation, increasing its signaling by inhibiting the protein tyrosine phosphatase; Ptp1b. This encourages the activation of insulin receptors and glucose uptake. Similarly, in preadipocytes, Nox4-derived H_2O_2 promotes insulin-mediated Akt activation and promotes preadipocyte differentiation into adipocytes [51].

Further, mitochondria have an efficient antioxidant system that neutralizes ROS in low concentrations. Mitochondria produces ATP via. Oxidative phosphorylation in conjunction with oxidation of metabolites via. Krebs' cycle and β-oxidation of fatty acids [52]. Under pathological conditions or when excess glucose is available, there is

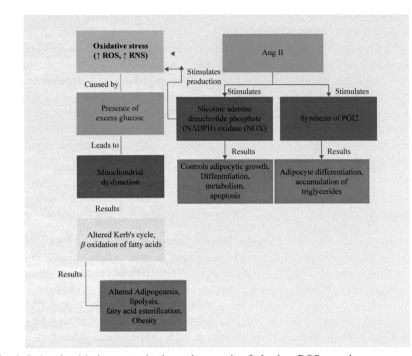

Fig. 1 Role of oxidative stress in the pathogenesis of obesity. *ROS*, reactive oxygen species; *Ang II*, angiotensin II; *NOX*, nicotine adenine dinucleotide phosphate oxidase; *RNS*, reactive nitrogen species; *PGI2*, prostaglandin I2.

an increased generation of ROS owing to the altered morphology and dynamics of mitochondria. ROS react with mitochondrial lipids, proteins, and DNA, and may induce alterations. They affect several constituents of the respiratory chain and Krebs cycle enzymes. This altogether may lead to altered adipogenesis, lipolysis, fatty acid esterification, and adiponectin production which results in obesity (Fig. 1) [52,53].

In addition, lipogenesis and lipolysis are medicated by enzymes such as lipoprotein lipase (LPL) and Hormone sensitive lipase (HSL), which are sensitive to ROS where lipogenesis is the process by which fatty acids are esterified with glycerol to form triglycerides which are stored as fat droplets in adipocyte and lipolysis is the breakdown stored triglycerides [52].

4   Role of angiogenesis in the development of obesity

Adipose tissue (AT) is composed of adipocytes, which are surrounded by various supportive tissues, including connective, vascular, and neural tissues. Additionally, non-adipocyte cells, collectively known as the stromal vascular fraction (SVF), are present in AT. These cells include immune cells (such as macrophages, natural killer cells, B-lymphocytes, and T-lymphocytes), endothelial cells (ECs), vascular and adipogenic progenitors, fibroblasts, and mesenchymal stem cells, all which are crucial for AT expansion and repair. The plasticity of AT allows it to expand greatly through hypertrophy and hyperplasia in adulthood, which is supported by neovascularization or vascular remodeling [54].

The expansion of adipose tissue relies heavily on the process of angiogenesis, which involves the formation of new blood vessels that supply nutrients and oxygen to the growing adipocytes. This increased vasculature not only provides oxygen but also creates an environment for multipotent progenitor cells to thrive. There is a direct correlation between the development of microvasculature and the proliferation of mesenchymal precursors. Studies have shown that promoting angiogenesis also promotes the proliferation of progenitor cells [55]. In animal models of obesity induced by a high-fat diet or genetic mutations, there is a significant increase in adipose tissue mass, along with a higher number of dividing endothelial cells, increased blood vessel volume, and higher blood counts compared to lean controls on a normal diet [56].

Angiogenesis in healthy adipose tissue includes the proliferation of primitive blood vessels in loose connective tissue followed by the accumulation of cytoplasmic microscopic fat vacuoles in mesenchymal cells, and finally, clear adipocytes and fat lobules are formed [57]. Key features in angiogenesis include:

I. Proliferation followed by migration of endothelial cells (EC) through extracellular matrix,
II. Formation of intercellular junctions and lumens,
III. Organization of perivascular support cells, and anastomosis with pre-existing vessels,
IV. Establishment of circulation.

4.1 Vascular endothelial growth factors (VEGF), placental growth factors (PlGF), notch signaling pathway in adipose tissue angiogenesis

The VEGF along with their receptors are the master regulator of the proliferation of EC. Deletion of VEGF in AT leads to reduced angiogenesis resulting in hypoxia and inflammation. The family of VEGF comprises six main growth factors ((VEGF-A)–(VEGF-F)). VEGF-A is the main regulator of angiogenesis during the expansion of adipose tissue and is associated with the differentiation of adipocytes and the proliferation of vascular smooth muscles [58]. VEGF-A interacts with a tyrosine kinase receptor (VEGF receptor 2) in a dose-dependent manner and exhibits potent mitogenic, angiogenic, and chemo-attractant signals to EC. VEGFR2 mediates most of the cellular responses to induce migration, survival, and proliferation of ECs (Fig. 2). Several studies show that deletion VEGF-A allele can result in impaired blood supply to the spinal cord and degeneration of motor neurons. Similarly, blocking VEGFR2 limits diet-induced AT expansion by decreasing angiogenesis and adipogenesis [59,60].

Dividing EC during angiogenesis acquires a specific phenotype (tip cell phenotype) marked by the development of numerous filopodia and subsequent vascular branches. Notch signaling pathway controls the response of EC to intracellular VEGF signals [61]. Delta like 4 (Dll4); a ligand for Notch, activates Notch signaling in tip cells, which suppresses the signaling of VEGF in neighboring cells. Hence there

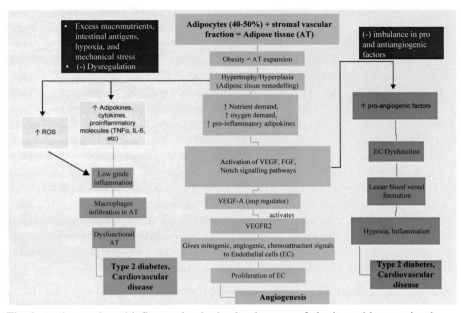

Fig. 2 Angiogenesis and inflammation in the development of obesity and its associated complications. *TNFα*, tumor necrosis factor α; *IL-6*, interleukin 6; *ROS*, reactive oxygen species; *VEGF*, vascular endothelial growth factor; *FGF*, fibroblast growth factor.

is high-level expression of the Notch ligand, Dll4 during EC proliferation. A continuous dynamic interaction between Dll4, Notch, and VEGF leads to the formation of angiogenic sprouts. Newly formed sprouts engage junctional transmembrane proteins such as VE-cadherin and matrix proteins that dynamically degrade and reorganize during vessel growth through interactions with smooth muscle cells and pericytes. It is stabilized and lumenised through a process that appears to be newly formed sprouts anastomose with existing vessels and expands tissue microcirculation [62]. VEGF overexpression is associated with increased insulin sensitivity and improved glucose tolerance due to angiogenesis. Overexpression of VEGF in WAT and BAT in mice led to increased number and size of blood vessels, increased insulin sensitivity, and improved glucose tolerance. PlGF has a 53% sequence identity with VEGF-A165 and is a homolog of VEGF which is responsible for pathological angiogenesis [57,62].

In the obese state, excess lipid accumulation in adipocytes triggers the secretion of proinflammatory adipokines, which in turn necessitates the formation of new blood vessels through the process of angiogenesis. Angiogenesis is regulated by the activity of growth factors such as VEGF and (Fibroblast growth factor) FGF, which stimulate the formation of new blood vessels and promote the expansion of adipose tissue. However, this process can also disrupt the balance between pro- and antiangiogenic factors, leading to dysfunction of endothelial cells (ECs) [59]. Impaired EC function can interfere with normal angiogenesis and reduce vascular density, which increases the risk of hypoxia and inflammation (Fig. 2). These factors can contribute to the development of metabolic diseases such as type 2 diabetes and cardiovascular disease [60,63].

It is important to note that, while angiogenesis is essential for the growth and expansion of adipose tissue, excessive angiogenesis can also lead to abnormal growth and development of blood vessels, which can contribute to the development of metabolic diseases. During hypoxia, the reduction of angiogenic growth factors like VEGF can impede the formation of new blood vessels in adipose tissue [64,65]. As a result, modulation of angiogenesis and vasculature has been suggested as a potential therapeutic strategy to mitigate the complications associated with obesity [66].

4.2 Hypoxia in adipose tissue angiogenesis

Adipocytes can store readily available nutrients as lipids in the form of lipid droplets in their cytoplasm. This results in adipocyte hypertrophy and a reduction in the supply of oxygen to AT causing the formation of the hypoxic state. This hypoxic status induces inflammation and requires remodeling of the extracellular matrix to increase blood supply and alleviate the condition. Several research publications suggest that a limited vascular supply can limit the growth and expansion of AT. Animal studies comprising HFD-induced obesity suggest growth of the vascular network is not parallel with that of AT growth resulting in hypoxia, causing inflammation and insulin resistance. In contrast, enhanced vascularization, reduced inflammation, and amelioration of HFD-induced insulin resistance occur when VEGF is overexpressed in adipose tissue. In obesity, with saturated storage capacity and malfunctioning adipocyte metabolism, there is a greater release of free fatty acids (FFAs) from adipocytes into the bloodstream [67].

5 Molecular mechanisms relating inflammation to obesity

Obesity is generally associated with low-grade inflammation. A few triggers of this inflammation are the presence of excess macronutrients, intestinal antigens, hypoxia, and mechanical stress. These triggers follow an acute, adaptive inflammatory response which is marked by an increase in systemic markers of inflammation. This contributes to the activation of the immune system leading to the development of obesity-associated pathologies like insulin resistance, altered glucose and lipid metabolism, atherosclerosis, hypertension, and metabolic syndrome (Fig. 2) [68].

AT is essentially involved in regulating energy balance and homeostasis. Their primary function is as a storage depot. They can sense energy storage and secrete hormones such as leptin which activate the sympathetic nervous system and reduce energy intake. In addition, they can activate β-adrenergic receptors on adipocytes, increasing the rate of lipolysis and thermogenic processes. Hypertrophy and hyperplasia are associated with hemostatic stress caused due to loss of energy balance and attaining a hyper-anabolic state. Disbalance in the energy initiates the release of various mediators leading to adaptive inflammatory response. There is a change in cytokine secretion, oxygen depletion, necrosis, increased recruitment of immune cells, and dysregulated fat metabolism. Adipose tissue expresses more than a dozen genes that encode inflammatory proteins, including IL-6, monocyte chemoattractant protein 1

(MCP-1), inducible nitric oxide synthase (iNOS), Matrix metalloproteinases (MMPs), and lipocalin [68,69]. Many of these proteins for example, IL-6 are associated with maintaining energy balance. The IL-6 receptor, expressed in the hypothalamus, regulates appetite and energy intake and plays a part in the regulation of energy homeostasis by inhibiting lipoprotein lipase activity [70,71].

Intestinal permeability is increased in obesity resulting in higher circulating levels of intestinal antigens like Lipopolysaccharide (LPS). LPS initiate pattern recognition receptors (PRRs), such as toll-like receptor 4 (TLR4) in adipocytes, and is associated with changes in gastrointestinal microbiota impacting body fat and insulin resistance. The TLR pathway, a family of surface receptors, is also activated by free fatty acids. Every cell has its own receptors. It controls the expression of several inflammatory genes and is crucial in vascular dysfunction and insulin resistance. Further, TLR2 and TLR4 upregulate the generation of ROS [68,71].

Excessive feeding and high free fatty acids activate nuclear factor kappa-light-chain-enhancer of activated B cells (NF-κB) and extracellular-signal-regulated kinase (ERK) signaling pathways. The NF-Bκ signaling regulates the activity of the inflammasomes pathway by stimulating the transcriptional expression of NOD-, LRR-, and pyrin domain-containing protein 3 (NLRP3) [68,71].

In stable energy homeostasis, immune cells, and AT coordinate to control energy storage and immobilization. In lean individuals, AT macrophages are <10%, whereas in the case of the obese individual, it can be as high as 40%–50% of total AT. It is also accompanied by abnormal cytokine and chemokine production. IL-33 regulates the Adipose tissue integrity and metabolism is regulated by IL-33. Eosinophils are activated by IL-33, which causes innate lymphoid cells to release IL-5 and IL-13. Eosinophils, in turn, release IL-4, which stimulates the polarization of M2 macrophages and causes the differentiation of beige adipocytes. The production of genes encoding antiinflammatory proteins such as IL-10, IL-1R, arginase 1 (ARG1), and chitinase-like protein 3 (CHIL3) is induced by the peroxisome proliferator-activated receptor-gamma (PPARγ) and PPAR. In addition to being antiinflammatory, IL-10 prevents lipolysis and maintains the insulin sensitivity of adipocytes [68].

6 Treatment for obesity

6.1 Lifestyle modifications to manage and prevent obesity

Lifestyle modifications are an effective way to prevent and manage obesity. These modifications include changes in diet, physical activity, and eating behaviors [72,73]. A balanced diet is essential for maintaining a healthy weight. It is recommended to consume a variety of fruits, vegetables, whole grains, lean proteins, and healthy fats. It is also important to limit the intake of processed foods, sugary drinks, and high-fat foods. Regular physical activity is also important, with experts recommending for at least 150 min of moderate-intensity exercise per week. This can include activities such as walking, cycling, swimming, or any other form of exercise that increases heart rate and burns calories [74]. In addition to diet and exercise, it is important to modify eating behaviors including eating slowly, avoiding distractions while eating, being mindful of portion sizes, and avoiding emotional eating. It is also important to limit sedentary behaviors such as watching TV or sitting for long periods. Keeping a food diary can also help identify areas for improvement [75]. Other lifestyle modifications include getting enough sleep, reducing stress, and avoiding smoking and excessive alcohol consumption. Seeking support from family, friends or a healthcare professional can also help in making lifestyle changes and managing obesity. Additionally, healthcare providers and health systems can implement strategies to prevent excess weight gain in children and adolescents, such as screening for obesity and offering or referring children with obesity to family-centered, comprehensive, intensive behavioral interventions to promote improvements in weight status [76].

6.2 Natural compounds to manage and prevent obesity

The metabolic and endocrine processes as well as the overall energy balance are highly impacted by diet. Most health guidelines place a strong emphasis on eating a diet comprising fruits and vegetables in major portions as they have a lower calorie content and higher nutrient density. Such diverse diets contain a variety of bioactive substances, which are known to have positive health benefits, especially in terms of antiinflammatory activity [77].

Inflammation, insulin resistance, and other obesity-related complications can be improved by dietary bioactive substances, such as polyphenols and specific fatty acids, which are known to decrease both systemic and adipose tissue inflammation [78].

Besides, there has been an increasing interest in the use of natural compounds and plant-derived extracts for the treatment of obesity. Several natural compounds, such as curcumin, resveratrol, and quercetin, have been shown to possess antiinflammatory and antioxidant properties, which can help reduce adipose tissue inflammation and oxidative stress [79].

6.2.1 Polyphenolic compounds

Polyphenolic compounds are natural compounds found in plant-based foods such as fruits, vegetables, whole grains, tea, and cocoa. These compounds have been extensively studied for their potential therapeutic effects against various diseases including obesity. Polyphenols have been shown to inhibit adipogenesis and reduce the accumulation of fat cells, which can help prevent obesity. They can lead to an increase in energy expenditure by increasing the activity of enzymes involved in fat metabolism, such as AMP-activated protein kinase (AMPK), which can help reduce body weight and can also suppress appetite by increasing the secretion of satiety hormones, such as glucagon-like peptide-1 (GLP-1) and peptide YY (PYY), which can help reduce food intake. Polyphenols exhibit antiinflammatory properties and can help reduce inflammation, which can improve metabolic health [80–82].

Some examples of polyphenolic compounds that have been studied for their potential therapeutic effects against obesity include resveratrol, catechins (tea-polyphenols), quercetin, isoflavones, epigallocatechin gallate (EGCG), and curcumin.

However, more research is needed to fully understand the mechanisms of action and potential therapeutic benefits of these compounds in the context of obesity.

6.2.2 Conjugated linoleic acid (CLA)

CLA is a group of naturally occurring fatty acids found in meat and dairy products. Studies have shown that CLA can help reduce body fat and improve body composition by increasing fat oxidation and reducing fat storage. Therefore, CLA has been studied as a potential nutraceutical for the management of obesity. Studies have shown that CLA supplementation can lead to modest reductions in body weight and body fat in humans [83], although the effects are generally small and inconsistent. Another proposed mechanism for the beneficial effects of CLA on body composition is its ability to modulate gene expression and promote adipocyte differentiation. Specifically, CLA has been shown to upregulate the expression of PPARγ, a transcription factor that plays a key role in adipogenesis and insulin sensitivity. By activating PPARγ, CLA may promote the differentiation of preadipocytes into adipocytes, which may lead to increased glucose uptake and improved insulin sensitivity. Additionally, CLA has been shown to have antiinflammatory and antioxidant properties, which may also contribute to its beneficial effects on body composition and insulin sensitivity [84]. CLA has been shown to reduce levels of proinflammatory cytokines such as TNF-α and IL-6 and to increase levels of antiinflammatory cytokines such as IL-10 through various pathways like TLRs suppression, activation of AMPK and PPARγ [85]. CLA also appears to have antioxidant effects, as it has been shown to increase levels of glutathione, a key antioxidant molecule. However, more research is needed to fully understand the potential therapeutic benefits of CLA for the management of obesity.

6.2.3 Probiotics

Probiotics are living microorganisms that confer a health benefit to the host when administered in adequate amounts. Several studies have investigated the effects of probiotics on obesity and related metabolic disorders, such as insulin resistance and dyslipidemia. One of the mechanisms by which probiotics may exert their antiobesity effects is through the modulation of gut microbiota. It has been shown that obese individuals have a different gut microbiota composition compared to lean individuals, characterized by a reduced microbial diversity and a higher proportion of Firmicutes bacteria. Probiotics can restore the gut microbiota balance by increasing the abundance of beneficial bacteria, such as Bifidobacterium and Lactobacillus, and reducing the abundance of harmful bacteria. Probiotics have also been shown to improve glucose homeostasis and insulin sensitivity [86]. This may be mediated by the production of short-chain fatty acids (SCFAs), which are the end products of bacterial fermentation of dietary fiber. SCFAs can stimulate the release of gut hormones, such as glucagon-like peptide-1 (GLP-1) and peptide YY (PYY), which can improve insulin sensitivity and reduce appetite. Moreover, probiotics can modulate the expression of genes involved in lipid metabolism, inflammation, and oxidative stress [87–89]. For example, *Lactobacillus rhamnosus* has been shown to decrease the expression of genes involved in lipogenesis and increase the expression of genes involved in lipolysis in adipose tissue [90]. This may lead to a reduction in adiposity and an improvement in lipid profile. However, the optimal strain, dose, and duration of probiotic administration need to be determined through further research.

6.2.4   Fiber supplements

Fiber is an essential nutrient that is found in plant-based foods such as fruits, vegetables, and whole grains. It is a type of carbohydrate that the body cannot digest or absorb. Soluble fiber supplements, such as psyllium, methylcellulose, and glucomannan, are commonly used to increase fiber intake and promote weight loss. One way that fiber supplements can promote weight loss is by increasing satiety, or feelings of fullness [91]. Fiber absorbs water in the digestive tract and expands, which can help to fill the stomach and reduce hunger. This can lead to a decrease in overall food intake and a reduction in calorie consumption. Additionally, fiber supplements can help to regulate blood sugar levels by slowing down the absorption of carbohydrates. This can prevent spikes in blood sugar levels that can lead to cravings and overeating. High fiber intake has also been associated with lower insulin levels and improved insulin sensitivity, which can help prevent the development of insulin resistance and type 2 diabetes [92]. Fiber supplements can also improve gut health by promoting the growth of beneficial bacteria in the gut. These bacteria can help to break down fiber into short-chain fatty acids, which have been shown to have several health benefits, including improved insulin sensitivity and reduced inflammation.

6.2.5   Aromadendrine and Kazinol B

Aromadendrine and Kazinol B are natural compounds found in plants that have been shown to have potential therapeutic effects on insulin resistance and glucose metabolism by activating PPARγ signaling pathways [93]. Aromadendrine is found in the leaves of the Chinese plant *Elaeagnus pungens* and has been shown to improve insulin sensitivity and glucose tolerance in animal models of obesity and diabetes. Kazinol B is a natural compound found in the plant Broussonetia Kozinski and has been shown to improve insulin sensitivity and reduce adipose tissue inflammation in obese states [93]. Specifically, they activate PPARγ by binding to the receptor and inducing a conformational change that allows it to form a heterodimer with retinoid X receptor (RXR). This heterodimer then binds to specific DNA sequences called PPAR response elements (PPREs) and regulates the expression of genes involved in glucose metabolism and adipogenesis [94].

6.2.6   Garcinia cambogia

Garcinia cambogia is a tropical fruit native to Southeast Asia and India, commonly used as a weight loss supplement. The active ingredient in Garcinia cambogia is hydroxycitric acid (HCA), which is believed to have potential weight loss properties. HCA works by inhibiting an enzyme called citrate lyase, which plays a key role in converting carbohydrates into fat. By inhibiting this enzyme, HCA is thought to prevent the accumulation of fat in the body and promote weight loss [95,96]. A detailed hepatic metabolic model was used to understand the underlying mechanism of HCA action on body weight, anthropometric parameters, and plasma lipid profile in obese individuals. HCA treatment led to significant reductions in body weight, fat mass, triceps, subscapular, and mid axillary measurements, as well as serum triglyceride, cholesterol, HDL, and LDL levels. Perturbation analysis of ATP citrate lyase activity indicated a net reduction in fatty acid, triglyceride, and cholesterol synthesis, along with increased protein synthesis fluxes under lipogenetic conditions [97]. It should also be noted that Garcinia cambogia and HCA supplements have been associated with some adverse effects, including digestive issues, headaches, and liver toxicity in some cases.

6.2.7   Capsaicin

Capsaicin's antiinflammatory action was demonstrated by the repressed expression of the proinflammatory gene iNOS. It also suppresses COX2 activity, thereby inhibiting PGE2 production, and inactivates NF-κB by blocking IκBα degradation (Kim et al., 2003). Further, capsaicin demonstrates antiinflammatory activity because it binds specifically to PPARγ, inactivates NF-κB, and inhibits the production of TNF-α [98].

Some other natural compounds that have been studied for their potential in managing and preventing obesity are summarized in Table 1.

6.3   Pharmacological therapy

Pharmacological treatment for weight loss should not be the first line of treatment for every obese patient. Obesity is a chronic disease that requires long-term treatment and antiobesity medication should be viewed as an adjunct to a continuous healthy lifestyle regime including increased daily activity and a calorie-deficit diet. Exercise, diet, and behavior modification should be the cornerstones of antiobesity treatment.

However, for individuals who fail to respond to lifestyle interventions after 6 months of treatment and have a BMI of $>30 \text{kg/m}^2$ or $>27 \text{kg/m}^2$ with weight-associated comorbidities, antiobesity drug treatment may be considered. It is important to note that weight reduction should not be the sole focus of treatment, but rather improving obesity-associated comorbidities such as hyperglycemia, hyperlipidemia, and hypertension. The efficacy of available antiobesity drugs is limited, and patients and healthcare providers should have realistic expectations regarding weight loss efficacy [131].

Table 1 An outline of natural compounds explored for therapeutics of obesity.

S. No.	Natural compound	Source	Properties	Specific target	Mechanism of action	References
1.	Omega-3 fatty acids	Fish oil	Polyunsaturated fatty acids	EPA and DHA	Reduces triglycerides and inflammation, improves endothelial function	[99–101]
2.	Curcumin	Turmeric root	Polyphenol	NF-kB and ROS	Reduces inflammation and oxidative stress	[102–105]
3.	Anthocyanins	Blueberries, blackcurrants, and cherries	Polyphenol	AMPK and ROS	Reduces inflammation and oxidative stress, improves lipid metabolism	[106,107]
4.	Ferulic acid	Rice bran, oats, and coffee	Polyphenol	PPARα, PPARγ, and ROS	Reduces inflammation and oxidative stress, improves lipid metabolism	[108,109]
5.	Resveratrol	Grapes, berries, and peanuts	Polyphenol	Sirtuins	Activates sirtuins, inhibit adipocyte glucose utilization, reduces inflammation and oxidative stress	[110]
6.	Epigallocatechin gallate (EGCG)	Green tea	Polyphenol	AMPK, PPARγ, and ROS	Reduces inflammation, oxidative stress, and improves lipid metabolism, inhibits adipogenesis	[111,112]
7.	Punicalagin	Pomegranate	Polyphenol	NF-kB and ROS	Reduces inflammation and oxidative stress, improves endothelial function	[113,114]
8.	Quercetin	Fruits and vegetables	Flavonoid	NF-kB and ROS	Reduces inflammation and oxidative stress	[115]
9.	Astaxanthin	Microalgae, yeast, salmon, trout,	Carotenoid	Nrf2 and ROS	Reduces inflammation and oxidative stress, increase HDL-C	[116,117]

Continued

Table 1 Continued

S. No.	Natural compound	Source	Properties	Specific target	Mechanism of action	References
10.	Vitamin E	krill, shrimp, crayfish Nuts, seeds, and vegetable oils	Fat-soluble vitamin	ROS	Antioxidant reduces oxidative stress	[118,119]
11.	Coenzyme Q10	Organ meats, fatty fish, and whole grains	Coenzyme	ROS	Antioxidant reduces oxidative stress, improves lipid metabolism, and reduces adipokine dysfunction	[120,121]
12.	N-Acetylcysteine	Whey protein, garlic, and onion	Antioxidant	ROS	Reduces oxidative stress and inflammation	[122,123]
13.	Piperine	Black pepper	Alkaloid	PPARα, PPARγ, and ROS	Reduces inflammation and oxidative stress, regulates the lipogenic and lipolytic genes, improves lipid metabolism	[124–126]
14.	Cannabidiol (CBD)	Hemp	Cannabinoid	CB1 and CB2 receptors, Nrf2, and ROS	Reduces inflammation, and oxidative stress, and improves glucose metabolism	[127,128]
15.	Ursolic acid	Apple peels, rosemary, and holy basil	Triterpenoid	AMPK, PPARα, and ROS	Reduces inflammation and oxidative stress, improves glucose and lipid metabolism	[129,130]

According to the Obesity Medicine Association, the most commonly prescribed weight loss medications include Qsymia (phentermine/topiramate), Contrave (bupropion/naltrexone), and Saxenda (liraglutide). A clinical practice guideline on pharmacological interventions for adults with obesity [132] also recommends the use of semaglutide 2.4 mg, liraglutide 3.0 mg, phentermine-topiramate extended-release (ER), naltrexone-bupropion ER, orlistat, phentermine, diethylpropion, and Gelesis 100 oral superabsorbent hydrogel. However, it is important to note that pharmacological therapy is recommended only for adults with overweight and obesity who have an inadequate response to lifestyle interventions alone.

7 Advanced drug therapies targeting angiogenesis and obesity

Despite lifestyle modifications, pharmacological interventions are necessary to treat obesity effectively. However, due to the limitations and nonspecificity of current antiobesity drugs, nanotechnology-based therapies present a promising alternative. Advanced drug delivery systems such as targeted nanosystems have gained much attention in recent years, due to their ability to improve tolerability, reduce side effects, and enhance bioavailability and efficacy compared to conventional therapies. Various nanocarriers (liposomes, polymeric, and gold nanoparticles) have been used to achieve these effects, indicating that targeted nanotherapy is a promising approach for combating obesity and its associated diseases [133]. Some of the strategies include targeting the inhibition of WAT angiogenesis, the transformation of WATs to BATs, photothermal lipolysis of WATs, and many more [134].

One approach is to use nanocarriers, such as liposomes, polymeric nanoparticles, and dendrimers, to deliver drugs that can modulate adipogenesis and angiogenesis. For instance, resveratrol-loaded liposomes have been shown to reduce adipocyte differentiation by inhibiting the expression of key transcription factors involved in adipogenesis and leading to enhanced browning of white adipocytes [135]. Similarly, polymeric nanoparticles loaded with antiangiogenic drugs have been demonstrated to reduce the formation of new blood vessels in adipose tissue, thereby reducing adipose tissue mass. Polymeric carriers, such as microspheres, nanoparticles, and hydrogels, can also be used to control drug release and target specific areas of the body. The controlled release of drugs can increase their efficacy while minimizing side effects. Additionally, the use of natural polymers in these carriers can potentially reduce toxicity [136].

Another approach proposed was to reduce subcutaneous fatty nano-encapsulating quercetin and resveratrol into sodium deoxycholate-elastic liposomes. This approach was found to inhibit adipogenesis and increase apoptosis in adipocytes. The liposomal formulation was stable and suitable for subcutaneous injection, and it reduced the use of phosphatidylcholine and cholesterol [137].

Some strategies use transdermal drug delivery systems (TDDS) that can specifically deliver drugs to adipose tissue. TDDS offers an effective and noninvasive method for treating obesity. TDDS offers advantages such as easy administration, sustained drug release, and reduced systemic exposure compared to other routes of drug administration [138].

Using stem/progenitor cells for therapeutic angiogenesis is a unique approach for treating severe ischemic disorders. Recent research in both humans and animals showed that autologous bone marrow mononuclear cell implantation successfully increased angiogenesis and collateral vessel development in ischemic skeletal muscles [139]. Autologous subcutaneous adipose tissue has been suggested as an intriguing cell source for therapeutic angiogenesis. Multipotent mesenchymal stem/progenitor cells, also known as adipose-derived stem/progenitor cells, are present in the stromal-vascular fraction of adipose tissue (ASCs). These cells can rebuild damaged tissues and can develop into several lineages, including fibroblasts, adipocytes, pericytes, osteoblasts, chondrocytes, and myocytes. Numerous angiogenesis-related growth factors, including VEGF, HGF, and chemokine stromal cell-derived factor-1, can also be released by ASCs (SDF-1). In a recent study, researchers successfully used ligand-coated nanoparticles to deliver trans-resveratrol to ASCs in obese mice. This led to the differentiation of ASCs into beige adipocytes and a subsequent 40% decrease in fat mass, improved glucose regulation, and reduced inflammation. The targeted delivery of browning agents to ASCs through nanoparticles could be a promising therapeutic approach to combat obesity with high efficacy and low toxicity [140].

Placental growth factor (PlGF) plays an important role in adipose tissue development and regulation and may be a potential target for the treatment of obesity-related metabolic disorders. Inhibition of PPAR-γ or VEGFR-2, as well as delivery of a placental growth factor (PlGF) neutralizing monoclonal antibody, can affect adipogenesis and de novo fat pad development. A study conducted in murine models demonstrated that PlGF deficiency leads to significantly reduced adipose tissue mass in both subcutaneous and gonadal adipose tissues in mice fed a standard diet. Furthermore, plasma leptin levels were significantly lower in PlGF-deficient mice and correlated with adipose tissue mass in both genotypes. On a high-fat diet, PlGF-deficient mice had a significantly lower body weight compared to wild-type mice [141]. Additionally, angiogenesis inhibitors like TNP-470 (a synthetic analog of fumagillin, which specifically inhibits endothelial cell growth by suppressing methionine aminopeptidase), angiostatin (a plasminogen fragment containing kringles), and endostatin (a COOH-terminal fragment of collagen XVIII) has demonstrated impaired growth of adipose tissue in mice [58].

Targeting prohibitin, a multifunctional membrane protein that is preferentially expressed in adipose tissue endothelial cells, may be another method to reduce fat accumulation through its vasculature. A synthetic peptide that binds to prohibitin and triggers apoptosis in adipose tissue blood vessels can be used to destroy fat tissue [142].

8   Conclusion and future perspective

The prevalence of overweight and obesity has become a global epidemic. Most current prevention and treatment strategies are individualistic and focus on "eat less, move more," placing blame solely on individual behavior. However, obesity is a multifactorial disease triggered by biological causes, genetics, and dysfunctional food systems. Obesity-related metabolic complications are linked to chronic inflammation and oxidative stress localized within adipose tissue. Various endocrine factors modulate different mechanisms in the body that contribute to the development and progression of metabolic diseases in obese states. Some of the prominent mechanisms that are modulated by endocrine factors in an obese state include insulin resistance, chronic low-grade inflammation, abnormal lipid metabolism, hormonal imbalances, and altered gut microbiome. Interventions that can target these mechanisms, such as weight loss, physical activity, and natural and pharmacological therapies, can help to improve the overall metabolic health of an individual in an obese state. There is a need for further research particularly in the development of more targeted and personalized approaches to obesity treatment. Advanced drug delivery systems hold great promise in the treatment of obesity by targeting the molecular pathways of faulty adipogenesis and obesity-related angiogenesis, oxidative stress, and inflammation. The development of targeted and efficient drug delivery systems for the treatment of obesity will be an essential step towards improving the efficacy and safety of obesity pharmacotherapy. A better understanding of the complex molecular mechanisms underlying obesity could lead to the development of new and more effective treatments that address the root causes of the condition, rather than simply managing symptoms.

Acknowledgments

I would like to express my sincere gratitude to UPES, and Dehradun for their valuable support. I would also like to thank the reviewers for their insightful comments and feedback, which greatly improved the quality of the chapter. Without their support, this chapter would not have become a reality.

References

[1] World Health Organization (WHO). World Obesity Day 2022 – Accelerating action to stop obesity, 2022. Available from: https://www.who.int/news/item/04-03-2022-world-obesity-day-2022-accelerating-action-to-stop-obesity. [Accessed 15 December 2022].
[2] NFHS-5. 2019-2021. India Report. Vol. 1. Mumbai: Ministry of Health and Family Welfare, Government of India. International Institute of Population Sciences. National Family Health Survey (NFHS-5); 2022.
[3] Bray GA, Kim KK, Wilding JPH, et al. Obesity: a chronic relapsing progressive disease process. A position statement of the world obesity federation. Obes Rev 2017;18:715–23. https://doi.org/10.1111/obr.12551.
[4] World Obesity Federation. Obesity is a disease, 2021. Available from: www.worldobesityday.org/assets/downloads/Obesity_Is_a_Disease.pdf. [Accessed 15 December 2022].
[5] ICD-11; World Health Organization. International statistical classification of diseases and related health problems. 11th ed; 2019.
[6] Deurenberg-Yap M, Chew SK, Lin VF, Tan BY, van Staveren WA, Deurenberg P. Relationships between indices of obesity and its co-morbidities in multi-ethnic Singapore. Int J Obes Relat Metab Disord 2001;25(10):1554–62.

[7] McKeigue PM, Shah B, Marmot MG. Relation of central obesity and insulin resistance with high diabetes prevalence and cardiovascular risk in South Asians. Lancet 1991;337 (8738):382–6.

[8] Misra A, Vikram NK. Insulin resistance syndrome (metabolic syndrome) and obesity in Asian Indians: evidence and implications. Nutrition 2004;20(5):482–91.

[9] Misra A, Chowbey P, Makkar BM, Vikram NK, Wasir JS, Chadha D, Joshi SR, Sadikot S, Gupta R, Gulati S, Munjal YP, Consensus group. Consensus statement for diagnosis of obesity, abdominal obesity and the metabolic syndrome for Asian Indians and recommendations for physical activity, medical and surgical management. J Assoc Phys India 2009;57:163–70.

[10] Chandalia M, Abate N, Garg A, Stray-Gundersen J, Grundy SM. Relationship between generalized and upper body obesity to insulin resistance in Asian Indian men. J Clin Endocrinol Metab 1999;84(7):2329–35.

[11] Saxena A, Tiwari P, Wahi N, Kumar A, Mathur SK. The common pathophysiologic threads between Asian Indian diabetic's 'Thin Fat Phenotype' and partial lipodystrophy: the peripheral adipose tissue transcriptomic evidences. Adipocyte 2020;9:253–63.

[12] Yajnik CS. Early life origins of insulin resistance and type 2 diabetes in India and other Asian countries. J Nutr 2004;134(1):205–10. https://doi.org/10.1093/jn/134.1.205.

[13] Ramachandran A, Snehalatha C, Shetty SA, Arun N, Susairaj P. Obesity in Asia – is it different from rest of the world. Diabetes Metab Res Rev 2012;28(Suppl 2):47–51.

[14] Adab P, Pallan M, Whincup PH. Is BMI the best measure of obesity? BMJ (Clin Res ed) 2018;360, k1274. https://doi.org/10.1136/bmj.k1274.

[15] World Health Organization (WHO)/NUT/NCD/98.1. Obesity: preventing and managing the global epidemic. Report of a WHO consultation. 894. Geneva: World Health Organization Technical Report Series; 2000. p. 1–253.

[16] WHO, IASO and IOTF. The Asia specific perspective: Redefining obesity and its treatment. Australia: International Diabetes Institute; 2000.

[17] Dludla PV, Nkambule BB, Jack B, Mkandla Z, Mutize T, Silvestri S, Orlando P, Tiano L, Louw J, Mazibuko-Mbeje SE. Inflammation and oxidative stress in an obese state and the protective effects of gallic acid. Nutrients 2019;11(1):23. https://doi.org/10.3390/nu11010023.

[18] Tencerova M, Ferencakova M, Kassem M. Bone marrow adipose tissue: role in bone remodeling and energy metabolism. Best Pract Res Clin Endocrinol Metab 2021;35 (4), 101545. https://doi.org/10.1016/j.beem.2021.101545.

[19] Sun W, von Meyenn F, Peleg-Raibstein D, Wolfrum C. Environmental and nutritional effects regulating adipose tissue function and metabolism across generations. Adv Sci (Weinh) 2019;6(11):1900275. https://doi.org/10.1002/advs.201900275.

[20] Kahn CR, Wang G, Lee KY. Altered adipose tissue and adipocyte function in the pathogenesis of metabolic syndrome. J Clin Invest 2019;129(10):3990–4000. https://doi.org/10.1172/JCI129187.

[21] Ha CWY, Martin A, Sepich-Poore GD, et al. Translocation of viable gut microbiota to mesenteric adipose drives formation of creeping fat in humans. Cell 2020;183(3): 666–683.e17. https://doi.org/10.1016/j.cell.2020.09.009.

[22] Stieber C, Malka K, Boucher JM, Liaw L. Human perivascular adipose tissue as a regulator of the vascular microenvironment and diseases of the coronary artery and aorta. J Cardiol Cardiovasc Sci 2019;3(4):10–5. https://doi.org/10.29245/2578-3025/2019/4.1174.

[23] Rosenwald M, Wolfrum C. The origin and definition of brite versus white and classical brown adipocytes. Adipocytes 2014;3:4–9.

[24] Longo M, Zatterale F, Naderi J, Parrillo L, Formisano P, Raciti GA, et al. Adipose tissue dysfunction as determinant of obesity-associated metabolic complications. Int J Mol Sci 2019;20:2358. https://doi.org/10.3390/ijms20092358.

[25] Cancello R, Clément K. Is obesity an inflammatory illness? Role of low-grade inflammation and macrophage infiltration in human white adipose tissue. Inter J Obstetr Gynaecol 2006;113(10):1141–7.

[26] Gesta S, Tseng Y-H, Kahn CR. Developmental origin of fat: tracking obesity to its source. Cell 2007;131(2):242–56.

[27] Mancuso P. The role of adipokines in chronic inflammation. Immunotargets Ther 2016;5:47–56. https://doi.org/10.2147/ITT.S73223.

[28] Huh JY, Park YJ, Ham M, Kim JB. Crosstalk between adipocytes and immune cells in adipose tissue inflammation and metabolic dysregulation in obesity. Mol Cell 2014;37:365–71. https://doi.org/10.14348/molcells.2014.0074.

[29] Zatterale F, Longo M, Naderi J, et al. Chronic adipose tissue inflammation linking obesity to insulin resistance and type 2 diabetes. Front Physiol 2020;10:1607. https://doi.org/10.3389/fphys.2019.01607.

[30] Fernandez-Sanchez A, Madrigal-Santillán E, Bautista M, Esquivel-Soto J, Morales-González A, Esquivel-Chirino C, Durante-Montiel I, Sánchez-Rivera G, Valadez-Vega C, Morales-González JA. Inflammation, oxidative stress, and obesity. Int J Mol Sci 2011;12:3117–32.

[31] Kim JE, Kim JS, Jo MJ, Cho E, Ahn SY, Kwon YJ, Ko GJ. The roles and associated mechanisms of adipokines in development of metabolic syndrome. Molecules 2022;27(2):334. https://doi.org/10.3390/molecules27020334.

[32] Chait A, den Hartigh LJ. Adipose tissue distribution, inflammation and its metabolic consequences, including diabetes and cardiovascular disease. Front Cardiovasc Med 2020;7:22. https://doi.org/10.3389/fcvm.2020.00022.

[33] Arner P, Rydén M. Human white adipose tissue: a highly dynamic metabolic organ. J Intern Med 2022;291(5):611–21. https://doi.org/10.1111/joim.13435.

[34] Glass CK, Olefsky JM. Inflammation and lipid signaling in the etiology of insulin resistance. Cell 2012;15:635–45.

[35] Geisler CE, Renquist BJ. Hepatic lipid accumulation: cause and consequence of dysregulated glucoregulatory hormones. J Endocrinol 2017;234(1):R1–R21. https://doi.org/10.1530/JOE-16-0513.

[36] Li Y, Xu S, Zhang X, Yi Z, Cichello S. Skeletal intramyocellular lipid metabolism and insulin resistance. Biophys Rep 2015;1:90–8. https://doi.org/10.1007/s41048-015-0013-0.

[37] Guzik TJ, Skiba DS, Touyz RM, Harrison DG. The role of infiltrating immune cells in dysfunctional adipose tissue. Cardiovasc Res 2017;113(9):1009–23. https://doi.org/10.1093/cvr/cvx108.

[38] Weisberg SP, McCann D, Desai M, Rosenbaum M, Leibel RL, Ferrante Jr AW. Obesity is associated with macrophage accumulation in adipose tissue. J Clin Investig 2003;112:1796–808.

[39] Kawai T, Autieri MV, Scalia R. Adipose tissue inflammation and metabolic dysfunction in obesity. Am J Physiol Cell Physiol 2021;320(3):C375–91. https://doi.org/10.1152/ajpcell.00379.2020.

[40] Han CY. Roles of reactive oxygen species on insulin resistance in adipose tissue. Diabetes Metab J 2016;40(4):272–9. https://doi.org/10.4093/dmj.2016.40.4.272.

[41] Corrales P, Vidal-Puig A, Medina-Gómez G. PPARs and metabolic disorders associated with challenged adipose tissue plasticity. Int J Mol Sci 2018;19(7):2124. https://doi.org/10.3390/ijms19072124.

[42] Mao Z, Zhang W. Role of mTOR in glucose and lipid metabolism. Int J Mol Sci 2018;19 (7):2043. https://doi.org/10.3390/ijms19072043.

[43] Hardy OT, Czech MP, Corvera S. What causes the insulin resistance underlying obesity? Curr Opin Endocrinol Diabetes Obes 2012;19(2):81–7. https://doi.org/10.1097/MED.0b013e3283514e13.

[44] Epingeac ME, Gaman MA, Diaconu CC, Gad MM, Gaman AM. The evaluation of oxidative stress levels in obesity. Revista de Chimie Bucharest 2019;70(6):2241–4.

[45] Hameister R, Kaur C, Dheen ST, Lohmann CH, Singh G. Reactive oxygen/nitrogen species (ROS/RNS) and oxidative stress in arthroplasty. J Biomed Mater Res Part B Biomaterials 2020;108(5):2073–87. https://doi.org/10.1002/jbm.b.34546.

[46] Pérez-Torres I, Castrejón-Téllez V, Soto ME, Rubio-Ruiz ME, Manzano-Pech L, Guarner-Lans V. Oxidative stress, plant natural antioxidants, and obesity. Int J Mol Sci 2021;22(4):1786. https://doi.org/10.3390/ijms22041786.

[47] Halliwell B. Free radicals and antioxidants: updating a personal view. Nutr Rev 2012;70 (5):257–65. https://doi.org/10.1111/j.1753-4887.2012.00476.x.

[48] Liu GS, Chan EC, Higuchi M, Dusting GJ, Jiang F. Redox mechanisms in regulation of adipocyte differentiation: beyond a general stress response. Cell 2012;1(4):976–93.

[49] Sart S, Song L, Li Y. Controlling redox status for stem cell survival, expansion, and differentiation. Oxid Med Cell Longev 2015; 105135. https://doi.org/10.1155/2015/105135.

[50] Vermot A, Petit-Härtlein I, Smith SME, Fieschi F. NADPH oxidases (NOX): an overview from discovery, molecular mechanisms to physiology and pathology. Antioxidants (Basel) 2021;10(6):890. https://doi.org/10.3390/antiox10060890.

[51] DeVallance E, Li Y, Jurczak MJ, Cifuentes-Pagano E, Pagano PJ. The role of NADPH oxidases in the etiology of obesity and metabolic syndrome: contribution of individual isoforms and cell biology. Antioxid Redox Signal 2019;31(10):687–709. https://doi.org/10.1089/ars.2018.7674.

[52] Bhatti JS, Bhatti GK, Reddy PH. Mitochondrial dysfunction and oxidative stress in metabolic disorders – a step towards mitochondria based therapeutic strategies. Biochim Biophys Acta Mol Basis Dis 2017;1863(5):1066–77. https://doi.org/10.1016/j.bbadis.2016.11.010.

[53] Li Q, Gao Z, Chen Y, Guan MX. The role of mitochondria in osteogenic, adipogenic and chondrogenic differentiation of mesenchymal stem cells. Protein Cell 2017;8(6):439–45. https://doi.org/10.1007/s13238-017-0385-7.

[54] Christiaens V, Lijnen HR. Angiogenesis and development of adipose tissue. Mol Cell Endocrinol 2010;318(1–2):2–9. https://doi.org/10.1016/j.mce.2009.08.006.

[55] Cheng R, Ma JX. Angiogenesis in diabetes and obesity. Rev Endocr Metab Disord 2015;16(1):67–75. https://doi.org/10.1007/s11154-015-9310-7.

[56] Voros G, Lijnen HR. Deficiency of thrombospondin-1 in mice does not affect adipose tissue development. J Thromb Haemost 2006;4(1):277–8.

[57] Corvera S, Solivan-Rivera J, Loureiro ZY. Angiogenesis in adipose tissue and obesity. Angiogenesis 2022;25(4):439–53.

[58] Lijnen HR. Angiogenesis and obesity. Cardiovasc Res 2008;78(2):286–93.

[59] Breier G, Chavakis T, Hirsch E. Angiogenesis in metabolic-vascular disease. Thromb Haemost 2017;117(7):1289–95.

[60] Carmeliet P, Moons L, Luttun A, Vincenti V, Compernolle V, De Mol M, Wu Y BF, Devy L, Beck H, Scholz D. Synergism between vascular endothelial growth factor and placental growth factor contributes to angiogenesis and plasma extravasation in pathological conditions. Nat Med 2001;7(5):575–83.

[61] Nijhawans P, Behl T, Bhardwaj S. Angiogenesis in obesity. Biomed Pharmacother 2020;126, 110103. https://doi.org/10.1016/j.biopha.2020.110103.

[62] Corvera S, Olga G. Adipose tissue angiogenesis: impact on obesity and type-2 diabetes. Biochim Biophys Acta (BBA) Mol Basis Dis 2014;1842(3):463–72.

[63] Eelen G, de Zeeuw P, Treps L, Harjes U, Wong BW, Carmeliet P. Endothelial cell metabolism. Physiol Rev 2018;98:3–58. https://doi.org/10.1152/physrev.00001.2017.

[64] Ye J. Adipose tissue vascularization: its role in chronic inflammation. Curr Diab Rep 2011;11:203–10.

[65] Hausman GJ, Richardson RL. Adipose tissue angiogenesis. J Anim Sci 2004;82: 925–34.

[66] Cao Y. Angiogenesis and vascular functions in modulation of obesity, adipose metabolism, and insulin sensitivity. Cell 2013;18:478–89.

[67] Herold J, Kalucka J. Angiogenesis in adipose tissue: the interplay between adipose and endothelial cells. Front Physiol 2021;11, 624903.

[68] Karczewski J, Śledzińska E, Baturo A, Jończyk I, Maleszko A, Samborski P, Begier-Krasińska B, Dobrowolska A. Obesity and inflammation. Eur Cytokine Netw 2018; 29(3):83–94. https://doi.org/10.1684/ecn.2018.0415.

[69] Ferrante AW. Obesity-induced inflammation: A metabolic dialogue in the language of inflammation. J Intern Med 2007;262(4):408–14. https://doi.org/10.1111/j.1365-2796. 2007.01852.x.

[70] Artemniak-Wojtowicz D, Pyrżak B, Kucharska AM. Obesity and chronic inflammation crosslinking. Central Eur J Immunol 2020;45(4):461–8. https://doi.org/10.5114/ CEJI.2020.103418.

[71] Ellulu MS, Patimah I, Khaza'ai H, Rahmat A, Abed Y. Obesity and inflammation: the linking mechanism and the complications. Arch Med Sci 2017;13(4):851–63. https://doi. org/10.5114/aoms.2016.58928.

[72] Wadden TA, Tronieri JS, Butryn ML. Lifestyle modification approaches for the treatment of obesity in adults. Am Psychol 2020;75(2):235–51. https://doi.org/10.1037/amp0000517.

[73] Wilson K. Obesity: lifestyle modification and behavior interventions. FP Essent 2020;492:19–24.

[74] World Health Organization (WHO). Physical activity: fact sheets, 2022. Available from: https://www.who.int/news-room/fact-sheets/detail/physical-activity. [Accessed 29 March 2023].

[75] Smethers AD, Rolls BJ. Dietary management of obesity: cornerstones of healthy eating patterns. Med Clin North Am 2018;102(1):107–24. https://doi.org/10.1016/j.mcna.2017. 08.009.

[76] Centers for Disease Control and Prevention (CDC). Healthcare strategies to prevent and treat childhood obesity. division of nutrition, physical activity, and obesity. National Center for Chronic Disease Prevention and Health Promotion; 2023. Available from: https://www.cdc.gov/obesity/strategies/healthcare/index.html. [Accessed 29 March 2023].

[77] Teodoro AJ. Bioactive compounds of food: their role in the prevention and treatment of diseases. Oxid Med Cell Longev 2019;2019, 3765986. https://doi.org/10.1155/2019/ 3765986.

[78] Pérez-Torres, Castrejón-Téllez V, Soto ME, Rubio-Ruiz ME, Manzano-Pech L, Guarner-Lans V. Oxidative stress, plant natural antioxidants, and obesity. Int J Mol Sci 2021;22 (4):1786. https://doi.org/10.3390/ijms22041786.

[79] Zhao Y, Chen B, Shen J, Wan L, Zhu Y, Yi T, Xiao Z. The beneficial effects of quercetin, curcumin, and resveratrol in obesity. Oxid Med Cell Longev 2017;1459497. https://doi.org/10.1155/2017/1459497.

[80] Aloo S-O, Ofosu FK, Kim N-H, Kilonzi SM, Oh D-H. Insights on dietary polyphenols as agents against metabolic disorders: obesity as a target disease. Antioxidants 2023;12 (2):416. https://doi.org/10.3390/antiox12020416.

[81] Balaji M, Ganjayi MS, Hanuma Kumar GE, Parim BN, Mopuri R, Dasari S. A review on possible therapeutic targets to contain obesity: the role of phytochemicals. Obes Res Clin Pract 2016;10(4):363–80. https://doi.org/10.1016/j.orcp.2015.12.004.

[82] Boix-Castejon M, Herranz-Lopez M, Perez Gago A, Olivares-Vicente M, Caturla N, Roche E, Micol V. Hibiscus and lemon verbena polyphenols modulate appetite-related biomarkers in overweight subjects: a randomized controlled trial. Food Funct 2018;9 (6):3173–84. https://doi.org/10.1039/c8fo00367j.

[83] Whigham LD, Watras AC, Schoeller DA. Efficacy of conjugated linoleic acid for reducing fat mass: a meta-analysis in humans. Am J Clin Nutr 2007;85(5):1203–11. https://doi.org/10.1093/ajcn/85.5.1203.

[84] Lehnen TE, da Silva MR, Camacho A, Marcadenti A, Lehnen AM. A review on effects of conjugated linoleic fatty acid (CLA) upon body composition and energetic metabolism. J Int Soc Sports Nutr 2015;12:36. https://doi.org/10.1186/s12970-015-0097-4.

[85] Siriwardhana N, Kalupahana NS, Cekanova M, LeMieux M, Greer B, Moustaid-Moussa N. Modulation of adipose tissue inflammation by bioactive food compounds. J Nutr Biochem 2013;24(4):613–23. https://doi.org/10.1016/j.jnutbio.2012.12.013.

[86] Abenavoli L, Scarpellini E, Colica C, Boccuto L, Salehi B, Sharifi-Rad J, Aiello V, Romano B, De Lorenzo A, Izzo AA, Capasso R. Gut microbiota and obesity: a role for probiotics. Nutrients 2019;11(11):2690. https://doi.org/10.3390/nu11112690.

[87] Larraufie P, Martin-Gallausiaux C, Lapaque N, Dore J, Gribble FM, Reimann F, Blottiere HM. SCFAs strongly stimulate PYY production in human enteroendocrine cells. Sci Rep 2018;8(1):74. https://doi.org/10.1038/s41598-017-18259-0.

[88] Psichas A, Sleeth M, Murphy K, et al. The short chain fatty acid propionate stimulates GLP-1 and PYY secretion via free fatty acid receptor 2 in rodents. Int J Obes (Lond) 2015;39:424–9. https://doi.org/10.1038/ijo.2014.153.

[89] Lu Y, Fan C, Li P, et al. Short chain fatty acids prevent high-fat-diet-induced obesity in mice by regulating G protein-coupled receptors and gut microbiota. Sci Rep 2016;6:37589. https://doi.org/10.1038/srep37589.

[90] Falcinelli S, Picchietti S, Rodiles A, Cossignani L, Merrifield DL, Taddei AR, Maradonna F, Olivotto I, Gioacchini G, Carnevali O. Lactobacillus rhamnosus lowers zebrafish lipid content by changing gut microbiota and host transcription of genes involved in lipid metabolism. Sci Rep 2015;5:9336. https://doi.org/10.1038/srep09336.

[91] Clark MJ, Slavin JL. The effect of fiber on satiety and food intake: a systematic review. J Am Coll Nutr 2013;32(3):200–11. https://doi.org/10.1080/07315724.2013.791194.

[92] Lattimer JM, Haub MD. Effects of dietary fiber and its components on metabolic health. Nutrients 2010;2(12):1266–89. https://doi.org/10.3390/nu2121266.

[93] Singh D, Sharma S, Choudhary M, Kaur P, Budhwar V. Role of plant-derived products through exhilarating peroxisome proliferator-activated receptor-γ (ppar-γ) in the amelioration of obesity induced insulin resistance. Curr Nutr Food Sci 2022;18(6):549–58.

[94] Bharti SK, Krishnan S, Kumar A, Kumar A. Antidiabetic phytoconstituents and their mode of action on metabolic pathways. Ther Adv Endocrinol Metab 2018;9(3): 81–100. https://doi.org/10.1177/2042018818755019.

[95] Hayamizu K, Ishii Y, Kaneko I, Shen M, Okuhara Y, Shigematsu N, Tomi H, Furuse M, Yoshino G, Shimasaki H. Effects of garcinia cambogia (hydroxycitric acid) on visceral fat accumulation: a double-blind, randomized, placebo-controlled trial. Curr Ther Res Clin Exp 2003;64(8):551–67. https://doi.org/10.1016/j.curtheres.2003.08.006.

[96] Sripradha R, Magadi SG. Efficacy of garcinia cambogia on body weight, inflammation and glucose tolerance in high fat fed male wistar rats. J Clin Diagn Res 2015;9(2): BF01–BF4. https://doi.org/10.7860/JCDR/2015/12045.5577.

[97] Tomar M, Rao RP, Dorairaj P, Koshta A, Suresh S, Rafiq M, Kumawat R, Paramesh R, Babu UV, Venkatesh KV. A clinical and computational study on anti-obesity effects of hydroxycitric acid. Roy Soc Chem 2019;9:18578–88. https://doi.org/10.1039/C9RA01345H.

[98] Leiherer A, Mündlein A, Drexel H. Phytochemicals and their impact on adipose tissue inflammation and diabetes. Vascul Pharmacol 2013;58(1–2):3–20. https://doi.org/10.1016/j.vph.2012.09.002.

[99] Skulas-Ray AC, Kris-Etherton PM, Harris WS, Vanden Heuvel JP, Wagner PR, West SG. Dose-response effects of omega-3 fatty acids on triglycerides, inflammation, and endothelial function in healthy persons with moderate hypertriglyceridemia. Am J Clin Nutr 2011;93(2):243–52. https://doi.org/10.3945/ajcn.110.003871.

[100] Young IE, Parker HM, Cook RL, O'Dwyer NJ, Garg ML, Steinbeck KS, Cheng HL, Donges C, Franklin JL, O'Connor HT. Association between obesity and omega-3 status in healthy young women. Nutrients 2020;12(5):1480. https://doi.org/10.3390/nu12051480.

[101] Zehr KR, Walker MK. Omega-3 polyunsaturated fatty acids improve endothelial function in humans at risk for atherosclerosis: a review. Prostaglandins Other Lipid Mediat 2018;134:131–40. https://doi.org/10.1016/j.prostaglandins.2017.07.005.

[102] Mokgalaboni K, Ntamo Y, Ziqubu K, et al. Curcumin supplementation improves biomarkers of oxidative stress and inflammation in conditions of obesity, type 2 diabetes and NAFLD: updating the status of clinical evidence. Food Funct 2021;12(24): 12235–49. https://doi.org/10.1039/d1fo02696h.

[103] Tabrizi R, Vakili S, Akbari M, et al. The effects of curcumin-containing supplements on biomarkers of inflammation and oxidative stress: a systematic review and meta-analysis of randomized controlled trials. Phytother Res 2019;33(2):253–62. https://doi.org/10.1002/ptr.6226.

[104] Maithilikarpagaselvi N, Sridhar MG, Swaminathan RP, Sripradha R. Preventive effect of curcumin on inflammation, oxidative stress and insulin resistance in high-fat fed obese rats. J Complement Integr Med 2016;13(2):137–43. https://doi.org/10.1515/jcim-2015-0070.

[105] Mokgalaboni K, Ntamo Y, Ziqubu K, et al. Curcumin supplementation improves biomarkers of oxidative stress and inflammation in conditions of obesity, type 2 diabetes and NAFLD: updating the status of clinical evidence. Food Funct 2021;12(24): 12235–49. https://doi.org/10.1039/d1fo02696h.

[106] van der Heijden RA, Morrison MC, Sheedfar F, et al. Effects of anthocyanin and flavanol compounds on lipid metabolism and adipose tissue associated systemic inflammation in diet-induced obesity. Mediators Inflamm 2016;2016:2042107. https://doi.org/10.1155/2016/2042107.

[107] Zhang H, Xu Z, Zhao H, Wang X, Pang J, Li Q, Yang Y, Ling W. Anthocyanin supplementation improves anti-oxidative and anti-inflammatory capacity in a dose-response manner in subjects with dyslipidemia. Redox Biol 2020;32, 101474. https://doi.org/10.1016/j.redox.2020.101474.

[108] Bumrungpert A, Lilitchan S, Tuntipopipat S, Tirawanchai N, Komindr S. Ferulic acid supplementation improves lipid profiles, oxidative stress, and inflammatory status in hyperlipidemic subjects: a randomized, double-blind, placebo-controlled clinical trial. Nutrients 2018;10(6):713. https://doi.org/10.3390/nu10060713.

[109] Ye L, Hu P, Feng LP, et al. Protective effects of ferulic acid on metabolic syndrome: a comprehensive review. Molecules 2022;28(1):281. https://doi.org/10.3390/molecules28010281.

[110] Carpéné C, Les F, Cásedas G, Peiro C, Fontaine J, Chaplin A, Mercader J, López V. Resveratrol anti-obesity effects: rapid inhibition of adipocyte glucose utilization. Antioxidants (Basel) 2019;8(3):74. https://doi.org/10.3390/antiox8030074.

[111] Li F, Gao C, Yan P, Zhang M, Wang Y, Hu Y, Wu X, Wang X, Sheng J. EGCG reduces obesity and white adipose tissue gain partly through AMPK activation in mice. Front Pharmacol 2018;9:1366. https://doi.org/10.3389/fphar.2018.01366.

[112] Liu J, Peng Y, Yue Y, Shen P, Park Y. Epigallocatechin-3-gallate reduces fat accumulation in *Caenorhabditis elegans*. Prev Nutr Food Sci 2018;23(3):214–9. https://doi.org/10.3746/pnf.2018.23.3.214.

[113] Kang B, Kim CY, Hwang J, et al. Punicalagin, a pomegranate-derived ellagitannin, suppresses obesity and obesity-induced inflammatory responses via the Nrf2/Keap1 signaling pathway. Mol Nutr Food Res 2019;63(22), e1900574. https://doi.org/10.1002/mnfr.201900574.

[114] Zou X, Yan C, Shi Y, Cao K, Xu J, Wang X, Chen C, Luo C, Li Y, Gao J, Pang W, Zhao J, Zhao F, Li H, Zheng A, Sun W, Long J, Szeto IM, Zhao Y, Dong Z, Zhang P, Wang J, Lu W, Zhang Y, Liu J, Feng Z. Mitochondrial dysfunction in obesity-associated non-alcoholic fatty liver disease: the protective effects of pomegranate with its active component punicalagin. Antioxid Redox Signal 2014;21(11):1557–70. https://doi.org/10.1089/ars.2013.5538.

[115] Salehi B, Machin L, Monzote L, Sharifi-Rad J, Ezzat SM, Salem MA, Merghany RM, El Mahdy NM, Kılıç CS, Sytar O, Sharifi-Rad M, Sharopov F, Martins N, Martorell M, Cho WC. Therapeutic potential of quercetin: new insights and perspectives for human health. ACS Omega 2020;5(20):11849–72. https://doi.org/10.1021/acsomega.0c01818.

[116] Choi HD, Kim JH, Chang MJ, Kyu-Youn Y, Shin WG. Effects of astaxanthin on oxidative stress in overweight and obese adults. Phytother Res 2011;25(12):1813–8. https://doi.org/10.1002/ptr.3494.

[117] Xia W, Tang N, Kord-Varkaneh H, et al. The effects of astaxanthin supplementation on obesity, blood pressure, CRP, glycemic biomarkers, and lipid profile: a meta-analysis of randomized controlled trials. Pharmacol Res 2020;161, 105113. https://doi.org/10.1016/j.phrs.2020.105113.

[118] Alcalá M, Sánchez-Vera I, Sevillano J, et al. Vitamin E reduces adipose tissue fibrosis, inflammation, and oxidative stress and improves metabolic profile in obesity. Obesity (Silver Spring) 2015;23(8):1598–606. https://doi.org/10.1002/oby.21135.

[119] Wong SK, Chin KY, Suhaimi FH, Ahmad F, Ima-Nirwana S. Vitamin E as a potential interventional treatment for metabolic syndrome: evidence from animal and human studies. Front Pharmacol 2017;8:444. https://doi.org/10.3389/fphar.2017.00444.

[120] Xu Z, Huo J, Ding X, Yang M, Li L, Dai J, Hosoe K, Kubo H, Mori M, Higuchi K, Sawashita J. Coenzyme Q10 improves lipid metabolism and ameliorates obesity by

regulating CaMKII-mediated PDE4 inhibition. Sci Rep 2017;7(1):8253. https://doi.org/10.1038/s41598-017-08899-7.

[121] Zhang P, Chen K, He T, Guo H, Chen X. Coenzyme Q10 supplementation improves adipokine profile in dyslipidemic individuals: a randomized controlled trial. Nutr Metab (Lond) 2022;19(1):13. https://doi.org/10.1186/s12986-022-00649-5.

[122] Dludla PV, Mazibuko-Mbeje SE, Nyambuya TM, et al. The beneficial effects of N-acetyl cysteine (NAC) against obesity associated complications: a systematic review of pre-clinical studies. Pharmacol Res 2019;146, 104332. https://doi.org/10.1016/j.phrs.2019.104332.

[123] Charron MJ, Williams L, Seki Y, Du XQ, Chaurasia B, Saghatelian A, Summers SA, Katz EB, Vuguin PM, Reznik SE. Antioxidant effects of N-acetylcysteine prevent programmed metabolic disease in mice. Diabetes 2020;69(8):1650–61. https://doi.org/10.2337/db19-1129.

[124] Du Y, Chen Y, Fu X, Gu J, Sun Y, Zhang Z, Xu J, Qin L. Effects of piperine on lipid metabolism in high-fat diet induced obese mice. J Funct Foods 2020;71, 104011.

[125] LiuC YY, Zhou J, et al. Piperine ameliorates insulin resistance via inhibiting metabolic inflammation in monosodium glutamate-treated obese mice. BMC Endocr Disord 2020;20:152. https://doi.org/10.1186/s12902-020-00617-1.

[126] Park UH, Jeong HS, Jo EY, et al. Piperine, a component of black pepper, inhibits adipogenesis by antagonizing PPARγ activity in 3T3-L1 cells. J Agric Food Chem 2012;60(15):3853–60. https://doi.org/10.1021/jf204514a.

[127] Ben-Cnaan E, Permyakova A, Azar S, et al. The metabolic efficacy of a cannabidiolic acid (CBDA) derivative in treating diet- and genetic-induced obesity. Int J Mol Sci 2022;23(10):5610. https://doi.org/10.3390/ijms23105610.

[128] Pinto JS, Martel F. Effects of cannabidiol on appetite and body weight: a systematic review. Clin Drug Investig 2022;42:909–19. https://doi.org/10.1007/s40261-022-01205-y.

[129] Jia Y, Kim S, Kim J, et al. Ursolic acid improves lipid and glucose metabolism in high-fat-fed C57BL/6J mice by activating peroxisome proliferator-activated receptor alpha and hepatic autophagy. Mol Nutr Food Res 2015;59(2):344–54. https://doi.org/10.1002/mnfr.201400399.

[130] Wang XT, Gong Y, Zhou B, et al. Ursolic acid ameliorates oxidative stress, inflammation and fibrosis in diabetic cardiomyopathy rats. Biomed Pharmacother 2018;97:1461–7. https://doi.org/10.1016/j.biopha.2017.11.032.

[131] Idrees Z, Cancarevic I, Huang L. FDA-approved pharmacotherapy for weight loss over the last decade. Cureus 2022;14(9), e29262. https://doi.org/10.7759/cureus.29262.

[132] Grunvald E, Shah R, Hernaez R, et al. AGA clinical practice guideline on pharmacological interventions for adults with obesity. Gastroenterology 2022;163(5):1198–225. https://doi.org/10.1053/j.gastro.2022.08.045.

[133] Sharma N, Behl T, Singh S, Kaur P, Zahoor I, Mohan S, Rachamalla M, Dailah HG, Almoshari Y, Salawi A, Alshamrani M. Targeting nanotechnology and nutraceuticals in obesity: an updated approach. Curr Pharm Des 2022;28(40):3269–88. https://doi.org/10.2174/1381612828666221003105619.

[134] Sibuyi NRS, Moabelo KL, Meyer M, Onani MO, Dube A, Madiehe AM. Nanotechnology advances towards development of targeted-treatment for obesity. J Nanobiotechnol 2019;17(1):122. https://doi.org/10.1186/s12951-019-0554-3.

[135] Zu Y, Overby H, Ren G, Fan Z, Zhao L, Wang S. Resveratrol liposomes and lipid nanocarriers: comparison of characteristics and inducing browning of white adipocytes.

Colloids Surf B Biointerfaces 2018;1(164):414–23. https://doi.org/10.1016/j.colsurfb. 2017.12.044.

[136] Huang D, Deng M, Kuang S. Polymeric carriers for controlled drug delivery in obesity treatment. Trends Endocrinol Metab 2019;30(12):974–89. https://doi.org/10.1016/j. tem.2019.09.004.

[137] Cadena PG, Pereira MA, Cordeiro RB, et al. Nanoencapsulation of quercetin and resveratrol into elastic liposomes. Biochim Biophys Acta 2013;1828(2):309–16. https://doi. org/10.1016/j.bbamem.2012.10.022.

[138] Li Z, Fang X, Yu D. Transdermal drug delivery systems and their use in obesity treatment. Int J Mol Sci 2021;22(23):12754. https://doi.org/10.3390/ijms222312754.

[139] Higashi Y, Kimura M, Hara K, et al. Autologous bone-marrow mononuclear cell implantation improves endothelium-dependent vasodilation in patients with limb ischemia. Circulation 2004;109(10):1215–8. https://doi.org/10.1161/01.CIR.0000121427.53291.78.

[140] Zu Y, Zhao L, Hao L, et al. Browning white adipose tissue using adipose stromal cell-targeted resveratrol-loaded nanoparticles for combating obesity. J Control Release 2021;333:339–51. https://doi.org/10.1016/j.jconrel.2021.03.022.

[141] Lijnen HR, Christiaens V, Scroyen I, et al. Impaired adipose tissue development in mice with inactivation of placental growth factor function. Diabetes 2006;55(10):2698–704. https://doi.org/10.2337/db06-0526.

[142] Kolonin MG, Saha PK, Chan L, Pasqualini R, Arap W. Reversal of obesity by targeted ablation of adipose tissue. Nat Med 2004;10(6):625–32. https://doi.org/10.1038/nm1048.

Molecular mechanism(s) of angiogenesis, inflammation, and oxidative stress in cancer

Swati Singh[a], Tapan Behl[b], and Dhruv Kumar[a]
[a]School of Health Sciences and Technology, UPES, Dehradun, Uttarakhand, India,
[b]Amity School of Pharmaceutical Sciences, Amity University, Mohali, Punjab, India

1   Introduction

Cancer poses a life-threatening danger because it has the potential to expand to nearby or faraway organs. Tumor cells can enter lymphatic or blood arteries, travel through the intravascular system, and then multiply at a different location [1]. Growth of the capillary network plays a crucial role in the metastatic dissemination of cancerous tissue. Angiogenesis and lymphangiogenic are the mechanisms that lead to the development of new blood and lymphatic arteries, respectively.

Worldwide, 1 in 5 people will develop cancer at some time in their life, and 1 in 8 men and 1 in 11 women will die from the disease, according to the International Agency for Research on Cancer (IARC). These most recent estimates indicate that more than 50 million people are still living 5 years after being diagnosed with cancer. The major causes of this increase continue to be socioeconomic risk factors and an aging world population.

The World Health Organization (WHO) states that, after cardiovascular illness, cancer is the second leading cause of mortality worldwide. Approximately 8.2 million people die from cancer each year, accounting for 13% of all fatalities [2]. Malignant cells that proliferate rapidly and unchecked outside of their normal borders become cancerous. Unlike healthy cells, malignant cells do not react to the signals that regulate their growth. As a result, they divide uncontrollably, infecting healthy tissues and organs and, in certain circumstances, eventually spreading throughout the body. This characteristic is exhibited in several cellular behaviors that set cancer cells apart from their healthy counterparts [3].

There are more than a 100 distinct forms of cancer [4]. Tumor cell types are used to categorize different types of cancer. Ninety percent of all cancer-related deaths in humans are caused by carcinomas, which are cancers of the breast, prostate, lung, pancreas, and colon epithelial cells. Lymphomas are cancer of the immune system, including the spleen, white blood cells, and lymph glands. Sarcomas are cancer of

Targeting Angiogenesis, Inflammation and Oxidative Stress in Chronic Diseases. https://doi.org/10.1016/B978-0-443-13587-3.00008-4

the fibrous connective tissues of bone, cartilage, fat tissue, muscle, and neurons. The ovaries and testicles both contained pluripotent stem cells [5].

One in four cancer diagnoses among women worldwide is breast cancer. Women are also more likely than males to get thyroid, lung, cervical, and colorectal cancer.

The International Agency for Research on Cancer's GLOBOCAN 2020 estimates of cancer incidence and death are used in this article to offer an update on the worldwide cancer burden. Globally, 19.3% million new instances of cancer are projected for 2020, 18.1 million of which will not be nonmelanoma skin cancer (NMSC), and more than 10 million cancer-related fatalities (9.9 million of which will not be nonmelanoma skin cancer). Female breast cancer is now more often diagnosed than lung cancer, with a projected 2.3 million new cases (11.7%). Cancers of the colon (10.0%), stomach (5.6%), lung (11.4%), and prostate (7.3%) are the following. Lung cancer was still the most prevalent kind of cancer, with an anticipated 1.8 million deaths (18%). Then came colon cancer (9.4%), stomach (7.7%), female breast (6.9%), and liver (8.3%). For both sexes, incidence was typically 2- to 3-fold higher in transitioned countries compared to transitional countries, whereas mortality varied by around 2-fold for men and less for women. However, the death rates for female cervix and breast cancers in transitional countries were much higher than those in transitioned countries (15.0 vs. 12.8 per 100,000 and 12.4 vs. 5.2 per 100,000, respectively). In 2040, there would be 28.4 million new instances of cancer globally, an increase of 47% from 2020. Due to demographic shifts, this rise will be more noticeable in transitioned (64%–95%) nations than in transitioned (32%–56%) countries, albeit this may be exacerbated by growing risk factors brought on by globalization and an expanding economy. The spread of cancer preventive strategies and the provision of cancer care in developing nations depend on sustained infrastructure development, according to the World Health Organization.

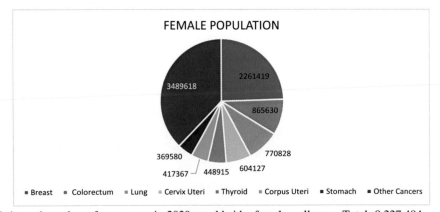

Estimated number of new cases in 2020, worldwide, females, all ages. Total: 9,227,484.

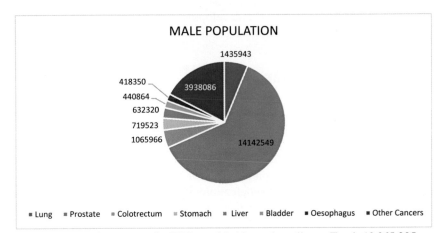

Estimated number of new cases in 2020, worldwide, males, all age. Total: 10,065,305.

2 Molecular role of angiogenesis in cancer

Angiogenesis is the process of new blood vessel formation from preexisting blood vessels. In cancer, angiogenesis plays a crucial role in tumor growth and metastasis. Tumors rely on angiogenesis to ensure a sufficient supply of oxygen and nutrients for their growth and progression.

Several signaling pathways, growth factors, and cellular elements interact intricately to form the molecular basis of angiogenesis in cancer. Vascular endothelial growth factor (VEGF), which is generated by cancer cells and activates the VEGF receptor (VEGFR) on endothelial cells lining the blood vessels, is the main factor in angiogenesis. This contact causes the (phosphatidylinositol-3-kinase) PI3K-Akt, Janus kinase-signal transducer and activator of transcription (JAK-STAT), and Ras-MAPK (mitogen-activated protein kinase) signaling pathways to be activated, which eventually leads to the development of new blood vessels. In addition to VEGF, other growth factors such as basic fibroblast growth factor (Bfgf), platelet-derived growth factor (PDGF), and transforming growth factor-beta (TGF-B) also contribute to the angiogenesis of cancer. These growth factors stimulate the proliferation and migration of endothelial cells, and they also promote the recruitment of other cells, including pericytes and smooth muscle cells, to stabilize the newly formed blood vessels.

Angiogenesis in cancer is a highly regulated process, and the balance between pro- and antiangiogenic factors determines the extent of blood vessels in tumors, which have been developed as a cancer treatment strategy. These therapies target VEGF or its receptor, as well as other signaling pathways involved in angiogenesis, such as the integrin and Notch pathways.

In summary, angiogenesis plays a critical role in tumor growth and metastasis by promoting the formation of new blood vessels. The molecular mechanism of angiogenesis involves the activation of multiple signaling pathways and the interaction between cancer cells and endothelial cells. Understanding the molecular basis of angiogenesis in cancer has led to the development of targeted therapies aimed at inhibiting this process and improving cancer treatment outcomes.

3   Inflammation in cancer

It has long been recognized that inflammation and the onset of cancer are related [6,7]. The inflammatory reaction coordinates the body's defenses against microbes and facilities tissue regeneration and healing, which can result from either infectious or noninfectious tissue injury. Epidemiological data suggested a link between chronic inflammation and a risk of developing cancer, or dysplasia, which is brought on by chronic inflammation. According to epidemiologic research, microbial infection contributes to about 15% of cancer prevalence globally [8].

At microbe-infected tumors or locations of protracted irritation, it is anticipated to see indications of inflammation, such as leukocyte infiltration. However, Virchow was the first to notice that many tumors, including mammary adenocarcinoma, exhibit a "lymphoreticular infiltrate," even though infection or inflammation may not always be a contributing factor. Many of these cancers have inflammatory gene expression signatures, as well as activated fibroblasts and macrophages. Inflammatory gene expression or quantitative wound healing features often show a negative correlation with cancer stage and prognosis [9–11]. Nonsteroidal antiinflammatory drugs (NSAIDs) have been shown to reduce inflammation by preventing the growth of spontaneous tumors in patients with familial adenomatous polyposis (FAP) [12].

Thus, epidemiology, histology, inflammatory profiles, and the effectiveness of antiinflammatory medications in prevention all link cancer to inflammation. These findings have sparked research hypotheses on the process and significance of the link between cancer and inflammation.

4   Oxidative stress in cancer

Oxidative stress is defined as an imbalance between the production of free radicals and reactive molecules, often referred to as oxidants or reactive oxygen species (ROS), and their elimination by defense mechanisms, sometimes referred to as antioxidants. This imbalance causes harm to vital cells and biomolecules, which could influence the entire body [13]. In reaction to changes in intracellular and extracellular environmental circumstances, ROS, which are byproducts of normal cellular metabolism, are crucial in stimulating signaling cascades in plant and animal cells [14]. The mitochondrial respiratory chain produces the majority of ROS in organisms [15]. According to antioxidant metabolism, oxidative stress is a physiological condition in which elevated amounts of ROS and free radicals are produced [16]. Free radicals and ROS are produced during normal cellular respiration, which is important for cell signaling networks [17]. The biggest powerhouse of cells, the mitochondria, mechanically produce ROS when they produce adenosine triphosphate (ATP), whereby oxygen (O_2) and electrons react to create the superoxide anion (O_2) [18]. Numerous studies have shown that oxidative stress and human metabolic illnesses may be fundamentally related [19–21]. Several cancers, including those of the breast, lung, liver, colon, prostate, ovary, and brain, are known to be affected by oxidative stress, which is well-known to be affected by oxidative stress, which is also well-known to change signaling cascades and harm the DNA molecule [22–28]. Furthermore, it has been claimed that the deoxyribose backbone of DNA, including the purine and pyrimidine bases, can bond with hydroxyl radicals throughout the entire DNA structure. 8-OH deoxyguanosine (8-OHdG) can be generated during these harmful processes, significantly raising the possibility of mutation [29]. The 8-OHdG molecules are generally acknowledged as an early detection instrument for the development of cancer [29,30]. They are also used as indicators to identify free radicals during DNA mutagenesis. Importantly, 8-OHdG can convert GC pairs to TA pairs during DNA replication, which, if oxidative defects are present, may cause mutagenesis and eventually lead to the onset of cancer [31].

5 Pathogenesis with emphasis on molecular pathways involved

Pathogenesis refers to the biological processes and mechanisms by which disease develops and progresses. The molecular pathways involved in pathogenesis can vary widely depending on the specific disease or condition. However, some general molecular pathways are commonly involved in many diseases.

Angiogenesis, inflammation, and oxidative stress are all important processes that can contribute to the progression and development of cancer.

1. Angiogenesis: Angiogenesis refers to the growth of new blood vessels, and it is an essential process for normal tissue growth and repair. However, in cancer, angiogenesis can become abnormal, resulting in the growth of new blood vessels that provide nutrients and oxygen to cancer cells, allowing them to grow and spread more easily.
2. Inflammation: Inflammation is a complex biological response to injury, infection, or tissue damage that involves a variety of molecules and cellular processes. Chronic inflammation is involved in the pathogenesis of many diseases including arthritis, atherosclerosis, diabetes, and cancer. Inflammatory cytokines such as interleukin-1 (IL-1) and tumor necrosis factor-alpha (TNF-alpha) are the key mediators of inflammation and contribute to the development of many diseases.
3. Oxidative stress: Oxidative stress is an imbalance between the production of ROS and the body's ability to detoxify them. ROS can damage cellular proteins, lipids, and DNA, and contribute to the pathogenesis of many diseases including cancer, neurodegenerative disease, and cardiovascular disease.

All these processes are interconnected, and they can contribute to the development and progression of cancer in complex ways. Understanding the role of angiogenesis, inflammation, and oxidative stress in cancer is important for developing new treatments and strategies for preventing and managing the disease.

5.1 *Relationship between oxidative stress, inflammation, and cancer*

See Fig. 1.

6 Drug targeting angiogenesis, inflammation, and oxidative stress in cancer treatment

Cancer is a complex disease that involves uncontrolled cell growth and proliferation. Angiogenesis, inflammation, and oxidative stress are the three key processes that contribute to the development and progression of cancer. In recent years, targeting these processes has emerged as a promising approach for the management and treatment of cancer. Here are some strategies for targeting angiogenesis, inflammation, and oxidative stress in cancer:

1. Targeting angiogenesis

Angiogenesis is the process of new blood vessel formation. Tumors require angiogenesis to grow and metastasize. Several drugs have been developed that target angiogenesis by blocking the activity of VEGF, which is a key regulator of angiogenesis. These drugs include bevacizumab, ramucirumab, and aflibercept. Additional angiogenic factors, such as FGF and PDGF, are also the focus of several additional medications in development.

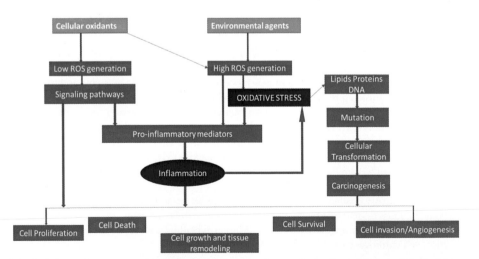

Fig. 1 Schematic representation of immune cells emits a variety of cytokines and chemokines during inflammation to draw additional immune cells to the area of oxidative stress or infection. Reflexively, immune cells near the site of inflammation produce more ROS, which leads to tissue damage and oxidative stress. To emphasize the relevance of "ROS-induced-inflammation" pathways in the development of various illnesses, the relationship between oxidative stress and inflammation is examined in this chapter.

2. Targeting inflammation

Inflammation is a key component of cancer development and progression. Chronic inflammation can contribute to DNA damage and mutations, which can lead to the development of cancer. Targeting inflammation in cancer treatment can involve the use of NSAIDs, which can inhibit cyclooxygenase-2 (COX-2) and reduce inflammation. Additionally, immune checkpoint inhibitors, such as pembrolizumab and nivolumab, can also target inflammation by blocking the activity of proteins that suppress the immune response.

3. Targeting oxidative stress

Oxidative stress is a condition in which there is an imbalance between the production of ROS and the ability of cells to detoxify them. ROS can cause DNA damage and promote tumor growth. Targeting oxidative stress in cancer treatment can involve the use of antioxidants, such as vitamins C and E, which can scavenge ROS and prevent oxidative damage. Additionally, drugs that target the production of ROS, such as inhibitors of (nicotinamide-adenine dinucleotide phosphate) NADPH oxidase (NOX), have also been developed.

6.1 Examples of drugs targeting these pathways

1. Antiangiogenic drugs: These drugs target the process of angiogenesis by blocking the signaling pathways involved in the formation of new blood vessels. Examples of antiangiogenic drugs include bevacizumab, sunitinib, and sorafenib.
2. Antiinflammatory drugs: These drugs target the chronic inflammation associated with cancer. Examples of antiinflammatory drugs include NSAIDs, such as aspirin and ibuprofen, and corticosteroids, such as prednisone.
3. Antioxidant drugs: These drugs target oxidative stress by neutralizing free radicals and reducing damage to DNA and other cellular components. Examples of antioxidant drugs include N-acetylcysteine, vitamin E, and coenzyme Q10.
4. Multitargeted drugs: These drugs target multiple pathways involved in cancer growth and survival, including angiogenesis, inflammation, and oxidative stress. Examples of multitargeted drugs include sunitinib, sorafenib, and pazopanib.

In conclusion, targeting angiogenesis, inflammation, and oxidative stress are promising approaches for the management and treatment of cancer. These strategies can complement traditional cancer therapies, such as chemotherapy and radiation therapy, and help to improve patient outcomes. It is important to note that targeted drugs can have side effects, and their use should be carefully monitored by a healthcare provider. Additionally, combination therapies that target multiple pathways may be more effective than targeting a single pathway alone. However, further research is needed to fully understand the mechanisms underlying these processes and to develop more effective and targeted drugs for cancer treatment (Table 1).

Please be aware that each study had its unique facts and results, and that since the knowledge cutoff date of September 2021, new research may have been done in the area. For the most latest details on nanotechnology-based methods for cancer management, it is crucial to consult contemporary scientific literature.

Table 1 An outline of nanoparticles explored for therapeutics of cancer.

Study	Nanoparticle type	Target	Findings	References
1.	Liposomes	Tumor vasculature	Liposomes loaded with anticancer drugs were shown to accumulate selectively in tumor blood vessels, leading to enhanced drug delivery and improved therapeutic outcomes	[32]
2.	Gold nanoparticles	Tumor cells	Gold nanoparticles conjugated with targeting ligands demonstrated high specificity for tumor cells, enabling selective drug delivery and minimizing off-target effects.	[33]
3.	Carbon nanotubes	Cancer imaging	For imaging methods like magnetic resonance imaging (MRI) and fluorescence imaging, functionalized carbon nanotubes were utilized as contrast agents, which improved sensitivity and resolution in identifying malignant tumors	[34]
4.	Dendrimers	Gene therapy	Targeted gene therapy methods for the treatment of different forms of cancer were made possible by the use of dendrimers to deliver therapeutic genes to cancer cells only	[35]
5.	Quantum dots	Sentinel lymph node mapping	Using quantum dots with special optical characteristics, sentinel lymph nodes were precisely mapped, assisting in the identification and staging of cancer metastasis	[36]
6.	Magnetic nanoparticles	Hyperthermia	An external magnetic field was used to heat magnetic nanoparticles, causing localized hyperthermia that killed cancer cells with the least amount of harm to adjacent healthy tissue	[37]
7.	Nanoemulsions	Drug combination therapy	Through the simultaneous administration of many therapeutic agents, nanoemulsions containing numerous anticancer medicines	[38]

8.	Polymeric nanoparticles	showed synergistic effects and enhanced therapeutic results
By delivering therapeutic compounds directly into drug-resistant cancer cells and avoiding efflux pumps, polymeric nanoparticles were created to circumvent drug resistance processes	[39]
	Drug resistance		
9.	Nanosensors	Researchers have created nanosensors that can identify cancer biomarkers in physiological fluids like blood or urine, potentially leading to earlier cancer detection and better patient outcomes	[40]
	Early cancer detection		
10.	Nanorobots	Nanorobots were developed to deliver medications directly to cancer cells while navigating through the circulation, therefore lowering systemic toxicity and improving therapeutic effectiveness	[41]
	Targeted drug delivery		

This table is describing a few previous study findings in the domain of nanotechnology-based cancer management strategies. Please be aware that the data are based on research completed as of September 2021, and that there may have been developments since then.

7 Strategic drug delivery approaches

To achieve and/or optimize the intended therapeutic effect(s) while minimizing any unfavorable effect(s), drug delivery [42] refers to numerous methods for administering a pharmaceutical substance in the human body [31,43]. Chemicals, peptides, antibodies, and vaccines are some examples of pharmaceutical molecules, in addition to gene-based medicines. Depending on how a drug is administered, multiple kinds of drug delivery systems may be identified. Novel drug delivery methods including targeted delivery and drug-device combinations are now garnering more and more interest in the drug development process, in addition to the conventional techniques like injectable, oral, transdermal, implants, inhalation, ophthalmic, suppository, and otic dosage forms.

The procedure of providing a pharmaceutical substance to have a therapeutic effect on animals or people is referred to as drug delivery. Drug delivery aims to deliver the medicine to the body's targeted site of action in high enough quantities to have the desired therapeutic effect while minimizing unwanted effects. To do this, several clever medication delivery strategies have been created:

1. Oral drug delivery: This is the most common way to give medication, and it entails taking pills, capsules, or syrups orally. Before entering systematic circulation, the medication is absorbed by the gastrointestinal tract and metabolized by the liver.
2. Injectable drug delivery: This entails giving the medication subcutaneously, intramuscularly, or intravenously. This technique avoids the digestive system and liver metabolism, resulting in a quicker and more effective medication administration.
3. Transdermal medication delivery: Using creams, gels, or patches, the medicine is applied directly to the skin. The medicine avoids the digestive system and liver metabolism by being absorbed via the skin and into systematic circulation.
4. Inhalation drug administration: This entails giving the medication by inhalation using tools like inhalers or nebulizers. The medication quickly reaches systematic circulation after being absorbed into the lungs.
5. Implantable drug delivery: In this method, a device that contains the medication is implanted into the body. Drug delivery over an extended period is sustained thanks to the drug's gradual release over time.
6. Targeted drug delivery: This method uses nanoparticles or other specialized delivery methods to deliver the medicine to a specific location in the body. By limiting the drug's exposure to tissues other than those it is intended to treat, this strategy lessens its negative effect.
7. Gene therapy: Viral vectors or other delivery mechanisms are used to introduce therapeutic genes that can make proteins that heal diseases or can fix genetic flaws.
8. Nanotechnology-based delivery: Nanoparticles and nanocarriers are used to encapsulate drugs, allowing for precise delivery, improved solubility, prolonged release, and targeting of specific tissues.

The area of medicine has undergone a revolution because of these clever medication delivery techniques that maximize effectiveness while minimizing negative effects.

It is difficult to classify drug delivery strategies without a specific criterion since they have been investigated so extensively to address pharmacology-related problems from so many different viewpoints. Drug distribution strategies typically do not entail chemically altering the active component, except the prodrug technique. These techniques include solubilization, permeability argumentation, modified release (MR), and other unique drug delivery systems such as targeted distribution, minimal local irritation, and drug-device integration. The pharmacologically active moiety is chemically modified during prodrug administration, which is covered in a different section.

Drug delivery strategies may affect a pharmaceutical drug's pharmacodynamic performance without altering the underlying pharmacodynamic characteristics of the chemical. A desirable drug target product profile (DTPP) has to be precisely specified to be used for drug development and subsequent regulatory review [6]. The following, among others, are clinical pharmacology-related concerns that could affect the expected therapeutic outcomes:

Dose, Dosing interval, Dose modifications, Characteristics of pharmacokinetics, Localized drug sensitivity, Patient cooperation, and Variation between patients.

Drug delivery refers to the methods or process of administrating a pharmaceutical compound to achieve a therapeutic effect in humans.

8 Conclusion and future perspectives

Natural products (NPs) have historically been important in the development of new medications, particularly for the treatment of infectious and cancerous diseases [6,7], as well as for other therapeutic conditions including cardiovascular disease (using statins, for example) and multiple sclerosis (using fingolimod, for example) [8–10].

Since the beginning of time, people have used natural remedies made from plants, animals, microorganisms, and marine organisms to treat and prevent disease. Humans have been employing plants as medicines for at least 60,000 years, according to fossil evidence [44,45]. It must have been rather difficult for early humans to use natural treatments as medicine. It is quite possible that early humans consumed toxic plants regularly while in quest for food, which led to comas, nausea, vomiting, diarrhea, or other fatal reactions—possibly even death. Even so, [46] early humans were able to learn about food plants and herbal cures in this way.

Traditional medicines (TMs) are very important and employ natural ingredients. Ayurveda, Kampo, traditional Korean medicine (TKM), traditional Chinese medicine (TCM), and unani are a few examples of medical practices that employ natural ingredients and have been used for hundreds or even thousands of years around the world. These procedures have developed into well-managed medical systems. They may have shortcomings in some of their numerous forms, but they nonetheless represent an essential repository of human knowledge [44,46,47].

Since natural products have evolved over millions of years, they have a specific chemical diversity that results in variation in their biological activities and drug-like properties. These products are now some of the most important raw materials for producing new lead compounds and scaffolds. Natural goods will be utilized often to meet the pressing requirement to create effective pharmaceuticals, and they will play a vital role in the development of medications for treating human ailments, particularly important disorders [48].

Natural products come in a broad variety of multidimensional chemical structures, and their potential for modifying biological functions has garnered a lot of interest. They have since been effectively applied in the search for novel medications and have had a significant influence on chemicobiology [49–51].

The treatment and prevention of several diseases have greatly benefited from the use of both natural and synthetic medications. Natural medicines—made from plants, animals, and other sources of the natural world—have been used for many years and served as the inspiration for many contemporary medications. However, laboratory-produced synthetic medications provide a more specialized and targeted approach to addressing particular illnesses.

Both natural and synthetic medicines have enormous potential for the future and will remain essential for the creation of novel cures and therapies. Here are a few instances:

1. Natural remedies: As more people become aware of the possible health advantages of natural goods, it is anticipated that the usage of natural remedies would rise. It has been discovered that many natural ingredients contain healing qualities, such as antiinflammatory, antioxidant, and antibacterial capabilities. Future research on natural medicines may make use of cutting-edge technology like metabolomics and genetics to increase their efficacy and safety.
2. Medicines made from synthetic materials: It is anticipated that this trend will continue in the future. In comparison to natural remedies, synthetic medications are more precise and have better control over the drug's physical characteristics, such as its solubility and stability. The creation of novel synthetic medications is anticipated to be facilitated by developments in drug discovery technologies including high-throughput screening and computer-aided drug design.
3. Future research is anticipated to place more emphasis on combination therapy, which includes both natural and artificial medications. Combination therapies have the potential to be more effective, less toxic, and more patient-compliant than single-agent treatments.
4. Advances in genomics and other technologies are anticipated to make it possible to create personalized medications that are adapted to a person's unique genetic profile and other characteristics. By increasing treatment effectiveness and minimizing side effects, personalized medicine has the potential to completely transform the healthcare industry.

In conclusion, drugs, both natural and artificial, will continue to be crucial in the creation of novel therapies and treatments. These medications have enormous potential for the future, and they will probably be combined with other therapy to provide patients with the greatest results possible.

8.1   Future burden of cancer in 2040

Different effects are predicted by experts for different cancer kinds, nations, and genders.

Given that the age gap between men and women in the population is expected to close in the future decades, the increase in both the number of new cases and the number of fatalities is predicted to affect males more severely than women. In 2040, males are predicted to experience 1.9 million new instances of cancer, compared to 1.5 million for women.

Due to the disparities in expected rates of demographic change between nations, there are also noticeable variances in the effects that population aging will have on various European nations. Between a 2% anticipated rise in new cancer cases in Bulgaria and a 65% increase in Ireland.

A country's population size and structure can change over time due to a variety of variables. Population forecasts made by Eurostat based on various scenarios of fertility, migration, and death provide the data used in the research. These predictions predict that by 2050, the proportion of adults 65 and older in Europe will rise from 20% to 30% of the entire population.

According to estimates, the most prevalent cancer forms will not change from 2020. A drop in testicular cancer, which largely affects younger males, and a rise in new occurrences of mesothelioma, gallbladder, and bladder cancer, which mostly affect older individuals, might be specifically caused by the changes in population size and structure.

To concentrate on the impact of population aging, the experts made these projections on the fictitious premise that cancer incidence and death levels will remain constant up to 2040. Over 40% of cancers are known to be preventable, though.

A projected 28.4 million new cancer cases globally (including NMSC but excluding basal cell carcinoma) are predicted to occur in 2040, a 47% increase from the similar 19.3 million cases in 2020, if country rates shown in 2020 remain unchanged. Low Human Development Index (HDI) nations (95%) and medium HDI countries (64%) show the greatest relative magnitude of growth. With 4.1 million more new cases in 2040 compared to 2020, the high HDI nations are predicted to suffer the largest rise in incidence. This prediction is primarily based on population increase and aging, and it might be made worse by the rising incidence of risk factors in various regions of the world (Fig. 2).

8.2 Summary

According to the GLOBOCAN 2020 projections discussed in this report, there would be 19.3 million new cases of cancer and about 10 million cancer-related deaths in the year 2020. Regardless of the degree of human development, illness is a significant source of morbidity and mortality everywhere on the globe. It bears repeating that the cumulative risk of female cancer deaths in Africa in 2020 is about equivalent to the risks seen among female cancer deaths in Northern America and the highest-income nations of Europe. Therefore, efforts to create a long-lasting infrastructure for the distribution of effective cancer preventive strategies and the provision of cancer care in developing nations are essential for the worldwide fight against cancer.

The tremendous diversity of cancer continues to provide hints about its underlying causes but also highlights the necessity of intensifying worldwide efforts to combat the illness. There are packages of efficient and resource-conscious preventative and curative interventions for cancer available [52,53] and their specialized integration

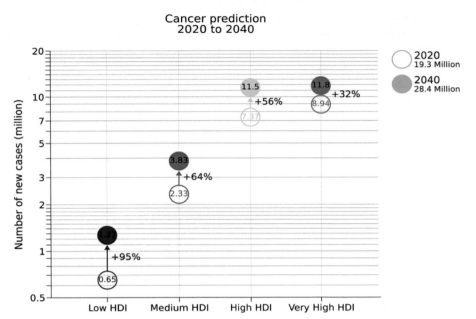

Fig. 2 Projected number of new cases for all cancers combined (both sexes combined) in 2040 according to the 4-Tier Human Development Index (HDI).
Source: GLOBOCAN 2020.

into national health planning can serve to lessen the burden and suffering caused by cancer in the future while reducing the clear cancer disparities between transitioning and transitioned countries currently observed.

Acknowledgments

The authors would like to thank School of Health Science, University of Petroleum and Energy Studies, Dehradun, Uttarakhand, India for providing facilities for the completion of this chapter.

References

[1] Folkman J. Tumor angiogenesis: therapeutic implications. New england journal of medicine 1971;285(21):1182–6.
[2] Cooper GM. The eukaryotic cell cycle. In: The cell: a molecular approach. 2nd ed. Sinauer Associates; 2000.
[3] Cairns J. Cancer: science and society. WH Freeman; 1978.
[4] National Research Council. Diet, nutrition, and cancer. National Academies Press; 1982.
[5] Harold F, Dvorak M. Tumors: wounds that do not heal. N Engl J Med 1986;315 (26):1650–9.
[6] Atanasov AG, Waltenberger B, Pferschy-Wenzig EM, Linder T, Wawrosch C, Uhrin P, Temml V, Wang L, Schwaiger S, Heiss EH, Rollinger JM. Discovery and resupply of

pharmacologically active plant-derived natural products: a review. Biotechnol Adv 2015;33(8):1582–614.

[7] Harvey AL. Natural products in drug discovery. Drug Discov Today 2008;13(19–20):894–901.

[8] Newman DJ, Cragg GM. Natural products as sources of new drugs over the nearly four decades from 01/1981 to 09/2019. J Nat Prod 2020;83(3):770–803.

[9] Waltenberger B, Mocan A, Šmejkal K, Heiss EH, Atanasov AG. Natural products to counteract the epidemic of cardiovascular and metabolic disorders. Molecules 2016;21(6):807.

[10] Tintore M, Vidal-Jordana A, Sastre-Garriga J. Treatment of multiple sclerosis—success from bench to bedside. Nat Rev Neurol 2019;15(1):53–8.

[11] Corley DA, Sedki M, Ritzwoller DP, Greenlee RT, Neslund-Dudas C, Rendle KA, Honda SA, Schottinger JE, Udaltsova N, Vachani A, Kobrin S. Cancer screening during the coronavirus disease-2019 pandemic: a perspective from the National Cancer Institute's PROSPR consortium. Gastroenterology 2021;160(4):999–1002.

[12] Kutikov A, Weinberg DS, Edelman MJ, Horwitz EM, Uzzo RG, Fisher RI. A war on two fronts: cancer care in the time of COVID-19. Ann Intern Med 2020;172(11):756–8.

[13] Ďuračková Z. Some current insights into oxidative stress. Physiol Res 2010;59(4):459–69.

[14] Jabs T. Reactive oxygen intermediates as mediators of programmed cell death in plants and animals. Biochem Pharmacol 1999;57(3):231–45.

[15] Clemmons DR. Role of IGF-I in skeletal muscle mass maintenance. Trends Endocrinol Metab 2009;20(7):349–56.

[16] Dinmohamed AG, Visser O, Verhoeven RH, Louwman MW, van Nederveen FH, Willems SM, Merkx MA, Lemmens VE, Nagtegaal ID, Siesling S. Fewer cancer diagnoses during the COVID-19 epidemic in the Netherlands. Lancet Oncol 2020;21(6):750–1.

[17] Alpay M, Backman LR, Cheng X, Dukel M, Kim WJ, Ai L, Brown KD. Oxidative stress shapes breast cancer phenotype through chronic activation of ATM-dependent signaling. Breast Cancer Res Treat 2015;151:75–87.

[18] Pisoschi AM, Pop A. The role of antioxidants in the chemistry of oxidative stress: a review. Eur J Med Chem 2015;97:55–74.

[19] Colombo R, Giustarini D, Milzani A. Biomarkers of oxidative damage in human disease. Clin Chem 2006;52:601–23.

[20] Tangvarasittichai S. Oxidative stress, insulin resistance, dyslipidemia, and type 2 diabetes mellitus. World J Diabetes 2015;6(3):456–80.

[21] Lee JD, Cai Q, Shu XO, Nechuta SJ. The role of biomarkers of oxidative stress in breast cancer risk and prognosis: a systematic review of the epidemiologic literature. J Women's Health 2017;26(5):467–82.

[22] Zhang L, Li L, Gao G, Wei G, Zheng Y, Wang C, Gao N, Zhao Y, Deng J, Chen H, Sun J. Elevation of GPRC5A expression in colorectal cancer promotes tumor progression through VNN-1 induced oxidative stress. Int J Cancer 2017;140(12):2734–47.

[23] Saijo H, Hirohashi Y, Torigoe T, Horibe R, Takaya A, Murai A, Kubo T, Kajiwara T, Tanaka T, Shionoya Y, Yamamoto E. Plasticity of lung cancer stem-like cells is regulated by the transcription factor HOXA5 that is induced by oxidative stress. Oncotarget 2016;7 (31):50043.

[24] Wang Z, Li Z, Ye Y, Xie L, Li W. Oxidative stress and liver cancer: etiology and therapeutic targets. Oxidative Med Cell Longev 2016;2016, 7891574.

[25] Oh B, Figtree G, Costa D, Eade T, Hruby G, Lim S, Elfiky A, Martine N, Rosenthal D, Clarke S, Back M. Oxidative stress in prostate cancer patients: a systematic review of case control studies. Prostate Int 2016;4(3):71–87.

[26] Saed GM, Diamond MP, Fletcher NM. Updates of the role of oxidative stress in the pathogenesis of ovarian cancer. Gynecol Oncol 2017;145(3):595–602.

[27] Jaroonwitchawan T, Chaicharoenaudomrung N, Namkaew J, Noisa P. Curcumin attenuates paraquat-induced cell death in human neuroblastoma cells through modulating oxidative stress and autophagy. Neurosci Lett 2017;636:40–7.

[28] Forcados GE, James DB, Sallau AB, Muhammad A, Mabeta P. Oxidative stress and carcinogenesis: potential of phytochemicals in breast cancer therapy. Nutr Cancer 2017;69 (3):365–74.

[29] Matsui A, Ikeda T, Enomoto K, Hosoda K, Nakashima H, Omae K, Watanabe M, Hibi T, Kitajima M. Increased formation of oxidative DNA damage, 8-hydroxy-2′-deoxyguanosine, in human breast cancer tissue and its relationship to GSTP1 and COMT genotypes. Cancer Lett 2000;151(1):87–95.

[30] Bradburn MJ, Clark TG, Love SB, Altman DG. Survival analysis part III: multivariate data analysis—choosing a model and assessing its adequacy and fit. Br J Cancer 2003;89 (4):605–11.

[31] Wen H, Jung H, Li X. Drug delivery approaches in addressing clinical pharmacology-related issues: opportunities and challenges. AAPS J 2015;17:1327–40.

[32] Blanco E, Shen H, Ferrari M. Principles of nanoparticle design for overcoming biological barriers to drug delivery. Nat Biotechnol 2015;33(9):941–51.

[33] Jeong HH, Choi E, Ellis E, Lee TC. Recent advances in gold nanoparticles for biomedical applications: from hybrid structures to multi-functionality. J Mater Chem B 2019;7 (22):3480–96.

[34] Chu DK, Kim LH, Young PJ, Zamiri N, Almenawer SA, Jaeschke R, et al. Mortality and morbidity in acutely ill adults treated with liberal versus conservative oxygen therapy (IOTA): a systematic review and meta-analysis. Lancet 2018;391(10131):1693–705.

[35] Franiak-Pietryga I, Ziemba B, Messmer B, Skowronska-Krawczyk D. Dendrimers as drug nanocarriers: the future of gene therapy and targeted therapies in cancer. In: Dendrimers: fundamentals and applications. IntechOpen; 2018. p. 7.

[36] Miao Q, Xie C, Zhen X, Lyu Y, Duan H, Liu X, Jokerst JV, Pu K. Molecular afterglow imaging with bright, biodegradable polymer nanoparticles. Nat Biotechnol 2017;35 (11):1102–10.

[37] Wilhelm S, Tavares AJ, Dai Q, Ohta S, Audet J, Dvorak HF, Chan WC. Analysis of nanoparticle delivery to tumours. Nat Rev Mater 2016;1(5):1–2.

[38] Peer E, Brandimarte L, Samat S, Acquisti A. Beyond the turk: alternative platforms for crowdsourcing behavioral research. J Exp Soc Psychol 2017;70:153–63.

[39] de Alcântara Rodrigues I, Ferrari RG, Panzenhagen PH, Mano SB, Conte-Junior CA. Antimicrobial resistance genes in bacteria from animal-based foods. Adv Appl Microbiol 2020;112:143–83.

[40] Gopalakrishnan V, Spencer CN, Nezi L, Reuben A, Andrews MC, Karpinets T, et al. Gut microbiome modulates response to anti–PD-1 immunotherapy in melanoma patients. Science 2018;359(6371):97–103.

[41] Yu W, Liu R, Zhou Y, Gao H. Size-tunable strategies for a tumor targeted drug delivery system. ACS Cent Sci 2020;6(2):100–16.

[42] Zhang L, Zhou H, Belzile O, Thorpe P, Zhao D. Phosphatidylserine-targeted bimodal liposomal nanoparticles for in vivo imaging of breast cancer in mice. J Control Release 2014; (183):114–23.

[43] Dawidczyk CM, Kim C, Park JH, Russell LM, Lee KH, Pomper MG, Searson PC. State-of-the-art in design rules for drug delivery platforms: lessons learned from FDA-approved nanomedicines. J Control Release 2014;(187):133–44.

[44] Yuan H, Ma Q, Ye L, Piao G. The traditional medicine and modern medicine from natural products. Molecules 2016;21:559.

[45] Fabricant DS. Approaches to drug discovery using higher plants. Environ Health Perspect 2001;109(1):69–75.

[46] Barboza RR, Souto WdMS, Mourão JdS. The use of zootherapeutics in folk veterinary medicine in the district of Cubati, Paraíba state, Brazil. J Ethnobiol Ethnomed 2007;3:1–4.

[47] Lorenz RT, Cysewski GR. Commercial potential for Haematococcus microalgae as a natural source of astaxanthin. Trends Biotechnol 2000;18(4):160–7.

[48] Hong J. Role of natural product diversity in chemical biology. Curr Opin Chem Biol 2011;15(3):350–4.

[49] Rosén J, Gottfries J, Muresan S, Backlund A, Oprea TI. Novel chemical space exploration via natural products. J Med Chem 2009;52(7):1953–62.

[50] Butler MS. Natural products to drugs: natural product-derived compounds in clinical trials. Nat Prod Rep 2008;25(3):475–516.

[51] Bray F, Laversanne M, Weiderpass E, Soerjomataram I. The ever-increasing importance of cancer as a leading cause of premature death worldwide. Cancer 2021;127(16):3029–30.

[52] Gelband H, Sankaranarayanan R, Gauvreau CL, Horton S, Anderson BO, Bray F, Cleary J, Dare AJ, Denny L, Gospodarowicz MK, Gupta S. Costs, affordability, and feasibility of an essential package of cancer control interventions in low-income and middle-income countries: key messages from disease control priorities. Lancet 2016;387(10033):2133–44.

[53] World Health Organization. Action plan for the prevention and control of non-communicable diseases in the WHO European region. World Health Organization. Regional Office for Europe; 2016.

Unraveling the enigma of rheumatoid arthritis: Exploring etiology, pathophysiology and its treatment

4

Diksha Chugh[a], Jyoti Upadhyay[b], Pooja Dhami[b], and Mukesh Nandave[a]
[a]Department of Pharmacology, Delhi Pharmaceutical Sciences and Research University (DPSRU), New Delhi, India, [b]School of Health Sciences and Technology, University of Petroleum and Energy Studies, Dehradun, Uttarakhand, India

1 Introduction

Rheumatoid Arthritis (RA) is a progressive disease that affects several organs, including the hands, knees, wrists, shoulders, and organs such as the eyes, lungs, and heart. It causes inflammation of the synovial joints and damage to the cartilage and bones [1]. Complexity in RA pathogenesis frequently causes missed diagnosis. In many rheumatoid arthritis patients, anticyclic citrullinated peptides (antiCCP) were identified in their serum samples [2]. Its sensitivity is like that of Rheumatoid Factor (RF) but has comparatively higher specificity than RF [3]. This disease may lead to low red blood cell count and inflammation in the surrounding area of the heart and lungs. Oxidative stress occurring as a result of inflammation in patients may affect the surrounding cells by converting them to free radicals. Low energy and fever may also be present as clinical manifestations. Most often, the symptoms may develop over a week to months [4]. Chronic inflammation causes functional impairment and erosive joints in most of the patient population [5]. The onset of rheumatoid arthritis varies from patient to patient depending upon the number, type, and pattern of joints involved. The course duration of RA may also be found to be different as per the presence or absence of different variables like genetics, presence of autoantibodies in serum, frequency of inflammation of joints, and severity of inflammation [6]. The etiology behind RA is unknown and is thought to believe from impaired immune responses. It can affect people of any age and can cause symptoms such as fatigue and prolonged stiffness during rest. As such, there are no drugs available to successfully treat RA, however, many are found to be effective and are increasingly available to treat this disease. These drugs can also prevent deformed joints. Along with medication and surgery, self-management techniques like exercises can cause a reduction in pain and stability. Globally, RA affects approx. 0.5% to 1% of the total population. Females are two to three times more prone to develop this disease as compared to males [7].

Targeting Angiogenesis, Inflammation and Oxidative Stress in Chronic Diseases. https://doi.org/10.1016/B978-0-443-13587-3.00017-5

2   Pathogenesis of rheumatoid arthritis

Posttranslational modification rate increases caused by genetic and environmental factors activate immune cells via Fc receptors on macrophages, leading to an inflammatory cascade caused by cytokine release [8]. It further differentiates B lymphocytes with the help of interleukin-6 (IL-6), leading to the accumulation of neutrophils in synovial fluid that involves Class II major histocompatibility complex (MHC) [9]. Fig. 1 explains the pathophysiology and pathogenesis of RA.

Pathogenesis of rheumatoid arthritis includes the following stages:

1. T cells presentation to antigen
2. Proliferation of T and B cells
3. Synovial membrane angiogenesis
4. Accumulation of neutrophils in synovial fluid
5. Proliferation of synovial cells
6. Synovitis polarization
7. Activation of chondrocytes
8. Degradation of cartilage
9. Cartilage invasion by pannus
10. Ligaments are stretched around joints [9]

2.1   Inflammation in rheumatoid arthritis

IL-6 performs a significant action in the pathophysiology of RA. It stimulates the differentiation and proliferation of T cells into T helper 17 (TH-17), which produces

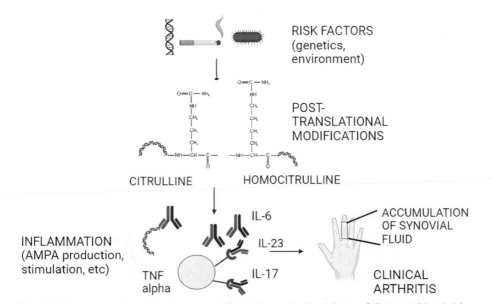

Fig. 1 Role of genetic and environmental factors in pathophysiology of rheumatoid arthritis.

IL-17, as well as beta-cell proliferation, leading to the mediation of chronic inflammation [10]. IL-6 performs a complex function of destroying joints by osteoclast activation, bone resorption, and pannus formation [10]. The formation of destructive pannus resulting in functional impairment and bone deformity is a consequence of the Rheumatoid Factor [11]. However, the pannus formation in RA is a very complex process and occurs over a period. A study conducted by Matsui et al. explains this pannus formation as an increasing accumulation of F labeled fluorodeoxyglucose (F-FDG) in swollen joints, which further worsens the situation in the presence of inflammatory cytokines [12]. Fig. 2 represents the role of IL-6 in RA.

2.2 Angiogenesis in rheumatoid arthritis

Angiogenesis, a process of developing new capillaries from preexisting ones, infiltrates inflammatory cytokines to joints that lead to synovial hyperplasia, tumor formation, and RA [13]. RA is defined as an "angiogenesis derived disease," with its significant role in damaging the cartilage and joints [12]. This synovial hyperplasia increases the oxygen demand of blood vessels and results in hypoxia, which greatly escalates the blood vessels count, infiltrated by inflammatory cytokines [14]. Angiogenesis in RA is characterized by the presence of several cell-surface bound mediators which are soluble that includes endogenous proteins like vascular endothelial growth factor (VEGF), hypoxia-inducible factors (HIFs), chemokines, cell adhesion molecules, and proteases [15]. The series of events includes (i) the prevascular phase, and (ii) the vascular phase showing an increase in vessel growth, are involved in neovascularization in RA patients [16]. Both these phases have great significance in pannus formation and inflammation at joints [16]. Table 1 depicts the phases of angiogenesis.

Factors that control angiogenesis in rheumatoid arthritis include [12]:

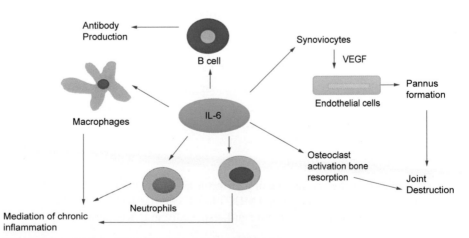

Fig. 2 Role of interleukin-6 (IL-6) in rheumatoid arthritis.

Table 1 Angiogenesis phases.

S. No.	Phases of angiogenesis	Description
1.	Prevascular phase	This phase correlates with the inflammation of the synovium; lymphocytes and macrophages infiltrate this inflamed synovium leading to pannus formation, which further increases the oxygen and nutrients demand at this site, supporting the arthritis progression and bone destruction [16].
2.	Vascular phase	This is characterized by synovial cell hyperplasia, infiltration of monocular cells, and tenderness in the joint [16].

2.2.1   Heparin-binding growth factors

Midkine (MK), a type of heparin-binding growth factor, is thought to possess a significant role in macrophage-like and fibroblast-like cellular expression in the synovial membrane of joints in RA patients [17]. It also acts on other growth factors like endothelial cell growth factor, acidic fibroblast growth factor, and eye and retina-derived growth factors [12]. MK, a 13-kd protein, is involved in two important steps of the development of RA, namely, inflammatory leukocytes migration and osteoclast differentiation, and can increase the migration of leukocytes and macrophages, thus involved in inflammation of joints and cartilage [17].

2.2.2   Angiogenin

Angiogenin, derived from cultured tumor cells, is a potent activator of neovascularization and has increased concentration in synovial fluid of RA patients [18]. Angiogenin is involved in the stimulation of angiogenesis and results in the cleavage of RNA in ribosomes [12].

2.2.3   Vascular permeability factor (VPF)

VPF/VEGF is a peptide secreted from tumor cells and a potent mitogen for endothelial cells [19,20]. It accumulates specifically in endothelial cells and causes tumor stroma formation along with angiogenesis [12,19].

2.2.4   Interferon-gamma

Several studies hypothesize the role of interferon-gamma in the progression of RA, as it is seen to activate macrophages and MHC class II cells [20]. However, they are present in very low amounts of synovial fluid, especially in RA patients [21].

2.2.5 TNF-α

TNF-α is among the most activated inflammatory cytokines during RA and induces the production of other cytokines, thereby accelerating cartilage damage [21]. In the in-vitro model, it also inhibits the mitogenesis of endothelial cells, causing procoagulant activity, thrombosis, hemorrhage, and small vessel disruption [12].

2.2.6 Interleukins

The various types of ILs that are implicated in the pathogenesis of RA include Interleukins-6, 1, 17, and 18 [21]. IL-1 alpha and IL-1 beta of the Interleukins class are involved in the inhibition of endothelial cell growth factor that induces the growth of endothelial cells [12]. ILs are major inflammatory cytokines that are involved in the proliferation and differentiation of T and B cells [20,21].

2.2.7 Prostaglandins

Prostaglandin D2 (PGD2) and J series are synthesized by prostaglandin D and have increased concentrations in tumors and inflammatory exudates [12,22]. The mechanism is unclear how prostaglandins act in angiogenesis; however, they may show inflammatory properties.

2.2.8 Fibrinogen

Alpha and beta chains of fibrin are identified as the potential target of antibody response along with the accumulation of citrullinated fibrinogen in synovial fluid of joints and cartilage [23]. However, the response may be short-lived because of the fibrin digestion caused by plasmin and trypsin [12,23]. Citrullinated fibrinogen is now recognized as an autoantigen, and these patients with increased fibrinogen also show a high amount of inflammatory cytokines such as ILs, TNFs, etc. [24,25]

Certain inhibitors of angiogenesis include exogenous and endogenous substances [26].

The exact cause of RA is not known; however, many recent studies indicate the role of genetics in developing RA, especially in patients with dominant Human Leukocyte Antigen (HLA)-DRB1*04 alleles [27]. The role in the destruction of cartilage is mainly by the enzymes secreted by neutrophils, synoviocytes, and chondrocytes, whereas the pannus destroys the bone in RA patients [27]. The pathogenesis of RA follows a hypoxic mechanism, which explains the shift of energy metabolism from aerobic to anaerobic glycolysis and blood flow stagnation [28]. Generally, in early-stage patients, there is a surge of Th17, which stimulates the release of inflammatory cytokines [29].

2.3   Oxidative stress in rheumatoid arthritis

Reactive oxygen species (ROS) and reactive nitrogen species (RNS) are the primary mediators of producing oxidative stress and cartilage damage in RA patients [30]. Oxidation of guanidino nitrogen of L-arginine produces nitric oxide (NO), which is responsible mainly for regulating inflammation, autoimmunity, and inflammation [30]. ROS, produced as a by-product in oxidative phosphorylation, includes peroxides, superoxides, and RNS [31]. However, the exact mechanism and consequences of ROS are not known, but they are expected to be produced as a result of inflammation and may release a variety of harmful enzymes and hormones [31]. Oxidative damage can also be caused by hypoxia induced by a rise in intracellular calcium, leading to the formation of ROS [32]. These generated ROS have a short biological half-life, which makes it very difficult for their detection [32]. The possible mechanism by which these ROS can create a cascade of reactions and produce inflammation in RA involves abstracting electrons from other compounds, thus converting them into free radicals [33]. The damage caused by these ROS is concentration dependent; as the concentration increases, they may damage the cellular components of the cell membrane as well as proteins and nucleic acids [33]. Today, novel approaches to treat RA by inhibiting the generation of ROS are followed, which is discussed further in the chapter.

2.4   Stages of rheumatoid arthritis

Prior detection of inflammation and its treatment in rheumatoid arthritis is critically important as it causes rapid damage to the tissue and joints and impairs their functioning. The progression of RA and types of joint involvement varies from individual to individual, as many patients with RA were found to have 70% joint destruction in hands and feet detected by X-ray [34]. Without proper treatment and management of RA, diagnosed after 20 years, significant functional impairment, that is, Stage III, was observed in more than 60% of patients. Patients with stage III RA need mobility aids and require joint replacement or sometimes require daily care and experienced loss of independence, that is, stage IV [34]. Table 2 represents the stages of RA.

American College of Rheumatology (ACR) and the European League Against Rheumatism (EULAR) proposed a new criterion to classify RA at its early stages in 2010. This criterion aimed to prevent disease progression and bone destruction using disease-modifying agents [36]. The most common presentation of RA is inflammation of large and small articulations accompanied by morning stiffness. As per this criterion, there are some changes in the clinical picture of this problem like in early diagnosis of the presence of seronegative monoarticular and oligoarthritis and the high rate of false positive diagnosis among patients with undifferentiated RA [37]. This criterion is not diagnostic but helps in identifying a disease with an increased likelihood of converting into a chronic form. A patient with RA is detected and classified on a score of six or greater [38]. This new classification has established a score value between 0 and 10.

Table 2 Stages of rheumatoid arthritis.

Stages	Stage description	Characteristics	Reference
Stage I	Early rheumatoid arthritis	Inflammation of synovial membrane, pain during joint movement, and swelling of joints. High cell density in the joints because of migrated immune cells. No joint destruction is evidenced by X-ray detection. Soft tissue swelling and some bone erosion are evidenced	[35]
Stage II	Moderate rheumatoid arthritis	Inflammation spreading across synovial joints may affect the cartilage joint cavity. Cartilage destruction occurs because of inflammation. Joint narrowing occurs at this stage	[35]
Stage III	Severe rheumatoid arthritis	Pannus formation in the synovium, cartilage and joint destruction, bone exposure below the cartilage, visible changes, and joint deformities are evidenced on X-ray detection	[35]
Stage IV	End stage or terminal rheumatoid arthritis	Fibrous tissue formation or fusing of the bones causes a halt in joint function. Subcutaneous nodule formation also occurs at this stage	[35]

2.5 Etiology of rheumatoid arthritis

RA, affects the synovial tissue of joints, bone, and cartilage, leading to inflammatory cytokine release at these locations, and is generally a consequence of epigenetic, genetic, and sometimes environmental factors that may include dust, smoke, cigarette, etc. [39] However, the exact etiology may be unknown, but certain elements may trigger the immune responses in the body and include RF, ACPA, pathogen-associated molecular patterns (PAMPs), and damage-associated molecular patterns (DAMPs) [39,40]. Robust immunity is seen in females against any pathogen; hormones such as estrogen and prolactin make them more susceptible to RA, as compared to males [39].

The most important factor leading to RA is known to be genetic, with a 50%–60% risk of developing RA, which follows a specific triad of gene-specific autoantibody disease, followed by environmental factors [40]. There are five known peptidyl arginine deiminases (PAD)-PAD 1, PAD 2, PAD 3, PAD 4, and PAD 5, with specific expressions of PAD 2 and PAD 4 in RA [40,41]. Citrullinated peptides in synovial tissue confirm the involvement of genes in the etiology of RA [41]. Fig. 1 represents the factors involved in RA development.

2.5.1 Environmental factors

Certain inhaled pollutants, such as tobacco smoke, can increase the probability of developing RA in individuals who have certain genetic risk factors. This is thought to be due to the effect of these pollutants on increasing levels of RF and ACPA, which are markers of RA [42,43]. Obesity, alcohol intake, increased body mass index (BMI), and hormonal factors are other environmental factors that increase the risk of RA [43]. Various biological interactions with smoking increase citrullinated proteins and inflammatory cytokines in the lungs along with joints, which could be related to early joint defects [44]. Fig. 3 represents the factors involved in the development of RA.

2.5.2 Genetic susceptibility

Genetics plays a key role in RA development, especially in first-degree relatives that have a specific set of alleles in the Major Histocompatibility complex, encoding amino acid sequences in HLA [44]. The two major loci of HLA proteins include Class I and Class II, which activate CD 8 and CD 4 T cells, respectively [45]. Class II molecules are majorly responsible for the production of cytokines that are involved in articular destruction [45]. There is a clear role of infectious agents in synovitis initiation and it

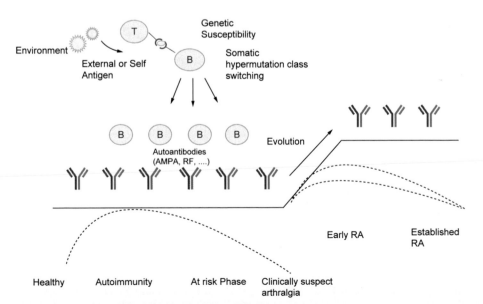

Fig. 3 Factors involved in the development of RA.

Table 3 Mechanism of action (MOA) by which microorganisms initiates synovitis.

Stages	Mechanism of action
I	Multiplication of an agent within a joint space. Examples-pyogenic bacteria and mycobacteria
II	Infectious agent localizes in the joint space and initiates an immune response. Example-Rubella in human beings
III	In some cases, infections caused by agents at a distant site can trigger an immune response that leads to the development of arthritis. Example-Rheumatic fever in which streptococcal antigens initiate an autoimmune response
IV	Arthritis is caused by toxins. Example-Arbovirus infections

is classified in four different ways [46]. Table 3 depicts this classification with examples of microorganisms [46].

2.6 Diagnosis of rheumatoid arthritis

Rheumatoid arthritis diagnosis is performed by physical examination, including signs and symptoms. Early diagnosis of RA within a period of 6 months with the onset of symptoms is important, ideally as it slows down or stops the beginning of disease progression. The RA diagnosis is made by detecting laboratory, clinical, and imaging characteristics. X-rays diagnosis of hands and feet are done for many patients with affected joints demonstrating juxta-articular osteopenia and swelling of soft tissues [47]. Subluxation and bony erosions occur with the advancement of RA. Ultrasound and magnetic resonance imaging (MRI) techniques are also useful in RA diagnosis [48]. MRI diagnosis helps detect radiographic images in RA and is capable of diagnosing synovial hypertrophy and cellular edema of bone marrow [49]. Ultrasonography techniques are also found to be important in guided joint injections and aspirations [50]. High-resolution sonograms allow the investigation of tendon sheaths, erosions, and vascularization of the synovial layer. Doppler ultrasound (color and power) identifies vascular signs of synovitis with the degree of inflammation. This diagnostic technique is important for detecting RA at early stages and acts to be the most precise biomarker of joint destruction in the future [51]. The laboratories' studies for the detection of RA usually fall under three categories, that is, inflammatory markers, hematological examination, and immunological parameters. These laboratory studies include the following tests:

1. Erythrocyte sedimentation rate (ESR)
2. Complete blood count
3. C-reactive protein level
4. RF assay
5. Antinuclear assay (ANA assay)
6. AntiCCP
7. AntiMCV (Antimutated citrullinated vimentin)
8. MicroRNA (miRNA)
9. Antifilaggrin antibodies (AFA)

3   Drugs targeting angiogenesis and inflammation in the management/treatment

After the initial evaluation of RA, the treatment should begin. Along with the management of RA, patient preference should also be considered [52]. The goal of antirheumatoid therapy is to minimize joint pain, prevent bone deformity (like ulnar deviation), swelling of joints, and bone erosions, maintain lifestyle, and regulate extraarticular symptoms [53]. Table 4 represents the drug therapy used in the treatment of angiogenesis and inflammation of RA. Five important drug classes are generally used in RA treatment, as shown in Table 4.

Table 4 Drug therapy is used in the treatment of rheumatoid arthritis.

S. No.	Drug classification	Treatment description
1	Analgesics	Used in mild to moderate arthritis, drugs included are tramadol, acetaminophen, narcotics, and capsaicin. They do not have antiinflammatory properties, so they are often used in combination with glucocorticoids, nonsteroidal antiinflammatory drugs (NSAIDs), anticytokine, and disease-modifying antirheumatic drugs DMARDs [53]
2	Disease-modifying antirheumatic drugs (DMARDs)	They are the drug of choice in rheumatoid arthritis treatment. They can be both biological and nonbiological. The biological agents include recombinant receptors and antibodies like monoclonal antibodies, which inhibit cytokines and promotion of inflammatory cascade [53]
	Methotrexate (MTX)	Methotrexate is the folic acid antagonist and acts within weeks in rheumatoid arthritis. Its exact mechanism of action is not known in RA. This drug has also been shown to reduce radiographic RA progression. It can also be used with other DMARDs for maintaining disease remission [54]
	Hydroxychloroquine, sulfasalazine, and leflunomide	Hydroxychloroquine or sulfasalazine can be recommended as monotherapy in RA patients with poor diagnostic features like nonerosive rheumatoid arthritis. They have very few side effects and are well tolerated. Leflunomide is a pyrimidine synthesis inhibitor used in combination with methotrexate. A triple-drug therapy combination of sulfasalazine, methotrexate, and hydroxychloroquine is found to be more effective and less toxic [55]

Table 4 Continued

S. No.	Drug classification	Treatment description
3	Anticytokine therapies (antiTNF (tumor necrosis factor))	These are the first-line biologic treatment for RA. They are also used in combination with MTX or other drug treatments. It improves endothelium-dependent vasodilation, which suggests a protective effect on vascular functioning [56]
4	Glucocorticoids	These are the class of steroids that binds to glucocorticoid receptors and are potent antiinflammatory agents. They are used as an adjunct to NSAIDS or DMARD treatment. Prednisolone is the most used glucocorticoid in RA treatment
5	Nonsteroidal antiinflammatory drugs	This class includes aspirin, naproxen, indomethacin, and ibuprofen. They are used in both pain and stiffness in RA patients. NSAIDs inhibit cyclooxygenase COX-1 and 2 enzymes that block prostaglandin synthesis. These drugs appear to have a lesser effect on patients' chronic condition of RA and thus do not come under first-line agents

Glucocorticoid drug therapy is the established therapy for inflammation and has been used in clinical practice for more than 60 years [57]. In RA, these drugs are used at large doses initially for short-term treatment, and for the next 6 months, the use of lower-dose therapy is recommended. The use of low-dose glucocorticoids for a longer duration of time is found to be a standard disease-modifying treatment as it reduces pain and improves a patient's quality of life. It also reduced joint erosion and suggested a disease-modifying action [58]. NSAIDs can be used with caution with patients having gastrointestinal, renal, and cardiac problems. The use of methotrexate and NSAIDs is found to be safe and effective if proper drug monitoring is done. The serious adverse drug reaction of antiTNF drug therapy includes demyelinating diseases, multiple sclerosis, transverse myelitis, and optic neuritis have also been reported with this therapy [58].

3.1 Drugs targeting oxidative stress in the management/ treatment

Reduction of oxidative stress in RA also becomes an important concern, with ROS being the main contributor, especially in chronic patients [59]. Antioxidants used for the same include majorly carotenes and minerals such as copper, iron, zinc, etc.

[59] Researchers have found the preventive role of polyphenols, a naturally occurring compound with antiinflammatory property also [59]. In many novels, synthetic approaches are adopted today to control oxidative damage in RA, with the role of Gold (I) thiolate drugs being used for many years [60]. The mechanism of action of these drugs is not very complex and lies in the innate ability of gold to block immunoglobulin G, a major thiol group involved in the potentiation of RA, by the formation of gold complexes that reduce the effects of ROS [60]. Although the exact MOA for these drugs is not precisely studied, further research will help us to know at what stage these drugs can act to reduce the destructive effects of ROS [60].

Many scientists are now aiming to target Nicotinamide Adenine Dinucleotide Phosphate Hydrogen (NADPH), which is responsible for the initiation of inflammation and the formation of superoxides and ROS [61]. Major classes of drugs used for its inhibition include angiotensin-converting enzyme (ACE) inhibitors and angiotensin receptor blockers, which potentially show scavenging effects as antioxidants [61]. A widely used drug for the management of breast cancer, Tamoxifen, is also known to reduce H_2O_2 production in neutrophils and thus exhibit antioxidant properties [61].

3.2   Strategic drug delivery approaches

The conventional drug delivery systems (DDS) include oral delivery of drugs, especially NSAIDs, that are the first-line of drug therapy used in RA treatment and Parenteral administration of drugs such as DMARDs also [62,63]. However, these conventional DDS have several disadvantages, such as gastrointestinal resorption and first-pass metabolism in oral therapy and the short plasma half-life and toxicity observed in systemic administration [63]. To overcome these, various novel drug delivery approaches are now used that lead to targeted drug delivery and better absorption of the drugs. Table 5 gives a brief outline of the novel drug delivery approaches for RA.

Table 5 Novel drug delivery approaches in the treatment of RA.

Delivery system	Mechanism of action and effects produced	Reference
Liposomes	EPR effect; angiogenesis	[63]
Nanoparticles	EPR effect; prolonged release	[63]
Dendrimers	Controlled release; increased drug loading	[64]
Lipogelosomes	Longer retention time	[64]
Microspheres	Controlled release; better encapsulation efficiency	[64]
Hydrogels	Increased antiinflammatory efficacy; prolonged release	[65]
Gene therapy	Silencing or downregulating the target proteins involved in inflammation in RA, with the aid of RNAi and siRNA	[66]

To date, the main aim of drug therapy has been to control and measure disease activity. However, developing suitable biomarkers that can predict the efficacy of specific drugs used and thereby administering them to only the appropriate patient population is still imperative [67]. The future prospects in drug therapy involve categorizing drugs and administering them to only suitable populations.

4 Conclusion and future perspectives

This chapter demonstrated the etiology, pathogenesis, diagnosis, and treatment of RA disease. Understanding the pathophysiology of RA can help researchers to discover a novel therapy for the treatment and management of RA. Diagnosis of RA is very important, and predictive biomarkers can help clinicians identify the different stages of RA. To conclude, many studies, preclinical and clinical, have confirmed the association of adverse drug reactions with conventional drugs. Early diagnosis of RA within a period of 6 months with the onset of symptoms is important, ideally as it slows down or stops the beginning of disease progression. Limited studies suggest that early detection of RA and monitoring the use of pharmacological agents and their combination in RA can help clinicians to manage this disease. It is also important to establish the optimum dosage bio-efficacy of drugs along with the duration of treatment. The future potential of medicines and drug therapy includes the involvement of suitable biomarkers for drug and patient targeting.

The novel drug delivery approaches, as compared to conventional ones, can show promising results while decreasing the potential disadvantages such as first-pass metabolism and nontargeted drug delivery. Worldwide, scientists are now focusing on gene therapy and the use of specific antibodies to develop a more efficient therapy to treat RA. A more focused approach also lies in reducing drug loading and associated toxicity. However, more thorough research is needed to prove these novel approaches are effective in the treatment and for effective commercialization.

References

[1] Smolen JS, Aletaha D, McInnes IB. Rheumatoid arthritis. Lancet 2016;388(10055): 2023–38.
[2] Nishimura K, Sugiyama D, Kogata Y, Tsuji G, Nakazawa T, Kawano S, Saigo K, Morinobu A, Koshiba M, Kuntz KM, Kamae I. Meta-analysis: diagnostic accuracy of anti-cyclic citrullinated peptide antibody and rheumatoid factor for rheumatoid arthritis. Ann Intern Med 2007;146(11):797–808.
[3] Geuskens GA, Burdorf A, Hazes JM. Consequences of rheumatoid arthritis for performance of social roles—a literature review. J Rheumatol 2007;34(6):1248–60.
[4] Brooks P. Rheumatoid arthritis: aetiology and clinical features. Medicine 2006;34: 379–82.
[5] Combe B. Progression in early rheumatoid arthritis. Best Pract Res Clin Rheumatol 2009;23:59–69.
[6] Gossec L, Combescure C, Rincheval N, Saraux A, Combe B, Dougados M. Relative clinical influence of clinical, laboratory, and radiological investigations in early arthritis on

the diagnosis of rheumatoid arthritis. Data from the French early arthritis cohort ESPOIR. J Rheumatol 2010;37:2486–92.

[7] Tedeschi SK, Bermas B, Costenbader KH. Sexual disparities in the incidence and course of SLE and RA. Clin Immunol 2013;149:211–8.

[8] Bennett JC. The infectious etiology of rheumatoid arthritis. Arthritis Rheum 1978; 21(5):531–8.

[9] Derksen VFAM, Huizinga TWJ, van der Woude D. The role of autoantibodies in the path-ophysiology of rheumatoid arthritis. Semin Immunopathol 2017;39(4):437–46.

[10] Epstein FH, Harris ED. Rheumatoid arthritis: pathophysiology and implications for therapy. N Engl J Med 1990;322(18):1277–89.

[11] Srirangan S, Choy EH. The role of interleukin 6 in the pathophysiology of rheumatoid arthritis. Ther Adv Musculoskelet Dis 2010;2(5):247–56.

[12] Matsui T, Nakata N, Nagai S, Nakatani A, Takahashi M, Momose T, et al. Inflammatory cytokines and hypoxia contribute to 18F-FDG uptake by cells involved in pannus forma-tion in rheumatoid arthritis. J Nucl Med 2009;50(6):920–6.

[13] White DHN. Rheumatoid arthritis and ankylosing spondylitis. In: Pietschmann P, editor. Principles of osteoimmunology. Vienna: Springer Vienna; 2012. p. 169–95. Available from: http://link.springer.com/10.1007/978-3-7091-0520-7_8.

[14] Elshabrawy HA, Chen Z, Volin MV, Ravella S, Virupannavar S, Shahrara S. The patho-genic role of angiogenesis in rheumatoid arthritis. Angiogenesis 2015;18(4):433–48.

[15] Colville-Nash PR, Scott DL. Angiogenesis and rheumatoid arthritis: pathogenic and ther-apeutic implications. Ann Rheum Dis 1992;51(7):919–25.

[16] Paleolog EM. Angiogenesis in rheumatoid arthritis. Arthritis Res Ther 2002;4(3):S81.

[17] Szekanecz Z, Besenyei T, Paragh G, Koch AE. Angiogenesis in rheumatoid arthritis. Autoimmunity 2009;42(7):563–73.

[18] Marrelli A, Cipriani P, Liakouli V, Carubbi F, Perricone C, Perricone R, et al. Angiogen-esis in rheumatoid arthritis: a disease specific process or a common response to chronic inflammation? Autoimmun Rev 2011;10(10):595–8.

[19] Maruyama K, Muramatsu H, Ishiguro N, Muramatsu T. Midkine, a heparin-binding growth factor, is fundamentally involved in the pathogenesis of rheumatoid arthritis. Arthritis Rheum 2004;50(5):1420–9.

[20] Strunk J, Rumbaur C, Albrecht K, Neumann E, Müller-Ladner U. Linking systemic angio-genic factors (VEGF, angiogenin, TIMP-2) and Doppler ultrasound to anti-inflammatory treatment in rheumatoid arthritis. Jt Bone Spine 2013;80(3):270–3.

[21] Fava RA, Olsen NJ, Spencer-Green G, Yeo KT, Yeo TK, Berse B, et al. Vascular perme-ability factor/endothelial growth factor (VPF/VEGF): accumulation and expression in human synovial fluids and rheumatoid synovial tissue. J Exp Med 1994;180(1):341–6.

[22] Kato M. New insights into IFN-γ in rheumatoid arthritis: role in the era of JAK inhibitors. Immunol Med 2020;43(2):72–8.

[23] Schurgers E, Billiau A, Matthys P. Collagen-induced arthritis as an animal model for rheu-matoid arthritis: focus on interferon-γ. J Interf Cytokine Res 2011;31(12):917–26.

[24] Fattahi MJ, Mirshafiey A. Prostaglandins and rheumatoid arthritis. Art Ther 2012; 2012:1–7.

[25] Zhao X, Okeke NL, Sharpe O, Batliwalla FM, Lee AT, Ho PP, et al. Circulating immune complexes contain citrullinated fibrinogen in rheumatoid arthritis. Arthritis Res Ther 2008;10(4):R94.

[26] Ho PP, Lee LY, Zhao X, Tomooka BH, Paniagua RT, Sharpe O, et al. Autoimmunity against fibrinogen mediates inflammatory arthritis in mice. J Immunol 2009;184 (1):379–90.

[27] Choy E. Understanding the dynamics: pathways involved in the pathogenesis of rheumatoid arthritis. Rheumatology 2012;51(Suppl 5):3–11.

[28] Rothschild BM, Masi AT. Pathogenesis of rheumatoid arthritis: a vascular hypothesis. Semin Arthritis Rheum 1982;12(1):11–31.

[29] Furst DE, Emery P. Rheumatoid arthritis pathophysiology: update on emerging cytokine and cytokine-associated cell targets. Rheumatology 2014;53(9):1560–9.

[30] Vasanthi P, Nalini G, Rajasekhar G. Status of oxidative stress in rheumatoid arthritis. Int J Rheum Dis 2009;12(1):29–33.

[31] Veselinovic M, Barudzic N, Vuletic M, Zivkovic V, Tomic-Lucic A, Djuric D, et al. Oxidative stress in rheumatoid arthritis patients: relationship to diseases activity. Mol Cell Biochem 2014;391(1–2):225–32.

[32] Mapp PI, Grootveld MC, Blake DR. Hypoxia, oxidative stress and rheumatoid arthritis. Br Med Bull 1995;51(2):419–36.

[33] Quiñonez-Flores CM, González-Chávez SA, Del Río Nájera D, Pacheco-Tena C. Oxidative stress relevance in the pathogenesis of the rheumatoid arthritis: a systematic review. Biomed Res Int 2016;2016, e6097417.

[34] Venables PJW, Maini RN. Clinical features of rheumatoid arthritis. In: O'Dell JR, Romain PR, editors. Up-to-date. Wolters Kluwer Health; 2023. www.uptodate.com. [Accessed 4 March 2023].

[35] Wheeless CR. Rheumatoid arthritis. In: Wheeless CR, Nunley JA, Urbaniak JR, editors. Wheeless' text of orthopaedics. Data Trace Internet Publishing, LLC; 2012. Available from: www.wheelessonline.com.

[36] Van Venrooij WJ, van Beers JJ, Pruijn GJ. Anti-CCP antibodies: the past, the present and the future. Nat Rev Rheumatol 2011;7(7):391–8. Available from: www.wheelessonline.com.

[37] De Hair MJ, Landewe RB, van de Sande MG. Smoking and overweight determine the likelihood of developing rheumatoid arthritis. Ann Rheum Dis 2013;72:1654–8.

[38] Aletaha D, Neogi T, Silman AJ, Funovits J, Felson DT, Bingham CO. Rheumatoid arthritis classification criteria: an American College of Rheumatology/European League against rheumatism collaborative initiative. Arthritis Rheum 2010;62(9):2569–81.

[39] Scherer HU, Häupl T, Burmester GR. The etiology of rheumatoid arthritis. J Autoimmun 2020;110, 102400.

[40] Alam J, Jantan I, Bukhari SNA. Rheumatoid arthritis: recent advances on its etiology, role of cytokines and pharmacotherapy. Biomed Pharmacother 2017;92:615–33.

[41] Wegner N, Lundberg K, Kinloch A, Fisher B, Malmström V, Feldmann M, et al. Autoimmunity to specific citrullinated proteins gives the first clues to the etiology of rheumatoid arthritis. Immunol Rev 2010;233(1):34–54.

[42] Suzuki A, Yamada R, Chang X, Tokuhiro S, Sawada T, Suzuki M, et al. Functional haplotypes of PADI4, encoding citrullinating enzyme peptidylarginine deiminase 4, are associated with rheumatoid arthritis. Nat Genet 2003;34(4):395–402.

[43] Alpízar-Rodríguez D, Finckh A. Environmental factors and hormones in the development of rheumatoid arthritis. Semin Immunopathol 2017;39(4):461–8.

[44] Sparks JA, Chen CY, Hiraki LT, Malspeis S, Costenbader KH, Karlson EW. Contributions of familial rheumatoid arthritis or lupus and environmental factors to risk of rheumatoid arthritis in women: a prospective cohort study: familial and environmental risk of RA. Arthritis Care Res 2014;66(10):1438–46.

[45] Deane KD, Demoruelle MK, Kelmenson LB, Kuhn KA, Norris JM, Holers VM. Genetic and environmental risk factors for rheumatoid arthritis. Best Pract Res Clin Rheumatol 2017;31(1):3–18.

[46] Taneja V. Cytokines pre-determined by genetic factors are involved in the pathogenesis of rheumatoid arthritis. Cytokine 2015;75(2):216–21.

[47] Baratelle AM, van der Heijde D. Radiographic imaging end points in rheumatoid arthritis trials. In: Clinical trials, clinical trials in rheumatoid arthritis and osteoarthritis. Springer; 2008. p. 201–21. 1–659.

[48] Cush JJ, Kavanaugh A, Weinblatt ME. Rheumatoid arthritis: early diagnosis and treatment. Professional Communications; 2010.

[49] Hoving JL, Buchbinder R, Hall S. A comparison of magnetic resonance imaging, sonography, and radiography of the hand in patients with early rheumatoid arthritis. J Rheumatol 2004;31:663–75.

[50] Wells AF, Haddad RH. Emerging role of ultrasonography in rheumatoid arthritis: optimizing diagnosis, measuring disease activity and identifying prognostic factors. Ultrasound Med Biol 2011;37(8):1173–84 [63].

[51] Schueller-Weidekamm C. Modern ultrasound methods yield stronger arthritis work-up. Diagn Imaging 2010;32:20–2.

[52] Deighton C, O'Mahony R, Tosh J, Turner C, Rudolf M, Guideline Development Group. Management of rheumatoid arthritis: summary of NICE guidance. BMJ 2009;338, b702.

[53] Wasserman. Diagnosis and management of rheumatoid arthritis. Am Fam Physician 2011;84(11):1245–52.

[54] Reddy DA, Trost LW, Lee T, Amir R, Baluch AR, Kaye AD. Rheumatoid arthritis: current pharmacologic treatment and anesthetic considerations. Middle East J Anesthesiol 2007;19(2):318.

[55] Weinblatt ME, Kavanaugh A, Genovese MC, Musser TK, Grossbard EB, Magilavy DB. An oral spleen tyrosine kinase (Syk) inhibitor for rheumatoid arthritis. N Engl J Med 2010;363:1303–12.

[56] Hurlimann D, Forster A, Noll G, Chenevard R, Distler O, Béchir M, Spieker LE, Neidhart M, Michel BA, Gay RE, Lüscher TF, Gay S, Ruschitzka F. Anti-tumor necrosis factor-alpha treatment improves endothelial function in patients with rheumatoid arthritis. Circulation 2002;106:2184–7.

[57] Kirwan JR. Links between radiological change, disability, and pathology in rheumatoid arthritis. J Rheumatol 2001;28(4):881–6.

[58] Myasoedova E, Crowson CS, Nicola PJ, et al. The influence of rheumatoid arthritis disease characteristics on heart failure. J Rheumatol 2011;38(8):1601–6 [115.Ofman JJ, Bad].

[59] Behl T, Upadhyay T, Singh S, Chigurupati S, Alsubayiel AM, Mani V, et al. Polyphenols targeting MAPK mediated oxidative stress and inflammation in rheumatoid arthritis. Molecules 2021;26(21):6570.

[60] Grootveld M, Blake DR, Sahinoglu T, Claxson AWD, Mapp P, Stevens C, et al. Control of oxidative damage in rheumatoid arthritis by gold(I)-thiolate drugs. Free Radic Res Commun 1990;10(4–5):199–220.

[61] Soomro S. Oxidative stress and inflammation. Open J Immunol 2019;9(1):1.

[62] Qindeel M, Ullah MH, Fakhar-ud-Din, Ahmed N, Rehman A. Recent trends, challenges and future outlook of transdermal drug delivery systems for rheumatoid arthritis therapy. J Control Release 2020;327:595–615.

[63] Tarner IH, Müller-Ladner U. Drug delivery systems for the treatment of rheumatoid arthritis. Expert Opin Drug Deliv 2008;5(9):1027–37.

[64] Thakur S, Riyaz B, Patil A, Kaur A, Kapoor B, Mishra V. Novel drug delivery systems for NSAIDs in management of rheumatoid arthritis: an overview. Biomed Pharmacother 2018;106:1011–23.

[65] Oliveira IM, Fernandes DC, Cengiz IF, Reis RL, Oliveira JM. Hydrogels in the treatment of rheumatoid arthritis: drug delivery systems and artificial matrices for dynamic in vitro models. J Mater Sci Mater Med 2021;32(7):74.

[66] Pirmardvand Chegini S, Varshosaz J, Taymouri S. Recent approaches for targeted drug delivery in rheumatoid arthritis diagnosis and treatment. Artif Cells Nanomed Biotechnol 2018;46(Suppl 2):502–14.

[67] Koga T, Kawakami A, Tsokos GC. Current insights and future prospects for the pathogenesis and treatment for rheumatoid arthritis. Clin Immunol 2021;225, 108680.

Inflammatory bowel diseases (IBDs)

Shuchi Upadhyay[a], Sanjay Kumar[b], Vinod Kumar[b,c], Indra Rautela[d], Shraddha Manish Gupta[e], and B.S. Rawat[f]

[a]Department of Allied Health Sciences, School of Health Science and Technology, University of Petroleum and Energy Studies UPES, Dehradun, Uttarakhand, India, [b]Department of Food Science and Technology, Graphic Era (Deemed to be University), Dehradun, Uttarakhand, India, [c]Peoples' Friendship University of Russia (RUDN University), Moscow, Russian Federation, [d]Department of Biotechnology, School of Applied and Life Sciences (SALS), Uttaranchal University, Dehradun, Uttarakhand, India, [e]Department of Pharmaceutical Sciences, School of Health Sciences and Technology, UPES, Dehradun, Uttarakhand, India, [f]Department of Physics, School of Applied and life Sciences, Uttaranchal University, Dehradun, India

1 Introduction

Inflammatory bowel disease (IBD) of the gastrointestinal system is a chronic functional disorder of the stomach. In IBD, generally, patients have episodic attacks with abdominal pain and disturbed bowel movement. There are either predominant irritable bowel syndrome (IBS-c), diarrhea-predominant IBS (IBS-d), or mixed IBS (IBS-m) [1]. The IBD is well-known worldwide problem. The global scenario represents 11% of IBD incidence worldwide [2]. The epidemiological causes are multifactorial with genetic immunological, and environmental factors which contribute to forming the disease. The disease is known for inflammation of the bowel thus it is known as inflammatory bowel disease. This is a holistic word that explains all diseases and disorders related to the digestive tract which is a long-standing (chronic) inflammation of tissues in human organs [3].

IBD is also known as a group of diseases with disorders that is a combination of an atypical immune reaction to the pathogen in a genetically designed host. The continuous severe infection in the form of inflammation along with several changes in the body and in the gastrointestinal route that reflect bowel disease outcomes of a disproportion of active lymphocytes and cytokines, which work like pro-inflammatory [4]. The large amount of reactive oxygen species (ROS) produced by different parts of cytokines and leukocytes with the activated part of macrophages of the immune system and responsible for oxidative stress (OS) disturbances. Several pathophysiological characteristics of IBD are represented by tissue injury (mucosal erosions) and fibrosis also connected with redox imbalance because of RDS generation, which decreases antioxidant molecule levels. Meanwhile, the effect of oxidative stress is destructive if it is uncontrolled inflammatory conditions, the body's antioxidant role

Targeting Angiogenesis, Inflammation and Oxidative Stress in Chronic Diseases. https://doi.org/10.1016/B978-0-443-13587-3.00006-0

work like a defenses and it can adverse the reaction of the results comes by an excess of reactive oxygen species.

The protective molecule in the direction of pro-oxidant molecules is antioxidant, the sources of antioxidants are plants herbs, vegan food, and different form of diet. IBD not only affect intestine but also the colon and other tissue of the digestive system that is distinguished by redness and tissue ruination. There are two major subtypes of IBD and well-known by their names, ulcerative colitis and Crohn's disease, both are the result of inflammation in the intestine. The symptoms of both diseases are similar to other bowel diseases like stomach cramps, gastric issues, and diarrhea. IBD is associated with bowel disturbance and discomfort [5]. It gives several attacks of stomach ache and disturbed bowel movement. If it is at a severe level, then it is known as IBS.

Children and adults are more affected by IBD worldwide and the percentage is raised to 5%–10% that is high in number. It affects both genders and most of the time it is identified in the severe stage. IBS and IBD in the first stage do not shorten any one life span but disturbs health and good life. It reduces physical health and creates a mental, physical, and economic burden in an individual's life. The diagnostic criteria of IBS and IBD determine the identification and treatment level. It is 11% around the world. The abnormal response of the human gut occurs in IBD, which creates the improper function of the immune system meaning over or underperformance of the immune system. Fig. 1 shows different forms and changes that occur in bacteria of microbiota that reduce infection and delayed bowel inflammation in several pathogen attacks. The body creates different responses in the natural stage to protect the body from any form of infection. The control of pain and sensation is regulated by the central nervous system to collect the right signals of bowel disease IBD. Chronic inflammation is characterized in two forms as mentioned in the above paragraph. If inflammation is extended over a long period, then it affects the GI tract and creates inflammation [6] in the long term GI tract results in the form of IBD, which is confused with a similar body.

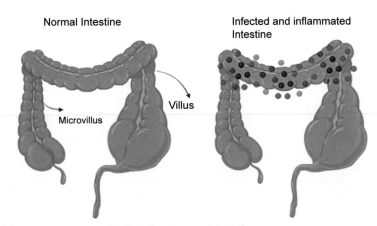

Fig. 1 Diagrammatic view of infected and normal intestine.

Symptoms: IBS has higher characteristics than IBD but, in general, both disorders had most of the common clinical symptoms such as diarrhea, stomach cramping, abdomen pain, etc. If the state of IBD is more critical, then the patient is experiencing fever, weight loss and blood in stool, this stage is alarming for the patient. The monitoring and check-up plan are entirely different in IBD in comparison to IBS and so care and diagnosis are important.

IBS can be segmented into those who indicate to have overbearing diarrhea or severe constipation. The rate of IBS is 25% as this is identified with a severe group of disorders over the last 50 years. There is a lot of evidence in several medical practice records for IBS that the association with occupation, behavior, and income group is partially available in any research area. The responsible factors such as genetic and dietary patterns are available in several research studies for IBS patients identified with IBS are more likely to have chances of other functional diseases and have cases of more treatment, and surgery than the general population. The disturbed motion and irritable colon are identified as the early stage of IBS and affect the spastic nerve. The changes in frequency and consistency of the bowel are the initial alarm and hallmark to think of IBS. Although the justification of IBS has not to date been fully interpreted as the real scenario [7]. There are several observations based on study and clinical data surveys that group of symptoms and results can present in one group and other groups with different symptoms, that is, because of visceral hypersensitivity neurotransmitters imbalance psychosocial factors, altered bowel motility, and infection. Psychosocial factors affect the person's hygienic behavior, contamination experience, affected person expectancies, and treatment outcome of IBS patients. As per several studies and reports of different case studies, the severity of IBS is yet not documented in any reports although it leads to other several complications in digestive organs that decrease the efficiency of the digestive tract. While irritable bowel syndrome in comparison to IBD is not considered a life-threatening disease or disorder. The disproportionate coordination of microbiota and a higher rate of comorbidity with different issues in IBS patients that include fibromyalgia, pelvic ache chronic fatigue, and psychiatric problems [8]. The muscle spasm and contractions were observed with a muscle spasm in several patients of IBS. These contractions of muscle could be slow or fast based on the stage of IBS. In several studies, extended sensitivity to stimuli triggers pain and abdomen discomfort observed. IBS gives a lot of socioeconomic burden on its society as it decreases the quality of life and increases medical expenses. Emerging technologies and innovative diagnostic areas can reduce the severity and the specific pharmacologic way might also support enhancing the affected person's life with care and decrease health care expenses and resources, thereby using doubtlessly decreasing the heavy burden this ailment presently bears. The yellow and red color of fruit and vegetables are rich in phytochemicals that provide antioxidant action. All provitamins are rich in extraordinary levels of antioxidant action that is similar to retinol-binding proteins. The body itself lowers the level of ROS if a sufficient amount of vitamin C and glucose is available in the body from different natural sources of fruits, vegetables, and plants. The vital minerals like zinc, copper, manganese, and iron take part in antioxidant enzymes for numerous isoforms of superoxide dismutase (SOD) activity. Studies reported that several cases of Crohn's disease (CD) and ulcerative colitis (UC) patients are documented with low levels of mentioned minerals and vitamins that are important for antioxidant activity. The nonvegan diet is a responsible factor for reducing the number of vitamins and minerals in such patients. Pain is the only signal in the body which triggers the patient to know about the disease's progress and therapy for controlling the symptoms. Aggravated pain and complications with ongoing intestinal inflammation are common causes of pain in IBD. Low residue diet and high-stress level in daily life create IBD and lead to Crohn's disease.

Bile-acid formation and inappropriate malabsorption can lead to diarrhea and stomach cramping that will usually respond to bile-acid sequestration. Gastrointestinal tract (GIT) can result in epithelial damage. Inflammatory bowel disorders (IBDs) are autoimmune, progressive persistent inflammatory illnesses. Fig. 2 covers the intestinal bacteria and food both make a substantial contribution to the large antigenic burden that exists there. Since they aid in digestion and thwart the colonization of pathogenic species, the majority of gut bacteria are helpful to health. IBD appears to manifest when there is an aberrant reaction to regular flora is present. The immune system attacks GIT cells as a result of an overreaction, which causes persistent inflammation and several other problems. The translocation of luminal antigens into the bowel, such as microorganisms from the commensal microbiome wall, is caused by changes in the epithelial barrier. If only one of these conditions exists, it is not enough for the disease to develop. Results from tests and genome-wide association studies revealed many immune systems disturbed and impaired pathways including several other pathways that affect pro-inflammatory rate mediated, diminished immunity-regulating process, and a barrier function deficit in the epithelium of the intestine [9]. This chapter covers all aspects of the disease including concept, diagnosis, recent studies on targeting

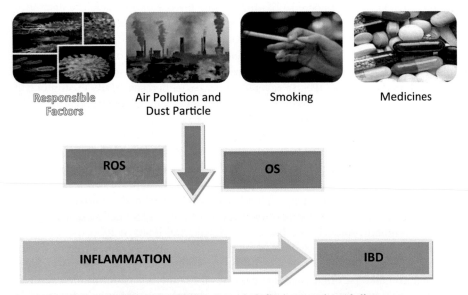

Fig. 2 Correlation of ROS and oxidative stress in inflammatory bowel disease.

angiogenesis, inflammation, and OS in IBD and the conventional to advance strategic drug delivery approaches with future perspectives.

2 Pathogenesis and responsible factors involved in IBD

In IBD, the intestinal barrier's strength and performance are both disrupted, resulting in a loss of resistance to typical food ingredients and/or an overactive immune response to infections, both of which exacerbate the inflammatory process as a whole.

It is generally accepted that a diverse range of factors [10], which include genetic mechanism and history, modifications in IBD, slow response of the immune system, and fast pace of intolerance by different organs even in microbiota and environment could create a level of oxidative stress. The OS and slow microbiota together contribute to the development of IBD [11].

Propels in sub-atomic hereditary qualities have shown that few IBD-related qualities associated with resistant pathways could address helpful restorative targets. Also, dysregulation of crosstalk among a few sub-atomic pathways in IBD needs further clarification. IBD patients' genetic variants have been thoroughly studied in recent years [12]. Different research studies identified 200 genetic uncertainty connected with IBS and IBD. TNFSF15, RIPK2, IFNGR1/2, TYK2, and CARD9 are just a few of the genes that have been linked to immunodeficiency by genome-wide association studies (GWAS). The process called autophagy is responsible for IBD, it is known to have a pathogenic process, activated to break down the cellular nucleus in response to cell deprivation.

There are persistent inflammatory (Fig. 3) alterations in the GI in many forms like CD and ulcerative colitis. Although they both result from T-cell activation, these are mediated by various immunologic mechanisms for each disease. Th1 cells, which coordinate the cell-mediated immune response, are assumed to be the clinical reason for Crohn's disease [13]. In Crohn's disease, the mucosa has higher levels of the cytokine IL-12. Increased Th1 response and interferon-gamma (IFN-γ) are the results of this. A loop of unchecked inflammation results from IFN-subsequent upregulation of macrophages interleukin-12 and Th1 responses in IBD.

To meet the demand for more efficient and secure treatments among IBD patients, several innovative techniques have been created. These methods include boosting Treg activity, preventing inflammatory cell trafficking, inhibiting pro-inflammatory cytokines, and strengthening epithelial barrier function. Loss of control of these excessively activated Th1 cells and macrophages also activates matrix metalloproteinase, causing tissue injury via IFN-γ and tumor necrosis factor-alpha (TNF-α). Th2-cells, which mediate B cell and antibody responses, are thought to control inflammation in ulcerative colitis,13 however, this has not been demonstrated. Although IL-5, a Th2 cytokine, has been proven to have enhanced expression, IL-4, another Th2 cytokine, has not. In ulcerative colitis, the rapid growth in IgG plasma cells, which are likely nourished by T cells, suggests that the Th2 contribution is aiding the antibody response.

Fig. 3 Responsible factors of inflammation bowel disease.

The study of IBD pathomechanisms have advanced quickly over the past 20 years. However, because of their major complications and side effects, existing IBD treatments are not quite ideal. Improved IBD management will evaluate based on a better understanding of the function of oxidative stress in IBD, particularly in relation to a combined drug regimen that uses both natural and synthetic antioxidant molecules [12]. OS is a situation carried out by a physiological response of the body to the inappropriate ability to detoxify wastes from body, these end products, and the formation or deposition of ROS in tissues and organs. ROS can and do carry out a variety of biological mechanisms.

Because of its nutrient-dense capability, which is essential to maintain balanced intestinal movement, the cell can adjust a few levels of ROS under physiological and pathological conditions. Furthermore, an increased load of OS resulting from increased ROS formation or reduced active processes can significantly increase membrane separation, change the response of inflammatory active signals, and cause DNA damage, apoptosis, lipid and protein alterations, and carcinogenesis.

After the leap forward of TNF-α bar by killing antibodies in both Cd and UC, numerous remedial specialists hindering the action of pro-inflammatory cytokines or supporting the activity of calming cytokines were assessed as treatment for IBD. Tragically, a considerable lot of them bombed in clinical examinations or had gainful impacts in subgroups of patients just, which highlight the thought that cytokine networks in human IBD are more mind-boggling than recently expected and may fluctuate among patients. This suggests the conversation starter of whether further and significant headway can be made by focusing on cytokines in IBD. If only one of these conditions exists, it is not enough for the disease to develop. Results from tests and genome-wide association studies revealed several impairments of inflammatory pathways, including pathways like pro-inflammatory mediation, reduce process and regulation of insulin resistance (IR), and a barrier function deficit in the epithelium of the intestine. In patients with inflammatory bowel disease, arthropathies are the most prevalent extraintestinal symptoms of IBD. The main clinical symptoms of IBD-associated arthropathy are back and joint discomfort. By accelerating chemical changes, various enzymes, like nicotinamide adenine dinucleotide phosphate (NADPH), xanthine, peroxidases, oxidases lipoxygenases, glucose oxidases, myeloperoxidases (MPOs), nitric oxide synthases, and cyclooxygenases (COXs), take part in the production of endogenous ROS. The role of several NOC complexes (nucleolar complexes) like 1 and 2 NOX and dual oxidase 2 among other mucosal NADPH oxidases (NOXs) have been identified as inflammatory risk factors, emphasizing the significance of disturbance in redox balance in blood for one of the pathological reasons of IBD. Additionally, extrinsic variables that are responsible for IBD in patients are continuous chemotherapy, long term use of smoking, unprotected radiation, luminal antigens, consumption of alcohol, frequent medications, and xenobiotics can also cause oxidative stress to be activated. High alcoholic beverages can harm the GI tract's mucosal barrier. Beyond that, it has been observed that a variety of medications and xenobiotic increase the accumulation of free radicals in the gastrointestinal tract. Most of the food contains *trans*-fatty acids (TFAs) in their processed and packed food [14], where acrylamide is available in many fast and processed food like chips and crackers. All processed foods are high in cholesterol [15] and intake of these foods, cause oxidative stress by boosting the release of ROS.

3   Drugs targeting angiogenesis of inflammatory bowel disease

The process of angiogenesis is the formation of new blood vessels from the vasculature that already exists. Numerous pathologic and physiological processes such as embryogenesis tissue growth, wound healing, and others. Angiogenesis may also play a role in the pathophysiology of conditions like cancer, psoriasis, tissue damage from reperfusion injury following ischemia or heart failure, diabetic retinopathy, and chronic inflammatory conditions of the joints or gastrointestinal tract. There is minimal relevant data to support the development of treatment approaches that target angiogenesis in chronic inflammatory diseases reactive oxygen species are created during metabolism of oxygen, which is important for survival of mammalian cells (ROS). Free radicals like alkoxyl, hydroxyl radicals, superoxide, peroxyl, and hydroperoxyl (HO_2) are examples of ROS [9]. Lipid hydroperoxides are another type of ROS, as are reactive nonradical substances like O_2, HOCl, chloramines, hydrogen peroxide, and ozone. Nitric oxide, nonradical molecules, nitrogen dioxide, peroxynitrite, and dinitrogen trioxide make up the majority of reactive nitrogen species (RNS) (N_2O_3). Because they have unpaired electrons, ROS and RNS are highly reactive and are both the main mediators of intracellular damage to CO_2, amino acids, fatty acids, lipids, and nucleic acids [16]. According to reports, reactive oxygen species and RNS increase the gene expression related to adaptive and innate immunity. By catalyzing chemical processes, the enzymes oxidase, nitric oxide synthase MPO, and COXs contribute to inside ROS production. The NOX2 complex, DUOX2, and NOX1, among other mucosal NOXs, are identified as risk factors IBD, highlighting the importance of the disproportion in redox equilibrium for the infectious of IBD. Endogenous ROS are constructed in the aqueous cytoplasm and extracellular matrix (ECM), similarly it is available in different organelles of the intracellular membrane such as the endoplasmic reticulum (ER), nucleus, mitochondria, and peroxisomes. A substantial portion of ROS is also created by the mitochondrial electron transport chain. The principal organelles in metabolic activities, mitochondria, are subjected to ROS. Impaired adenosine triphosphate (ATP) synthesis, inhibition of the intracellular electron transport chain (ETC), and DNA damage in mitochondria are all effects of excessive ROS generation. If this condition persists then affected Bioenergy mitochondrial and homeostasis convert into cell death. Instead of being viewed as an adverse reaction of ongoing inflammation in the intestinal mucosa, oxidative stress has become recognized to be effective pathogens and an important part of the development, severity, growth, and progression of IBD [17]. However, the underlying workings of these processes are still far from being fully understood. According to laboratory studies and clinical trials, a growing range of various antioxidants, the inhibitors of ROS generation, different hormones, numerous artificial substances, different polyphenols, various extracts of medicinal plants, and different functional foods can control inflammation.

4 Oxidative stress management in inflammatory bowel disease

The dairy and nondairy probiotics, water- and fat-soluble vitamins, have several metabolic roles for smooth function of the gut and metabolism for maintaining of oxidative stress. However, limited research has been done, and further is required to determine the targeted action of these promising substances as well as the most effective dosages and routes of administration. IBD and other diseases may benefit more from the therapeutic effects of new antioxidant-enhancing medicines when combined with traditional medications. There are numerous ways when inflammation promotes increased angiogenesis [18,19].

5 Probiotic and antioxidant defenses mechanism for treatment of IBD disease and other GI tract disorders

IBD causes the intestinal microbiota to become abnormal. Malfunction activity bacterial stimulation may cause inflammation in the mucosal lining and upset the equilibrium between pro-inflammatory chemicals and T-cell activity [19]. Aberrant

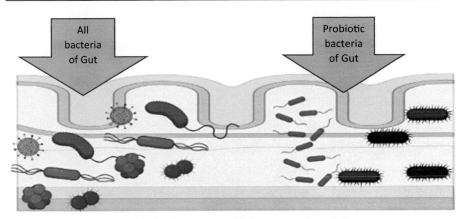

Fig. 4 Involvement of probiotics for control of inflammatory bowel disease.

microflora (Fig. 4) and mucosal immune function disorders are closely related, in contrast to healthy controls, bifidobacteria counts were significantly lower in patients, however, bacteroides and lactobacilli levels stayed constant. Additionally, bacterial enzyme activity, particularly b-D-galactosidase, decreased in bifidobacteria counts were observed in fecal extracts from patients with Crohn's disease. In another study, the related microbiota of the mucosal lining of different colonies in patients with various levels of ulcer and colitis were examined. Patients with active ulcerative colitis showed a noteworthy down in the series of anaerobic bacteria, anaerobic gram negatives, and lactobacillus [20]; however, there were no changes in the numbers of aerobic bacteria.

However, there was no discernible difference between patients with inactive ulcerative colitis and healthy conditions in the microbiota associated with the intestinal mucosa. SODs, glutathione peroxidase (GPX), catalase (CAT), glutathione, an intracellular nonenzymatic antioxidant, and extracellular antioxidants such vitamins, minerals, ceruloplasmin, and uric acid consist of the majority of the endogenous antioxidant system. Various probiotic microorganisms have been proven in studies to reduce or even prevent intestinal inflammation in some animal models. Additionally, supporting the use of probiotics in IBD is clinical evidence. It is generally known that in Crohn's disease, changing the fecal stream tends to result in mucosal healing. The concept that bacteria play a critical role in gut inflammation and that probiotic bacteria may modify the host-microbe interaction in a way that is directly advantageous to patients is well supported by evidence. Probiotic therapy in several means useful in different nutraceutical and pharmaceutical products for the treatment of IBD and associated diseases.

A significant nonenzymatic intracellular antioxidant is glutathione. It is more common in its reduced form, GSH, a soluble antioxidant that is highly expressed in the cytoplasm, nucleus, and mitochondria. An antioxidant barrier is formed in the gut mucosa by the enzymes GPX, glutathione reductase (GSR), and glutathione-S-transferase (GST). GSH has been used as a biomarker for both (Fig. 5) oxidative stress and inflammation. Strong antioxidant melatonin (MEL) was first synthesized

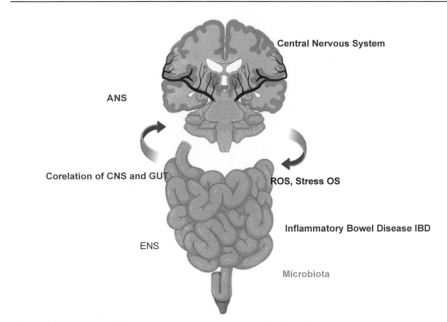

Fig. 5 Mechanism of the central nervous system and microbiota.

by mammals in their pineal gland. By overcoming physiological barriers like the mitochondrial membrane, it can lower oxidative stress in both lipid and aqueous cell settings. MEL acts as a protective factor in both the early and late phases of numerous disorders involving ROS metabolites such as IBD. It acts as an antioxidant for peroxyl and hydroxyl radicals.

6   Strategic drug delivery approaches in IBD

Numerous approaches have been used to control IBD, antiinflammatory drugs must be used often and in high quantities in regular practice [21] to treat inflammatory bowel disease. This causes the drugs to be absorbed from the small intestine, which in the long term can have serious side effects. Therefore, a variety of strategic approaches have been used such as the creation of prodrugs that specifically deliver medications to the large intestine after specific bacterial enzymes in the colon separate the active component from the hydrophilic carrier. IBD treatment has progressed significantly in the beyond 20 years with the utilization of biological specialists, for the most part, hostile to TNF-α-based treatments. The helpful requirements of IBD are still neglected, and the viability of these new treatments is restricted in certain gatherings of patients. To meet the demand for more efficient and secure treatments among IBD patients [22], several innovative techniques have

Table 1 Different drug delivery approaches and effects.

Drug delivery system	Mode of action	Effect	References
Nanoparticulate dual scavenger	Oral dose	Therapeutic activity has seen in both ulcerative colitis and Crohn's disease models	[23,24]
Enzymatic antioxidant defense mechanism	Oral dose	Superoxide dismutase (SOD) and catalase (CAT) activities	[25]
Immune modulators	Oral	Control symptoms and reduce further growth	[26]
Antiinflammatory drugs (corticosteroids)	Intravenous/ oral	Reduce pain and severity	[26]
Biodegradable nanoparticles	Oral administration	Therapeutic efficiency higher than the conventional	[1]

been created. These methods include boosting activity, preventing inflammatory cell trafficking, inhibiting pro-inflammatory cytokines, and strengthening epithelial barrier function (Table 1).

As well as the development of solid dose forms that allow the medicine to be released in the colon according to the physiological environment. Administration of drugs by rectal route is also currently used. However, it is useless if the swollen tissues are not located in the upper parts of the colon. Numerous research recently unveiled that, drug carrier systems with a size larger than 200 μm are subjected to diarrhea symptoms, resulting in a decreased gastrointestinal transit period and hence a sharp decrease in performance.

7 Conventional drug delivery approach (CDDA) for IBD

Existing medical treatment of inflammatory bowel diseases determines the severity of symptoms. The orally administered routes are the most popular and, through these, the medicines reach systemic levels. When searching for local effects, parenteral or rectal foams with about the same active substances are used. However, there is no standard therapy for these pathologies and patients experience incidences throughout their lifetime, which substantially limits their standard of health. But often the severity of symptoms is suppressed for the patient.

8   Colonic drug delivery systems approach (CDDSA) for IBD

CDDS approaches have received much interest over the last decade, since it was identified that the several benefits in the colon area as a local target of medicines [26], to increase the health benefits for various medications and pathological treatments. The CDDSA approach covers protein- and peptide-based drug delivery to treat IBD. The CDDS drug delivery systems must cover safety from different enzymes like proteolytic activity-based enzymes that are accessible to the colon and other parts of the large intestine of the stomach, where it finds a suitable inner environment for intense absorption and for targeting the colon against deformity or diseases that needs local targeted treatment for reducing the systemic adverse effect of several diseases like inflammatory bowel diseases, colorectal cancer, and irritable bowel syndrome. The most often used routes are dietary or parenteral, and through them, the medicines are absorbed into the body. When seeking for local effects, suppositories or rectal foams with the same active ingredients are utilized. However, there is no permanent cure for many diseases. Patients endure breakouts throughout the duration of their lives, which severely reduces their quality of life. Sometimes a patient's symptoms are so severe that they are incapacitating [25]. Additionally, a sizeable a certain percentage of individuals require hospitalization due to an outbreak, and they are treated with a harsh medication cocktail or possibly surgery. Inflammatory diseases can be treated with both should advance, and numerous research teams are working to provide new treatments.

9   Advance drug delivery approach (ADDA) for IBD

9.1   Nanotechnological delivery approach (NA)

Nanoparticles are meant to regulate medicine upon oral ingestion, the creation of such, in particular, using nanoparticles to decrease the dosage regimen. Neutrophils and natural killer cells are generally more prevalent [27]. Mast cells, regulatory T cells [28], and cells, all of which play a crucial role inflammatory bowel disease pathophysiology disease, additionally, reports suggest that microspheres and macrophages are capable of effectively absorbing nanoparticles. The colon delivery by nanocarriers increases the digestibility and lessens the systemic complications seen in various administrations like oral and intravenous.

Therefore, it may be assumed that the storage of the individual carrier system in the intended area will be facilitated by the uptake of different particles into cells of immune-related or by the breakdown of the colon barrier, especially intestine function. The control amount and dosage are possible since nanoparticles are assumed to have a high stability period than recent drug delivery approaches and methods.

As a result, the use of any medications needs to be carefully regulated as part of a customized treatment schedule for IBD. Immune modulators can have cytotoxic effects that may exacerbate other diseases or infections; however, they are (Fig. 6) frequently used in conjunction with antiinflammatory medications. Some of these

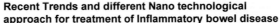

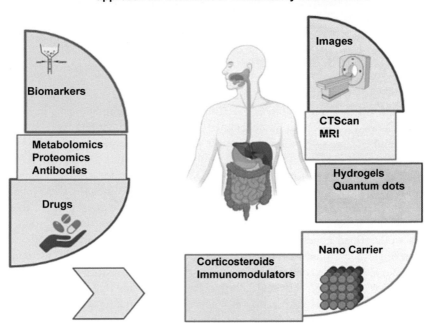

Fig. 6 Trends and nanotechnological approach for the treatment of IBD.

medications can directly scavenge free radicals, whereas others have indirect antioxidant effects through TNF.

9.2   Nutritional therapy in IBD

The association between fatty acids and intestinal inflammation is of significant interest. Permanent settle cells that are known as a resident in the damaged tissue create PIM (pro-inflammatory mediators) in soluble stage, like derived of eicosanoids from x6-arachidonic acid, during acute inflammation. This controls the early stages of the responses of the inflammatory disease and encourages the entrance of blood white blood cell (WBC) neutrophils, the immune system, and their phagocytic role from WBC. Leukocytes' selflimited response to infection and any inflammation is an active-controlled process that results in a temporal switch in the production of lipid mediators within the eicosanoids family, from pro- to antiinflammatory prostaglandins and leukotrienes as well as their free fatty acid (FFA) precursors [29]. To prevent the recurrence of the disease, different nutritional therapies used a comprehensive combination of vitamins and amino acids, involving the antioxidant [30] different vitamins D, E, A, and k for the treatment of inflammatory disease. Food sources rich in vitamin C [31] are intentionally designed [15] to add value to food products and meet the requirements. Several food industries [32] and nutraceutical [33] industries are aligned toward the nutritional need of micromineral for nutritional meet.

10 Conclusion and future perspectives

Immune cells such as leukocytes, monocytes, and neutrophils generate the ROS during the metabolism of prostaglandins, leukotrienes, and respiration during inflammation, which creates additional tissue damage. IBD patients, including those suffering from CD and UC, are often more susceptible to developing colorectal cancer (CRC). In fact, along with the inborn disorders of familial adenomatous polyposis (FAP) and nonFAP are hereditary colorectal cancer, IBD is included in the common three risk conditions for developing colorectal cancer. The risk of IBD-based CRC emerges to be a more strongly connected disease related to gastrointestinal mucosa than to other apparent genetic predisposition, in distinction to the latter two disorders that have a well-defined genetic etiology. Immense infiltration of leukocytes of the gut is a sign of chronic inflammation in IBD. These cells produce an unrestricted number of ROS and RNS species along with an extensive range of advanced stages of inflammation in cytokines when activated. It is important to mention that the expression and development of IBD are significantly influenced by the dramatic and persistent modification of redox equilibrium that is available in the gut mucosa and create an excess of oxidative reactions that condition is called OS. By triggering redox-sensitive pathways of different signals and transcription factors, oxidative stress keeps intestinal mucosal inflammation active.

It is essential to specify that various medications could create comparative helpful results in regard to the adjustment of provocative pathways. Numerous studies have tested new viewpoints on the revelation of atomic focuses for conclusive IBD treatment. It is observed that exploratory information does not necessarily compare to illness in clinical settings and that numerous enhancements and refinements of our sub-atomic way to deal with IBD still cannot seem to be finished. New examinations evaluating the mind-boggling safe marks because of various treatments are required. It is still controversial whether intestinal inflammation is a cause of or a product of damage to the transmembrane proteins that create junctions.

Endocytosis of junctional molecules rises in inflammatory conditions, such as the ones experienced by IBD patients, and intracellular subdivision may lead to a collapse of the protective barrier. The current chapter on targeting angiogenesis, inflammation, and oxidative stress in chronic diseases is outlined in the IBD section. The main factors influencing the pathophysiology and development of IBD comprise high production ROS and low antioxidant activity. The gastrointestinal way epithelium is constantly exposed to a variety of stimuli including food consumed, local microbiota or other illnesses, stomach acid, and other pro-oxidants. The fast development of biomedical devices of a new era with extensive international cooperation gives new areas for the identification and medicament of IBD.

Acknowledgments

Author acknowledges UPES R&D/SHODH/ 202111 for financial and laboratory support and School of Health Sciences and Technology, UPES Dehradun, Uttarakhand, India for providing facilities for the completion of this book chapter.

References

[1] Lamprecht A, Ubrich N, Yamamoto H, Schäfer U, Takeuchi H, Maincent P, et al. Biodegradable nanoparticles for targeted drug delivery in treatment of inflammatory bowel disease. J Pharmacol Exp Ther 2001;299(2):775–81.

[2] Guan Q, Zhang J. Recent advances: the imbalance of cytokines in the pathogenesis of inflammatory bowel disease. Mediators Inflamm 2017;2017.

[3] Belali N, Wathoni N, Muchtaridi M. Advances in orally targeted drug delivery to colon. J Adv Pharm Technol Res 2019;10(3):100–6.

[4] Ambalam P, Raman M, Purama RK, Doble M. Probiotics, prebiotics and colorectal cancer prevention. Best Pract Res Clin Gastroenterol 2016;30(1):119–31.

[5] Lemmens B, De Hertogh G, Sagaert X. Inflammatory bowel diseases [Internet]. In: Pathobiology of human disease: a dynamic encyclopedia of disease mechanisms. Elsevier Inc; 2014. p. 1297–304. https://doi.org/10.1016/B978-0-12-386456-7.03806-5.

[6] Winter MW, Weinstock JV. Inflammatory bowel disease. Autoimmune Dis 2019;871–94.

[7] Lunney PC, Leong RWL. Review article: ulcerative colitis, smoking and nicotine therapy. Aliment Pharmacol Ther 2012;36(11–12):997–1008.

[8] Donnelly TM, Fox GJ, Marini RP. Ferrets: inflammatory bowel disease. Clin Vet Advis Birds Exot Pets 2012;465–7.

[9] Tian T, Wang Z, Zhang J. Pathomechanisms of oxidative stress in inflammatory bowel disease and potential antioxidant therapies. Oxid Med Cell Longev 2017;2017.

[10] Roudsari NM, Lashgari NA, Momtaz S, Farzaei MH, Marques AM, Abdolghaffari AH. Natural polyphenols for the prevention of irritable bowel syndrome: molecular mechanisms and targets; a comprehensive review. DARU J Pharm Sci 2019;27(2):755–80.

[11] Akbar A, Walters JRF, Ghosh S. Review article: visceral hypersensitivity in irritable bowel syndrome: molecular mechanisms and therapeutic agents. Aliment Pharmacol Ther 2009;30(5):423–35.

[12] Mahurkar-Joshi S, Chang L. Epigenetic mechanisms in irritable bowel syndrome. Front Psych 2020;11(August):1–15.

[13] Mishima Y, Ishihara S. Molecular mechanisms of microbiota-mediated pathology in irritable bowel syndrome. Int J Mol Sci 2020;21(22):1–25.

[14] Dziąbowska-Grabias K, Sztanke M, Zając P, Celejewski M, Kurek K, Szkutnicki S, et al. Antioxidant therapy in inflammatory bowel diseases. Antioxidants 2021;10(3):1–18.

[15] Kumar S, Krishali V, Purohit P, Saini I, Kumar V, Singh S, et al. Physicochemical properties, nutritional and sensory quality of low-fat Ashwagandha and Giloy-fortified sponge cakes during storage. J Food Process Preserv 2022;46(2):1–14.

[16] Itzkowitz SH, Yio X. Inflammation and cancer—IV. Colorectal cancer in inflammatory bowel disease: the role of inflammation. Am J Physiol Gastrointest Liver Physiol 2004;287:150–1.

[17] Hatoum OA, Heidemann J, Binion DG. The intestinal microvasculature as a therapeutic target in inflammatory bowel disease. Ann N Y Acad Sci 2006;1072:78–97.

[18] Biasi F, Leonarduzzi G, Oteiza PI, Poli G. Inflammatory bowel disease: mechanisms, redox considerations, and therapeutic targets. Antioxid Redox Signal 2013;19 (14):1711–47.

[19] Piechota-Polanczyk A, Fichna J. Review article: the role of oxidative stress in pathogenesis and treatment of inflammatory bowel diseases. Naunyn Schmiedebergs Arch Pharmacol 2014;387(7):605–20.

[20] Singh S, Gupta R, Chawla S, Gauba P, Singh M, Tiwari RK, et al. Natural sources and encapsulating materials for probiotics delivery systems: recent applications and challenges in functional food development. Front Nutr 2022;9.

[21] Singh S, Sanwal P, Bhargava S, Behera A, Upadhyay S, Akhter M, et al. Smart advancements for targeting solid tumors via protein and peptide drug delivery (PPD). Curr Drug Deliv 2023.

[22] Liu TC, Stappenbeck TS. Genetics and pathogenesis of inflammatory bowel disease. Annu Rev Pathol Mech Dis 2016;11(314):127–48.

[23] Kotla NG, Singh R, Baby BV, Rasala S, Rasool J, Hynes SO, et al. Inflammation-specific targeted carriers for local drug delivery to inflammatory bowel disease. Biomaterials [Internet] 2022;281(March 2021), 121364. https://doi.org/10.1016/j.biomaterials. 2022.121364.

[24] Hua S, Marks E, Schneider JJ, Keely S. Advances in oral nano-delivery systems for colon targeted drug delivery in inflammatory bowel disease: selective targeting to diseased versus healthy tissue. Nanomed Nanotechnol Biol Med [Internet] 2015;11(5): 1117–32. https://doi.org/10.1016/j.nano.2015.02.018.

[25] Teruel AH, Gonzalez-Alvarez I, Bermejo M, Merino V, Marcos MD, Sancenon F, et al. New insights of oral colonic drug delivery systems for inflammatory bowel disease therapy. Int J Mol Sci 2020;21(18):1–30.

[26] Singh D, Srivastava S, Pradhan M, Kanwar JR, Singh MR. Inflammatory bowel disease: pathogenesis, causative factors, issues, drug treatment strategies, and delivery approaches. Crit Rev Ther Drug Carrier Syst 2015;32(3):181–214.

[27] Zhang M, Merlin D. Nanoparticle-based oral drug delivery systems targeting the colon for treatment of ulcerative colitis. Inflamm Bowel Dis 2018;24(7):1401–15.

[28] Gaurav N. Current scenario and future perspectives of nanotechnology in sustainable agriculture and food production. Plant Cell Biotechnol Mol Biol 2021;22(11–12):99–121.

[29] Kechagia M, Basoulis D, Konstantopoulou S, Dimitriadi D, Gyftopoulou K, Skarmoutsou N, et al. Health benefit of probiotic: a review. Hindawi Publ Corp 2013;2013:1–7.

[30] Kumar S, Jain I, Khare A, Kumar V, Singh S, Gautam P, et al. Numerical optimization of microwave treatment and impact of storage period on bioactive components of cumin, black pepper and mustard oil incorporated coriander leave paste. J Food Meas Charact 2022;16(3):2071–85.

[31] Kohli D, Kumar A, Kumar S, Upadhyay S. Waste utilization of Amla pomace and germinated finger millets for value addition of biscuits. Curr Res Nutr Food Sci J 2019;7 (1):272–9.

[32] Kaur N, Bains A, Kaushik R, Dhull SB, Melinda F, Chawla P. A review on antifungal efficiency of plant extracts entrenched polysaccharide-based nanohydrogels. Nutrients 2021;13(6):2055.

[33] Mittal M, Thakur A, Kaushik R, Chawla P. Physicochemical properties of Ocimum sanctum enriched herbal fruit yoghurt. J Food Process Preserv 2020;44(12), e14976.

Natural and synthetic agents targeting angiogenesis, oxidative stress, and inflammation in psoriasis

6

Deepika Sharma[a], Sudeep Pukale[b], and Shraddha Manish Gupta[a]
[a]Department of Pharmaceutical Sciences, School of Health Sciences and Technology, UPES, Dehradun, Uttarakhand, India, [b]Lupin Research Park, Nande, Maharashtra, India

1 Introduction

Psoriasis is a chronic, noncontagious skin disorder that results in severe hyperkeratosis and hyperplasia, erythema, scaling, and thickness, as well as persistent pain, swelling, and bleeding. Symptom relief, not a complete cure, is the main objective of therapeutic treatments. People of all ages (both men and women) are negatively impacted, and it has a serious negative impact on patient's lives (most commonly at the age of 50–69). Between 0.09% and 11.43% of the world's population suffer from psoriasis, making it a significant condition that affects at least 100 million people. It mostly impacts the skin and nails and is linked to several cooccurring ailments, including depression, inflammatory bowel disease, metabolic syndrome, arthritis, and cardiovascular diseases. Due to erroneous or delayed diagnosis and a lack of available therapies, many psoriasis patients endure needless suffering 1.3%–34.7% of people with psoriasis advance psoriatic arthritis, a disabling, chronic inflammatory disease that affects the joints, while 4.2%–69% of people with the condition experience nail changes. It often affects older people and is more common in countries with robust economies. Psoriasis significantly reduces the quality of life in emotional, physical, and social ways that harm productivity, and mental health, and create social alienation and discontent [1,2].

With significant levels of hyperkeratosis and hyperplasia in the skin lesions, as well as chronic itching, scaling, thickness, oedema, pain, inflammation, and bleeding, psoriasis is now recognized as an autoimmune skin disease. When T cells are activated, keratinocytes hyper proliferate, and the rate of turnover of epidermal keratinocytes is dramatically reduced. Even if a genetic predisposition is the root of the problem, the cause itself is still not known [3–7]. Various factors, including stress, sunburn, mild trauma, systemic drugs, and infections, as well as internal and external factors, might lead to the development of this condition, according to research. Recent research has revealed that the primary cause is the predominance of prooxidants like reactive

Targeting Angiogenesis, Inflammation and Oxidative Stress in Chronic Diseases. https://doi.org/10.1016/B978-0-443-13587-3.00005-9

oxygen species (ROS) over the skin's natural antioxidant defense mechanism, which results in severe dermal inflammation [8,9].

Oral, systemic, topical, and photodynamic therapies are currently assessable treatment techniques. The initial treatment for psoriasis in the clinic is topical corticosteroids. Second-line treatments, such as calcipotriene, anthralins, and corticosteroids, are given to patients when initial treatments fail (intralesional injection). These medicines are often used with keratolytics, moisturizers, and climatotherapy to improve their therapeutic effects. Other treatments include sulfasalazine, leflunomide, vitamin D analogs, sulfasalazine, cyclosporine, retinoids, fumaric acid esters, methotrexate, mycophenolate mofetil, hydroxyurea, and cyclosporine [10]. Topical antioxidant treatment was also found to be effective in reducing psoriatic inflammation, which may have resulted from the elimination of free radicals and other reactive oxygen species. In addition to other antioxidant molecules, berberine has been associated with improvement in psoriasis [11,12] epigallocatechin-3-gallate [13], resveratrol [14], mangeferin [8], glabridin [15], propylthiouracil [16], quercetin [9], proanthocyanidins, etc. All the aforementioned drugs work by inhibiting cell proliferation in one way or another, decreasing the inflammatory immune response and relieving symptoms.

2 Clinical types of psoriasis

It is classified into several subtypes based on etiology, lesion characteristics, and treatment outcomes. Ninety percent of all psoriasis cases are due to plaque psoriasis, the most prevalent form. Plaques emerge unexpectedly on the head, the extremities, and the trunk, starting as small papules that quickly spread and become itchy, silvery-white, scaly, and red [17–19]. When Group A Streptococcus infects the tonsils and spreads rapidly, it causes guttate psoriasis, which is triggered by an overactive immune response and responds well to UV therapy. Plaque psoriasis affects nearly a third of persons with this disorder in adulthood. Eventually, the pink papule will have a scaly texture [20,21]. Erythrodermic psoriasis is a more severe and altered type of this ailment that can arise on its own or emerge as a result of the exacerbation of any earlier existing variety of the condition. It causes a significant drop in protein stores, making it difficult to regulate temperature and fluid balance, and ultimately requiring medical intervention in the form of hospitalization. The treatment may exacerbate preexisting conditions such as growth retardation, staphylococcus infections, arthropathy, and pustulosis [17]. Those affected by pustular psoriasis have skin that is covered in a large number of pustules that have merged together. This condition is subdivided into the Acrodermatitis continuum of Hallopeau (ACS) and psoriasis pustulosa palmoplantaris (PPP) based on phenotypic observations. Hands and feet, as examples of extremities, are in the middle of the two types. In contrast to PPP, which is limited to the palms and soles, ACS often manifests at more distant locations, such as the tips of the toes and fingers, where it produces structural alterations in the nails. Localized pustular psoriasis is challenging to treat and has little systemic effects, although the lesions might reoccur. Fever and electrolyte imbalance are common symptoms of relapsing psoriasis of any type, including the widespread pustular type [22].

3   Role of oxidative stress and inflammation in psoriasis

Oxidative stress plays an important role in psoriasis [23], and recent research reveals that there is a redox imbalance that is present in the blood and skin of psoriasis patients [24], which is a key factor in the etiology of the condition. Malondialdehyde (MDA), erythrocyte MDA, lipid hydroperoxides, protein carbonyl, and nitric oxide end products are examples of oxidative stress markers that were considerably elevated in psoriatic individuals [25–32]. The etiology of psoriasis is mostly mediated by dendritic cells (DCs) [33]. The inflammatory DCs express inflammatory molecules and produce mediators that play a role in the activation of T cells and keratinocytes (KCs) [34] in a downstream manner [35].

According to studies, oxidative stress increases the production of TNF-α and IL-8 by DCs [36] and promotes them to express several DC surface markers that interact with T cells, including costimulatory molecules and MHC Class II molecules [37,38]. DCs that have been exposed to ROS enhance T-cell proliferation more potently than DCs that have not been exposed [37]. Oxidative stress can increase the ability of DCs to induce the production of proinflammatory cytokines. Importantly, TNF-α is regarded as a significant mediator of inflammation in psoriasis and is one of the cytokines whose production is increased by oxidative stress [39–41].

Several cytokines contribute to psoriasis-induced KC proliferation. The signals are ultimately sent directly or indirectly to transcription factors such as STAT or NF-κB. ROS are primarily responsible for activating transcription factors in KCs [42]. Regarding the etiology of psoriasis, MAPK, JAK-STAT, and NF-κB signaling pathways are key inflammatory pathways impacted by reactive oxygen species [43].

4   Role of angiogenesis in psoriasis

Angiogenesis in psoriasis is triggered by a comparatively high concentration of vascular endothelial growth factor (VEGF), specifically the VEGF121 isoform [44,45]. It's interesting to note that this isoform is more prevalent in psoriatic skin than in skin from healthy donors [46,47]. Moreover, VEGF121 significantly dilates the blood vessels and increases their permeability, but it does not cause the formation of new blood vessels [48]. Data from few studies support the fact that VEGF121 is the major isoform and may be the key regulator of angiogenesis in psoriatic scales observed in psoriatic lesions [49,50]. Many compounds found in high concentrations within scales, including epidermal growth factor (EGF), transforming growth factor-1 (TGF-1), and TNF-α, in addition to genetic predispositions, increase the production of VEGF. It is crucial to comprehend the mechanisms underlying VEGF secretion in psoriasis patients' scales [51–55]. Role of oxidative stress, inflammation, and angiogenesis in psoriasis has been depicted in Fig. 1.

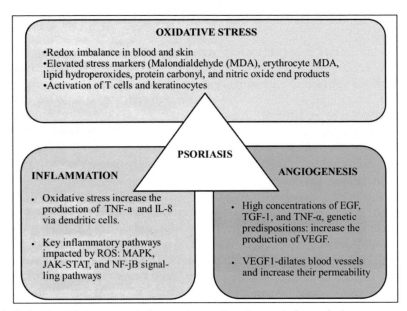

Fig. 1 Role of oxidative stress, inflammation, and angiogenesis in psoriasis.

5 Currently accessible treatments

Systemic, phototherapy, and topical treatments are currently offered to treat psoriasis. The phototherapy method (PUVA and UVB irradiation) is the most expensive and is not recommended for long-term use [56]. A crucial component of the psoriasis treatment plan has been the combination of systemic and topical treatments. In 20% of cases, systemic therapies are still required even when topical therapeutic strategies are helpful in reducing the symptoms of psoriasis. The issue with using systemic therapy (calcineurin inhibitors and immunosuppressive medicines) is the severity of their side effects, which include cancer from cyclosporine and PUVA, teratogenicity from retinoids, and nephrotoxicity and hepatotoxicity from methotrexate and cyclosporine. These formulations also reach the targeted intended site (deeper dermal layers) and widespread diffusion to undesirable organs, resulting in toxic effects. Moreover, the total amount of medication reaching the site of action is extremely modest due to significant plasma protein binding. These worries make it difficult to maintain using this therapeutic technique over the long term.

Drugs used in topical therapies are only absorbed by the skin, the site of the ailment, rather than being absorbed by other organs, making these treatments safer [57]. Many OTC topical treatments for psoriasis contain corticosteroids, retinoids, salicylic acid, coal tar, vitamin D analogs, and anthralin; however, these ingredients are linked to undesirable reactions such as thinning of the skin, dilation of blood vessels, and irritation of the skin making them inappropriate for long-term use [58]. The bulk of recognized pharmacological molecules used in modern medicine are hydrophobic, which complicates drug dispersion. Its mainstream preparations (gels, creams, ointments, etc.) are less effective because of rapid localized drug release and systemic drug leaching, leading to side effects like skin atrophy, infections, stretch marks, and redness, as well as systemic toxicity like downregulation of the hypothalamic-pituitary-adrenal (HPA) axis [59]. To prevent these toxins and boost therapeutic success, we need new carriers that are both cost-effective and capable of penetrating deeper into healthy epidermis without systemic absorption and releasing the medicament locally at even a sustained speed over an extended period of time.

6 Strategic drug delivery approaches: Topical drug delivery based on nano-carrier

Innovative formulation advances increase patient quality of life by boosting in vivo efficacy and minimizing conventional pharmaceutical adverse effects. The similar outcome could be accomplished by restricting medication administration to the desired site (only deeper dermal regions) and subjecting other healthy organs to relatively low drug concentration. The focus has now shifted to nanotherapeutics to tackle these problems. According to several reports, nanocarriers are preferred for their potential delivery of therapeutic agents to specific areas of the skin surface because of their small size, elevated skin deposition, improved surface characteristics, enhanced drug release characteristics, and enhanced penetration via biological barriers, which have proven beneficial in the treatment of psoriasis [10,32,34,60]. Solid lipid nanoparticles (SLNs), nanostructure lipid carriers (NLCs), vesicular systems, micro- or nanoemulsions, and other lipidic and polymeric nanocarriers are examples of these nanosystems (nanoparticles, micelles, nano-conjugate, dendrimers, etc.). In comparison to conventional delivery systems, these nanosystems have many advantages, such as a prolonged drug release profile, protection of the active principle from the harmful bioenvironment, side effect reduction from lower doses, drug targeting at the active site, and modification of dermatokinetic parameters. Although substantial literature exists on numerous benefits describing both polymeric and lipidic nanoparticles, but these systems are also associated with few drawbacks [61].

6.1 Nanoliposystems

6.1.1 Vesicular systems

Since liposomes are bilayered vesicular systems that are made up of cholesterol, phospholipids, and long-chain fatty acids, they are essential members of this class. Large unilamellar vesicles are 0.10 mm in diameter, while multilamellar vesicles are 0.05 mm, while small unilamellar vesicles are 0.01 mm in diameter (0.025–0.05 mm). These systems scan deliver hydrophilic as well as lipophilic drugs via the cutaneous route as their lipid content is comparable to that of skin, making them simpler to permeate through the epidermal barrier than other standard formulations [61,62]. For instance, when administered as multilamellar liposomes made from cholesterol and dipalmitoyl phosphatidylcholine, for delivery of triamcinolone acetonide. The results revealed that liposomal formulations provided a comparatively high dermal concentration (4.8-fold) of loaded drug with decreased systemic leaching when compared with the conventional formulation (2.1-fold). Another study found that betamethasone dipropionate-containing liposomes significantly reduced the scaling and erythema that accompany atopic eczema when compared to commercial formulations. Knudsen et al. found that topical delivery of calcipotriol loaded liposomes was significantly increased when delivered in a liquid state as opposed to a gel state, with obtained values of 62.58.2% and 49.89.9%, respectively. According to their findings, the extent to which a topical delivery is achieved depends vastly on the mobility of the vesicular vehicle [63]. In another study, when the liposomal formulation of calcipotriol was prepared using poly (ethylene glycol)-distearoylphosphoethanolamine lipopolymer;

The poly(ethylene glycol)-distearoylphosphoethanolamine lipopolymer improved colloidal stability without affecting drug delivery or skin penetration when added to the liposomal formulation of calcipotriol. Additionally, it was discovered that the size of this vesicular system significantly influenced as small unilamellar vesicles perform better at skin penetration than bigger multilamellar vesicles. Oleic acid (OA) and phosphatidylcholine (PC), produced at various ratios 1:2 and 1:9, respectively, were used to make deformable liposomes carrying MTX, and their performance was compared to two standard liposomal systems made from cholesterol (CH) and PC prepared at varied ratios. Sizes between 80 and 140 nm and a narrow PDI were present in all formulations, which is ideal for effective topical distribution. When compared to deformable liposomes (PC2.5:OA1), the drug loading, entrapment efficiency, and colloidal stability of traditional liposomes (PC9:CH1 and PC2:CH1) were shown to be much higher. This occurred because OA has elastic and deformable features. A good penetration-enhancing activity is also claimed for OA [55]. In a different investigation, Gizaway et al., made betamethasone di-propionate-loaded transferosomes via thin-film hydration. These transferosomes showed excellent chemical and physical stability (drug release and colloidal stability) [64].

Ethosomes, another version of the liposomal system with diameters between 30 nm and a few m, have higher alcohol concentrations. Similar to liposomes, they are able to carry both hydrophilic and lipophilic drug molecules and improve the distribution of drugs to deeper skin layers by enhancing penetration and creating new channels through the skin. The formulation that contained 40% (v/v) ethanol and 5% (w/v) lipid S 100 was the most effective, and it may be able to increase the efficacy of PUVA therapy while lowering toxicity. Both ethosomes and psoralen-loaded conventional liposomes were created by Zhang et al. using Lipoid S 100 and various concentrations of ethanol [65].

6.1.2 Nanoemulsions

The oil phase is dispersed in the water phase (o/w type) in these dispersed systems, and the globules are stabilized by surfactants and reside in the nano-range (20–500 nm). In a study, a nanoemulsion was prepared using tween 20, corticosteroid clobetasol propionate, ethanol, eucalyptus oil, and distilled water. The nanoemulsion significantly decreased edema up to 84.15% at 12 h, in contrast to the Glevate cream (a commercial product), which only managed to decrease edema up to 40.99% at the same period, upon in vivo evaluation study on Wistar rats [66].

Exopolysaccharides from the bacterium *Bacillus amyloliquefaciens* were utilized as the emulsifier in a different study to make a nanoemulsion of the medication calcipotriol, which significantly decreased psoriatic inflammation thanks to improved drug deposition at the target site and less adverse effects [67]. In another study, exopolysaccharides from *B. amyloliquefaciens* were employed as the emulsifier to produce a nanoemulsion of calcipotriol. This reduced psoriatic inflammation effectively since there was more medication deposition at the target region and fewer adverse effects. This shows that exopolysaccharides have been employed in medicine for dermatological purposes [68]. To create synergistic antipsoriatic efficacy, medicines that have functional qualities have also been combined with excipients. For instance, a tacrolimus nanoemulsion produced with Kalonji oil (a multifunctional excipient) demonstrated a synergistic antipsoriatic effect. The resulting nanoemulsion demonstrated a protracted and biphasic profile of drug release with improved local accumulation (4.33-folds), less systemic diffusion, and much higher efficacy in comparison to the marketed formulation by Tacroz Pharmaceuticals. This approach has made it possible to explore a new field of functional excipients for dermatological care utilizing a nanotechnology-based methodology [69]. Furthermore, TPGS-based microemulsion (ME) with tacrolimus showed significantly better topical drug distribution and better antipsoriatic activity in comparison to Protopic. The vehicle's critical significance as a potential adjuvant that, when used in conjunction with the medicine, seems to have a synergistic effect in psoriasis treatment [70].

6.1.3 Delivery systems based on lipids

Nanocarriers systems based on lipids have emerged as a viable alternative to the traditionally used emulsions and vesicular systems, allowing for more precise regulation of the drug release profile and delivery phase. Solid lipid nanoparticles (SLNs) are a type of dispersion in which the oil globules in a nanoemulsion are completely replaced by solid lipids; SLNs, like nanoemulsions, are stabilized by surfactants with a size (10–1000 nm) and a spherical geometry. The hydrophobic core lipid can be used to dissolve additional hydrophobic drug molecules. Enhanced skin permeability of these nanosystems can be attributed to several factors, including extended drug residence time, higher skin moisture, and the interaction of the lipid part of SLNs with the skin structure [71–74]. Nanostructured lipid carriers (NLCs) are nanocarriers in which liquid lipid (oil) partially replaces the distributed solid lipid portion of SLNs [75,76]. The inclusion of oil in NLCs lowers the crystallinity of the core matrix and imparts fluidity, allowing them to better load and preserve the drug compared to SLNs [77–80].

Likewise, NLCs are also known to improve dermal distribution and systemic leaching of fluocinolone acetonide and triamcinolone acetonide [81,82]. Another study showed improved skin permeability of all-trans retinoic acid upon delivery via NLCs and SLNs [83]. Similarly, Ridolfi et al. loaded tretinoin (TRE) in SLNs to improve stability and reduce local side effects. Further addition of chitosan into these SLNs improved antibacterial activity and was found to be noncytotoxic to HaCaT cells [84].

Similarly, thermosensitive SLNs loaded with tacrolimus (Tac) have also shown better results in a reduction of skin irritation in erythema due to enhanced distribution of Tac into dermal layers of skin in comparison to the reference product [85]. Tac-NLC and Tac-SLN, have also shown the potential to go into deeper skin layers and hence can be utilized in the treatment of psoriasis as deeper skin layers are the target location for drug delivery in this skin disease [86].

Advantages of lipidic nanosystems include reduced production costs, improved encapsulation efficiencies, fewer steps, and the absence of toxic byproducts. These systems include SLNs, NLCs, vesicular systems, and micro- or nanoemulsions. Burst release, lack of chemical modification options, drug partitioning, high polydispersity index, and drug ejection are all disadvantages associated with these lipid-based carriers [87,88]. Various lipid-based nanosystems that have been explored to deliver therapeutic moiety to psoriatic skin are displayed in Table 1.

Table 1 Various lipid-based nanosystems have been explored to deliver therapeutic moiety to psoriatic skin.

Delivery system/ Carrier	Drug/ Therapeutic agent/extract	Effects/Benefits	References
Transferosomes	*Berberis aristata* extracts	Improved antiinflammatory activity	[89]
Cerosomes	Tazarotene	Increased entrapment of tazarotene, enhanced deposition in the skin	[90]
Liposomes	Omiganan	Controlled release and a better permeation profile than conventional formulations	[91]
Liposomes	Cyclosporine	Reduced levels of key psoriatic cytokines (tumor necrosis factor-α, IL-17, and IL-22)	[92]
Squarticles	Clobetasol propionate	Enhanced permeation, Increased skin retention, and enhanced effect of a drug	[93]
Deformable liposome	Methotrexate	enhanced permeability of Methotrexate	[94]
Niosomes	Celastrol	Celastrol was mainly accumulated in the skin instead of exposure to blood or the lymphatic system	[95]
Lipospheres	Thymoquinone	Deeper skin penetration, slow-release, and skin compatibility	[96]

Table 1 Continued

Delivery system/ Carrier	Drug/ Therapeutic agent/extract	Effects/Benefits	References
Liposomes	Fusidic acid	Percent skin permeation (>75%), enhanced cellular uptake	[97]
Ethosomes and liposomes	Psoralen	Higher psoralen transdermal flux and skin deposition in via ethosomes	[98]
Proposomes	Tofacitinib citrate	Enhanced skin permeation and deposition of TC	[99]

6.1.4 Polymeric nanosystems

Polymeric nanosystems are yet another category of drug delivery system that is prepared using different polymers. Among various biocompatible and biodegradable polymers that are used in the formulation of polymeric nanosystems; poly(-caprolactone) (PCL) and poly(lactic-*co*-glycolic acid) (PLGA) have received extensive attention due to their numerous useful properties and are GRAS. Cyclosporine A (CsA) loaded in PEG-b-poly(d,L-lactide-*co*-glycolide) nanoparticles showed better efficacy and safety profile upon comparison with free drug (IC50 value 30 ng/mL) [100]. In addition, a study has shown that CsA loaded polymeric nanoparticles made up of PLGA can also prevent systemic leaching, as the deposition of CsA in the dermal layer and stratum corneum is much higher when administered as PLGA nanoparticles, in comparison to free drug delivery [101]. Nanoparticles made of PLGA and nanoparticles made of PLGA-PEG-CsA were also examined with regard to skin penetration and drug release of CsA. Both nanosystems were composed of PLGA (triblock copolymer), and their sizes and PDIs were roughly the same (∼30 nm and ∼0.2, respectively). Upon comparison of CsA loaded PLGA nanoparticles with CsA loaded PLGA (triblock copolymer), the study demonstrated that nanoparticles developed using triblock copolymer showed a rapid drug release while nanoparticles developed using PLGA demonstrated a sustained release of CsA [102]. These nanosystems can be used as a promising technique to solve the issues of rapid degradation and toxicity associated with psoralens [103]. The benzo psoralen encapsulated PLGA nanoparticles were designed to specifically target neutrophils, eosinophils, and macrophages, the "workhorses" of the target cell population, and to trigger apoptosis aggressively, leading to improved therapeutic efficacy in conjugation with UV (ultraviolet) treatment [104].

Likewise, PCL polymer has also been utilized to improve efficacy and reduce toxicity shown by drugs [105]. For example, hydrocortisone acetate's toxicity and systemic leaching were reduced in another study when the drug was administered in the form of PCL nanoparticles with the following characteristics: particle size

(190–230 nm), zeta potential value (3–5 mV), and encapsulation efficiency (≈62%) [106].

Since the overall natural charge of skin tissues is negative, a positive charge on the nanoparticles has been utilized for the interaction of skin with drug delivery systems for its beneficial effects in topical drug administration. The degradation of retinol was eliminated by formulating it into cationic nanoparticles made of Eudragit RS 100, which also improved the drug's stability for up to 2 months and significantly increased drug residence duration [107].

6.1.5 Micelles

Micelles are selfassembled nanostructures (10–100 nm), that display higher cutaneous deposition (deeper dermal layers) and have superior colloidal stability. Micelles have been widely used in topical drug delivery systems. An example of a drug that has been delivered via micelles is Tacrolimus. It was successfully delivered through polymeric micelles, which addressed the drug's poor skin permeability. This topical drug delivery system enhanced the antipsoriatic efficacy [108]. In an additional study, Jin et al. combined zinc phthalocyanine with a Brij 58 PEG chain to create core-shell type selfassembled micelles. Average zeta potential (15 mV) value of these micelles indicated that these micelles were stable. When combined with UV treatment and supported by extensive histological evidence, these micelles reduced the inflammation in the psoriasis-induced animal model (guinea pig) [109]. They can be employed as a promising method for treating psoriasis.

6.1.6 Dendrimers

Dendrimers are three-dimensional hyperbranched macromolecular nano-sized structures consisting of a central core surrounded by concentric shells. Therapeutic drug molecules can be placed into hydrophobic empty areas or covalently bonded to peripheral functions. They provide enhanced penetration of hydrophobic drugs across the skin [110,111]. In a study, penetration of anthralin across skin was enhanced via a fifth-generation polypropylene imine (PPI) dendrimer dendrimer-based system [112]. In another study, G3.5 and G2.5 poly(amidoamine) dendrimers were used to improve 8-Methoxypsoralene skin penetration [113]. These systems have poor encapsulation and burst release efficiency than others. Hyperbranched structures were altered to improve entrapment efficiency and burst release of two model drugs methotrexate (MTX) and polyester-co-polyether (PEPE). Despite advancements, encapsulation was inefficient [114].

6.1.7 Lipid-polymer hybrid nanoparticles

Lipid-polymer hybrid nanosystems are developed to get the combined benefits of polymeric nanocarriers and lipids. These nanosystems have controlled drug release properties, good drug-loading capacities, better cellular uptake, and biocompatibility. Based on their structure, these nanosystems are of various types. Monolithic systems have uniformly dispersed lipids that form a drug-loading core [115]. Core-shell hybrid systems have a polymeric core surrounded by lipid layers. These systems are also modified to have a lipid-based hollow core with a polymeric shell and a lipid shell [116,117]. Covering the RBC membrane onto the polymeric core results in biomimetic lipid-polymer nanoparticles [118]. Liposomes are caged with polymers. Their pharmaceutical applications include vaccine adjuvants, gene delivery, siRNA delivery, and delivery of diagnostic agents. Based on these applications, lipid-polymer hybrid nanoparticles are now being utilized widely for delivering therapeutic agents [119]. For example, amoxicillin-loaded monolithic lipid-polymer hybrid nanoparticles, with an average diameter of 200 nm have displayed sustained and complete drug release and were successful elimination *H. pylori* [115]. Similarly, Zhao et al. codelivered gemcitabine and Hypoxia-inducible factor 1 (HIF1) siRNA via a hybrid drug delivery system for pancreatic cancer treatment. This hybrid system was formulated using the double emulsion method with mPEG-PLGA, PEGylated lipid bilayers, and poly-lysine. Cationic charge on poly-lysine complexed efficiently with anionic charge on the surface (si-HIF1). Gemcitabine was encapsulated in the hydrophilic inner core, while the outer lipid bilayer was PEGylated to prevent aggregation of nanoparticles [120]. These hybrid systems showed good tumor metastasis suppression in an orthotopic tumor model. Lipid-polymer hybrid nanoparticles that mimic biological processes have attracted a lot of interest for use in creating nanosystems with much longer in vivo residence times. Various polymeric nanosystems have been explored as topical drug delivery systems in psoriasis management. List of various polymeric delivery systems explored topically for psoriasis management is given in Table 2.

Table 2 List of various polymeric delivery systems explored topically for psoriasis management.

Delivery system/ Carrier	Drug/ Therapeutic agent/extract	Benefits/Effects	References
Polymeric nanoparticle incorporated in the gel	Curcumin	Gel with curcumin-loaded nanoparticles had a greater therapeutic impact than curcumin alone	[121]
poly(ε-caprolactone)-poly(ethylene glycol)-poly(ε-caprolactone) (PCEC) copolymers based micelles	Tacrolimus (FK506)	Much slower and sustained release of FK506-loaded micelles	[122]
poly(D,L-lactic/glycolic acid) (PLGA) or poly(D, L-lactic acid) (PLA) nanoparticles	Betamethasone phosphate	Encapsulated betamethasone phosphate has a higher efficiency	[123]

Continued

Table 2 Continued

Delivery system/ Carrier	Drug/ Therapeutic agent/extract	Benefits/Effects	References
Micelle	Tacrolimus	Targeted tacrolimus administration to the hair follicle, selective cutaneous absorption	[108]
Nanogel	Acitretin and Aloe-emodin	Higher acitretin and aloe emodin deposition in the epidermal and dermal layers	[124]

6.1.8 Polymeric hybrid nanoparticles based on lipids

Polymeric hybrid nanoparticles based on lipids have been developed using nanoprecipitation and emulsion solvent evaporation techniques, which present challenges during formulation processes. Additionally, probe sonicator and high-shear homogenizers, which are generally employed to prepare these hybrid nanosystems are limited to lab-scale production only. Among different nano formulation methods, high-pressure homogenization is cost-effective method due to its scalable [125]. High-pressure homogenizers are more efficient than high-shear homogenizers and ultrasonicator, as in earlier one principle behind size reduction is impact instead of shear stress and cavitation in the case of ultrasonicators [126].

7   Conclusion and future perspectives

Topical therapy is a more appropriate drug delivery method than systemic approaches and phototherapy as it delivers therapeutic substances directly to the site of action, preventing toxicity in unintended body organs. Poor penetration across biological skin layers, especially the stratum corneum, is a difficulty for topical drug delivery. Nanoparticle-based techniques improve skin permeability, minimize dose, improve efficacy, reduce side effects, and limit the systemic leaching of topically administered therapeutic molecules. The literature divides nanocarriers into lipidic and polymeric kinds. Both types have advantages over conventional systems but also suffer from a few limitations. Lipidic nanosystems have various limitations like instability, burst release, a high polydispersity index, drug expulsion, drug partitioning, and limited chemical modifications, etc. While however, polymeric nanocarriers have poor drug entrapment efficiency, the preparation method is a multistep process, and the cost of manufacturing and scalability is high. A carrier is needed that combines the benefits of both systems while removing their drawbacks. This hybrid system combines the benefits of polymeric nanocarriers and lipids, with better-controlled drug release, good drug-loading capacity, enhanced cellular uptake, and is biocompatible in nature. A monolithic system is simple and scalable for lipid-polymer hybrid nanoparticles, yet it can achieve hybrid benefits. Most studies employed emulsion solvent evaporation and nanoprecipitation methods to prepare these hybrid systems, via probe sonicators and high-shear homogenizers. Among different methods, based on the number of benefits being offered by high-pressure homogenization, this method is lucrative for preparing nanoformulations because this method is easily scalable.

References

[1] Danielsen K. Increased risk of death in patients with psoriasis: disease or lifestyle? Br J Dermatol 2019;180(1):3–4. https://doi.org/10.1111/bjd.17141.
[2] Parisi R, Iskandar IY, Kontopantelis E, Augustin M, Griffiths CE, Ashcroft DM. National, regional, and worldwide epidemiology of psoriasis: systematic analysis and modelling study. BMJ 2020;369. https://doi.org/10.1136/bmj.m1590.
[3] Garg T, Rath G, Goyal AK. Nanotechnological approaches for the effective management of psoriasis. Artif Cells Nanomed Biotechnol 2016;44(6):1374–82. https://doi.org/10.3109/21691401.2015.1037885.
[4] Rahman M, Alam K, Zaki Ahmad M, Gupta G, Afzal M, Akhter S, Kazmi I, Jalees Ahmad F, Anwar F. Classical to current approach for treatment of psoriasis: a review. Endocr Metab Immune Disord Drug Targets 2012;12(3):287–302. https://doi.org/10.2174/187153012802002901.
[5] Rahman M, Zaki Ahmad M, Kazmi I, Akhter S, Beg S, Gupta G, Afzal M, Saleem S, Ahmad I, Shaharyar A, Jalees AF. Insight into the biomarkers as the novel anti-psoriatic drug discovery tool: a contemporary viewpoint. Curr Drug Discov Technol 2012;9(1):48–62. https://doi.org/10.2174/157016312799304516.
[6] Peters BP, Weissman FG, Gill MA. Pathophysiology and treatment of psoriasis. Am J Health Syst Pharm 2000;57(7):645–59. https://doi.org/10.1093/ajhp/57.7.645.
[7] Lowes MA, Bowcock AM, Krueger JG. Pathogenesis and therapy of psoriasis. Nature 2007;445(7130):866–73. https://doi.org/10.1038/nature05663.
[8] Li P, Li Y, Jiang H, Xu Y, Liu X, Che B, Tang J, Liu G, Tang Y, Zhou W, Zhang L. Glabridin, an isoflavan from licorice root, ameliorates imiquimod-induced psoriasis-like inflammation of BALB/c mice. Int Immunopharmacol 2018;59:243–51. https://doi.org/10.1016/j.intimp.2018.04.018.
[9] Lai R, Xian D, Xiong X, Yang L, Song J, Zhong J. Proanthocyanidins: novel treatment for psoriasis that reduces oxidative stress and modulates Th17 and Treg cells. Redox Rep 2018;23(1):130–5. https://doi.org/10.1080/13510002.2018.1462027.
[10] Bhat M, Pukale S, Singh S, Mittal A, Chitkara D. Nano-enabled topical delivery of antipsoriatic small molecules. J Drug Deliv Sci Technol 2021;62, 102328. https://doi.org/10.1016/j.jddst.2021.102328.
[11] Sun S, Zhang X, Xu M, Zhang F, Tian F, Cui J, Xia Y, Liang C, Zhou S, Wei H, Zhao H. Berberine downregulates CDC6 and inhibits proliferation via targeting JAK-STAT3 signaling in keratinocytes. Cell Death Dis 2019;10(4):274. https://doi.org/10.1038/s41419-019-1510-8.

[12] Zhang S, Liu X, Mei L, Wang H, Fang F. Epigallocatechin-3-gallate (EGCG) inhibits imiquimod-induced psoriasis-like inflammation of BALB/c mice. BMC Complement Altern Med 2016;16(1):1. https://doi.org/10.1186/s12906-016-1325-4.

[13] Khurana B, Arora D, Narang RK. QbD based exploration of resveratrol loaded polymeric micelles based carbomer gel for topical treatment of plaque psoriasis: in vitro, ex vivo and in vivo studies. J Drug Deliv Sci Technol 2020;59, 101901. https://doi.org/10.1016/j.jddst.2020.101901.

[14] Pleguezuelos-Villa M, Diez-Sales O, Manca ML, Manconi M, Sauri AR, Escribano-Ferrer E, Nácher A. Mangiferin glycethosomes as a new potential adjuvant for the treatment of psoriasis. Int J Pharm 2020;573, 118844. https://doi.org/10.1016/j.ijpharm.2019.118844.

[15] Utaş S, Köse K, Yazici C, Akdaş A, Keleştimur F. Antioxidant potential of propylthiouracil in patients with psoriasis. Clin Biochem 2002;35(3):241–6. https://doi.org/10.1016/S0009-9120(02)00294-1.

[16] Chen H, Lu C, Liu H, Wang M, Zhao H, Yan Y, Han L. Quercetin ameliorates imiquimod-induced psoriasis-like skin inflammation in mice via the NF-κB pathway. Int Immunopharmacol 2017;48:110–7. https://doi.org/10.1016/j.intimp.2017.04.022.

[17] de Jong EM. The course of psoriasis. Clin Dermatol 1997;15(5):687–92. https://doi.org/10.1016/S0738-081X(97)00023-0.

[18] Ortonne JP, Chimenti S, Luger T, Puig L, Reid F, Trüeb RM. Scalp psoriasis: European consensus on grading and treatment algorithm. JEADV 2009;23(12):1435–44. https://doi.org/10.1111/j.1468-3083.2009.03372.x.

[19] Nestle FO, Kaplan DH, Schon MP, Barker J. Psoriasis. N Engl J Med 2009;361(17):496–509. https://doi.org/10.1056/NEJMra0804595.

[20] Van Onselen J. Psoriasis in general practice. Nurs Stand (through 2013) 1998;12 (30):32. https://doi.org/10.7748/ns.12.30.32.s46.

[21] Ko H-C, JWA S-W, Song M, Kim M-B, Kwon K-S. Clinical course of guttate psoriasis: long-term follow-up study. J Dermatol 2010;37(10):894–9. https://doi.org/10.1111/j.1346-8138.2010.00871.x.

[22] Navarini AA, Burden AD, Capon F, Mrowietz U, Puig L, Köks S, Kingo K, Smith C, Barker JN, ERASPEN Network, Bachelez H. European consensus statement on phenotypes of pustular psoriasis. J Eur Acad Dermatol Venereol 2017;31(11):1792–9. https://doi.org/10.1111/jdv.14386.

[23] Wagener FA, Carels CE, Lundvig DM. Targeting the redox balance in inflammatory skin conditions. Int J Mol Sci 2013;14(5):9126–67. https://doi.org/10.3390/ijms14059126.

[24] Barygina VV, Becatti M, Soldi G, Prignano F, Lotti T, Nassi P, Wright D, Taddei N, Fiorillo C. Altered redox status in the blood of psoriatic patients: involvement of NADPH oxidase and role of anti-TNF-α therapy. Redox Rep 2013;18(3):100–6. https://doi.org/10.1179/1351000213Y.0000000045.

[25] Houshang N, Reza K, Masoud S, Ali E, Mansour R, Vaisi-Raygani A. Antioxidant status in patients with psoriasis. Cell Biochem Funct 2014;32(3):268–73. https://doi.org/10.1002/cbf.3011.

[26] ŞikarAktürk A, Özdoğan HK, Bayramgürler Dİ, Çekmen MB, Bilen N, Kıran R. Nitric oxide and malondialdehyde levels in plasma and tissue of psoriasis patients. J Eur Acad Dermatol Venereol 2012;26(7):833–7. https://doi.org/10.1111/j.1468-3083.2011.04164.x.

[27] Kadam DP, Suryakar AN, Ankush RD, Kadam CY, Deshpande KH. Role of oxidative stress in various stages of psoriasis. Indian J Clin Biochem 2010;25:388–92. https://doi.org/10.1007/s12291-010-0043-9.

[28] Ferretti G, Bacchetti T, Campanati A, Simonetti O, Liberati G, Offidani A. Correlation between lipoprotein (a) and lipid peroxidation in psoriasis: role of the enzyme

paraoxonase-1. Br J Dermatol 2012;166(1):204–7. https://doi.org/10.1111/j.1365-2133.2011.10539.x.

[29] Rocha-Pereira P, Santos-Silva A, Rebelo I, Figueiredo A, Quintanilha A, Teixeira F. Dislipidemia and oxidative stress in mild and in severe psoriasis as a risk for cardiovascular disease. Clinicachim Acta 2001;303(1–2):33–9. https://doi.org/10.1016/S0009-8981(00)00358-2.

[30] Woźniak A, Drewa G, Krzyżyńska-Malinowska E, Czajkowski R, Protas-Drozd F, Mila-Kierzenkowska C, Rozwodowska M, Sopońska M, Czarnecka-Żaba E. Oxidant-antioxidant balance in patients with psoriasis. Med Sci Monit 2006;13(1):CR30–3.

[31] Gabr SA, Al-Ghadir AH. Role of cellular oxidative stress and cytochrome c in the pathogenesis of psoriasis. Arch Dermatol Res 2012;304:451–7. https://doi.org/10.1007/s00403-012-1230-8.

[32] Drewa G, Krzyzyńska-Malinowska E, Woźniak A, Protas-Drozd F, Mila-Kierzenkowska C, Rozwodowska M, Kowaliszyn B, Czajkowski R. Activity of superoxide dismutase and catalase and the level of lipid peroxidation products reactive with TBA in patients with psoriasis. Med Sci Monit Int Med J Exp Clin Res 2002;8(8):BR338–43.

[33] Zaba LC, Fuentes-Duculan J, Eungdamrong NJ, Abello MV, Novitskaya I, Pierson KC, Gonzalez J, Krueger JG, Lowes MA. Psoriasis is characterized by accumulation of immunostimulatory and Th1/Th17 cell-polarizing myeloid dendritic cells. J Investig Dermatol 2009;129(1):79–88. https://doi.org/10.1038/jid.2008.194.

[34] Johnson-Huang LM, Lowes MA, Krueger JG. Putting together the psoriasis puzzle: an update on developing targeted therapies. Dis Model Mech 2012;5(4):423–33. https://doi.org/10.1242/dmm.009092.

[35] Sheng KC, Pietersz GA, Tang CK, Ramsland PA, Apostolopoulos V. Reactive oxygen species level defines two functionally distinctive stages of inflammatory dendritic cell development from mouse bone marrow. J Immunol 2010;184(6):2863–72. https://doi.org/10.4049/jimmunol.0903458.

[36] Verhasselt V, Goldman M, Willems F. Oxidative stress up-regulates IL-8 and TNF-α synthesis by human dendritic cells. Eur J Immunol 1998;28(11):3886–90. https://doi.org/10.1002/(SICI)1521-4141(199811)28:11<3886::AID-IMMU3886>3.0.CO;2-M.

[37] Rutault K, Alderman C, Chain BM, Katz DR. Reactive oxygen species activate human peripheral blood dendritic cells. Free Radic Biol Med 1999;26(1–2):232–8. https://doi.org/10.1016/S0891-5849(98)00194-4.

[38] Kantengwa S, Jornot L, Devenoges C, Nicod LP. Superoxide anions induce the maturation of human dendritic cells. Am J Respir Crit Care Med 2003;167(3):431–7. https://doi.org/10.1164/rccm.200205-425OC.

[39] Kyriakou A, Patsatsi A, Vyzantiadis TA, Sotiriadis D. Serum levels of TNF-α, IL-12/23p40, and IL-17 in plaque psoriasis and their correlation with disease severity. J Immunol Res 2014;2014. https://doi.org/10.1155/2014/467541.

[40] Brotas AM, Cunha JM, Lago EH, Machado CC, Carneiro SC. Tumor necrosis factor-alpha and the cytokine network in psoriasis. An Bras Dermatol 2012;87:673–83. https://doi.org/10.1590/s0365-05962012000500001.

[41] Simplex EB. Reactive oxygen species in tumor necrosis factor-a-activated primary human keratinocytes: implications for psoriasis and inflammatory skin disease. J Investig Dermatol 2009;129:1838. https://doi.org/10.1038/jid.2008.122.

[42] Bito T, Nishigori C. Impact of reactive oxygen species on keratinocyte signaling pathways. J Dermatol Sci 2012;68(1):3–8. https://doi.org/10.1016/j.jdermsci.2012.06.006.

[43] Armstrong AW, Voyles SV, Armstrong EJ, Fuller EN, Rutledge JC. Angiogenesis and oxidative stress: common mechanisms linking psoriasis with atherosclerosis. J Dermatol Sci 2011;63(1):1–9. https://doi.org/10.1016/j.jdermsci.2011.04.007.

[44] Henno A, Blacher S, Lambert CA, Deroanne C, Noël A, Lapière C, de la Brassinne M, Nusgens BV, Colige A. Histological and transcriptional study of angiogenesis and lymphangiogenesis in uninvolved skin, acute pinpoint lesions and established psoriasis plaques: an approach of vascular development chronology in psoriasis. J Dermatol Sci 2010;57(3):162–9. https://doi.org/10.1016/j.jdermsci.2009.12.006.

[45] Zhang Y, Matsuo H, Morita E. Vascular endothelial growth factor 121 is the predominant isoform in psoriatic scales. Exp Dermatol 2005;14(10):758–64. https://doi.org/10.1111/j.1600-0625.2005.00356.x.

[46] Detmar M, Brown LF, Claffey KP, Yeo KT, Kocher O, Jackman RW, Berse B, Dvorak HF. Overexpression of vascular permeability factor/vascular endothelial growth factor and its receptors in psoriasis. J Exp Med 1994;180(3):1141–6. https://doi.org/10.1084/jem.180.3.1141.

[47] Henno A, Blacher S, Lambert C, Colige A, Seidel L, Noël A, Lapière C, De La Brassinne M, Nusgens BV. Altered expression of angiogenesis and lymphangiogenesis markers in the uninvolved skin of plaque-type psoriasis. Br J Dermatol 2009;160(3):581–90. https://doi.org/10.1111/j.1365-2133.2008.08889.x.

[48] Keyt BA, Berleau LT, Nguyen HV, Chen H, Heinsohn H, Vandlen R, Ferrara N. The carboxyl-terminal domain (111–165) of vascular endothelial growth factor is critical for its mitogenic potency. J Biol Chem 1996;271(13):7788–95. https://doi.org/10.1074/jbc.271.13.7788.

[49] Küsters B, de Waal RM, Wesseling P, Verrijp K, Maass C, Heerschap A, Barentsz JO, Sweep F, Ruiter DJ, Leenders WP. Differential effects of vascular endothelial growth factor A isoforms in a mouse brain metastasis model of human melanoma. Cancer Res 2003;63(17):5408–13.

[50] Young HS, Bhushan M, Griffiths CE, Summers AM, Brenchley PE. Single-nucleotide polymorphisms of vascular endothelial growth factor in psoriasis of early onset. J Investig Dermatol 2004;122(1):209–15. https://doi.org/10.1046/j.0022-202X.2003.22107.x.

[51] Wongpiyabovorn J, Yooyongsatit S, Ruchusatsawat K, Avihingsanon Y, Hirankarn N. Association of the CTG (− 2578/− 460/+ 405) haplotype within the vascular endothelial growth factor gene with early-onset psoriasis. Tissue Antigens 2008;72(5):458–63. https://doi.org/10.1111/j.1399-0039.2008.01134.x.

[52] Kakurai M, Demitsu T, Umemoto N, Kobayashi Y, Inoue-Narita T, Fujita N, Ohtsuki M, Furukawa Y. Vasoactive intestinal peptide and inflammatory cytokines enhance vascular endothelial growth factor production from epidermal keratinocytes. Br J Dermatol 2009;161(6):1232–8. https://doi.org/10.1111/j.1365-2133.2009.09439.x.

[53] Kuroda K, Sapadin A, Shoji T, Fleischmajer R, Lebwohl M. Altered expression of angiopoietins and Tie2 endothelium receptor in psoriasis. J Investig Dermatol 2001;116(5):713–20. https://doi.org/10.1046/j.1523-1747.2001.01316.x.

[54] Simonetti O, Lucarini G, Campanati A, Goteri G, Zizzi A, Marconi B, Ganzetti G, Minardi D, Di Primio R, Offidani A. VEGF, survivin and NOS overexpression in psoriatic skin: critical role of nitric oxide synthases. J Dermatol Sci 2009;54(3):205–8. https://doi.org/10.1016/j.jdermsci.2008.12.012.

[55] Coto-Segura P, Coto E, Mas-Vidal A, Morales B, Alvarez V, Díaz M, Alonso B, Santos-Juanes J. Influence of endothelial nitric oxide synthase polymorphisms in psoriasis risk. Arch Dermatol Res 2011;303:445–9. https://doi.org/10.1007/s00403-011-1129-9.

[56] Sala M, Elaissari A, Fessi H. Advances in psoriasis physiopathology and treatments: up to date of mechanistic insights and perspectives of novel therapies based on innovative skin drug delivery systems (ISDDS). J Control Release 2016;239:182–202. https://doi.org/10.1016/j.jconrel.2016.07.003.

[57] Pukale SS, Sharma S, Dalela M, Kumar Singh A, Mohanty S, Mittal A, Chitkara D. Multi-component clobetasol-loaded monolithic lipid-polymer hybrid nanoparticles ameliorate imiquimod-induced psoriasis-like skin inflammation in Swiss albino mice. Acta Biomater 2020;115:393–409. https://doi.org/10.1016/j.actbio.2020.08.020.

[58] Menter A, Griffiths CE. Current and future management of psoriasis. Lancet 2007;370 (9583):272–84. https://doi.org/10.1016/S0140-6736(07)61129-5.

[59] Horn EJ, Domm S, Katz HI, Lebwohl M, Mrowietz U, Kragballe K, International Psoriasis Council. Topical corticosteroids in psoriasis: strategies for improving safety. J Eur Acad Dermatol Venereol 2010;24(2):119–24. https://doi.org/10.1111/j.1468-3083.2009.03358.x.

[60] Perera GK, Di Meglio P, Nestle FO. Psoriasis. Annu Rev Pathol 2012;7:385–422. https://doi.org/10.1146/annurev-pathol-011811-132448.

[61] Sharma N, Zahoor I, Sachdeva M, Subramaniyan V, Fuloria S, Fuloria NK, Naved T, Bhatia S, Al-Harrasi A, Aleya L, Bungau S. Deciphering the role of nanoparticles for management of bacterial meningitis: an update on recent studies. Environ Sci Pollut Res 2021;1:1–8. https://doi.org/10.1007/s11356-021-16570-y.

[62] Van Tran V, Moon JY, Lee YC. Liposomes for delivery of antioxidants in cosmeceuticals: challenges and development strategies. J Control Release 2019;300:114–40. https://doi.org/10.1016/j.jconrel.2019.03.003.

[63] Knudsen NØ, Jorgensen L, Hansen J, Vermehren C, Frokjaer S, Foged C. Targeting of liposome-associated calcipotriol to the skin: effect of liposomal membrane fluidity and skin barrier integrity. Int J Pharm 2011;416(2):478–85. https://doi.org/10.1016/j.ijpharm.2011.03.014.

[64] El Gizaway SA, Fadel MA, Mourad BA, Elnaby FE. Betamethasone dipropionate gel for treatment of localized plaque psoriasis. Int J Pharm Pharmaceut Sci 2017;9:173–82. https://doi.org/10.22159/ijpps.2017v9i8.18571.

[65] Zhang H, Zhang K, Li Z, Zhao J, Zhang Y, Feng N. In vivo microdialysis for dynamic monitoring of the effectiveness of nano-liposomes as vehicles for topical psoralen application. Biol Pharm Bull 2017;40(11):1996–2000. https://doi.org/10.1248/bpb.b17-00302.

[66] Alam MS, Ali MS, Alam N, Siddiqui MR, Shamim M, Safhi MM. In vivo study of clobetasol propionate loaded nanoemulsion for topical application in psoriasis and atopic dermatitis. Drug Invent Today 2013;5(1):8–12. https://doi.org/10.1016/j.dit.2013.02.001.

[67] Song B, Song R, Cheng M, Chu H, Yan F, Wang Y. Preparation of calcipotriol emulsion using bacterial exopolysaccharides as emulsifier for percutaneous treatment of psoriasis vulgaris. Int J Mol Sci 2019;21(1):77. https://doi.org/10.3390/ijms21010077.

[68] Kaur A, Katiyar SS, Kushwah V, Jain S. Nanoemulsion loaded gel for topical co-delivery of clobitasol propionate and calcipotriol in psoriasis. Nanomed Nanotechnol Biol Med 2017;13(4):1473–82. https://doi.org/10.1016/j.nano.2017.02.009.

[69] Sahu S, Katiyar SS, Kushwah V, Jain S. Active natural oil-based nanoemulsion containing tacrolimus for synergistic antipsoriatic efficacy. Nanomedicine 2018;13(16):1985–98. https://doi.org/10.2217/nnm-2018-0135.

[70] Wan T, Pan J, Long Y, Yu K, Wang Y, Pan W, Ruan W, Qin M, Wu C, Xu Y. Dual roles of TPGS based microemulsion for tacrolimus: enhancing the percutaneous delivery and antipsoriatic efficacy. Int J Pharm 2017;528(1–2):511–23. https://doi.org/10.1016/j.ijpharm.2017.06.050.

[71] Dingler A, Blum RP, Niehus H, Muller RH, Gohla S. Solid lipid nanoparticles (SLNTM/LipopearlsTM) a pharmaceutical and cosmetic carrier for the application of vitamin E in dermal products. J Microencapsul 1999;16(6):751–67. https://doi.org/10.1080/026520499288690.

[72] Jenning V, Gysler A, Schäfer-Korting M, Gohla SH. Vitamin A loaded solid lipid nanoparticles for topical use: occlusive properties and drug targeting to the upper skin. Eur J Pharm Biopharm 2000;49(3):211–8. https://doi.org/10.1016/S0939-6411(99) 00075-2.

[73] Wissing SA, Müller RH. The influence of solid lipid nanoparticles on skin hydration and viscoelasticity—in vivo study. Eur J Pharm Biopharm 2003;56(1):67–72. https://doi.org/ 10.1016/S0939-6411(03)00040-7.

[74] Geszke-Moritz M, Moritz M. Solid lipid nanoparticles as attractive drug vehicles: composition, properties and therapeutic strategies. Mater Sci Eng C 2016;68:982– 94. https://doi.org/10.1016/j.msec.2016.05.119.

[75] Behl T, Singh S, Sharma N, Zahoor I, Albarrati A, Albratty M, Meraya AM, Najmi A, Bungau S. Expatiating the pharmacological and nanotechnological aspects of the alkaloidal drug berberine: current and future trends. Molecules 2022;27(12):3705. https://doi. org/10.3390/molecules27123705.

[76] Müller RH, Radtke M, Wissing SA. Solid lipid nanoparticles (SLN) and nanostructured lipid carriers (NLC) in cosmetic and dermatological preparations. Adv Drug Deliv Rev 2002;54:S131–55. https://doi.org/10.1016/S0169-409X(02)00118-7.

[77] Garcês A, Amaral MH, Lobo JS, Silva AC. Formulations based on solid lipid nanoparticles (SLN) and nanostructured lipid carriers (NLC) for cutaneous use: a review. Eur J Pharm Sci 2018;112:159–67. https://doi.org/10.1016/j.ejps.2017.11.023.

[78] Schäfer-Korting M, Mehnert W, Korting HC. Lipid nanoparticles for improved topical application of drugs for skin diseases. Adv Drug Deliv Rev 2007;59(6):427–43. https:// doi.org/10.1016/j.addr.2007.04.006.

[79] Ganesan P, Narayanasamy D. Lipid nanoparticles: different preparation techniques, characterization, hurdles, and strategies for the production of solid lipid nanoparticles and nanostructured lipid carriers for oral drug delivery. Sustain Chem Pharm 2017;6:37– 56. https://doi.org/10.1016/j.scp.2017.07.002.

[80] Pardeike J, Hommoss A, Müller RH. Lipid nanoparticles (SLN, NLC) in cosmetic and pharmaceutical dermal products. Int J Pharm 2009;366(1–2):170–84. https://doi.org/ 10.1016/j.ijpharm.2008.10.003.

[81] Pradhan M, Singh D, Murthy SN, Singh MR. Design, characterization and skin permeating potential of Fluocinolone acetonide loaded nanostructured lipid carriers for topical treatment of psoriasis. Steroids 2015;101:56–63. https://doi.org/10.1016/j. steroids.2015.05.012.

[82] Pradhan M, Singh D, Singh MR. Fabrication, optimization and characterization of triamcinolone acetonide loaded nanostructured lipid carriers for topical treatment of psoriasis: application of box Behnken design, in vitro and ex vivo studies. J Drug Deliv Sci Technol 2017;41:325–33. https://doi.org/10.1016/j.jddst.2017.07.024.

[83] Charoenputtakhun P, Opanasopit P, Rojanarata T, Ngawhirunpat T. All-trans retinoic acid-loaded lipid nanoparticles as a transdermal drug delivery carrier. Pharm Dev Technol 2014;19(2):164–72. https://doi.org/10.3109/10837450.2013.763261.

[84] Ridolfi DM, Marcato PD, Justo GZ, Cordi L, Machado D, Durán N. Chitosan-solid lipid nanoparticles as carriers for topical delivery of tretinoin. Colloids Surf B Biointerfaces 2012;93:36–40. https://doi.org/10.1016/j.colsurfb.2011.11.051.

[85] Kang JH, Chon J, Kim YI, Lee HJ, Oh DW, Lee HG, Han CS, Kim DW, Park CW. Preparation and evaluation of tacrolimus-loaded thermosensitive solid lipid nanoparticles for improved dermal distribution. Int J Nanomedicine 2019;5381–96. https://doi.org/10.2147/ IJN.S215153.

[86] Jain S, Addan R, Kushwah V, Harde H, Mahajan RR. Comparative assessment of efficacy and safety potential of multifarious lipid based tacrolimus loaded nanoformulations. Int J Pharm 2019;562:96–104. https://doi.org/10.1016/j.ijpharm.2019.03.042.

[87] Hadinoto K, Sundaresan A, Cheow WS. Lipid–polymer hybrid nanoparticles as a new generation therapeutic delivery platform: a review. Eur J Pharm Biopharm 2013;85 (3):427–43. https://doi.org/10.1016/j.ejpb.2013.07.002.

[88] Mandal B, Bhattacharjee H, Mittal N, Sah H, Balabathula P, Thoma LA, Wood GC. Core–shell-type lipid–polymer hybrid nanoparticles as a drug delivery platform. Nanomed Nanotechnol Biol Med 2013;9(4):474–91. https://doi.org/10.1016/j.nano.2012.11.010.

[89] Nimisha, Rizvi DA, Fatima Z, Neema, Kaur CD. Antipsoriatic and anti-inflammatory studies of Berberis aristata extract loaded nanovesicular gels. Pharmacogn Mag 2017;13(Suppl 3):S587–94. https://doi.org/10.4103/pm.pm_210_17.

[90] Abdelgawad R, Nasr M, Moftah NH, Hamza MY. Phospholipid membrane tubulation using ceramide doping "Cerosomes": characterization and clinical application in psoriasis treatment. Eur J Pharm Sci 2017;101:258–68. https://doi.org/10.1016/j.ejps.2017.02.030.

[91] Javia A, Misra A, Thakkar H. Liposomes encapsulating novel antimicrobial peptide Omiganan: characterization and its pharmacodynamic evaluation in atopic dermatitis and psoriasis mice model. Int J Pharm 2022;624, 122045. https://doi.org/10.1016/j. ijpharm.2022.122045.

[92] Walunj M, Doppalapudi S, Bulbake U, Khan W. Preparation, characterization, and in vivo evaluation of cyclosporine cationic liposomes for the treatment of psoriasis. J Liposome Res 2020;30(1):68–79. https://doi.org/10.1080/08982104.2019.1593449.

[93] Dadwal A, Mishra N, Rawal RK, Narang RK. Development and characterisation of clo-betasol propionate loaded Squarticles as a lipid nanocarrier for treatment of plaque psoriasis. J Microencapsul 2020;37(5):341–54. https://doi.org/10.1080/02652048.2020. 1756970.

[94] Srisuk P, Thongnopnua P, Raktanonchai U, Kanokpanont S. Physico-chemical character-istics of methotrexate-entrapped oleic acid-containing deformable liposomes for in vitro transepidermal delivery targeting psoriasis treatment. Int J Pharm 2012;427(2):426–34. https://doi.org/10.1016/j.ijpharm.2012.01.045.

[95] Qiu F, Xi L, Chen S, Zhao Y, Wang Z, Zheng Y. CelastrolNiosome hydrogel has anti-inflammatory effect on skin keratinocytes and circulation without systemic drug exposure in psoriasis mice. Int J Nanomedicine 2021;16:6171–82. https://doi.org/10.2147/IJN. S323208.

[96] Jain A, Pooladanda V, Bulbake U, Doppalapudi S, Rafeeqi TA, Godugu C, Khan W. Liposphere mediated topical delivery of thymoquinone in the treatment of psoriasis. Nanomedicine 2017;13(7):2251–62. https://doi.org/10.1016/j.nano.2017.06.009.

[97] Wadhwa S, Singh B, Sharma G, Raza K, Katare OP. Liposomal fusidic acid as a potential delivery system: a new paradigm in the treatment of chronic plaque psoriasis. Drug Deliv 2016;23(4):1204–13. https://doi.org/10.3109/10717544.2015.1110845.

[98] Zhang YT, Shen LN, Wu ZH, Zhao JH, Feng NP. Comparison of ethosomes and liposomes for skin delivery of psoralen for psoriasis therapy. Int J Pharm 2014;471(1–2):449–52. https://doi.org/10.1016/j.ijpharm.2014.06.001.

[99] Kathuria H, Nguyen DTP, Handral HK, Cai J, Cao T, Kang L. Proposome for transdermal delivery of tofacitinib. Int J Pharm 2020;585, 119558. https://doi.org/10.1016/j. ijpharm.2020.119558.

[100] Tang L, Azzi J, Kwon M, Mounayar M, Tong R, Yin Q, Moore R, Skartsis N, Fan TM, Abdi R, Cheng J. Immunosuppressive activity of size-controlled PEG-PLGA

nanoparticles containing encapsulated cyclosporine A. J Transplant 2012;2012. https://doi.org/10.1155/2012/896141.

[101] Jain S, Mittal A, Jain K, A. Enhanced topical delivery of cyclosporin-A using PLGA nanoparticles as carrier. Curr Nanosci 2011;7(4):524–30. https://doi.org/10.2174/157341311796196835.

[102] Takeuchi I, Kagawa A, Makino K. Skin permeability and transdermal delivery route of 30-nm cyclosporin A-loaded nanoparticles using PLGA-PEG-PLGA triblock copolymer. Colloids Surf A Physicochem Eng Asp 2020;600, 124866. https://doi.org/10.1016/j.colsurfa.2020.124866.

[103] Gomes AJ, Lunardi LO, Caetano FH, Machado AE, Oliveira-Campos AM, Bendhack LM, Lunardi CN. Biodegradable nanoparticles containing benzopsoralens: an attractive strategy for modifying vascular function in pathological skin disorders. J Appl Polym Sci 2011;121(3):1348–54. https://doi.org/10.1002/app.33427.

[104] Gomes AJ, Faustino AS, Lunardi CN, Lunardi LO, Machado AE. Evaluation of nanoparticles loaded with benzopsoralen in rat peritoneal exudate cells. Int J Pharm 2007;332(1–2):153–60. https://doi.org/10.1016/j.ijpharm.2006.09.035.

[105] Frušić-Zlotkin M, Soroka Y, Tivony R, Larush L, Verkhovsky L, Brégégère FM, Neuman R, Magdassi S, Milner Y. Penetration and biological effects of topically applied cyclosporin A nanoparticles in a human skin organ culture inflammatory model. Exp Dermatol 2012;21(12):938–43. https://doi.org/10.1111/exd.12051.

[106] Rosado C, Silva C, Reis CP. Hydrocortisone-loaded poly (ε-caprolactone) nanoparticles for atopic dermatitis treatment. Pharm Dev Technol 2013;18(3):710–8. https://doi.org/10.3109/10837450.2012.712537.

[107] Goudon F, Clément Y, Ripoll L. Controlled release of retinol in cationic co-polymeric nanoparticles for topical application. Cosmetics 2020;7(2):29. https://doi.org/10.3390/cosmetics7020029.

[108] Lapteva M, Mondon K, Möller M, Gurny R, Kalia YN. Polymeric micelle nanocarriers for the cutaneous delivery of tacrolimus: a targeted approach for the treatment of psoriasis. Mol Pharm 2014;11(9):2989–3001. https://doi.org/10.1021/mp400639e.

[109] Jin Y, Zhang X, Zhang B, Kang H, Du L, Li M. Nanostructures of an amphiphilic zinc phthalocyanine polymer conjugate for photodynamic therapy of psoriasis. Colloids Surf B Biointerfaces 2015;128:405–9. https://doi.org/10.1016/j.colsurfb.2015.02.038.

[110] Zahoor I, Singh S, Sharma N, Behl T, Wani SN. Dendrimers: versatile and revolutionary nanocarriers for infectious diseases. ECS Trans 2022;107(1):8619. https://doi.org/10.1149/10701.8619ecst.

[111] Sharma N, Zahoor I, Singh S, Behl T, Antil A. Expatiating the pivotal role of dendrimers as emerging nanocarrier for management of liver disorders. J Integr Sci Technol 2023;11 (2):489.

[112] Agrawal U, Mehra NK, Gupta U, Jain NK. Hyperbranched dendritic nano-carriers for topical delivery of dithranol. J Drug Target 2013;21(5):497–506. https://doi.org/10.3109/1061186X.2013.771778.

[113] Borowska K, Wołowiec S, Głowniak K, Sieniawska E, Radej S. Transdermal delivery of 8-methoxypsoralene mediated by polyamidoamine dendrimer G2. 5 and G3. 5—in vitro and in vivo study. Int J Pharm 2012;436(1–2):764–70. https://doi.org/10.1016/j.ijpharm.2012.07.067.

[114] Dhanikula RS, Hildgen P. Influence of molecular architecture of polyether-co-polyester dendrimers on the encapsulation and release of methotrexate. Biomaterials 2007;28 (20):3140–52. https://doi.org/10.1016/j.biomaterials.2007.03.012.

[115] Cai J, Huang H, Song W, Hu H, Chen J, Zhang L, Li P, Wu R, Wu C. Preparation and evaluation of lipid polymer nanoparticles for eradicating H. pylori biofilm and impairing antibacterial resistance in vitro. Int J Pharm 2015;495(2):728–37. https://doi.org/10.1016/j.ijpharm.2015.09.055.

[116] Hu Y, Hoerle R, Ehrich M, Zhang C. Engineering the lipid layer of lipid–PLGA hybrid nanoparticles for enhanced in vitro cellular uptake and improved stability. Acta Biomater 2015;28:149–59. https://doi.org/10.1016/j.actbio.2015.09.032.

[117] Shi J, Xiao Z, Votruba AR, Vilos C, Farokhzad OC. Differentially charged hollow core/shell lipid-polymer-lipid hybrid nanoparticle for small interfering RNA delivery. Angew Chem Int Ed Engl 2011;50(31):7027. https://doi.org/10.1002/anie.201101554.

[118] Hu CM, Zhang L, Aryal S, Cheung C, Fang RH, Zhang L. Erythrocyte membrane-camouflaged polymeric nanoparticles as a biomimetic delivery platform. Proc Natl Acad Sci 2011;108(27):10980–5. https://doi.org/10.1073/pnas.1106634108.

[119] Aoki I, Yoneyama M, Hirose J, Minemoto Y, Koyama T, Kokuryo D, Bakalova R, Murayama S, Saga T, Aoshima S, Ishizaka Y. Thermoactivatable polymer-grafted liposomes for low-invasive image-guided chemotherapy. Transl Res 2015;166(6):660–73. https://doi.org/10.1016/j.trsl.2015.07.009.

[120] Zhao X, Li F, Li Y, Wang H, Ren H, Chen J, Nie G, Hao J. Co-delivery of HIF1α siRNA and gemcitabine via biocompatible lipid-polymer hybrid nanoparticles for effective treatment of pancreatic cancer. Biomaterials 2015;46:13–25. https://doi.org/10.1016/j.biomaterials.2014.12.028.

[121] Mao KL, Fan ZL, Yuan JD, Chen PP, Yang JJ, Xu J, ZhuGe DL, Jin BH, Zhu QY, Shen BX, Sohawon Y, Zhao YZ, Xu HL. Skin-penetrating polymeric nanoparticles incorporated in silk fibroin hydrogel for topical delivery of curcumin to improve its therapeutic effect on psoriasis mouse model. Colloids Surf B Biointerfaces 2017;160:704–14. https://doi.org/10.1016/j.colsurfb.2017.10.029.

[122] Wang Y, Wang C, Fu S, Liu Q, Dou D, Lv H, Fan M, Guo G, Luo F, Qian Z. Preparation of tacrolimus loaded micelles based on poly(ε-caprolactone)-poly(ethylene glycol)-poly(ε-caprolactone). Int J Pharm 2011;407(1–2):184–9. https://doi.org/10.1016/j.ijpharm.2011.01.018.

[123] Ishihara T, Izumo N, Higaki M, Shimada E, Hagi T, Mine L, Ogawa Y, Mizushima Y. Role of zinc in formulation of PLGA/PLA nanoparticles encapsulating betamethasone phosphate and its release profile. J Control Release 2005;105(1–2):68–76. https://doi.org/10.1016/j.jconrel.2005.02.026.

[124] Divya G, Panonnummal R, Gupta S, Jayakumar R, Sabitha M. Acitretin and aloe-emodin loaded chitin nanogel for the treatment of psoriasis. Eur J Pharm Biopharm 2016;107:97–109. https://doi.org/10.1016/j.ejpb.2016.06.019.

[125] Durán-Lobato M, Enguix-González A, Fernández-Arévalo M, Martín-Banderas L. Statistical analysis of solid lipid nanoparticles produced by high-pressure homogenization: a practical prediction approach. J Nanopart Res 2013;15:1–4. https://doi.org/10.1007/s11051-013-1443-6.

[126] Tang PL, Sudol ED, Silebi CA, El-Aasser MS. Miniemulsion polymerization—a comparative study of preparative variables. J Appl Polym Sci 1991;43(6):1059–66. https://doi.org/10.1002/app.1991.070430604.

Unveiling the pharmacological and nanotechnological aspects for the management of hypertension: State-of-the-art and future perspectives

7

Neelam Sharma[a], Sonam Grewal[a], Sukhbir Singh[a], Ikmeet Kaur Grewal[b], and Ishrat Zahoor[a]
[a]Department of Pharmaceutics, MM College of Pharmacy, Maharishi Markandeshwar (Deemed to be University), Mullana-Ambala, Haryana, India, [b]Department of Pharmacy, Government Medical College, Patiala, Punjab, India

1   Introduction

Due to its widespread prevalence around the world, hypertension is a significant public health issue. In the world, high blood pressure is responsible for over 7.5 million deaths per year, or 12.8% of all fatalities, and in 2025, 1.56 billion adults would have hypertension, according to predictions [1]. A condition of high arterial blood pressure is referred to as hypertension (or HTN) also known as high blood pressure. A systolic blood pressure of greater than 140 mmHg and/or a diastolic blood pressure of less than 90 mmHg are considered to be signs of hypertension. Prehypertension is the gray region between a systolic blood pressure reading of 120 to 139 and a diastolic blood pressure reading of 80 to 89 [2]. Despair is brought on by the fact that hypertension is, quantitatively, the biggest risk factor for cardiovascular diseases that it is increasing in prevalence and that it is almost poorly controlled. Hope is brought on by the fact that prevention is possible, and that almost all patients can be effectively treated, leading to significant drops in stroke and heart attack rates [3]. The most frequent modifiable risk factor for cardiovascular disease and death is hypertension, meanwhile treatment with antihypertensive medications that lower both BP and related target organ damage can significantly minimize the elevated risk associated with blood pressure rise [4]. Only 10% of patients have a recognized reason of their hypertension (referred to as essential or primary hypertension) and in nearly 90% of cases, the cause is unknown (secondary hypertension). Essential hypertension can be managed even though it cannot be cured, and it is commonly believed that essential hypertension has unknown causes; this is only

Targeting Angiogenesis, Inflammation and Oxidative Stress in Chronic Diseases. https://doi.org/10.1016/B978-0-443-13587-3.00013-8

partially true because we know very little about genetic variations, genes that are overexpressed or underexpressed, and the intermediate phenotypes that these genes regulate to cause high blood pressure. Among the factors that cause blood pressure to rise are physical activity, obesity, insulin resistance, a high salt and alcohol diet, aging, sedentary living, stress, inadequate potassium and calcium intakes, and physical inactivity [5]. Nearly two-thirds of persons having age of 60 years or older are at risk because it is the most prevalent risk factor for cardiovascular disease. Uncontrolled hypertension is thought to be the cause of 7.5 million deaths annually worldwide and costs the US economy about $47 billion annually in lost productivity due to employee absences and medical expenses [6]. Hypertension is an important public health problem in both economically developed and developing nations. Systolic blood pressure increased in economically emerging regions like Oceania, East Africa, South and Southeast Asia, and lowered in economically developed regions like Australia, North America, and Western Europe [7]. In terms of national prevalence, Taiwan, South Korea, Japan, and a few western European nations had the lowest rates of hypertension in 2019 for women, while various low- and middle-income nations had the lowest rates for men. Canada and Peru had the lowest rates for both men and women. Fewer than 24% of women and less than 25% of males in each of these nations experienced age-standardized prevalence in 2019 [8]. Epidemiological refers to the rise that is caused by industrialization, urbanization, and associated lifestyle modifications. Over 10.5 million people pass away each year in India, and it was noted that 20.3% of the deaths among males and 16.9% of the fatalities among women were attributed to cardiovascular disease. According to RGI data for the years 2010 to 2013, the proportion of fatalities from cardiovascular disease increased to 23% of all deaths and 32% of deaths among adults. Mortality ranges from 10% in rural areas of less developed states to >35% in urban areas with higher levels of development [9]. The process of inflammation is intricate and involves many different cells and released substances, several of which have been linked to hypertension. In both animal models of essential hypertension and dendritic cells from hypertensive patients, reactive oxygen species production is increased. This sets off a series of actions that cause T-cells to become activated. One potential autoimmune mechanism for hypertension is the hypothesized direct cellular harm caused by CD8+ "killer" T-cells [10]. Angiogenesis, which relieves tissue ischemia by creating new microvessels and microvascular networks from already existing ones, is a tightly controlled process that occurs in response to hypoxia and other stimuli. In order to have an effect, angiogenic growth factors like vascular endothelial growth factor and fibroblast growth factor produce NO. Angiogenesis is either induced by effectors molecules and associated receptors of the RAAS (Bradykinin, Angiotensin II) and inhibits the angiogenesis [11]. Vascular relaxation results from the endothelium's release of nitric oxide and oxygen-derived free radical superoxide anion quickly break down nitric oxide. Reduced NO bioavailability and elevated oxidative stress are both related to hypertension. Reduced levels of glutathione, reactive oxygen species or reactive nitrogen species scavengers like catalase and/or superoxide dismutase, as well as vitamins C and E, also contribute to oxidative stress [12].

2 Pathogenesis of hypertension

Essential hypertension cause is unknown; however, it usually manifests in the fifth or sixth decade of life, is frequently linked to obesity and excess salt consumption, and has a close relationship to family history, suggesting a possible genetic susceptibility. On the other hand, secondary hypertension is accompanied with definable causes such adrenal disorders, chronic kidney disease, sleep apnea, and renal artery stenosis [13]. The derangement of various mechanism involved in the maintenance of normal blood pressure is the shared phenomenon in both situations, and as a result, the sympathetic nervous system, the renin-angiotensin-aldosterone system, endothelial function, as well as sodium and water retention, all involved in the development of the hypertension [14]. The efferent and afferent pathways of the renal sympathetic nervous system, which both impact blood pressure, are important in the development and maintenance of hypertension. The efferent pathway transports signals from the SNS to the kidney, where it stimulates renin release, activates the RAAS system, and increases sodium and water retention, all of which lead to increased circulation volumes and thus, increase blood pressure [15]. The pathophysiology of hypertension is described in Fig. 1.

3 Role of inflammation in hypertension

A defensive reaction to damage or infection is inflammation. Inflammatory cells identify the injured tissue first, followed by the recruitment of leukocytes into the tissue, the removal of the causative agent, and the healing of the wound. Cell surfaces,

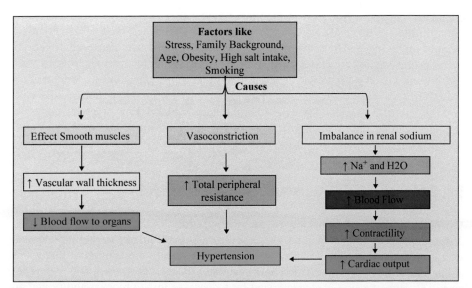

Fig. 1 The factors involved and their description in pathogenesis of hypertension.

extracellular matrix, and proinflammatory mediators must interact for inflammation to occur [16]. Numerous cell types and released substances are involved in the complex process of inflammation, which has been linked to high blood pressure in many instances. In animal models of essential hypertension as well as in dendritic cells from hypertensive patients, the production of reactive oxygen species is increased [17]. Spleen function highlights the significance of the interaction between innate and adaptive immune cells in hypertension. It has been demonstrated that the innate and adaptive immune systems' cells are crucial in the development of hypertension [18]. C-reactive protein, an acute phase protein, plays several roles in innate immune responses, including stimulating the signaling pathway and promoting phagocytosis. In addition to causing endothelial cells to express intracellular adhesion molecule (ICAM)-1 and vascular cell adhesion molecule (VCAM)-1, CRP can cause monocytes to release proinflammatory cytokines like interleukin-6 (IL-6) and interleukin-1 beta (IL-1β) and tumor necrosis factor alpha (TNF-α) [19].

4 Role of angiogenesis in hypertension

Potentially, genetic predisposition, inadequate placental and embryonic vascular development which result in low birth weight as well as reduced postembryonic vascular growth in general and in target organs may all contribute to defective angiogenesis in people predisposed to hypertension. Thus, defective angiogenesis or the inadequate growth of new blood vessels may be inherently linked to hypertension and the development of hypertension-dependent target organ damage [20]. Angiogenesis is the process through which new blood vessels are created from preexisting microvessels. It supports the provision of nutrients, oxygen, and waste disposal. Endothelial cells and pericytes make up capillary blood vessels. These two cell types transmit information to create new capillary networks, branches, and tubes. Hypoxia brought on by inadequate tissue perfusion is only stimulus that causes angiogenesis by releasing angiogenic molecules [21]. NO is essential for angiogenesis in vivo and growth factor-mediated endothelial tube creation in vitro, and its reduced bioavailability is a major contributor in the development of arterial hypertension. By controlling vascular permeability, vasodilation, tumor blood flow, platelet adhesion and aggregation, among other processes, NO operates as a mediator in the vascular, neurological, and immunological systems [22].

5 Role of oxidative stress in hypertension

Various molecules that belong to the ROS family all affect cellular function differently. A significant number of these behaviors are linked to the pathological alterations seen in cardiovascular disease. Redox-sensitive modulation of many signaling molecules and second messengers mediates the effects of reactive oxygen species [23]. Hypertension is characterized by a decrease in NO bioavailability and an increase in oxidative stress. These results are generally based on elevated levels of oxidative stress and lipid peroxidation indicators. Increased amounts of reactive oxygen species scavengers, such as glutathione and vitamins E and C, as well as decreased antioxidant activity (superoxide dismutase and catalase) may potentially contribute to oxidative stress [24]. NADPH oxidase is the ROS source with the highest characterization. The main biological generator of ROS, especially superoxide, in the vasculature is NADPH oxidase. Rich sources of NADPH oxidase-derived ROS are the kidney and vasculature, which play a significant role in renal failure and vascular damage under pathological circumstances [25].

6   Pharmacology investigations related to management of hypertension

Several pharmacological investigations have been performed for examining the role of various medications in management of hypertension and have been described in Table 1.

7   Strategies for management of hypertension

The objective of the therapy is to provide the patient with such a symptomatic relief as is clinically feasible and also preventing any long-term potential risks. The most common approaches that are all used in the treatment are all used in the treatment of hypertension. The classification of drugs used for the management of hypertension is given in Fig. 2, and their mechanism of action is given in Fig. 3.

7.1   Lifestyle modification

When someone has a risk factor for cardiovascular disease such obesity, hypertension, dyslipidemia, or diabetes, lifestyle modification is the cornerstone of preventative therapy. Before beginning drug therapy for hypertensive patients, lifestyle changes can be suggested as a first line of treatment, and when medication is already being used, they can be used as an addition to it [43]. Because everyone is at risk, and it is impossible to determine who will develop hypertension as they get older. However, there are a few significant contributing variables for critical hypertension. They include obesity, higher sodium intake, dietary fat and alcohol intake, and a lack of exercise. Consuming too little whole grain food or fruit and vegetable intake has also been linked [44].

7.2   ACE and ARBs inhibitors

Both ACE inhibitors and ARBs have a similar and distinct mechanism of action. Although both classes of drugs act on the renin-angiotensin-aldosterone system, ACE inhibitors inhibit the formation of angiotensin II and consequently the downstream effects through the angiotensin II type 1 (AT1) receptor (vasoconstriction, cell

Table 1 Description of in-vivo studies investigated for the management of hypertension.

Route of administration/ Drug	Dose/Duration	Animal model	Outcomes	References
i.v./Captopril	3 mg/kg/day for 4 weeks	Rats	Study showed effect in the protection of endothelial dysfunction and played a crucial role for treatment of hypertension.	[26]
p.o./Captopril	10 mg/kg Body Weight	Rats	Significant approach for the treatment of hypertension as well as to increase efficacy of drug in combination with celery.	[27]
Intraabdominal/ Captopril	30 mg/kg	Rabbits	Study showed that arterial pressure was reduced	[28]
p.o./Captopril and L-Carnitine	80 mg/kg/day for 12 weeks	Rats	Inhibition of the RAAS, modulation of the factor NF-Kb, and reduced mechanical stress.	[29]
p.o./Captopril	40 or 80 mg/kg for 5 weeks	Rats	Study demonstrated that the blood pressure reduced	[30]
i.v./Captopril	10 mg/kg	Rats	Study showed that arterial pressure was reduced	[31]
i.p./Lisinopril	10 mg/kg	Rats	Reduced the oxidative stress-induced lipid peroxidation level	[32]
Ramipril	1 mg/kg/day for 2 weeks	Rats	Investigated that the myocardial infarction was reduced	[33]
p.o./Candesartan cilexitel	0.1 mg/kg	Rats	Decreased myocardial fibrosis and mRNA expression of TGF-β1	[34]
p.o./Quinapril	0.5 mg/kg/day for 14 days	Rats	Blocked the increased blood pressure and injury against RAAS.	[35]
p.o./Irbesartan	20 mg/kg for 8 weeks	Rats	Significant in lowered the diastolic pressure	[36]
p.o./Telmisartan	10 mg/kg/day	Rats	Suppressed inflammation and help in lowering B.P.	[37]
p.o./Telmisartan	5 mg/kg/day for 8 weeks	Rats	Reduced the oxidative stress and meanwhile reduced the cardiovascular problems in diabetic patients.	[38]
p.o./Bisoprolol	5 mg/kg for 4 weeks	Rats	Helpful in lowered the diastolic and systolic pressure.	[39]
p.o./Bisoprolol	100 mg/kg for 3 weeks	Rats	Helpful in protection of heart during antihypertensive Treatment.	[40]
p.o./Atenolol	2 mg/kg	Rats	Showed the synergistic effects and helpful in lowering B.P.	[41]
Amlodipine	10 mg/kg/day for 19 weeks	Rats	Showed that high dose of drug shows the antihypertensive effects.	[42]

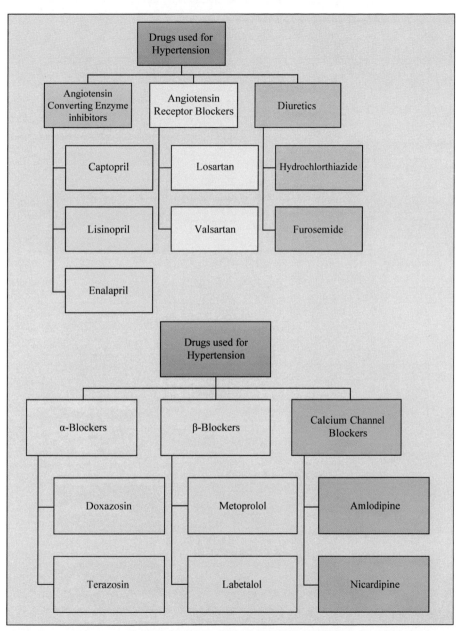

Fig. 2 Classification of drugs used for the management of hypertension.

growth, sodium and water retention sympathetic activation) and the angiotensin II type 2 (AT2) receptor. One disadvantage of ACE inhibitors is that the presence of non-ACE pathways results in continued low-level production of angiotensin II despite the inhibition of ACE [45]. Both systemic and local factors influence the renin-angiotensin system. The systemic process is triggered by the kidney's response

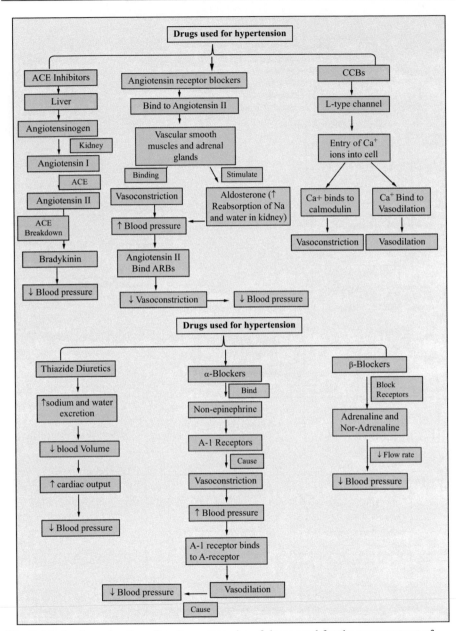

Fig. 3 Mechanism of action of various categories of drugs used for the management of hypertension.

to decreased effective blood volume and begins with the secretion of renin from the renal cortex. Angiotensinogen is broken down by renin after release to produce angiotensin I. Angiotensin-converting enzyme, which is primarily formed in the pulmonary vasculature, converts this product into angiotensin II. Research demonstrates that ACE inhibition attenuates the potent vasoconstrictor's effects on the tissues and systems it affects [46].

7.3 Calcium channel blockers

Although all members of the drug class have the same ability to interact with L-type voltage-dependent transmembrane calcium channels, CCBs exhibit a number of significant differences from the perspectives of pharmacokinetics and pharmacodynamic as well as selectivity and duration of pharmacological action [47]. Calcium channel blockers (CCBs) stop extracellular calcium from flowing via cell wall-spanning ion-specific channels. Although other varieties of these channels have been found, L-type channels in humans are inhibited by the CCBs that are now on the market. Inhibiting inward calcium flow causes vascular smooth muscle cells to relax, which lowers blood pressure and causes vasodilation (BP) [48]. CCBs are a diverse class of medications that can be categorized into two main groups depending on their predominate physiologic effects: (1) Dihydropyridines (DHPs), which bind L-type calcium preferentially and (2) non-DHPs (verapamil and diltiazem), which have equipotent effects on L-type calcium channels in the heart and the vasculature which preferentially bind calcium channels at the sinoatrial and atrioventricular node, which causes vasodilatation and lowers blood pressure (BP) [49].

7.4 α-Blockers

In the management of hypertension, α1-blockers offer a number of therapeutic benefits. First, adrenergic predominance of the sympathetic nervous system, which contributes to hypertension, is combated with α1-blockers. Second, α1-blockers have positive effects on insulin resistance and glucose tolerance, which in turn lowers blood sugar and cholesterol levels [50]. Despite their therapeutic utility, α1-blockers have not been used appropriately or enough in the treatment of hypertension due to safety concerns regarding heart failure and other potential side effects like orthostatic hypotension [51].

7.5 β-Blockers

Patients with hypertension who also have concurrent ischemic heart disease, heart failure, obstructive cardiomyopathy, or certain arrhythmias should be treated with b-blockers. The first b-blocker to be approved for use as an oral antihypertensive was propranolol. Pheochromocytoma was treated with propranolol as an additional therapy to phentolamine, α-adrenergic blocker [52]. Those with combined systolic and diastolic hypertension as well as the majority of those with isolated systolic hypertension will have lower blood pressure with b-blockers alone or in conjunction with other antihypertensive [53].

7.6   Diuretics

Diuretics were prescribed to 12% of US people in 2012, and the relative increase in prescriptions from 1999 to 2012 was 1.4, matching the rise in antihypertensive medications as a whole. Thiazide and thiazide-like diuretics are being prescribed at even higher rates (relative increase 1.7). The second most popular class of antihypertensive drugs given is thiazide and thiazide-like diuretics [54]. Resistant hypertension, which affects 5% of all people and is a significant cause of morbidity and death, must be managed with the use of diuretics. Diuretics that save potassium are probably underused. The SPRINT study supports an SBP aim of less than 120 mmHg in many patients since there are more obese and elderly patients who are salt-sensitive patients [55].

8   Nanotechnology as a boon for the management of hypertension

The nanotechnology approaches like solid lipid nanoparticles, nanostructured lipid carriers, and micelles have been investigated for the management of hypertension in recent years. Table 2 provides the detailed description about method of preparation, excipients used, in-vivo details, and outcomes and clinical significance of these nanocarriers.

8.1   Solid lipid nanoparticles

Solid lipid nanoparticles (SLNs) are introduced as an effective carrier approach for correcting dynamic medicine and water-soluble medicines. Colloidal particles between 10 and 1000 nm in size are considered nanoparticles. They are made of synthetic, unique polymers with improved drug delivery and decreased lethality as their main goals [75]. SLNs are colloidal carrier systems made up of an aqueous surfactant on top of a solid core of high melting point lipid. BCS Classes II and IV are the categories of drugs utilized in SLNs. Lipids include triglycerides, partial glycerides, fatty acids, hard fats, and waxes in a broad sense. A clear advantage of SLN is the fact that the lipid matrix is made up of physiological lipids, which lowers the risk of both acute and long-term toxicity [76]. Water, emulsifiers, coemulsifiers, and lipids (matrix materials) are frequently utilized elements in the creation of SLNs. To meet the demands of stability and targeting aspects, charge modifiers, stealthy chemicals that enhance extended circulation time and targeting ability are also used [77]. The hot homogenization technique and the cold homogenization technique are the two primary SLN production processes. In both methods, the medication is dissolved or solubilized in the lipid by heaing at its melting point.   For the hot homogenization method, a hot aqueous surfactant solution of the same temperature is used to stir a drug-containing melt into a uniform mixture. When the heated O/W nanoemulsion

Table 2 The nanotechnology-based approaches investigated for the management of hypertension.

Method of preparation	Excipients	Dose/Animal	Outcomes and clinical significance	References
Solid Lipid NPs				
Hot homogenization method	Glyceryl monostearate and Tween 80	1.023 mg/kg/rats	Study showed that solid lipid nanoparticles revealed higher C_{max} of 1610 ng/mL, higher AUC of 15,492.50 ng/mL and increased relative bioavailability by almost 2.3-fold compared to marketed formulation	[56]
Hot high-shear homogenization	Tween 80 and pluronic	10 mg/kg/rats	Effective in increasing the oral BA by 1.8 fold and also effective for sustained release	[57]
Ultrasonication method	Glyceryl monooleate, span 20		Study revealed that entrapment efficiency of SLN dispersion was found to be in the range of 72.50% to 86.90% and increased the poor water solubility of drug	[58]
Ultrasonication method	Poloxamer 188 and egg lecithin E80	10 mg/kg/rats	The pharmacodynamic research of SLNs in hypertensive rats revealed a drop in systolic blood pressure for 48h, whereas suspension exhibited a decrease in systolic blood pressure for just 2h, according to the study. The oral bioavailability was enhanced by 2.75-fold after inclusion into SLNs.	[59]
Homogenization method	Compritol, poloxamer and lecithin	240 mg/kg/rats	Study estimated that absolute oral bioavailability of nanoparticulate systems ranged from 10% to 90% and the relative bioavailability related to the AUC of NB was found to be 150.06% in PEG4-S19NB compared to commercial tablet.	[60]
Hot homogenization	Stearic acid, glyceryl monostearate	10 mg/kg/rats	Study showed that SLNs prepared with GMS having size of 188.6 ± 3.6 nm, PDI of 0.273 ± 0.052, ZP of −21.8 ± 2.7 mV with entrapment efficiency of 86.86 ± 0.75% which increase the poor aqueous solubility	[61]

Continued

Table 2 Continued

Method of preparation	Excipients	Dose/Animal	Outcomes and clinical significance	References
Film homogenization technique	Tween 80 and glycerol monostearate	10 mg/kg/rats	Study investigated that nanometer-sized spherical particles with high entrapment efficiency of 91.33% and increased oral bioavailability of lipophilic drug by 12-fold.	[62]
Ultrasonication method	Eudragit L100, soya lecithin		In comparison to ONps and drug solution, the oral bioavailability of ISR in ONbp was 3.21 and 4.45 times higher, respectively. The pharmacokinetic investigation revealed that coated SLN with Ru had an oral bioavailability that was between 3.2 and 4.7 times greater than that of a coated formulation without Ru and a normal drug solution. On the other hand, it was found that the coated SLN without Ru had a 1.00- to 1.11-fold higher bioavailability than drug solution.	[63]
Solvent evaporation	Poloxamer-188 and Compritol 888	10 mg/kg/rats	Results obtained from the oral pharmacokinetic study and showed that C_{max} values were found to increase significantly by 1.57 fold increase from 363.6 ± 23.5 ng/mL in free CND to 572.4 ± 25.3 ng/mL while the AUC0–∞ values increased by 2.41-fold	[64]
Homogenization-ultrasonication technique	Gelucire solution and stearylamine	1 mg/kg/rabbits	Prolonged the release for more than 23 h and enhanced the oral bioavailability by more than 2-fold	[65]
Nanostructured Lipid Carriers				
Melt emulsification technique	Precirol and PVA		Study demonstrated effective long-term release and high entrapment efficiency	[66]
Homogenization-ultrasonication and microemulsion	Pluronic F68, Tween 80	10 mg/kg/rats	Study showed AUC was 20% more than corresponding suspension also the complete release and higher stability rate than SLNs.	[67]

Method	Components	Dose	Findings	Ref.
Ultrasonication method	Glyceryl monostearate, oleic acid, Tween 20	3.53 mg/kg/rats	According to a pharmacokinetic investigation, the relative bioavailability of the telmisartan-loaded NLC was 2.17 and 3.46 times higher than that of the commercial formulation and pure drug solution.	[68]
Hot high-shear homogenization	Compritol ATO 888, oleic acid, Tween 80, poloxamer 188		Investigated that verapamil cellular uptake from the VER-9 formulation was 10.90- and 134.91-fold higher than the verapamil solution and verapamil-dextran complex which played significant role in improving the absorption of the drug.	[69]
Hot homogenization-ultrasonication technique	Lutrol F127 and Cremophor	10 mg/kg/rats	Cellular uptake research suggests that NLCs should be taken up by enterocytes and then delivered into the bloodstream. The pharmacokinetic data showed that encapsulation significantly increased the oral bioavailability by more than twofold	[70]
Probe sonication technique	Tween 80 and PEG4000	20 mg/kg/rats	Study showed EE and loading in between 69.45% and 88.56% and 9.58%–12.56% while relative bioavailability increased by 3.95-fold as compared to the suspension	[71]
Micelles				
Solvent evaporation	Poly(ethylene glycol) methyl ether-*block*-poly(ε-caprolactone)		Study showed that cumulative release of amlodipine-loaded micelles using the larger membrane resulted in a ~2.5-fold increase in release rate, compared to free amlodipine and higher solubility achieved with micellar encapsulation will likely lead to increased bioavailability	[72]
Direct dissolution method	Monomethoxy poly(ethylene glycol), tetrahydrofuran		Valsartan was loaded into empty micelles with a high encapsulation effectiveness of up to 84% and in vitro drug release is characterized by a burst release of up to 52% in 12h followed by a gradual release.	[73]
Thin-film hydration method	Pluronic F127 and P123		Study showed a good stability and the highest release efficiency of the drug.	[74]

is cooled to room temperature after being homogenized with a piston-gap homogenizer (such as the Micron LAB40), the lipid recrystallizes and forms solid lipid nanoparticles [78].

8.2 Nanostructured lipid carriers

It was discovered that adding a liquid lipid to the solid matrix of the nanoparticle increased the number of defects in the solid matrix, making it easier to incorporate more medication while maintaining the physical stability of the nanocarriers. As nanostructured lipid carriers, this new unstructured-matrix SLN was created. The number of papers using NLC formulations has significantly expanded since NLCs were first described as freshly created drug delivery methods [79]. NLCs adopt combinations of liquid and solid lipids and maintain their solid state by regulating the liquid lipid concentration. NLCs, as opposed to emulsions, can more effectively immobilize drugs and stop the particle from coalescing because of the solid matrix. The benefits of SLNs such as low toxicity, biodegradation, drug protection, gradual release, and avoiding the use of organic solvents in manufacturing are also present in NLCs [80]. Several techniques including high-pressure homogenization, emulsion/solvent evaporation, phase inversion, and microemulsion have been used to create SLNs and NLCs [81]. For the administration of medications via oral, parenteral, ophthalmic, pulmonary, topical, and transdermal routes, NLCs have emerged as a viable carrier system. Lately, NLCs have also been used in the delivery of cosmeceuticals, nutraceuticals, chemotherapy, gene therapy, and food products [82].

8.3 Micelles

The size of PMs, which are spherical colloidal particles, typically ranges from 10 to 100 nm. They are made of hydrophobic and hydrophilic amphiphilic copolymers with unique building units. Amphiphilic triblock (hydrophilic-hydrophobic-hydrophilic) or diblock (hydrophilic-hydrophobic-hydrophilic) polymers are the most common types of these copolymers. PEG, which has a molecular weight range of 2 to 15 kDa, is the most often used hydrophilic building block, whereas polyesters, polyether or polyamino acids like poly(l-aspartic acid), PLA, poly(-caprolactone) (PCL), and poly(propylene oxide) are generally employed as hydrophobic building blocks [83]. The core-shell structure of polymeric micelles, which are nanosized drug delivery devices, was created via the self-assembly of amphiphilic block copolymers in aqueous solution. Amphiphilic molecules coexist separately in diluted aqueous solutions, and they serve as surfactants that lower surface tension at the air-water interface. As more chains are introduced to the system, the adsorption at the interface increases until unimers aggregate because the bulk solution is saturated [84]. Micelles are self-assembling, hydrophobic-cored, hydrophilic-shelled nanoscale colloidal particles. For poorly water soluble medicines, which make up about 25% of traditional, commercially available treatments and close to 50% of candidates found through screening methods, micelles unique structure makes them appropriate carriers [85]. The hydrophilic shell of the micelle has significantly increased the solubility of medications, and due to the tunable size of micelles, pharmaceuticals can be targeted to areas with increased permeability, such as tumor and inflammatory tissue. Furthermore, micelles can achieve greater tissue selectivity and cellular absorption when modified by functional molecules that sense chemical cues particular to damaged areas [86].

9   Conclusion and future perspectives

For therapy of hypertension, a deep clinical evaluation and study are needed. For the treatment of hypertension, β-blockers, ACE-inhibitors, ARBs, α-blockers, CCBs, and diuretics are part of the treatment. Nanotechnology due to its small particle size, high specific surface area, many active centers, higher surface reactivity, and superior adsorption capacity is used to improve the efficacy and bioavailability. By enabling the precise delivery of nanodrugs into target cells, it avoids interfering with the physiological processes in other organs. The problem of enhancing the bioavailability and efficiency of further drugs whose nanoformulations are not formed awaits a solution.

Acknowledgments

The authors would like to thank Department of Pharmaceutics, MM College of Pharmacy, Maharishi Markandeshwar (Deemed to be University), Mullana-Ambala, Haryana, India 133207.

References

[1] Singh S, Shankar R, Singh GP. Prevalence and associated risk factors of hypertension: a cross-sectional study in urban Varanasi. Int J Hypertens 2017;2017. https://doi.org/10.1155/2017/5491838.

[2] Kumar MR, Shankar R, Singh S. Hypertension among the adults in rural Varanasi: a cross sectional study on prevalence and health seeking behaviour. Indian J Prev Soc Med 2016;47(1–2):6.

[3] Kaplan NM. Kaplan's clinical hypertension. Lippincott Williams & Wilkins; 2010.

[4] Oparil S, Schmieder RE. New approaches in the treatment of hypertension. Circ Res 2015;116(6):1074–95.

[5] Ilyas M. Hypertension in adults: part 1. Prevalence, types, causes and effects. South Sudan Med J 2009;2(3):9–10.

[6] Delacroix S, Chokka RG, Worthley SG. Hypertension: pathophysiology and treatment. J Neurol Neurophysiol 2014;5(6):1–8. https://doi.org/10.4172/2155-9562.1000250.

[7] Kishore J, Gupta N, Kohli C, Kumar N. Prevalence of hypertension and determination of its risk factors in rural Delhi. Int J Hypertens 2016;2016.

[8] Zhou B, Carrillo-Larco RM, Danaei G, Riley LM, Paciorek CJ, Stevens GA, Gregg EW, Bennett JE, Solomon B, Singleton RK, Sophiea MK. Worldwide trends in hypertension prevalence and progress in treatment and control from 1990 to 2019: a pooled analysis of 1201 population-representative studies with 104 million participants. Lancet 2021;398(10304):957–80. https://doi.org/10.1016/S0140-6736(21)01330-1.

[9] Gupta R, Mohan I, Narula J. Trends in coronary heart disease epidemiology in India. Ann Glob Health 2016;82(2):307–15. https://doi.org/10.1016/j.aogh.2016.04.002.

[10] Patrick DM, Van Beusecum JP, Kirabo A. The role of inflammation in hypertension: novel concepts. Curr Opin Physio 2021;19:92–8. https://doi.org/10.1016/j.cophys.2020.09.016.

[11] Kiefer FN, Neysari S, Humar R, Li W, Munk VC, Battegay EJ. Hypertension and angiogenesis. Curr Pharm Des 2003;9(21):1733–44.

[12] Baradaran A, Nasri H, Rafieian-Kopaei M. Oxidative stress and hypertension: possibility of hypertension therapy with antioxidants. J Res Med Sci 2014;19(4):358.

[13] Weber MA, Schiffrin EL, White WB, Mann S, Lindholm LH, Kenerson JG, Flack JM, Carter BL, Materson BJ, Ram CV, Cohen DL. Clinical practice guidelines for the management of hypertension in the community: a statement by the American Society of Hypertension and the International Society of Hypertension. J Clin Hypertens 2014;16 (1):14. https://doi.org/10.1097/HJH.0000000000000065.

[14] Carretero OA, Oparil S. Essential hypertension: part II: treatment. Circulation 2000;101 (4):446–53. https://doi.org/10.1161/01.cir.101.4.446.

[15] Sarzani R, Salvi F, Dessì-Fulgheri P, Rappelli A. Renin–angiotensin system, natriuretic peptides, obesity, metabolic syndrome, and hypertension: an integrated view in humans. J Hypertens 2008;26(5):831–43. https://doi.org/10.1097/HJH.0b013e3282f624a0.

[16] Keane MP, Strieter RM. Chemokine signaling in inflammation. Crit Care Med 2000;28 (4):N13–26.

[17] Norlander AE, Madhur MS, Harrison DG. The immunology of hypertension. J Exp Med 2018;215(1):21–33. https://doi.org/10.1084/jem.20171773.

[18] Kirabo A, Fontana V, De Faria AP, Loperena R, Galindo CL, Wu J, Bikineyeva AT, Dikalov S, Xiao L, Chen W, Saleh MA. DC isoketal-modified proteins activate T cells and promote hypertension. J Clin Invest 2014;124(10):4642–56. https://doi.org/10.1172/JCI74084.

[19] Pasceri V, Willerson JT, Yeh ET. Direct proinflammatory effect of C-reactive protein on human endothelial cells. Circulation 2000;102(18):2165–8.

[20] Levy BI, Ambrosio G, Pries AR, Struijker-Boudier HA. Microcirculation in hypertension: a new target for treatment? Circulation 2001;104(6):735–40. Le Noble FA, Stassen FR, Hacking WJ, Struilker Boudier HA. Angiogenesis and hypertension. J Hypertens 1998;16:1563–72.

[21] Carmeliet P. Mechanisms of angiogenesis and arteriogenesis. Nat Med 2000;6(4):389–95.

[22] Chung NA, Lydakis C, Belgore F, Blann AD, Lip GY. Angiogenesis in myocardial infarction. An acute or chronic process? Eur Heart J 2002;23(20):1604–8.

[23] Kimura S, Zhang GX, Nishiyama A, Shokoji T, Yao L, Fan YY, Rahman M, Abe Y. -Mitochondria-derived reactive oxygen species and vascular MAP kinases: comparison of angiotensin II and diazoxide. Hypertension 2005;45(3):438–44.

[24] Ceriello A. Possible role of oxidative stress in the pathogenesis of hypertension. Diabetes Care 2008;31(Suppl 2):S181–4.

[25] Touyz RM. Reactive oxygen species, vascular oxidative stress, and redox signaling in hypertension: what is the clinical significance? Hypertension 2004;44(3):248–52.

[26] Luo HL, Zang WJ, Lu J, Yu XJ, Lin YX, Cao YX. The protective effect of captopril on nicotine-induced endothelial dysfunction in rat. Basic Clin Pharmacol Toxicol 2006;99 (3):237–45.

[27] Siska S, Munim A, Bahtiar A, Suyatna FD. Effect of *Apium graveolens* extract administration on the pharmacokinetics of captopril in the plasma of rats. Sci Pharm 2018;86 (1):6. https://doi.org/10.3390/scipharm86010006.

[28] Wu PC, Huang YB, Chang JJ, Chang JS, Tsai YH. Evaluation of pharmacokinetics and pharmacodynamics of captopril from transdermal hydrophilic gels in normotensive rabbits and spontaneously hypertensive rats. Int J Pharm 2000;209(1–2):87–94. https://doi.org/ 10.1016/s0378-5173(00)00557-3.

[29] Miguel-Carrasco JL, Monserrat MT, Mate A, Vázquez CM. Comparative effects of captopril and l-carnitine on blood pressure and antioxidant enzyme gene expression in the heart of spontaneously hypertensive rats. Eur J Pharmacol 2010;632(1–3):65–72. https://doi.org/10.1016/j.ejphar.2010.01.017.

[30] Sharma JN, Kesavarao U. Effect of captopril on urinary kallikrein, blood pressure and myocardial hypertrophy in diabetic spontaneously hypertensive rats. Pharmacology 2002;64(4):196–200. https://doi.org/10.1159/000056171.

[31] Zicha J, Dobešová Z, Kuneš J. Antihypertensive mechanisms of chronic captopril or N-acetylcysteine treatment in L-NAME hypertensive rats. Hypertens Res 2006;29(12): 1021–7.

[32] Morsy MA. Protective effect of lisinopril on hepatic ischemia/reperfusion injury in rats. Indian J Pharmacol 2011;43(6):652.

[33] Ji X, Tan BK, Zhu YC, Linz W, Zhu YZ. Comparison of cardioprotective effects using ramipril and DanShen for the treatment of acute myocardial infarction in rats. Life Sci 2003;73(11):1413–26. https://doi.org/10.1016/s0024-3205(03)00432-6.

[34] Shirai K, Watanabe K, Ma M, Wahed MI, Inoue M, Saito Y, Suresh PS, Kashimura T, Tachikawa H, Kodama M, Aizawa Y. Effects of angiotensin-II receptor blocker candesartan cilexetil in rats with dilated cardiomyopathy. Mol Cell Biochem 2005;269(1): 137–42.

[35] Watanabe K, Juan W, Narasimman G, Ma M, Inoue M, Saito Y, Wahed MI, Nakazawa M, Hasegawa G, Naito M, Tachikawa H. Comparative effects of angiotensin II receptor blockade (candesartan) with angiotensin-converting enzyme inhibitor (quinapril) in rats with dilated cardiomyopathy. J Cardiovasc Pharmacol 2003;41:S93–7.

[36] Cerbai E, De Paoli P, Sartiani L, Lonardo G, Mugelli A. Treatment with irbesartan counteracts the functional remodeling of ventricular myocytes from hypertensive rats. J Cardiovasc Pharmacol 2003;41(5):804–12.

[37] Sukumaran V, Veeraveedu PT, Gurusamy N, Yamaguchi KI, Lakshmanan AP, Ma M, Suzuki K, Kodama M, Watanabe K. Cardioprotective effects of telmisartan against heart failure in rats induced by experimental autoimmune myocarditis through the modulation of angiotensin-converting enzyme-2/angiotensin 1-7/mas receptor axis. Int J Biol Sci 2011;7(8):1077.

[38] Goyal BR, Parmar K, Goyal RK, Mehta AA. Beneficial role of telmisartan on cardiovascular complications associated with STZ-induced type 2 diabetes in rats. Pharmacol Rep 2011;63(4):956–66.

[39] Andika M, Arifin H, Rivai H. Effect of bisoprolol against reduction of systolic and diastolic blood pressure in hypertension white rat with hypercholesterolemia complications. World J Pharm Pharm Sci 2020;9:122.

[40] Mougenot N, Médiani O, Lechat P. Bisoprolol and hydrochlorothiazide effects on cardiovascular remodeling in spontaneously hypertensive rats. Pharmacol Res 2005;51(4): 359–65. https://doi.org/10.1016/j.phrs.2004.10.010.

[41] Xu LP, Shen FM, Shu H, Miao CY, Jiang YY, Su DF. Synergism of atenolol and amlodipine on lowering and stabilizing blood pressure in spontaneously hypertensive rats. Fundam Clin Pharmacol 2004;18(1):33–8. https://doi.org/10.1111/j.1472-8206.2004. 00200.x.

[42] Nishikawa N, Masuyama T, Yamamoto K, Sakata Y, Mano T, Miwa T, Sugawara M, Hori M. Long-term administration of amlodipine prevents decompensation to diastolic heart failure in hypertensive rats. J Am Coll Cardiol 2001;38(5):1539–45.

[43] Ghezelbash S, Ghorbani A. Lifestyle modification and hypertension prevention. ARYA Atheroscler J 2012;8(Special Issue):1–6.

[44] Samadian F, Dalili N, Jamalian A. Lifestyle modifications to prevent and control hypertension. Iran J Kidney Dis 2016;10(5).

[45] Messerli FH, Bangalore S, Bavishi C, Rimoldi SF. Angiotensin-converting enzyme inhibitors in hypertension: to use or not to use? J Am Coll Cardiol 2018;71(13):1474–82.

[46] Bicket DP. Using ACE inhibitors appropriately. Am Fam Physician 2002;66(3):461.

[47] Tocci G, Battistoni A, Passerini J, Musumeci MB, Francia P, Ferrucci A, Volpe M. Calcium channel blockers and hypertension. J Cardiovasc Pharmacol Ther 2015;20 (2):121–30.

[48] Elliott WJ, Ram CV. Calcium channel blockers. J Clin Hypertens 2011;13(9):687.

[49] Basile J. The role of existing and newer calcium channel blockers in the treatment of hypertension. J Clin Hypertens 2004;6(11):621–9.

[50] Li H, Xu TY, Li Y, Chia YC, Buranakitjaroen P, Cheng HM, Van Huynh M, Sogunuru GP, Tay JC, Wang TD, Kario K. Role of α1-blockers in the current management of hypertension. J Clin Hypertens 2022;24(9):1180–6.

[51] Williams B, Mancia G, Spiering W, Agabiti Rosei E, Azizi M, Burnier M, Clement DL, Coca A, De Simone G, Dominiczak A, Kahan T. 2018 ESC/ESH guidelines for the management of arterial hypertension: the task force for the management of arterial hypertension of the European Society of Cardiology (ESC) and the European Society of Hypertension (ESH). Eur Heart J 2018;39(33):3021–104.

[52] Frishman WH, Saunders E. β-adrenergic blockers. J Clin Hypertens 2011;13(9):649.

[53] Dodson PM. Hypertension and diabetes. Curr Med Res Opin 2002;18(sup1):s48–57.

[54] Roush GC, Sica DA. Diuretics for hypertension: a review and update. Am J Hypertens 2016;29(10):1130–7.

[55] Krakoff LR. Diuretics for hypertension. Circulation 2005;112(10):e127–9.

[56] Pandya NT, Jani P, Vanza J, Tandel H. Solid lipid nanoparticles as an efficient drug delivery system of olmesartan medoxomil for the treatment of hypertension. Colloids Surf B Biointerfaces 2018;165:37–44. https://doi.org/10.1016/j.colsurfb.2018.02.011.

[57] Cirri M, Mennini N, Maestrelli F, Mura P, Ghelardini C, Mannelli LD. Development and in vivo evaluation of an innovative "hydrochlorothiazide-in cyclodextrins-in solid lipid nanoparticles" formulation with sustained release and enhanced oral bioavailability for potential hypertension treatment in pediatrics. Int J Pharm 2017;521(1–2):73–83. https://doi.org/10.1016/j.ijpharm.2017.02.022.

[58] Ekambaram P, Sathali AA. Formulation and evaluation of solid lipid nanoparticles of ramipril. J Young Pharm 2011;3(3):216–20. https://doi.org/10.4103/0975-1483.83765.

[59] Dudhipala N, Veerabrahma K. Candesartan cilexetil loaded solid lipid nanoparticles for oral delivery: characterization, pharmacokinetic and pharmacodynamic evaluation. Drug Deliv 2016;23(2):395–404.

[60] Üstündağ-Okur N, Yurdasiper A, Gündoğdu E, Homan GE. Modification of solid lipid nanoparticles loaded with nebivolol hydrochloride for improvement of oral bioavailability in treatment of hypertension: polyethylene glycol versus chitosan oligosaccharide lactate. J Microencapsul 2016;33(1):30–42. https://doi.org/10.3109/02652048.2015. 1094532.

[61] Thirupathi G, Swetha E, Narendar D. Role of isradipine loaded solid lipid nanoparticles on the pharmacodynamic effect in rats. Drug Res 2017;67(03):163–9.

[62] Zhang Z, Gao F, Bu H, Xiao J, Li Y. Solid lipid nanoparticles loading candesartan cilexetil enhance oral bioavailability: in vitro characteristics and absorption mechanism in rats. Nanomed Nanotechnol Biol Med 2012;8(5):740–7. https://doi.org/10.1016/j.nano.2011.08.016.

[63] Kumar V, Chaudhary H, Kamboj A. Development and evaluation of isradipine via rutin-loaded coated solid–lipid nanoparticles. Intervention Med Appl Sci 2018;10(4):236–46. https://doi.org/10.1556/1646.10.2018.45.

[64] Diwan R, Ravi PR, Pathare NS, Aggarwal V. Pharmacodynamic, pharmacokinetic and physical characterization of cilnidipine loaded solid lipid nanoparticles for oral delivery optimized using the principles of design of experiments. Colloids Surf B Biointerfaces 2020;193, 111073. https://doi.org/10.1016/j.colsurfb.2020.111073.

[65] El-Say KM, Hosny KM. Optimization of carvedilol solid lipid nanoparticles: an approach to control the release and enhance the oral bioavailability on rabbits. PloS One 2018;13(8), e0203405. https://doi.org/10.1371/journal.pone.0203405.

[66] Nafee N, Makled S, Boraie N. Nanostructured lipid carriers versus solid lipid nanoparticles for the potential treatment of pulmonary hypertension via nebulization. Eur J Pharm Sci 2018;125:151–62. https://doi.org/10.1016/j.ejps.2018.10.003.

[67] Cirri M, Maestrini L, Maestrelli F, Mennini N, Mura P, Ghelardini C, Di Cesare ML. Design, characterization and in vivo evaluation of nanostructured lipid carriers (NLC) as a new drug delivery system for hydrochlorothiazide oral administration in pediatric therapy. Drug Deliv 2018;25(1):1910–21. https://doi.org/10.1080/10717544.2018.1529209.

[68] Thapa C, Ahad A, Aqil M, Imam SS, Sultana Y. Formulation and optimization of nano-structured lipid carriers to enhance oral bioavailability of telmisartan using Box–Behnken design. J Drug Deliv Sci Technol 2018;44:431–9. https://doi.org/10.1016/j.jddst.2018.02.003.

[69] Khan AA, Abdulbaqi IM, Abou Assi R, Murugaiyah V, Darwis Y. Lyophilized hybrid nanostructured lipid carriers to enhance the cellular uptake of verapamil: statistical optimization and in vitro evaluation. Nanoscale Res Lett 2018;13(1):1–6.

[70] Anwar W, Dawaba HM, Afouna MI, Samy AM, Rashed MH, Abdelaziz AE. Enhancing the oral bioavailability of candesartan cilexetil loaded nanostructured lipid carriers: in vitro characterization and absorption in rats after oral administration. Pharmaceutics 2020;12(11):1047.

[71] Mishra A, Imam SS, Aqil M, Ahad A, Sultana Y, Ameeduzzafar AA. Carvedilol nano lipid carriers: formulation, characterization and in-vivo evaluation. Drug Deliv 2016;23(4):1486–94. https://doi.org/10.3109/10717544.2016.1165314.

[72] Di Trani N, Liu HC, Qi R, Viswanath DI, Liu X, Chua CY, Grattoni A. Long-acting tunable release of amlodipine loaded PEG-PCL micelles for tailored treatment of chronic hypertension. Nanomed Nanotechnol Biol Med 2021;37, 102417.

[73] Zhu Q, Zhang B, Wang Y, Liu X, Li W, Su F, Li S. Self-assembled micelles prepared from poly (D, L-lactide-co-glycolide)-poly (ethylene glycol) block copolymers for sustained release of valsartan. Polym Adv Technol 2021;32(3):1262–71. https://doi.org/10.1002/pat.5175.

[74] El-Gendy MA, El-Assal MI, Tadros MI, El-Gazayerly ON. Olmesartan medoxomil-loaded mixed micelles: preparation, characterization and in-vitro evaluation. Future J Pharm Sci 2017;3(2):90–4. https://doi.org/10.1016/j.fjps.2017.04.001.

[75] Nikdouz A, Namarvari N, Shayan RG, Hosseini A. Comprehensive comparison of theranostic nanoparticles in breast cancer. Am J Clin Exp Immunol 2022;11(1):1.

[76] Yadav N, Khatak S, Sara US. Solid lipid nanoparticles-a review. Int J Appl Pharm 2013;5 (2):8–18.

[77] Manjunath K, Reddy JS, Venkateswarlu V. Solid lipid nanoparticles as drug delivery systems. Methods Find Exp Clin Pharmacol 2005;27(2):127–44.

[78] Müller RH, Mäder K, Gohla S. Solid lipid nanoparticles (SLN) for controlled drug delivery – a review of the state of the art. Eur J Pharm Biopharm 2000;50(1):161–77.

[79] Beloqui A, Solinís MÁ, Rodríguez-Gascón A, Almeida AJ, Préat V. Nanostructured lipid carriers: promising drug delivery systems for future clinics. Nanomed Nanotechnol Biol Med 2016;12(1):143–61.

[80] Jaiswal P, Gidwani B, Vyas A. Nanostructured lipid carriers and their current application in targeted drug delivery. Artif Cells Nanomed Biotechnol 2016;44(1):27–40.

[81] Duong VA, Nguyen TT, Maeng HJ. Preparation of solid lipid nanoparticles and nanostructured lipid carriers for drug delivery and the effects of preparation parameters of solvent injection method. Molecules 2020;25(20):4781.

[82] Jaiswal P, Gidwani B, Vyas A. Nanostructured lipid carriers and their current application in targeted drug delivery. Artif Cells Nanomed Biotechnol 2016;44(1):27–40.

[83] Sutton D, Nasongkla N, Blanco E, Gao J. Functionalized micellar systems for cancer targeted drug delivery. Pharm Res 2007;24:1029–46.

[84] Ghezzi M, Pescina S, Padula C, Santi P, Del Favero E, Cantù L, Nicoli S. Polymeric micelles in drug delivery: an insight of the techniques for their characterization and assessment in biorelevant conditions. J Control Release 2021;332:312–36.

[85] Zhang N, Wardwell PR, Bader RA. Polysaccharide-based micelles for drug delivery. Pharmaceutics 2013;5(2):329–52.

[86] Xiong XB, Falamarzian A, Garg SM, Lavasanifar A. Engineering of amphiphilic block copolymers for polymeric micellar drug and gene delivery. J Control Release 2011;155(2): 248–61.

Decrypting the role of angiogenesis, inflammation, and oxidative stress in pathogenesis of congestive heart failure: Nanotechnology as a boon for the management of congestive heart failure

8

Neelam Sharma[a], Shahid Nazir Wani[b,c], Sukhbir Singh[a], Ishrat Zahoor[a], Tapan Behl[d], and Irfan Ahmad Malik[c]
[a]Department of Pharmaceutics, MM College of Pharmacy, Maharishi Markandeshwar (Deemed to be University), Mullana-Ambala, Haryana, India, [b]Chitkara College of Pharmacy, Chitkara University, Rajpura, Punjab, India, [c]Aman Pharmacy College, Jhunjhunu, Rajasthan, India, [d]Amity School of Pharmaceutical Sciences, Amity University, Mohali, Punjab, India

1 Introduction

Congestive heart failure (CHF) is a complicated clinical condition that emerges as a result of a structural or functional heart disorder that impedes the capacity of the ventricles to either fill with blood or to evacuate blood into the systemic circulation. This causes the ventricles to be unable to pump blood effectively, which leads to the development of heart failure (HF). The distinguishing aspect of the condition is that it is characterized by an inability to meet the demands that are imposed on the system by circulation [1]. A clinical state in which the quantity of blood pumped by the heart was inadequate to fulfil the needs of the different organ systems was the traditional definition of heart failure throughout a significant portion of medical history [2]. Due to anatomical and/or functional problems of the heart, heart failure is now described in a more comprehensive manner as a clinical illness that is marked by dyspnea and tiredness, whether at rest or with effort. In order for there to be a symptom known as heart failure, the heart itself has to have some kind of underlying issue [2]. Heart failure continues to be a disorder that is very common all over the globe and has a significant morbidity and death rate. Its prevalence is estimated to be 26 million individuals all over the globe, and it adds to the rise in the expense of healthcare everywhere [1]. Heart failure may be caused by a number of conditions or illnesses. However, the

Targeting Angiogenesis, Inflammation and Oxidative Stress in Chronic Diseases. https://doi.org/10.1016/B978-0-443-13587-3.00009-6

majority of therapy suggestions are centered on the existence of heart failure alone, instead of the root cause. This is because the etiology of heart failure affects the treatment plan to some degree, but only to a limited degree. There is not a single cause of heart failure; rather, the pathophysiology of the condition is rather complicated. Heart failure may be brought on by a variety of distinct forms of heart disease. The majority of cases of heart failure can be traced back to well-established etiologic variables; nevertheless, even individuals who seem to be in good condition can be hiding risk factors for the eventual development of heart failure [2]. Reduced left ventricular myocardial function is the most prevalent cause of HF; however, malfunction of the pericardium, myocardium, endocardium, heart valves, or great vessels alone or in combination may also be linked with HF [3]. Increased hemodynamic overload, ischemia-related dysfunction, ventricular remodeling, excessive neurohumoral stimulation, abnormal myocyte calcium cycling, excessive or inadequate proliferation of the extracellular matrix, accelerated apoptosis, and genetic mutations are some of the major pathogenic mechanisms that lead to heart failure [3,4]. Other major pathogenic mechanisms include genetic mutations. More than one million individuals are hospitalized to hospitals in the United States each year because of the condition known as CHF [5,6]. This makes CHF the disorder that leads to the highest hospitalizations of adults aged 65 and over in the United States. Despite advances in diagnosis and treatment, CHF is still responsible for over 300,000 deaths per year in the United States. In addition, individuals admitted to the hospital with an acute exacerbation of CHF have a mortality rate that may reach as high as 12% during their stay [6]. Patients who have acute decompensated heart failure (ADHF) regularly visit the emergency department for examination. A significant number of these patients are in a serious condition and need prompt medical attention. The development and severity of heart failure are connected to the morbidity and mortality that comes with it. As a result, therapeutic strategies must to be developed not only to alleviate symptoms but also to slow down the course of heart failure. A malfunction in the left ventricle's systolic pumping action is what starts off the course of heart failure. It would seem that a certain minimum level of left ventricular dysfunction is necessary in order to start the remodeling process in the left ventricle [7]. The gradual expansion of the left ventricular size that is linked with a rise in end-diastolic and end-systolic volumes and a reduction in the left ventricular ejection fraction are the defining characteristics of left ventricular remodeling. The threshold degree of left ventricular systolic dysfunction that commences ventricular remodeling in individual patients may vary; however, in general, it seems that left ventricular remodeling is triggered in most patients if the left ventricular ejection fraction is lowered to 40% or less [7]. Because of this, it has a significant amount of promise for use in the management of HF in order to address the primary physiological abnormalities that are brought on by HF. This promise might be further enhanced by a delivery strategy based on nanotechnology, which would allow for the targeted delivery of payloads to diseased regions. In addition, the pathophysiological changes that separate disorders from healthy parts of the body might act as biological triggers to "switch on" nanomedicine in the body. This ensures that the nanomedicine is only able to give its therapeutic effect at the problematic site [8].

2   Classification of heart failure

Heart failure can be classified based on several factors, including side of the heart affected, course of the disease, and ejection fraction. Here are some common ways that HF is classified in Fig. 1.

2.1   Based on the side of the heart affected

2.1.1   Left heart failure

Left heart failure (LHF) is a condition where the left side of the heart is unable to pump blood effectively to the rest of the body. The left side of the heart is responsible for receiving oxygen-rich blood from the lungs and pumping it out to the rest of the body. When LHF occurs, the heart is unable to meet the body's demand for oxygenated blood, leading to a variety of symptoms [9].

There are two types of LHF: systolic heart failure and diastolic heart failure.

Systolic Heart Failure: Systolic heart failure occurs when the left ventricle of the heart is unable to contract effectively, resulting in reduced blood flow to the body. This can be caused by a number of factors, including coronary artery disease, hypertension, and cardiomyopathy. When the heart is unable to contract properly, it is unable to pump enough blood to the body, which can lead to a buildup of fluid in

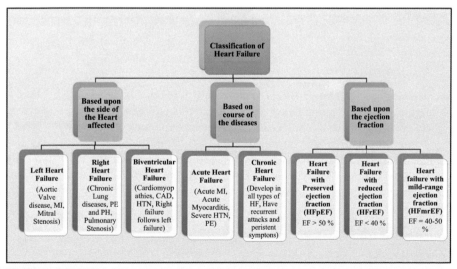

Fig. 1 Classification of congestive heart failure and associated conditions. *MI*, myocardial infarction; *PE*, pulmonary embolism; *PH*, pulmonary hypertension; *CAD*, coronary artery disease; *HTN*, hypertension; *HFpEF*, heart failure with preserved ejection fraction; *HFrEF*, heart failure with reduced ejection fraction; *HFmrEF*, heart failure with mild range ejection fraction.

the lungs, a condition known as pulmonary edema. Symptoms of systolic heart failure may include: shortness of breath, especially during exercise or when lying down, fatigue and weakness, swelling in the legs and ankles, rapid or irregular heartbeat, persistent coughing or wheezing, reduced ability to exercise or do physical activity [10].

Diastolic Heart Failure: Diastolic heart failure occurs when the left ventricle of the heart is unable to relax properly, making it difficult for the heart to fill with enough blood to pump to the body. This can be caused by conditions such as hypertension, diabetes, and obesity. When the heart is unable to relax properly, it is unable to fill with enough blood to pump to the body, which can lead to a buildup of fluid in the lungs. Symptoms of diastolic heart failure may include: shortness of breath, especially during exercise or when lying down, fatigue and weakness, swelling in the legs and ankles, rapid or irregular heartbeat, and difficulty breathing during physical activity [10].

2.1.2   Right heart failure

Right heart failure (RHF) is a condition where the right side of the heart is unable to pump blood effectively. RHF can be acute or chronic, and is typically caused by underlying medical conditions such as pulmonary hypertension, chronic obstructive pulmonary disease (COPD), or LHF. Symptoms of RHF may include shortness of breath, fatigue, swelling in the legs and abdomen, and reduced ability to exercise. Treatment for RHF typically involves treating the underlying medical condition, as well as medications to improve heart function and manage symptoms. In severe cases, surgical interventions such as heart transplant or mechanical circulatory support may be necessary [11,12].

2.1.3   Biventricular heart failure (BiHF)

Biventricular heart failure (BiHF) is a condition where both the left and right sides of the heart are unable to pump blood effectively, and is typically caused by underlying medical conditions such as coronary artery disease, hypertension, or cardiomyopathy [13]. Symptoms of BiHF may include shortness of breath, fatigue, swelling, rapid or irregular heartbeat, and reduced ability to exercise. Treatment for BiHF typically involves medications, lifestyle changes, and in severe cases, surgical interventions such as heart transplant or mechanical circulatory support.

2.2   Based on the course of the disease

2.2.1   Acute heart failure

Acute heart failure (AHF) is a sudden onset of heart failure symptoms that usually require urgent medical attention. It can occur in individuals with no prior history of heart disease or in those with underlying chronic heart failure. AHF can present in several different forms, including pulmonary edema, cardiogenic shock, and ADHF. Symptoms of AHF may include shortness of breath, rapid heartbeat, swelling in the legs and ankles, chest pain, fainting, fatigue, and persistent coughing or wheezing. Treatment for AHF typically involves medications to improve heart function and manage symptoms, such as diuretics and vasodilators, as well as interventions to remove excess fluid from the body. In severe cases, mechanical circulatory support may be necessary. It is important to seek immediate medical attention if you are experiencing symptoms of AHF, as prompt treatment can improve outcomes and prevent complications.

AHF can have a significant impact on an individual's quality of life, and it is important for individuals to work closely with their healthcare providers to manage their condition and prevent future episodes. Lifestyle changes such as a heart-healthy diet, regular exercise, and avoiding smoking and excessive alcohol consumption can also help improve outcomes for individuals with AHF [14].

2.2.2   Chronic heart failure

Chronic heart failure, also referred to as congestive heart failure or heart failure, is a continuous condition characterized by the heart's inability to pump enough blood throughout the body, leading to insufficient oxygen supply. Risk factors for the condition include advanced age, diabetes, high blood pressure, and obesity. While the name of the condition may be confusing, it does not indicate that the heart has stopped or is entirely failing, but that it is weaker than a healthy heart. Symptoms range from mild to severe and can include shortness of breath, fatigue, chest pain, cough, and swelling of the legs and ankles. It is most common in older adults and those with pre-existing heart conditions and is typically managed through lifestyle changes, medication, and diet modifications. Chronic heart failure is a progressive condition that worsens over time and cannot be cured, but its symptoms can be effectively managed. In contrast to acute heart failure, which develops rapidly, chronic heart failure is a long-term condition [14].

2.3   Based on the ejection fraction

2.3.1   Heart failure with preserved ejection fraction (HFpEF)

The term "heart failure with preserved ejection fraction" (HFpEF) refers to a specific type of heart failure in which the ejection fraction (EF) is normal, defined as greater than 50%. Even though the heart muscle may be healthy and robust, it loses its ability to relax and expand adequately after contracting and ejecting blood. This results in an inadequate filling of the heart with blood, leading to insufficient blood being pumped out into the body, particularly during strenuous physical activities [15]. EF is frequently more than 50% in patients diagnosed with HFpEF, the majority of whom are female and older people; the volume of the left ventricular (LV) cavity is normally normal, but the LV wall is thickened and stiff; as a result, the ratio of LV mass to end-diastolic volume is high. This means that while the heart is pumping normally, it is not filling up with enough blood, which causes blood to back up in the lungs and other organs, leading to symptoms such as shortness of breath, fatigue, and swelling. HFpEF is typically caused by underlying medical conditions such as hypertension, diabetes, and obesity [15,16].

2.3.2   Heart failure with reduced ejection fraction (HFrEF)

Progressive left ventricular enlargement and deleterious heart remodeling characterize HFrEF, which is diagnosed when the left ventricular ejection fraction (LVEF) is 40% or less. Individuals who have HFrEF often have an enlarged left ventricular cavity, and the ratio of LV mass to end-diastolic volume is either normal or decreased in these patients. At the level of the cell, both the width of the cardiomyocytes and the volume of the myofibrils are larger in HFpEF than they are in HFrEF [17].

2.3.3   Heart failure with mild range ejection fraction (HFmrEF)

Heart failure with mid-range ejection fraction (HFmrEF) is a type of heart failure where the ejection fraction (EF) falls between 40% and 49%. This type of heart failure shares characteristics of both heart failure with reduced ejection fraction (HFrEF) and heart failure with preserved ejection fraction (HFpEF) [18]. People with HFmrEF may experience symptoms such as shortness of breath, fatigue, and swelling in the legs or ankles.

3   Epidemiology

It is believed that 64.3 million individuals throughout the globe are now living with heart failure [19]. The incidence of known heart failure is commonly considered to be between 1% and 2% of the overall adult population in developed nations [20]. Even taking into account the fact that different diagnostic criteria might provide different results, the majority of research have indicated that more than half of all patients with heart failure in the general population have a preserved LVEF, and that this fraction is rising [20,21]. Between the years 2013 and 2016, there were about 6.2 million persons in the United States who were clinically diagnosed with HF [22,23]. According to some statistics, the incidence rate has reached a peak; yet, the prevalence is growing as a greater number of patients undergo treatment. Patients who suffer from heart failure have not seen either an improvement in their quality of life or a reduction in the number of times they have been hospitalized as a result of this. The Global Health Data Exchange registry estimates that there are now 64.34 million instances of CHF affecting people all over the globe [23]. This amounts to 9.91 million years lost because of disability and 346.17 billion USD in healthcare expenditures in the United States. Age is one of the most important factors in determining HF. It doesn't matter what causes heart failure or what criterion is used to categorize people who have heart failure (HF), the occurrence of HF rises rapidly with age. A propensity for race is also noted by the registry, with a frequency of HF that is 25% greater in patients of African-American heritage than in Caucasians [23]. According to the American Heart Association, heart failure is still the leading cause of hospitalization among the senior population. Furthermore, heart failure is responsible for 8.5% of fatalities in the United States that are attributable to cardiovascular disease [24]. According to the findings of the survey, Hispanic Americans, African Americans, Native Americans, and recent immigrants from developing countries had a much greater incidence and prevalence of heart failure. The Candesartan in Heart Failure Assessment of Reduction in Mortality and Morbidity (CHARM program) found that the occurrence of heart failure is considerably higher in patients of younger age. The researchers believed that the reason for this was due to the fact that obese people tend to be younger. [25] After the age of 65, the risk of heart failure in males increases by a factor of three for every 10 years that pass after that age, but the risk for women in the same age cohort is three times higher. [24] The data gathered internationally about the epidemiology of HF are comparable to those gathered in the United States. The incidence rises substantially with age, and important risk factors include metabolic risk factors in addition to leading a sedentary lifestyle. In underdeveloped nations, ischemic cardiomyopathy is one of the primary contributors of HF, along with hypertension [26]. A greater incidence of isolated right heart failure is the most striking variation based on a study of small cohort studies from these countries. One possible explanation for this phenomenon is that there is a greater incidence of lung illness, pericardial disease, and tuberculosis in this population. There is a lack of reliable evidence to support these claims.

4 Pathophysiology of congestive heart failure

Heart failure is a complex condition that is difficult to diagnose and treat due to the numerous compensatory mechanisms present at every level of the body. Failure of the heart to pump blood effectively only occurs when this network of adaptations is overloaded [27–29]. The Frank-Starling process is one such mechanism that describes how an enhanced preload assists in maintaining heart function. Myocardial hypertrophy, the growth of the heart's contractile tissue, may occur with or without ventricular chamber dilatation. Activation of neurohumoral systems [30] is another compensatory mechanism that occurs in response to heart failure. The discharge of norepinephrine by adrenergic cardiac nerves increases myocardial contractility and activates the sympathetic nervous system (SNS), renin-angiotensin-aldosterone system (RAAS), and other neurohumoral modifications that sustain arterial pressure and the perfusion of essential organs [31]. Acute heart failure causes the heart's limited adaptive mechanisms to become maladaptive while trying to maintain appropriate cardiac function, ultimately resulting in remodeling, most often of the eccentric type [32]. Reducing wall stress has been a long-standing approach in the treatment of heart failure. A cascade of hemodynamic and neurohormonal derangements is set into motion when cardiac output is reduced as a result of myocardial injury [33]. The secretion of epinephrine and norepinephrine, along with vasoactive substances such as endothelin-1 (ET-1) and vasopressin, induces vasoconstriction, leading to a rise in calcium afterload and cytosolic calcium entry via an increase in cyclic adenosine monophosphate (cAMP). This contributes to enhanced myocardial contractility and impaired myocardial relaxation, hindering lusitropy. Calcium toxicity may cause abnormal heart rhythms, ultimately resulting in death. Elevated afterload and myocardial contractility, along with impaired lusitropy, contribute to a rise in myocardial energy expenditure, causing myocardial cell death, also known as apoptosis [34]. Activation of the RAAS causes a retention of salt and water, resulting in a rise in preload and additional enhancements in myocardial energy expenditure. Increases in renin cause a reduction in the supply of chloride to the macula densa and an increase in the activation of beta1-adrenergic receptors, leading to a rise in levels of angiotensin II (Ang-II), which in turn leads to an elevation in levels of aldosterone, which causes the release of aldosterone to be stimulated [30]. Together with ET-1, Ang-II plays an essential role in the successful maintenance of intravascular homeostasis, achieved by vasoconstriction and aldosterone-induced water and salt retention processes [35]. In heart failure, the fast increase in myocyte loss overwhelms the replacement mechanism, resulting in an imbalance between hypertrophy and mortality against regeneration. Increasing rates of myocyte turnover have been seen during periods of pathological stress [34]. The concept that the heart may function as an organ that regenerates itself is a very recent one, sparking an entire area of study devoted to enhancing cardiac regeneration.

4.1 Role of oxidative stress and inflammation

Typically, reactive oxygen species (ROS) is necessary to maintain a cell's physiological function, but too much ROS can harm important cell components like DNA, lipids, and proteins, ultimately leading to cell death [36]. This is known as oxidative stress, which is linked to many diseases including cardiovascular disease. Specifically, oxidative stress can damage the microvessels in the heart, leading to cardiovascular problems [37]. Excessive ROS can thicken the capillary basement membrane, increase endothelial permeability, and disrupt vascular smooth muscle cell function [38]. In heart failure, mitochondrial dysfunction and increased levels of oxidative stress and inflammation markers are commonly observed and are associated with poor prognosis. Excessive ROS release from mitochondria can trigger further ROS generation by various mechanisms, including uncoupling of NOS mitochondrial isoform and the conversion of xanthine dehydrogenase to its ROS-producing form, xanthine oxidase [39]. These reactions can generate highly reactive oxidative species, such as peroxynitrite and superoxide anions, that can harm cell components and lead to alterations in calcium regulation, cardiomyocyte hypertrophy, apoptosis, fibrosis, and inflammatory responses, all of which are key elements in the development of heart failure and are shown in Fig. 2 [40]. Oxidative stress and inflammation are closely linked to each other, both in the immediate result of a heart attack and during long-term heart remodeling. During ischemia-reperfusion injury, there is an increase in the production of ROS, which activates the inflammatory response [41]. ROS recruits inflammatory cells and fibroblast progenitors through various mechanisms, such as chemokine upregulation, neutrophil integrin activation, and endothelial cell adhesion molecule expression [42]. Dysfunctional cardiomyocytes with elevated ROS levels experience severe oxidative DNA damage, which stimulates nuclear enzyme poly(ADP-ribose) polymerase. Overactivation of this enzyme impairs several cellular metabolic pathways and promotes the expression of inflammatory mediators,

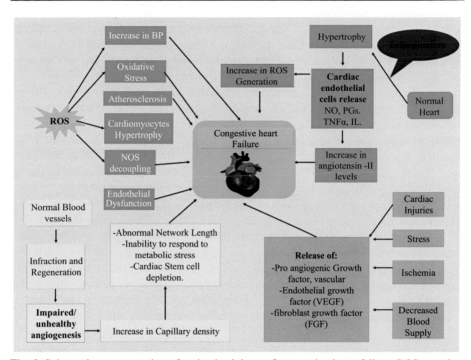

Fig. 2 Schematic representation of pathophysiology of congestive heart failure. *ROS*, reactive oxygen species; *BP*, blood pressure; *PGs*, prostaglandins; *TNF-α*, tumor necrosis factor-α; *NOS*, nitric oxide synthetase.

leading to subclinical inflammation that significantly contributes to cardiac remodeling and failure [43]. Heart failure is characterized by a systemic proinflammatory state, as evidenced by high levels of circulating inflammatory mediators like tumor necrosis factor (TNFα) and interleukin (IL)-6 [44]. TNFα overexpression damages mitochondrial DNA, inhibits antioxidant factors, and alters the activity of mitochondrial complex III, which leads to increased ROS generation. Conversely, in heart failure animal models, inhibiting TNFα in vivo stabilizes oxidative balance and reduces apoptosis [39,41].

4.2 Role of angiogenesis

The molecular pathways and mechanisms underlying the role of angiogenesis in the pathophysiology of CHF are complex and involve multiple signaling pathways and cell types. After cardiac injury or stress, such as a heart attack or hypertension, ischemia or decreased blood supply triggers the release of proangiogenic growth factors, including vascular endothelial growth factor (VEGF) and fibroblast growth factor (FGF) (Fig. 2) [45,46]. These growth factors bind to receptors on endothelial cells, which line the interior surface of blood vessels, and activate downstream signaling pathways that promote cell proliferation, migration, and tube formation [47].

This leads to the formation of new blood vessels in the heart, a process known as angiogenesis. However, in CHF, angiogenesis becomes dysregulated and excessive, leading to the formation of abnormal, leaky blood vessels that are prone to rupture and bleeding [48]. This abnormal vessel formation is driven by a number of factors, including chronic inflammation, oxidative stress, and increased levels of proangiogenic growth factors [49]. In addition to abnormal vessel formation, angiogenesis also contributes to the development of fibrosis, or scarring, in the heart muscle [50]. This is because the new blood vessels that form are often accompanied by the growth of fibroblasts, cells that produce collagen and other extracellular matrix proteins. Excessive fibrosis can stiffen the heart muscle and impair its ability to pump blood effectively. The molecular mechanisms underlying this fibrosis involve the activation of various signaling pathways, including transforming growth factor beta (TGF-β) and connective tissue growth factor (CTGF) [51]. These pathways stimulate the production of extracellular matrix proteins by fibroblasts and promote the differentiation of fibroblasts into myofibroblasts, which are specialized cells that play a key role in wound healing and tissue repair. Overall, while angiogenesis is an important response to cardiac injury and stress, dysregulated angiogenesis can contribute to the pathophysiology of CHF by promoting abnormal vessel formation and fibrosis through the activation of multiple signaling pathways and cell types [52].

4.3 Other predisposing factors

4.3.1 Diastolic and systolic failure

Both diastolic and systolic heart failure may lead to a reduction in stroke volume in the patient [53,54]. This results in the stimulation of central and peripheral baroreflexes as well as chemoreflexes, both of which are able to cause significant improvements in the volume of traffic carried by sympathetic nerves. In spite of the fact that the neurohormonal responses to a reduced stroke volume have certain similarities with one another, the neurohormone-mediated processes which follow have been most thoroughly investigated in people who suffer from systolic heart failure [53]. The subsequent spike in plasma norepinephrine has a direct correlation with the degree of cardiac dysfunction, and this correlation has substantial prognostic consequences. While norepinephrine is directly toxic to cardiac myocytes, it is also responsible for a variety of signal-transduction abnormalities. These abnormalities include the downregulation of beta1-adrenergic receptors, the uncoupling of beta 2-adrenergic receptors, and the increased activity of inhibitory G-protein. Alterations in beta1-adrenergic receptors cause overexpression and contribute to the development of cardiac hypertrophy [53,54].

4.3.2 Atrial natriuretic peptide and B-type natriuretic peptide

Atrial natriuretic peptide (ANP) and B-type natriuretic peptide (BNP) are peptides that are produced naturally in the body and become active in response to upsurge in the volume or pressure of the atrium or ventricle. ANP and BNP are both secreted by the heart's atria and ventricles, respectively, and they both help to dilate blood vessels and cause the kidneys to excrete more sodium [30]. Their effects on blood flow are made possible by decreases in the pressures exerted by the ventricles during filling. These pressure decreases are caused by reductions in the preload and afterload placed on the heart. In particular, BNP causes selective vasodilation of the afferent arteriolar capillaries and blocks salt reabsorption in the proximal convoluted tubule. In addition to this, it suppresses the release of renin and aldosterone, and as a result, adrenergic activation. Chronic heart failure is associated with higher levels of both ANP and BNP. BNP in particular has potentially substantial implications for diagnostics, therapy, and prognosis [55].

4.3.3   Angiotensin II

Research shows that the local production of Ang II in the heart (which lowers lusitropy, enhances inotropy, and raises afterload) results in a higher rate of myocardial energy use. Ang-II has also been demonstrated to enhance the rate of apoptosis in myocytes in both in vitro and in vivo studies [56]. In this way, the acts that Ang II carries out in heart failure are comparable to those that norepinephrine carries out. Additionally, Ang-II is responsible for the mediation of myocardial cellular hypertrophy and may contribute to the gradual loss of cardiac function. The neurohumoral variables described above cause myocyte hypertrophy and interstitial fibrosis, which in turn leads to an enhancement in myocardial volume as well as an increase in myocardial mass, in addition to the loss of myocytes. Because of this, the architecture of the heart alters, which, in turn, causes an even greater rise in the volume and mass of the myocardium [57].

4.3.4   Heart failure with preserved ejection fraction

The relatively similar pathophysiologic mechanisms that result in reduced cardiac output in systolic heart failure occur in diastolic heart failure (heart failure with preserved ejection fraction [HFpEF]). However, these pathophysiologic mechanisms occur in response to a unique set of circulatory and hemodynamic environmental factors that decrease cardiac output [58]. In HFpEF, an elevation in ventricular afterload causes abnormal relaxation and enhanced rigidity of the ventricle (owing to delayed calcium absorption by the myocyte sarcoplasmic reticulum and delayed calcium efflux from thymocyte) (pressure overload). In individuals with HFpEF, the poor relaxation of the ventricle causes reduced diastolic filling of the left ventricle (LV). This dysfunction may be linked to the similar fibrotic events that impair the subendocardial layer of the LV and, to a lesser degree, pressure overload in the right ventricular (RV). It may have a role in the symptomatology of HFpEF patients [59].

4.3.5   Myocytes and myocardial remodeling

Increased myocardial volume in the failing heart is characterized by larger myocytes toward the end of their life cycle [34]. As more myocytes are lost, a heavier load is imposed on the remaining myocardium, and this adverse environment is passed on to the progenitor cells responsible for replacing lost myocytes. As the underlying pathological process worsens and cardiac failure accelerates, the effectiveness of progenitor cells diminishes. Myocardial remodeling is characterized by an increase in myocardial volume and mass, as well as a net loss of myocytes [60]. This remodeling process results in early adaptive mechanisms, like an expansion of the stroke volume (Frank-Starling mechanism) and a reduce in wall stress (Laplace law), followed by maladaptive mechanisms, like an increment in myocardial oxygen demand, impaired contractility, myocardial ischemia, and arrhythmogenesis. As heart failure progresses, the vasodilatory actions of endogenous vasodilators such as nitric oxide (NO), ANP, bradykinin (BK), prostaglandins (PGs), and BNP decrease [61]. This decrease occurs with an elevation in vasoconstrictor chemicals from the RAAS and adrenergic system, which promotes additional improvements in vasoconstriction and, therefore, preload and afterload [61]. This leads to cellular proliferation, adverse myocardial remodeling, and antinatriuresis, as well as an increase in total body fluid and a worsening of heart failure symptoms.

4.3.6   Left ventricle chamber stiffness

An increment in LV chamber rigidity can be the result from any one of the three possible processes, or a combination of two or more of them: an increase in filling pressure, a transition to a steeper ventricular pressure-volume curve, or a reduction in ventricular contractility. An increase in filling pressure is the movement of the ventricle up along its pressure-volume curve to a steeper portion, which can take place in circumstances like as volume overload caused by acute valvular regurgitation or acute LV failure due to myocarditis [62]. The most frequent cause of a transition toward a steeper ventricular pressure-volume curve are a significant rise in ventricular mass and wall thickness (as seen in aortic stenosis and long-standing hypertension), as well as infiltrative condition (such as amyloidosis), myocardial ischemia and endomyocardial fibrosis. A reduction in ventricular contractility is often described to as a parallel upward shift of the diastolic pressure-volume curve. This phenomenon occurs as the ventricle becomes less distensible. Extrinsic compression of the ventricles is often the culprit in cases like these [63].

4.3.7   Arrhythmias

Although potentially fatal rhythms are much more prevalent in ischemic cardiomyopathy, arrhythmia imposes a substantial load on all types of heart failure [64]. Some arrhythmias may further exacerbate cardiac failure. Among all rhythms linked to cardiac failure, life-threatening ventricular arrhythmias are the most prominent [65]. Regardless of the underlying reason, typical structural substrates for ventricular arrhythmias in heart failure comprise ventricular dilatation, myocardial hypertrophy, and myocardial fibrosis. Myocytes may be subjected to enhanced stretch, wall strain, catecholamines, ischemia, and electrolyte imbalance at the cellular level. The combination of these variables increases the risk of arrhythmogenic sudden cardiac death in heart failure patients [66].

4.3.8   Concentric LV hypertrophy

Pressure overload resulting in concentric left ventricle hypertrophy (LVH), as seen in aortic stenosis, hypertrophic cardiomyopathy and hypertension, shifts the diastolic pressure-volume curve to the left along its volume axis [67]. As a consequence of this, the diastolic pressure in the ventricles is abnormally raised, despite the fact that the chamber stiffness may or may not be affected [67,68]. Enhancement in diastolic pressure results in an increase in myocardial energy expenditure, remodeling of the ventricle, enhanced myocardial oxygen demand, myocardial ischemia, and eventually the progression of the maladaptive mechanisms of the heart that eventually leads to decompensated heart failure [68].

4.3.9   Other vasoactive systems

Other vasoactive systems involved in the development of heart failure involve the endothelin receptor system, the adenosine receptor system, vasopressin, and tumor necrosis factor alpha (TNF-alpha) [69]. ET, a chemical generated by the endothelium of the blood vessels, may have a role in the control of myocardial function, vascular tone, and peripheral resistance in heart failure. The degree of cardiac failure is significantly correlated with elevated ET-1 levels [70]. ET-1 is a powerful vasoconstrictor with enhanced vasoconstrictor effects in the renal vasculature, resulting in a decrease in renal plasma blood flow, sodium excretion, and glomerular filtration rate (GFR). TNF-alpha has been associated with a variety of infectious and inflammatory disorders. Consistently elevated TNF-alpha levels have been seen in patients with heart failure and seem to correspond with the severity of myocardial dysfunction [71]. Several studies indicate that local generation of TNF-alpha may have deleterious effects on the myocardium, hence impairing systolic and diastolic cardiac function. Therefore, in patients with systolic dysfunction, neurohormonal responses to reduced stroke volume lead in a transitory increase in systolic blood pressure and tissue perfusion. In all cases, however, the available data support the hypothesis that these neurohormonal responses contribute to the evolution of cardiac dysfunction over time [72].

5   Conventional therapies

The purpose of chronic CHF therapy is to enhance symptomatic management and quality of life, reduce hospitalizations, and reduce the disease's overall mortality rate [73]. The objective of pharmacologic treatment is to provide all prescribed medicines rather than a single agent since the cumulative impact of various therapies is superior to that of any single agent administered alone. The main combination treatment for HFrEF consists of a renin-angiotensin system inhibitor (like as an angiotensin-converting enzyme (ACE) inhibitor, angiotensin receptor neprilysin inhibitor (ARNI), or an angiotensin II receptor blocker (ARB)), diuretics, and a beta-blocker [74]. If ACE inhibitor, ARNI, and ARB treatments are contraindicated, the combination of nitrate and hydralazine provides an alternative to angiotensin system blockers for main therapy. The hydralazine and nitrate combination is also advised to minimize morbidity and mortality in African American patients undergoing appropriate pharmacological treatment for symptomatic HFrEF [75]. In comparison to ACE inhibitors alone, the combined treatment of ARB-ARNI substantially decreased cardiovascular mortality and HF hospitalizations [76]. Mineralocorticoid receptor antagonists, like as spironolactone or eplerenone, are recommended for patients having NYHA functional class II to IV with an LVEF of 35% or less. Individuals with symptomatic HF following a myocardial infarction (MI) with an LVEF of less than 40% are also candidates for these medications. In individuals with a recent MI and a poor EF who had symptoms of HF, however, these medicines demonstrated minimal benefit [77]. Ivabradine has been shown to suppress the funny current (I-f) in a selective manner in the sino-atrial node. Patients who have consistently symptomatic heart failure with an EF of less than or equal to 35% in sinus rhythm are considered ideal candidates for treatment with ivabradine, as recommended by the AHA and ACC. Despite the use of goal-directed beta-blocker medication, the resting heart rate should be larger than 70 beats per minute [78]. The powerful vasodilator vericiguat is a medication that activates the intracellular receptor for endogenous NO. This results in increased blood flow across the body. It was recently authorized by the FDA in 2021 to lower the risk of death and HF hospitalizations in individuals hospitalized with HF exacerbation who had chronic symptomatic HF and an EF of less than 45%. This was the first time that the FDA has given its blessing to a drug for this purpose [79]. Digoxin may be explored in symptomatic patients with sinus rhythm despite sufficient goal-directed treatment in order to lower the risk of hospitalizations from all causes, but its usefulness is restricted. Implantable cardioverter-defibrillators (ICDs) are suggested for the primary prevention of sudden cardiac death in patients with heart failure (HF) who have an LVEF of less than or equal to 35% and a NYHA functional class of II to III while receiving goal-directed medication treatment. It is also recommended for patients with NYHA functional class II and an EF of less than or equal to 30% despite effective medical treatment [80]. Cardiac resynchronization treatment (CRT) with biventricular pacing is recommended for patients with HFrEF, NYHA functional class II to IV, LVEF less than or equal to 35%, and QRS length more than 150 ms [80]. According to the European Society of Cardiology (ESC), CRT is not advised for individuals with a QRS length of less than 130 milliseconds since many studies have shown the possibility of injury. The European Society of Cardiology (ESC) advises CRT for patients with nonleft bundle branch block (LBBB) morphology who match the criteria for CRT; however, the ACC/AHA recommendations restrict it to individuals with LBBB morphology on ECG. Whether QRS shape or QRS duration should be the major criterion for the selection of CRT is the subject of continuing discussion [81].

6　Role of nanoparticles in congestive heart failure

The introduction of nanotechnology and the subsequent creation of biocompatible nanomaterials has marked the beginning of dramatic change in the realm of biomedicine. These nanomaterials have seen extensive usage in a wide range of biological applications, one of which is the treatment of cardiovascular diseases (CVDs), which has led to increased therapeutic advantages all around [8,82]. Diseases of the cardio-vascular system have maintained their position as the major cause of death and disability throughout the globe. Because adult cardiomyocytes have a limited capacity for proliferation, the heart is unable to regenerate new myocardium following an incident that causes myocardial ischemia. This causes the heart to become progressively weaker over time, which may lead to heart failure and possibly death [82]. Administration of cardioprotective medicines as soon as possible after the event may assist in protecting the heart from additional cell death and enhance cardiac function; however, delivery techniques and the potential adverse effects of these treatments may be a challenge. Researchers have been able to improve their medication targeting capabilities due to recent developments in nanotechnology, notably nanoparticles for drug delivery [83]. This has resulted to an increase in the overall effectiveness of treatment. Nanomedicine has been the subject of extensive research over the last several decades for a wide range of disease types, including cancer, CVDs, and diabetes [84,85]. Nanomedicine refers to the application of nanotechnology to accomplish disease therapy at the nanometer ($\sim 10^{-9}$ nm)-size scale. In contrast to traditional medicine, it modifies the characteristics of materials so that they are suited for usage at the nanometric scale. The physics, chemistry, and biology of a substance at a greater size are often different from the same thing at a smaller size. In addition, the dimensions of various biological systems found inside the human body may be described using the nanometric scale. This permits nanoparticles (NPs) to effectively traverse natural barriers so that they can access the new delivery sites and interact with DNA or small proteins at different levels, whether in the blood or within organs, tissues, or cells. This is because of the fact that the size of numerous biological systems found in the human body corresponds to the nanometric scale. At the moment, nanomedicine encompasses a variety of subspecialties within the medical industry, such as targeted nanotherapeutics, medical imaging diagnostics, vaccinations, and regenerative therapies [85–87]. Specifically, targeted nanotherapeutics have the potential to boost therapeutic outcomes while simultaneously lowering the risk of off-target negative effects. In spite of the development of a wide range of research studies using nanomedicine, the application of nanomedicine to CVD is still difficult. It is difficult to correctly deliver therapeutic cargos to particular blood vessel lesions as a result of the diverse variety of lesion locations that are caused by typical vascular disorders. Therefore, it is anticipated that a targeted nanomedicine system that is capable of correctly delivering a therapeutic medication to a desired region in a desired dosage may prove to be a potential therapy technique for vascular disease [85].

6.1 Polymeric nanoparticles

Biodegradable polymeric nanoconstructs have been widely studied as drug carriers. This kind of nanocarrier is comprised of copolymers, each of which consists of two or more blocks that have a distinct level of hydrophobicity. These copolymers often form micellar structures when they are exposed to environments containing water. These structures are made up of a hydrophilic shell and a hydrophobic core. While the hydrophilic corona may be further changed to allow for the attachment of water-soluble drugs or targeting moieties, water-insoluble medications can be housed within the hydrophobic core of the molecule [88]. This is not always the case, though. To date, several polymeric nanostructures with very diverse architectures have been produced. Polymers, for instance, may produce hydrogels, network-like scaffolds, microparticles, nanospheres, and nanoshells that are appropriate for the release of both hydrophobic and hydrophilic medicines, proteins, genes, nucleic acids, and genes [88–93]. The regulated release of encapsulated pharmaceuticals is based on time or the surrounding environment. The kinetics of the release of pharmaceuticals bound to nanoparticles are dependent on the nanoparticles' residence time in the organism and the properties of their microenvironment. In general, biodegradable polymer formulations keep the required therapeutic drug concentration in the target tissue longer than other nanocarriers. This characteristic makes them appropriate delivery vehicles for extremely poisonous, poorly soluble, and unstable medicines. Poly (D, L-lactic acid), poly (D, L-glycolic acid) (PLGA) is a copolymer extensively utilized in medical applications due to its low toxicity, good biocompatibility and biodegradability, and FDA approval. In this manner, several research groups have concentrated their efforts on producing nanomedicine materials based on PLGA. Chang and colleagues, for example, demonstrated the manufacture of insulin-like growth factor-1 (IGF-1) complexed PLGA NPs for early cardioprotection after acute MI. It has been discovered that IGF-1 is essential for regulating heart function and promotes cardiomyocyte proliferation. IGF-1 therapy improves heart function following infarction in humans. Electrostatic interactions allowed anionic IGF-1 to form complexes with PLGA in order to sustain its biological activity. In order to produce PLGA-IGF-1 complexation and take use of the fact that PLGA itself has a negative charge, polyethylenimine conjugation was used. This turned the surface of PLGA into a cationic layer. After synthesizing NPs of several sizes (1 m, 200 nm, and 70 nm), researchers discovered that the smallest ones (70 nm) had the highest binding capacity and triggered the most Akt phosphorylation in cultured cardiomyocytes. This was the case regardless of the size of the NPs. In mice, a single intramyocardial injection of IGF-1-PLGA NPs prolonged IGF-1 retention time to at least 24 h, inhibited cardiomyocyte apoptosis, lowered infarct size, and prevented ventricular dilation and wall thinning 21 days after a MI. All of these effects were observed after a single intramyocardial injection. These findings highlight attention to the possibility of using nanomedicine in the treatment of cardiovascular disorders and stimulate the field's implementation in clinical settings [94]. In both animal models and human patients with acute myocardial infarction, the administration of statins during reperfusion did not have any therapeutic benefits. Despite this, Nagaoka and his colleagues discovered that Pitavastatin given by nanoparticles had cardioprotective benefits. They generated Pitavastatin-PLGA NPs, which owing to their improved vascular permeability deposited in the infarcted area of the myocardium of a rat model of MI. Pitavastatin nanoparticles, when given a single intravenous injection at the time of reperfusion, were shown to minimize the size of the infarct after 24 h and enhance left ventricular function. In addition to this, they were successful in inhibiting inflammation and cardiomyocyte necrosis in the infarcted heart, in contrast to the free pitavastatin medication, which did not demonstrate any cardioprotective effects.

In addition, pitavastatin nanoparticles (NPs) have successfully completed phase I clinical trials at a university hospital, indicating that this nanoparticle-based technology has the potential to serve as an innovative therapeutic therapy for ischemia damage [95]. Therapeutic angiogenesis has emerged as a promising clinical approach for the recovery of ischemic tissues in the conditions of coronary artery disease (CAD) and peripheral artery disease (PAD). This approach involves the direct administration of pro-angiogenic growth factors to target tissues in order to promote the synthesis of growth factors by those target tissues. In this respect, a novel angiogenic factor known as adrenomedullin-2 (ADM-2) has recently been discovered as being capable of regulating both blood flow and the function of the cardiovascular system. Because of its short biological half-life and poor stability in plasma, ADM-2 has only a restricted number of uses at this point. As a result of this, Quadros and his colleagues attempted to encase this peptide inside PLGA nanoformulations in the hopes that these formulations may act as a delivery method for heart repair. PLGA-ADM-2 NPs were manufactured using a double emulsion technique, and the resulting particles exhibited a size range of 300–350 nm, a negative surface charge, and high colloidal stability. Furthermore, the PLGA-ADM-2 NPs did not exhibit any toxicity in cardiomyocyte cells [96]. As determined by ELISA, the entrapment effectiveness of ADM-2 reached around 70%. In vitro peptide release indicated a 60% release in the first 3 days, followed by a slower release rate in the next 3 weeks. With vitro investigations in EA.hy926 endothelial cells also shown increased cell proliferation (1.4 times greater), making AMD-2-PLGA NPs an innovative method for therapeutic angiogenesis in CVDs [96]. Nguyen et al. described a technique for potentially treating heart failure after a MI that is based on the intravenous delivery of enzyme responsive nanoparticles. This study was groundbreaking since it was the first of its kind. These nanomaterials have the capacity to change their structure from discrete micellar nanostructures to network-like scaffolds in response to the enzymatic conditions that are present in the heart after an acute MI. This occurs when the nanomaterials are exposed to the enzymatic circumstances [97]. It is important to point out that these nanosystems, which consist of brush peptide-polymer amphiphiles based on a polynorbornene backbone with peptide targeting sequences specific for recognition by matrix metalloproteinases MMP-2 and MMP-9, have the capacity to remain stable during blood circulation conditions until they reach the infarct region via the vascular leakage that occurs after a MI. This is something that should be a t this location, their structure is changed, and it has been discovered that they stay on the injured tissue for up to 28 days after injection even after they have been injected. In earlier research, measurements using superresolution fluorescence microscopy demonstrated that the responsive nanoparticles are activated by MMP-9 and that they remain assembled into scaffolds at the injection site for a period of 7 days. On the other hand, the nonresponsive particles were inactive in the presence of MMP-9 and were removed within 1 h. In addition, the effectiveness of the system was also examined in rats with MI, and the results showed that the responding particles were activated in the presence of MMPs and remained aggregated in network-like scaffolds for 6 days. This was in contrast to the reduced aggregation that was observed in nonresponsive particles [91]. After injecting responsive and nonresponsive particles into the tail veins of healthy rats and 24 h post-MI rats, a 28-day buildup of responsive particles in the damaged heart tissue was observed. MMP-responsive nanoparticles leverage the EPR effect for initial passive targeting and propose a highly promising strategy for the effective and regulated delivery of medicines to the heart following MI [91].

6.2 Liposomes

In both diagnostics and treatment, the utilization of liposomes aims to accomplish the delivery of a drug to the affected region. Liposomes are spherical structures that are self-closed and made of one or more concentric lipid bilayers [97]. There is an aqueous phase both within and between the lipid bilayers of a liposome. Because of their potential to entrap different water-soluble compounds within the inner aqueous phase and lipophilic agents between liposomal bilayers, liposomes have been shown to be effective for the delivery of various types of drugs as well as for carrying diagnostic agents in all imaging modalities, including gamma-scintigraphy, magnetic resonance imaging (MRI), computed tomography (CT) imaging, and sonography. Liposomal modification using polyethylene glycol (PEG) broadens the applications for these particles by extending the length of time they remain in circulation and allowing for the attachment of antibodies or other types of targeting moieties to their surfaces [97]. These modified liposomes and immunoliposomes can be regarded for intravascular drug delivery through using cells and noncellular components (like as endothelial cells, subendothelial structures, and blood components) as the targeted sites for diagnosing and treating the most important cardiac pathologies, including MI, coronary thrombosis, and atherosclerosis and heart failure [97]. Verma et al. revealed that optimized and targeted liposomes loaded with ATP may generate cardioprotective effects ex vivo in the Langendorff isolated rat heart model [98]. After 30 min of reperfusion, the ischemic myocardium received considerable protection from ATP-liposomes (ATP-L) injections that were administered 1 min before the beginning of a period of global ischemia. Compared to the Krebs-Henseleit (KH) buffer, the use of ATP-L at the end of reperfusion resulted in a 61% decrease in the left ventricular end diastolic pressure [98]. Because of their unique characteristics, like as their biodegradability, minimal toxicity, great carrying capacity, and simplicity of production, liposomes are a good choice for the encapsulation of drugs, which enables them to be administered through injection with far less discomfort. Istaroxime, a potential and safe therapy for both acute and chronic heart failure, was administered using a drug delivery system composed of vesicles synthesized with phospholipids and PEG-HS, an excipient designed to alter the characteristics of the bilayer [99]. Coenzyme Q10, also known as CoQ10, has been suggested as a possible therapy and preventative measure for ischemia-reperfusion damage, CAD, hypertension, and hyperlipidemia [100]. The solubility of berberine in buffer was improved by the use of liposomal encapsulation, and the ejection percent was maintained following MI. This demonstrates that introduction of berberine-loaded liposomes greatly enhances its therapeutic availability and suggests berberine-loaded liposomes as a possible therapy of unfavorable remodeling after MI [101].

6.3 Niosomes

Niosomes are able to encapsulate a wider variety of pharmaceuticals than liposomes and polymersomes, which enables them to improve both the medications' stability and their therapeutic value. As a result, niosomes have the potential to serve as an alternative to liposomes and polymersomes. Liposomes, polymersomes, and niosomes are all capable of being loaded with both hydrophilic and hydrophobic medicines, in contrast to other nanoparticles. Structurally, these three types of particles have many similarities [102]. Because of this, they were able to codeliver medicines that were both hydrophilic and hydrophobic in a single vesicle. Niosomes are far more stable than lipids, in terms of their both physical and chemical stabilities, due to the fact that the ingredients that go into their formation—nonionic surfactants—are themselves more stable [102]. Because the lipid bilayer can only maximally tolerate about 5%–6% mol% of PEG, which may induce some stability issues like the lysis of liposomes at high concentrations, the PEG on the surface of liposomes, which could prolong the half-life after being administered, was limited. This was because PEG could potentially lengthen the half-life after being administered. Because of the niosomes' high degree of stability, the processing of the formulation was made simpler. Niosomes also have a lower production cost than liposomes do [102–104]. A niosome is made up of many components, including medicines, cholesterol or cholesterol derivatives, nonionic surfactants, and sometimes ionic amphiphiles. Niosomes are able to enclose the medications, regardless of whether they are hydrophilic or hydrophobic. Medications that are hydrophilic are encased in the corresponding core, while drugs that are hydrophobic are encased in the hydrophobic area of the bilayer. Because cholesterol interacts with nonionic surfactants, the correct quantity of cholesterol must be added to the niosomes in order to produce the most stable formulation possible [105]. Cholesterol is not only forming the structure of the bilayer; nonetheless, it is capable of mixing with the bilayer membrane and performing the function of controlling the structure and flexibility of the membrane while acting as a reliable buffer. Niosomes have a number of applications in the delivery of chemical medications, one of which is the use of this formulation to increase oral bioavailability. Carvedilol is a kind of clinical medicine that is used extensively in the therapy of CHF and coronary artery disorders [102]. The first-pass metabolism and the relatively short half-life after administration both contribute to the restricted systemic availability of the drug. Numerous researches have been conducted to investigate the possibility of producing novel formulations that would increase the bioavailability of carvedilol. Niosome is one candidate for a solution because of its ability to prevent the degradation of the loaded medication, maintain control over the drug's releasing profiles via the optimization of its constituent parts, and bypass the first-pass metabolism [106]. It has been reported that carvedilol niosomes may be generated by a film hydration approach with a minimum size of 167 nm (PDI 0.6) and the greatest encapsulation rate of 77.7% in a variety of formulations. It has been demonstrated that the release of all formulations is capable of reaching approximately one 100% after 20 h [102], with no significant differences.

6.4 Polymeric micelles

Micelles are nanoscale colloidal particles that contain a hydrophilic exterior and a hydrophobic center. Micelles are specialized carriers that are effective for transporting poorly soluble medicines. There are several uses for surface-modified polymeric micelles in the fields of drug delivery and theranostics [107]. These micelles are largely made up of amphiphilic macromolecules. The formation of polymeric micelles has been accomplished with the help of a wide variety of polysaccharides. A PEGylated-based polycationic blocks' copolymeric micellar assembly was designed in order to target vascular injury in the rabbit carotid artery and deliver genes by utilizing this technique. [108]. Peters et al. showed that an anticoagulant medication could be delivered to a comparable targeted region using a PEG-based lipid micellar technology that was coloaded with a fluorophore as an imaging agent [109]. In other investigations, PEG-based lipid micelles with surface-modified scavenger receptor-based antibodies and gadolinium (Gd) complex accumulated well in atherosclerotic arterial locations [107]. These micelles were able to do this because they were encapsulated with gadolinium. At atherosclerotic aortic locations, Gd-encapsulated micelles functionalized with anti-CD36 antibodies might identify macrophages in human atherosclerotic aortic tissues acquired at autopsy. Ding and colleagues evaluated a variety of likely micelle-based techniques aimed for improperly functioning endothelia that comprises a large portion of thrombotic or atherosclerotic tissues. Their findings demonstrated a substantial potential of micellar site-specific delivery and effectiveness for the therapy of CVDs [107,110]. Lipid-based micelles may passively collect in the infarct location with a shorter latency than liposomes and are more likely to concentrate in cardiomyocytes rather than in immune cells or arteries, even when delivered 1 week after acute IRI [111]. In a same fashion, polymeric micelles comprised of amphiphilic PEG and phosphatidyl-ethanolamine (PE) likewise concentrate at the damaged heart but not at the distant healthy tissue [112]. Micelles may also be targeted to various cell types by modifying the surface makeup. This is the case with CCR antibodies that when linked to lipid micelles may diminish monocytic infiltration at the damage site [113]. Micelles allow for the solubilization of hydrophobic medicines, and as a result, enhance their bioavailability, diminish unfavorable off-target effects, and amplify their potential to traverse biological barriers [114]. These features make micelles an ideal DDS in the cardiovascular system. Recently, the clinically licensed medication rapamycin, administered chronically as a micellar nanoformulation (Rapatar), was able to decrease MI size in diabetic rats [114]. The use of polymeric micelles to supply hydrogen disulfide donors, the release of which is dependent on endocytosis, resulted in substantial protective role on cardiomyocytes that were hypoxic [115]. Micelles have been investigated for their potential use in the passive distribution of a wide range of naturally occurring antioxidants that, in their unaltered forms, are insoluble in water. In H9c2 cells that had been treated with doxorubicin, the presence of polymeric pluronic micelles that concurrently transported curcumin and resveratrol resulted in a decrease in effector caspase action and ROS concentrations [116]. According to research carried out by the same team, coadministration of quercetin and resveratrol had comparable effects in H9c2 cells, while in vivo, the combination was shown to be cardioprotective [117]. In line with these findings, ginsenoside that was supplied as part of a pluronic micelles formulation was able to minimize the cardiac dysfunction that was caused by doxorubicin [118]. This was likely accomplished by protecting the integrity of the mitochondria. An additional plant byproduct called tilianin, which is a natural flavonoid, was encapsulated in polymeric micelles and put through a hypoxia-reoxygenation test preparation. After sustaining the simulated damage, the survival of cultured cardiomyoblasts was considerably increased when either tilianin on its own or encased in micelles was applied. Despite this, the water solubility of micelles was shown to be much higher than that of flavonoid on its own [119]. Isoflavone puerarin, which was encased in PEG-PE micelles, was shown to be effective in inhibiting the proapoptotic actions of isoprenaline in H9c2 cells. In fact, in terms of its natural hemolytic activity, the conjugate demonstrated a superior safety profile when compared to puerarin on its own [120]. The antiapoptotic effect of puerarin was further increased by the addition of triphenylphosphonium cation, and confocal imaging showed that the cation pushed the micelles directly to the mitochondria [121].

6.5 Gold nanoparticles

Gold nanoparticles offer appropriate delivery vehicles in the realm of nanotechnology, particularly to treat CVDs because of their less cytotoxicity, good stability, and their biocompatible nature. They offer a wide range of uses in biomedical research, such as molecular labeling and drug administration, as well as a high degree of stability in the systemic circulation and the capacity to breakdown at the target location [107]. Due to their enormous surface area, AuNPs may be functionalized with a diversity of biological ligands, including medicines, various proteins, antibodies as well as genes. The imaging of CVDs using AuNPs with potential applications in atherosclerotic abrasions and inflammation is under progress [122]. Using the characteristics of Au for the fabrication of nanostructures encapsulating vascular drugs, photothermal therapeutic techniques have also been created. The subsequent reaction may contribute to the thermal instability of NPs, allowing the loaded medicines to be released at their respective target location. Researchers are evaluating the involvement of different macrophages in vascular abrasions and their capacity to absorb AuNPs in order to provide increased intravascular photoimaging of heart problems. In addition, AuNPs possess potent antioxidant characteristics that may be advantageous for managing CVDs. The photothermal property of AuNPs was put to use in the diagnostic process of photothermal revascularization of occluded arteries [123], where it was used in a plaque-specific manner for delivery. Lisinopril was used as the active agent that Ghann and his colleagues encapsulated in order to generate AuNP-based CT contrast agents. Citrate-coated AuNPs were subjected to a ligand exchange process, which resulted in the development of AuNPs that were loaded with pure lisinopril, decreased thioctic lisinopril. X-ray CT was employed to test the targeting of ACE, and the greater stability features of thioctic lisinopril AuNPs were put to use in this evaluation [124]. The images displayed a strong contrast in the area of the heart and lungs, which clearly indicated that the targeting of ACE was associated with the progression of pulmonary and cardiac fibrosis. Additionally, overexpression of ACE was associated with the targeting of ACE. Utilizing CT imaging, this innovative technique may prove to be a helpful tool for monitoring the pathophysiology of the cardiovascular system. AuNPs, either on their own or in combination with other substances, have the potential to provide large-scale therapeutic and diagnostic techniques for CVDs [124]. Spivak and his colleagues were successful in creating gold nanoparticles having a diameter of 30 nm, and their AuNPs-Simdax combination showed promise in terms of its biocompatibility and biosafety both in vitro and in vivo. When examined on rats with doxorubicin induced heart failure, the cardioprotective effects of AuNPs-Simdax and AuNPs were shown to be comparable and much greater than those of Simdax. According to all of the factors that were investigated, intrapleural administration (which is local) is preferable above intravenous delivery (which is systemic). It has been shown that sonoporation may improve the distribution of gold nanoparticles to cardiac cells in vivo [125].

6.6   Dendrimers

The dendrimers are a category of sphere-shaped polymers having numerous branches and topological characteristics. Researchers in several important domains are becoming more interested in their distinctive features [126]. A typical dendrimer contains three different components, which bestow several unique advantages: a central fundamental component of a single atom or atomic group; these atomic groups act as building blocks made of repeated units delivering from the center like branches, and various peripheral functional groups [127]. Numerous functional groups on the periphery of dendrimers may interact with the outer environment, therefore dictating their macroscopic features [128]. The site selective alteration and functionalization of the terminal groups of dendrimers is made possible by the sequential synthesis of dendrimers. For the purpose of producing targeted delivery systems, specific chemicals may be coupled to dendrimers either in a covalent or noncovalent fashion. It is now common practice to couple dendrimers with multiple copies of specific ligands, like antibodies, and functional moieties, such as medications and dyes, in order to make use of novel materials that are beneficial for both the diagnosis and treatment of sickness [129–131]. The polyamidoamines represent the most common kind of dendrimer that incorporates nitric oxide (NO) as a free radical for use in CVDs because of their substantial relaxing effect on vascular smooth muscle [132,133]. Suppression of ATIR expression in cardiomyocytes in vitro was accomplished by the use of siRNA-based oligo-arginine-conjugated dendritic delivery techniques which demonstrate excellent preservation of cardiac functions in rats with ischemia reperfusion damage [134,135]. Dendritic assembly in artery walls led to the regulated release and therapeutic administration of NO [136,137]. Lysine-based dendrimer designs were also used to combat atherosclerosis. Dendritic structures that are biocompatible and functionalized with nonlinear RGD peptides that contain entrapped 76Be have been applied site specifically in mice in order to perform PET imaging. It has been discovered that streptokinase-modified dendrimers made with different types of polymers (PLGA or chitosan or PEG, glycol chitosan) sustain the circulatory distribution of antithrombotic medications [136,138].

6.7 Quantum dots

Quantum dots, also known as QDs, are nanocrystals that have a diameter of 10 nm on average and are classified as semiconductors. They have fluorescence features that are size-dependent, as well as a wide range of quantized energy levels. Their characteristics shine with brilliance. Recent research has been conducted with the goal of developing less hazardous QDs that do not contain heavy metals [139]. This is a response to the fact that QDs have certain limitations regarding the clinical safety as a result of their use of heavy metals, which are known to induce toxicity in addition to other adverse effects. As a result of this use of heavy metals, QDs have certain limitations regarding the clinical safety. It has been discovered that polar polymeric biomolecules may be coated on QDs made of cadmium selenite that have been encapsulated with zinc sulfide [107]. Because of their increased sensitivity, quantum dots have found widespread usage in a variety of biological applications, including imaging, targeted medication administration, and in vitro diagnostics. In the field of biomarker analysis, in vitro diagnostic lateral flow-based assay methods are becoming more popular. This is mostly due to the ease with which the test may be performed and the speed with which the findings can be obtained. N-terminal natriuretic peptide (NT-proBNP) is a biomarker for heart failure that has been investigated [140]. It has been investigated whether or not it is possible to use water-soluble carboxylic ($-$COOH) functionalized photoluminescent cadmium telluride quantum dots (CdTe) nanoparticles for lateral flow-based detection of the biomarker.

6.8 Carbon nanotubes

The architectural construction of CNTs may be compared to grapheme sheets that have been shaped or coiled. The specified structure of CNTs confers considerable physiochemical capabilities, increased surface area, good optical qualities, robust mechanical stability, and increased electrical conductivity [107]. Numerous studies have shown that CNTs are nontoxic when used to transport a range of biologically active compounds in vitro and in vivo for medicinal applications. For instance, it has been illustrated that applying CNTs to cultured cardiomyocytes can enhance their survival and proliferation as well as induce their maturation. This was discovered when assessing the electrophysiological attributes of the cardiomyocytes. On the other hand, these systems can help endothelial cells in vascular system in providing oxygen to the heart tissue [141,142]. Xu and his colleagues have come up with a multipurpose power support system that relies upon carbon nanotubes (CNTs) that mimics the beating of the heart (MPS). The multipoint pacing system (MPS) provides a multipoint pacing function in addition to an electric-driven power-assist function for a damaged heart in a manner that does not entail contact with blood [143]. This is accomplished by enclosing the heart while it is still in place. Through the process of tissue engineering, conductive superaligned carbon nanotube sheets, also known as SA-CNTs, have been used to repair the shape and function of damaged myocardium. Additionally, these sheets have been produced as effective cardiac pacing electrodes. A pacing device that is constructed utilizing SA-CNTs is capable of facilitating the transition from epithelial to the mesenchymal cells and also of directing the movement of proregenerative epicardial cells. The power-assist unit in the meantime exhibits an exceptional frequency respond to fluctuating voltage, which imitates the natural systolic and diastolic amplitudes of the heart. This technology also exhibits excellent pacing when connected to a rabbit heart's surface, and both in vitro and in vivo tests show it to be highly biocompatible [143]. Table 1 summarizes the nanoparticles investigated for treatment of congestive heart failure.

7   Current challenges of nanocardiology

In spite of the fact that over the course of the last decade, NPs have been produced for use as DDSs, and the majority of multifunctional NPs have been applied to the study of cancer. Patients only have access to therapeutic nanoparticles (NPs) as a last resort when other treatments have been unsuccessful [150]. In clinical research, the use of therapeutic nanoparticles (NPs) as a first-line therapy is unusual. In addition, the majority of the instances that have been effectively transferred into clinical settings to treat conditions other than cancer are quite limited. In the next part, we will discuss some of the challenges that NP delivery systems confront in the field of CVD and cardioprotection [151].

7.1   Biological challenges

The heart, which serves as both the body's primary organ and its primary pump, is both very sensitive and highly resistant to damage from external factors. Because of this, it is more difficult to administer medications directly into the heart muscle. After the administration of nanomedicine, the majority of the time there is very little biodistribution in the heart. The coronary microvasculature becomes leaky in acute myocardial IRI. This is a well-known phenomenon in malignancies and is referred to as the increased permeability and retention effect [152]. Because of the increases permeability and retention effect, nanoparticles (NPs) are able to transport therapeutic agents into the ischemic heart. One more obstacle comes from the immune system, which is easily triggered by an injection with NPs, causing rapid clearance from the systemic circulation. This makes it difficult to treat some diseases. It is possible for NPs to be PEGylated or PEG-terminated in order to prevent them from being eliminated from the body by immune cells and to enable them to retain in the body for a longer period of time. In a similar manner, applying a hydrophilic sugar coating, such as dextran, on iron-oxide nanoparticles produces the same result [151]. Decorating biological proteins, such as albumin, is another technique to lengthen the half-life before immune detection. In terms of acute myocardial IRI, it is particularly important to safeguard the NPs during the acute phase, when the majority of IR damage occurs. This necessitates that DDSs must concentrate in the heart swiftly and effectively.

Table 1 Summary of nanoparticles explored for treatment of congestive heart failure.

Drug (technique)	Excipients	Dosage form	Outcome and significance	Reference
Digoxin (solvent displacement method)	Tetrahydrofuran, PLGA	Polymeric nanoparticles	↑ Encapsulation efficiency and sustained release over 48h as well as ↑ permeability across BeWo cell layers of digoxin-loaded nanoparticles when compared with free digoxin.	[144]
Curcumin-doxorubicin	NVA622 polymer, EDCI	Polymeric nanoparticles	Composite DOX-curcumin nanoparticle that overcomes both MDR-based DOX chemoresistance and DOX-induced cardiotoxicity	[145]
ATP (Freezing-thawing method)	Phosphatidylcholine, cholesterol, 1,2-distearoyl-sn-glycero-3-phosphoethanolamine-N-[methoxy(polyethyleneglycol)-2000]	Liposomes	Significantly lower left ventricular end diastolic pressure after ischemia/reperfusion than controls.	[98]
Candesartan Cilexetil (film hydration method combined with sonication)	Span 60, Pluronic L64 & 127, dicetyl phosphate, stearyl amine, cholesterol	Niosomes	The drug-loaded niosomes exhibited entrapment efficiency, particle size, and zeta potential of these niosomes varied within the range of 99.06 ± 1.74 to 36.26 ± 2.78, 157.3 ± 3.3 to 658.3 ± 12.7 nm, and -14.7 ± 2.8 to -44.5 ± 1.5 mV, respectively, and the in vitro drug release from niosomes was improved after niosomal entrapment compared to pure candesartan cilexetil.	[146]

Continued

Table 1 Continued

Drug (technique)	Excipients	Dosage form	Outcome and significance	Reference
Nerium oleander (Thin-film hydration technique)	Tween 60, cholesterol	Niosomes	The methanol extract of oleander roots (MOE) showed higher polyphenolic content and exhibits a better antioxidant activity in compared to the hydromethanol (20% methanol) extract (MOWE). Encapsulation efficiencies of the vesicles were found as 16.2% for MON (contain MOE) and 13.24% for MWON (contain MOWE).	[147]
Paeonia emodi	HAuCl4	Gold nanoparticles	The plant subfraction *Paeonia emodi* (Pe. EA) and its gold NPs significantly reduced the serum levels of alanine aminotransferase (ALT), aspartate aminotransferase (AST), lactate dehydrogenase (LDH), creatine phosphokinase (CPK) to 66.07 ± 1.54, 77.08 ± 1.79, 84.86 ± 1.34 and $265.34 \pm 4.34 \mathrm{IU/L}$ respectively as compared to ISO treated group. Pe. EA 40-AuNPs (40 mg/kg) reduced the levels of ALT, AST, CPK and LDH to 60.74 ± 2.79, 75.47 ± 1.67, 80.48 ± 2.64 and $247.54 \pm 5.57 \mathrm{IU/L}$ respectively.	[148]
Polyoxalate containing vanillyl alcohol (PVAX) Polymer (Solvent Evaporation method)	Dichloromethane (DCM), polyvinyl alcohol (PVA)	Polymeric nanoparticles	Injecting PVAX nanoparticles at a dose of 4 mg/kg/day into mice treated with DOX-reduced apoptosis in the liver and heart by suppressing PARP-1 and caspase-3 activation. PVAX treatment also prevented DOX-induced cardiac dysfunction and significantly improved the survival rate compared to the vehicle-treated group.	[149]

The fundamental complexity of IRI pathophysiology presents an additional obstacle. Since it is recognized that numerous separate mechanisms lead to the pathophysiology of IRI, it is crucial to emphasize the necessity for a multitarget strategy to heart protection [153]. NPs might be laden with a mix of multitarget payloads designed to address IRI from a therapeutic standpoint. Consequently, additive cardioprotection may be accomplished by targeting two or more signaling mechanisms, the same route at various locations, or numerous cell types, such as endothelial cells, platelets, fibroblasts cardiomyocytes, and macrophages [154].

7.2 Toxicity challenges

One of the most significant challenges which needs to be first overcome prior nanomedicine can be successfully used in the clinical settings is the problem of patient safety. Toxicology studies on nanomedicine, which are relatively new as a kind of treatment, have been mostly ignored throughout the course of the previous several decades [155]. There is presently no standard that can be used to assess and classify the different nanomaterials' degrees of toxicity, despite the growing interest in the safety aspects of diverse nanomedicine. As a result, a consistent procedure for systematic toxicity testing need to be developed. The utilization of the material in the heart necessitates paying particular focus to the toxicity at cellular level and the biological destiny of the materials, particularly in situations when the goal of cardioprotection is to preserve cell life. A technique for translational nanomedicine is to manufacture NPs out of materials that have been certified safe by the FDA and are biodegradable. For cardiac applications, nondegradable materials, such as inorganic nanoparticles (NPs), need to have their physical clearance thoroughly explored [155].

7.3 Technological challenges

In the laboratory, some of the technological difficulties presented by NPs involve scale up production, optimization of performance, as well as projections of future performance. In case of preclinical research, the production of NPs takes place in relatively smaller batches, and it may not always be possible to scale up the process to generate considerable amounts due to the batch size. Nanoparticle preparations which show the best results in animal studies have the most potential in human trials, but despite this, they are rarely systematically improved upon. The goal of performance optimization is to maximize the effectiveness of the nanoparticles. The data from animals and people need to be correlated in order to make accurate forecasts since there are both numerous differences and many similarities between the two [155]. Therefore, this creates a barrier for forecasting the efficacy of nanoparticles in case of preclinical and clinical settings. It is essential that the attempts to correlate the data be put into action as quickly as is practically practicable in order to uncover any general patterns, provided that this is even conceivable for better outcomes. Due to the fact that the majority of quantitative procedures require organ isolation or tissue harvesting, there are significant challenges involved in assessing and identifying whether or not the nanomedicine is beneficial in people. When it comes to researching time-dependent biodistribution, noninvasive imaging methods could be the solution. Imaging a beating heart with high spatial resolution remains a formidable challenge [155].

7.4 Administration routes

Injection into an intravenous vein is the most typical route of administration for NPs that are intended to treat AMI. Intramyocardial injection with enough retention is preferred in the case of hydrogels and self-assembling peptides. This allows for more exact local distribution of the substance. Despite this, the intrusive approach only has a limited use in clinical settings and has a higher risk. Other strategies for the delivery of NPs to the heart, such as oral administration, inhalation, and intraperitoneal injection, have been studied as potential alternatives to intravenous administration [156–158]. It is necessary to do more study in order to discover whether or not these procedures are suitable for the treatment of AMI with other kinds of NPs [155].

8 Future perspectives

The various existing medical issues for the treatment of CVDs include the creation of clear, comprehensive clinical diagnostic judgments and the frequent monitoring of treatment responses. Several advancements have been made to overcome these difficulties. Nanosensors have the potential to improve the diagnostic tools' specificity while maintaining their sensitivity. The creation of biosensors that make use of biomarkers is an essential component in the development of CVD diagnosis. In order to detect various disease markers and the correctly diagnosing the heart disease, it is of the utmost importance to build biosensor platforms that are both highly selective and sensitive, making use of surface chemistry and of nanomaterials too. The instantaneous detection of various biomarkers in a single experimentation utilizing small quantity of blood sample considerably improves the system's validity and reduces diagnostic costs for disease stage evaluation. Combining current techniques like proteomics, microfluidics, and polymer science with the identification of biomarkers and the creation of biosensors may provide miniaturized, user-friendly, precise, and cost-effective biosensing equipment. While major breakthroughs have already been achieved in the nanotechnology for diagnosing the CVDs, early stage diagnosis remains tough due to ambiguous symptoms and the relatively low expression of early stage cardiac biomarkers, making nanotechnological testing difficult. A significant worry is the biomarker's sensitivity and specificity. Earlier study indicated that might be insufficient single indicators, resulting in a lack of sensitivity and specificity for the proper diagnosis of CVD. Because of the ambiguity, heterogeneity, and unpredictability of etiology in different groups, it is impossible to get definitive diagnostic findings with a single biomarker. In addition, the course of illnesses could cause distinct biomarkers to be controlled in different ways. On the other hand, the growing use of nanomaterials in a diverse array of biological applications has led to a rise in concerns over their toxicity. It is widely held that the physicochemical as well as morphological properties of nanomaterials that are advantageous for use in biomedical applications may play a significant part in determining the degree to which these materials are toxic to a variety of organs, including the liver, kidneys, skin, brain, and heart, among other organs. The toxicity of nanoparticles can be decreased or eradicated by designing their surfaces with different kinds of natural or synthetic polymers or other chemicals, despite the fact that nanomaterials have considerable adverse side effects in several experimental scenarios. While there are numerous researches on the biological uses of nanomedicine and their toxicity level, there is no comprehensive information on all of these elements.

9   Conclusion

The applications of nanomaterials may greatly increase the number of methods accessible to doctors for combating life-threatening disorders such as CVDs. In the realm of conventional materials, the prospective applications and effectiveness of nanoparticles such as liposomes, CNTs, and polymeric NPs are thought to be revolutionary. With the growth of nanoscience, new tactics and methods will emerge. As additional nanomaterials fulfil their potential, the predictability of therapies for CVDs may be enhanced and modified. The advancements in NPs make their use intriguing for delivering diagnostics and treatments to ischemic heart after AMI. However, there are various obstacles to bringing the nanomedicine into clinical practice, and main concern is related to safety issues. In order to overcome these challenges and ensure the continued success of translational nanomedicine in CVD and cardioprotection, it is essential to do further research into the design of NPs in relation to their safety and their efficacy.

Acknowledgments

The authors would like to thank Department of Pharmaceutics, MM College of Pharmacy, Maharishi Markendeshwar (Deemed to be University), Mullana-Ambala, Haryana, India 133207 for providing facilities for the completion of this review.

References

[1] Savarese G, Lund LH. Global public health burden of heart failure. Card Fail Rev 2017;3 (1):7. https://doi.org/10.15420/cfr.2016:25:2.
[2] Francis GS, Tang WW. Pathophysiology of congestive heart failure. Rev Cardiovasc Med 2003;4(S2):14–20.
[3] Inamdar AA, Inamdar AC. Heart failure: diagnosis, management and utilization. J Clin Med 2016;5(7):62. https://doi.org/10.3390/jcm5070062.
[4] Dassanayaka S, Jones SP. Recent developments in heart failure. Circ Res 2015;117(7): e58–63. https://doi.org/10.1161/CIRCRESAHA.115.305765.
[5] Roger VL. Epidemiology of heart failure. Circ Res 2013;113(6):646–59. https://doi.org/ 10.1161/CIRCRESAHA.113.300268.
[6] Bui AL, Horwich TB, Fonarow GC. Epidemiology and risk profile of heart failure. Nat Rev Cardiol 2011;8(1):30–41. https://doi.org/10.1038/nrcardio.2010.165.

[7]   Chatterjee K. Congestive heart failure. Am J Cardiovasc Drugs 2002;2(1):1–6. https://doi.org/10.2165/00129784-200202010-00001.

[8]   Lin C, Gao H, Ouyang L. Advance cardiac nanomedicine by targeting the pathophysiological characteristics of heart failure. J Control Release 2021;(337):494–504. https://doi.org/10.1016/j.jconrel.2021.08.002.

[9]   Rosenkranz S, Gibbs JS, Wachter R, De Marco T, Vonk-Noordegraaf A, Vachiery JL. Left ventricular heart failure and pulmonary hypertension. Eur Heart J 2016;37(12):942–54. https://doi.org/10.1093/eurheartj/ehv512.

[10] McMurray JJ. Systolic heart failure. N Engl J Med 2010;362(3):228–38. https://doi.org/10.1056/NEJMcp0909392.

[11] Bogaard HJ, Abe K, Noordegraaf AV, Voelkel NF. The right ventricle under pressure: cellular and molecular mechanisms of right-heart failure in pulmonary hypertension. Chest 2009;135(3):794–804. https://doi.org/10.1378/chest.08-0492.

[12] Mehra MR, Park MH, Landzberg MJ, Lala A, Waxman AB. Right heart failure: toward a common language. Pulm Circ 2013;3(4):963–7. https://doi.org/10.1086/674750.

[13] Ansalone G, Giannantoni P, Ricci R, Trambaiolo P, Fedele F, Santini M. Biventricular pacing in heart failure: back to basics in the pathophysiology of left bundle branch block to reduce the number of nonresponders. Am J Cardiol 2003;91(9):55–61. https://doi.org/10.1016/S0002-9149(02)03339-8.

[14] MacIver DH, Dayer MJ, Harrison AJ. A general theory of acute and chronic heart failure. Int J Cardiol 2013;165(1):25–34. https://doi.org/10.1016/j.ijcard.2012.03.093.

[15] Borlaug BA, Paulus WJ. Heart failure with preserved ejection fraction: pathophysiology, diagnosis, and treatment. Eur Heart J 2011;32(6):670–9. https://doi.org/10.1093/eurheartj/ehq426.

[16] Dunlay SM, Roger VL, Redfield MM. Epidemiology of heart failure with preserved ejection fraction. Nat Rev Cardiol 2017;14(10):591–602. https://doi.org/10.1038/nrcardio.2017.65.

[17] Chen YT, Wong LL, Liew OW, Richards AM. Heart failure with reduced ejection fraction (HFrEF) and preserved ejection fraction (HFpEF): the diagnostic value of circulating microRNAs. Cell 2019;8(12):1651. https://doi.org/10.3390/cells8121651.

[18] Savarese G, Stolfo D, Sinagra G, Lund LH. Heart failure with mid-range or mildly reduced ejection fraction. Nat Rev Cardiol 2022;19(2):100–16. https://doi.org/10.1038/s41569-021-00605-5.

[19] James SL, Abate D, Abate KH, Abay SM, Abbafati C, Abbasi N, Abbastabar H, Abd-Allah F, Abdela J, Abdelalim A, Abdollahpour I. Global, regional, and national incidence, prevalence, and years lived with disability for 354 diseases and injuries for 195 countries and territories, 1990–2017: a systematic analysis for the Global Burden of Disease Study 2017. Lancet 2018;392(10159):1789–858. https://doi.org/10.1016/S0140-6736(18)32279-7.

[20] Groenewegen A, Rutten FH, Mosterd A, Hoes AW. Epidemiology of heart failure. Eur J Heart Fail 2020;22(8):1342–56. https://doi.org/10.1002/ejhf.1858.

[21] Gerber Y, Weston SA, Redfield MM, Chamberlain AM, Manemann SM, Jiang R, Killian JM, Roger VL. A contemporary appraisal of the heart failure epidemic in Olmsted County, Minnesota, 2000 to 2010. JAMA Intern Med 2015;175:996–1004 [go to original source. 2015].

[22] Virani SS, Alonso A, Benjamin EJ, Bittencourt MS, Callaway CW, Carson AP, Chamberlain AM, Chang AR, Cheng S, Delling FN, Djousse L. Heart disease and stroke statistics—2020 update: a report from the American Heart Association. Circulation 2020;141(9):e139–596. https://doi.org/10.1161/CIR.0000000000000757.

[23] Malik A, Brito D, Vaqar S, Chhabra L, Doerr C. Congestive heart failure (nursing). Trea-sure Island, FL: StatPearls Publishing; 2022.

[24] Benjamin EJ, Blaha MJ, Chiuve SE, Cushman M, Das SR, Deo R, de Ferranti SD, Floyd J, Fornage M, Gillespie C, et al. Heart disease and stroke statistics-2017 update: a report from the American Heart Association. Circulation 2017;135:e146–603. https://doi.org/10.1161/CIR.0000000000000485.

[25] Wong CM, Hawkins NM, Jhund PS, MacDonald MR, Solomon SD, Granger CB, Yusuf S, Pfeffer MA, Swedberg K, Petrie MC, McMurray JJ. Clinical characteristics and out-comes of young and very young adults with heart failure: the CHARM programme (Can-desartan in Heart Failure Assessment of Reduction in Mortality and Morbidity). J Am Coll Cardiol 2013;62(20):1845–54. https://doi.org/10.1016/j.jacc.2013.05.072.

[26] Yusuf S, Joseph P, Rangarajan S, Islam S, Mente A, Hystad P, Brauer M, Kutty VR, Gupta R, Wielgosz A, AlHabib KF. Modifiable risk factors, cardiovascular disease, and mortality in 155 722 individuals from 21 high-income, middle-income, and low-income countries (PURE): a prospective cohort study. Lancet 2020;395(10226):795–808. https://doi.org/10.1016/S0140-6736(19)32008-2.

[27] Greyson CR. Pathophysiology of right ventricular failure. Crit Care Med 2008;36(1):S57–65. https://doi.org/10.1097/01.CCM.0000296265.52518.70.

[28] Haddad F, Doyle R, Murphy DJ, Hunt SA. Right ventricular function in cardio-vascular disease, part II: pathophysiology, clinical importance, and management of right ventricular failure. Circulation 2008;117(13):1717–31. https://doi.org/10.1161/CIRCULATIONAHA.107.653584.

[29] Onwuanyi A, Taylor M. Acute decompensated heart failure: pathophysiology and treatment. Am J Cardiol 2007;99(6):S25–30. https://doi.org/10.1016/j.amjcard.2006.12.017.

[30] Boieşan R. Heart failure – a major public health problem. Anul 2011;XVI(2):187.

[31] Cody RJ. Hormonal alterations in heart failure. In: Congestive heart failure: pathophys-iology, diagnosis and comprehensive approach to management. Philadelphia: Lippincott Williams & Wilkins; 2000. p. 199–212.

[32] Gheorghiade M, Pang PS. Acute heart failure syndromes. J Am Coll Cardiol 2009;53(7):557–73. https://doi.org/10.1016/j.jacc.2008.10.041.

[33] Konstam MA, Udelson JE, Anand IS, Cohn JN. Ventricular remodeling in heart failure: a credible surrogate endpoint. J Card Fail 2003;9(5):350–3. https://doi.org/10.1054/j.cardfail.2003.09.001.

[34] Kajstura J, Leri A, Finato N, Di Loreto C, Beltrami CA, Anversa P. Myocyte proliferation in end-stage cardiac failure in humans. Proc Natl Acad Sci 1998;95(15):8801–5. https://doi.org/10.1073/pnas.95.15.8801.

[35] Anversa P, Nadal-Ginard B. Myocyte renewal and ventricular remodelling. Nature 2002;415(6868):240–3. https://doi.org/10.1038/415240a.

[36] Senoner T, Dichtl W. Oxidative stress in cardiovascular diseases: still a therapeutic tar-get? Nutrients 2019;11(9):2090. https://doi.org/10.3390/nu11092090.

[37] Hou J, Yuan Y, Chen P, Lu K, Tang Z, Liu Q, Xu W, Zheng D, Xiong S, Pei H. Pathological roles of oxidative stress in cardiac microvascular injury. Curr Probl Car-diol 2022; 101399. https://doi.org/10.1016/j.cpcardiol.2022.101399.

[38] Petrie JR, Guzik TJ, Touyz RM. Diabetes, hypertension, and cardiovascular disease: clin-ical insights and vascular mechanisms. Can J Cardiol 2018;34(5):575–84. https://doi.org/10.1016/j.cjca.2017.12.005.

[39] Rosca MG, Hoppel CL. Mitochondrial dysfunction in heart failure. Heart Fail Rev 2013;18:607–22. https://doi.org/10.1007/s10741-012-9340-0.

[40] Burgoyne JR, Mongue-Din H, Eaton P, Shah AM. Redox signaling in cardiac physiology and pathology. Circ Res 2012;111(8):1091–106. https://doi.org/10.1161/CIRCRESAHA.111.255216.

[41] Aimo A, Castiglione V, Borrelli C, Saccaro LF, Franzini M, Masi S, Emdin M, Giannoni A. Oxidative stress and inflammation in the evolution of heart failure: from pathophysiology to therapeutic strategies. Eur J Prev Cardiol 2020;27(5):494–510. https://doi.org/10.1177/2047487319870344.

[42] Frangogiannis NG. The inflammatory response in myocardial injury, repair, and remodelling. Nat Rev Cardiol 2014;11(5):255–65. https://doi.org/10.1038/nrcardio.2014.28.

[43] Pacher P, Szabó C. Role of poly (ADP-ribose) polymerase 1 (PARP-1) in cardiovascular diseases: the therapeutic potential of PARP inhibitors. Cardiovasc Drug Rev 2007;25 (3):235–60. https://doi.org/10.1111/j.1527-3466.2007.00018.x.

[44] Paulus WJ, Tschöpe C. A novel paradigm for heart failure with preserved ejection fraction: comorbidities drive myocardial dysfunction and remodeling through coronary microvascular endothelial inflammation. J Am Coll Cardiol 2013;62(4):263–71. https://doi.org/10.1016/j.jacc.2013.02.092.

[45] Pandya NM, Dhalla NS, Santani DD. Angiogenesis—a new target for future therapy. Vasc Pharmacol 2006;44(5):265–74. https://doi.org/10.1016/j.vph.2006.01.005.

[46] Testa U, Pannitteri G, Condorelli GL. Vascular endothelial growth factors in cardiovascular medicine. J Cardiovasc Med 2008;9(12):1190–221. https://doi.org/10.2459/JCM.0b013e3283117d37.

[47] Khan S, Villalobos MA, Choron RL, Chang S, Brown SA, Carpenter JP, Tulenko TN, Zhang P. Fibroblast growth factor and vascular endothelial growth factor play a critical role in endotheliogenesis from human adipose-derived stem cells. J Vasc Surg 2017;65 (5):1483–92. https://doi.org/10.1016/j.jvs.2016.04.034.

[48] Mühleder S, Fernandez-Chacon M, Garcia-Gonzalez I, Benedito R. Endothelial sprouting, proliferation, or senescence: tipping the balance from physiology to pathology. Cell Mol Life Sci 2021;78(4):1329–54. https://doi.org/10.1007/s00018-020-03664-y.

[49] Kim YW, West XZ, Byzova TV. Inflammation and oxidative stress in angiogenesis and vascular disease. J Mol Med 2013;91:323–8. https://doi.org/10.1007/s00109-013-1007-3.

[50] Johnson A, DiPietro LA. Apoptosis and angiogenesis: an evolving mechanism for fibrosis. FASEB J 2013;27(10):3893. https://doi.org/10.1096/fj.12-214189.

[51] Schirone L, Forte M, Palmerio S, Yee D, Nocella C, Angelini F, Pagano F, Schiavon S, Bordin A, Carrizzo A, Vecchione C. A review of the molecular mechanisms underlying the development and progression of cardiac remodeling. Oxidative Med Cell Longev 2017;2017. https://doi.org/10.1155/2017/3920195.

[52] Zhang Q, Kandic I, Kutryk MJ. Dysregulation of angiogenesis-related microRNAs in endothelial progenitor cells from patients with coronary artery disease. Biochem Biophys Res Commun 2011;405(1):42–6. https://doi.org/10.1016/j.bbrc.2010.12.119.

[53] Henes J, Rosenberger P. Systolic heart failure: diagnosis and therapy. Curr Opin Anesthesiol 2016;29(1):55–60. https://doi.org/10.1097/ACO.0000000000000270.

[54] Nicoara A, Jones-Haywood M. Diastolic heart failure: diagnosis and therapy. Curr Opin Anesthesiol 2016;29(1):61–7. https://doi.org/10.1097/ACO.0000000000000276.

[55] Kjær A, Hesse B. Heart failure and neuroendocrine activation: diagnostic, prognostic and therapeutic perspectives. Clin Physiol 2001;21(6):661–72. https://doi.org/10.1046/j.1365-2281.2001.00371.x.

[56] Leri A, Claudio PP, Li Q, Wang X, Reiss K, Wang S, Malhotra A, Kajstura J, Anversa P. Stretch-mediated release of angiotensin II induces myocyte apoptosis by activating p53 that enhances the local renin-angiotensin system and decreases the Bcl-2-to-Bax protein ratio in the cell. J Clin Invest 1998;101(7):1326–42. https://doi.org/10.1172/JCI316.

[57] Rohini A, Agrawal N, Koyani CN, Singh R. Molecular targets and regulators of cardiac hypertrophy. Pharmacol Res 2010;61(4):269–80. https://doi.org/10.1016/j.phrs.2009.11.012.

[58] Gary R, Davis L. Diastolic heart failure. Heart Lung 2008;37(6):405–16. https://doi.org/10.1016/j.hrtlng.2007.12.002.

[59] Morris DA, Gailani M, Pérez AV, Blaschke F, Dietz R, Haverkamp W, Özcelik C. Right ventricular myocardial systolic and diastolic dysfunction in heart failure with normal left ventricular ejection fraction. J Am Soc Echocardiogr 2011;24(8):886–97. https://doi.org/10.1016/j.echo.2011.04.005.

[60] Mann DL, Bristow MR. Mechanisms and models in heart failure: the biomechanical model and beyond. Circulation 2005;111(21):2837–49. https://doi.org/10.1161/CIRCULATIONAHA.104.500546.

[61] Volpe M, Carnovali M, Mastromarino V. The natriuretic peptides system in the pathophysiology of heart failure: from molecular basis to treatment. Clin Sci 2016;130(2):57–77. https://doi.org/10.1042/CS20150469.

[62] Leite-Moreira AF. Current perspectives in diastolic dysfunction and diastolic heart failure. Heart 2006;92(5):712–8. https://doi.org/10.1136/hrt.2005.062950.

[63] Nagueh SF, Appleton CP, Gillebert TC, Marino PN, Oh JK, Smiseth OA, Waggoner AD, Flachskampf FA, Pellikka PA, Evangelisa A. Recommendations for the evaluation of left ventricular diastolic function by echocardiography. Eur J Echocardiogr 2009;10(2):165–93. https://doi.org/10.1093/ejechocard/jep007.

[64] Silvestri NJ, Ismail H, Zimetbaum P, Raynor EM. Cardiac involvement in the muscular dystrophies. Muscle Nerve 2018;57(5):707–15. https://doi.org/10.1002/mus.26014.

[65] Hernandez-Madrid A, Paul T, Abrams D, Aziz PF, Blom NA, Chen J, Chessa M, Combes N, Dagres N, Diller G, Ernst S. Arrhythmias in congenital heart disease: a position paper of the European Heart Rhythm Association (EHRA), Association for European Paediatric and Congenital Cardiology (AEPC), and the European Society of Cardiology (ESC) Working Group on Grown-up Congenital Heart Disease, endorsed by HRS, PACES, APHRS, and SOLAECE. EP Europace 2018;20(11):1719–53. https://doi.org/10.1093/europace/eux380.

[66] McElwee SK, Velasco A, Doppalapudi H. Mechanisms of sudden cardiac death. J Nucl Cardiol 2016;23(6):1368–79. https://doi.org/10.1007/s12350-016-0600-6.

[67] Chinnaiyan KM, Alexander D, Maddens M, McCullough PA. Curriculum in cardiology: integrated diagnosis and management of diastolic heart failure. Am Heart J 2007;153(2):189–200. https://doi.org/10.1016/j.ahj.2006.10.022.

[68] Mandinov L, Eberli FR, Seiler C, Hess OM. Diastolic heart failure. Cardiovasc Res 2000;45(4):813–25. https://doi.org/10.1016/S0008-6363(99)00399-5.

[69] Feldman AM, Combes A, Wagner D, Kadakomi T, Kubota T, You Li Y, McTiernan C. The role of tumor necrosis factor in the pathophysiology of heart failure. J Am Coll Cardiol 2000;35(3):537–44. https://doi.org/10.1016/S0735-1097(99)00600-2.

[70] Spieker LE, Noll G, Ruschitzka FT, Lüscher TF. Endothelin receptor antagonists in congestive heart failure: a new therapeutic principle for the future? J Am Coll Cardiol 2001;37(6):1493–505. https://doi.org/10.1016/s0735-1097(01)01210-4.

[71] Bartekova M, Radosinska J, Jelemensky M, Dhalla NS. Role of cytokines and inflammation in heart function during health and disease. Heart Fail Rev 2018;23(5):733–58. https://doi.org/10.1007/s10741-018-9716-x.

[72] Convertino VA, Cooke WH, Holcomb JB. Arterial pulse pressure and its association with reduced stroke volume during progressive central hypovolemia. J Trauma Acute Care Surg 2006;61(3):629–34. https://doi.org/10.1097/01.ta.0000196663.34175.33.

[73] Ekman I, Chassany O, Komajda M, Böhm M, Borer JS, Ford I, Tavazzi L, Swedberg K. Heart rate reduction with ivabradine and health related quality of life in patients with chronic heart failure: results from the SHIFT study. Eur Heart J 2011;32(19):2395–404. https://doi.org/10.1093/eurheartj/ehr343.

[74] Braunwald E. The path to an angiotensin receptor antagonist-neprilysin inhibitor in the treatment of heart failure. J Am Coll Cardiol 2015;65(10):1029–41. https://doi.org/10.1016/j.jacc.2015.01.033.

[75] Ferdinand KC, Elkayam U, Mancini D, Ofili E, Piña I, Anand I, Feldman AM, McNamara D, Leggett C. Use of isosorbide dinitrate and hydralazine in African-Americans with heart failure 9 years after the African-American Heart Failure Trial. Am J Cardiol 2014;114(1):151–9. https://doi.org/10.1016/j.amjcard.2014.04.018.

[76] Bhatt AS, Vaduganathan M, Claggett BL, Liu J, Packer M, Desai AS, Lefkowitz MP, Rouleau JL, Shi VC, Zile MR, Swedberg K. Effect of sacubitril/valsartan vs. enalapril on changes in heart failure therapies over time: the PARADIGM-HF trial. Eur J Heart Fail 2021;23(9):1518–24. https://doi.org/10.1002/ejhf.2259.

[77] Beygui F, Cayla G, Roule V, Roubille F, Delarche N, Silvain J, Van Belle E, Belle L, Galinier M, Motreff P, Cornillet L. Early aldosterone blockade in acute myocardial infarction: the ALBATROSS randomized clinical trial. J Am Coll Cardiol 2016;67 (16):1917–27. https://doi.org/10.1016/j.jacc.2016.02.033.

[78] Su Y, Ma T, Wang Z, Dong B, Tai C, Wang H, Zhang F, Yan C, Chen W, Xu Y, Ye L. Efficacy of early initiation of ivabradine treatment in patients with acute heart failure: rationale and design of SHIFT-AHF trial. ESC Heart Fail 2020;7(6):4465–71. https://doi.org/10.1002/ehf2.12997.

[79] Ezekowitz JA, Zheng Y, Cohen-Solal A, Melenovský V, Escobedo J, Butler J, Hernandez AF, Lam CS, O'Connor CM, Pieske B, Ponikowski P. Hemoglobin and clinical outcomes in the vericiguat global study in patients with heart failure and reduced ejection fraction (VICTORIA).   Circulation   2021;144(18):1489–99.   https://doi.org/10.1161/CIRCULATIONAHA.121.056797.

[80] van der Meer P, Gaggin HK, Dec GW. ACC/AHA versus ESC guidelines on heart failure: JACC guideline comparison. J Am Coll Cardiol 2019;73(21):2756–68. https://doi.org/10.1016/j.jacc.2019.03.478.

[81] Yancy CW, Jessup M, Bozkurt B, Butler J, Casey DE, Drazner MH, Fonarow GC, Geraci SA, Horwich T, Januzzi JL, Johnson MR. 2013 ACCF/AHA guideline for the management of heart failure: a report of the American College of Cardiology Foundation/American Heart Association Task Force on Practice Guidelines. J Am Coll Cardiol 2013;62 (16):e147–239. https://doi.org/10.1161/CIR.0b013e31829e8776.

[82] Prajnamitra RP, Chen HC, Lin CJ, Chen LL, Hsieh PC. Nanotechnology approaches in tackling cardiovascular diseases. Molecules 2019;24(10):2017. https://doi.org/10.3390/molecules24102017.

[83] Haba MŞ, Şerban DN, Şerban L, Tudorancea IM, Haba RM, Mitu O, Iliescu R, Tudorancea I. Nanomaterial-based drug targeted therapy for cardiovascular diseases: ischemic heart failure and atherosclerosis. Crystals 2021;11(10):1172. https://doi.org/10.3390/cryst11101172.

[84] Matoba T, Egashira K. Nanoparticle-mediated drug delivery system for cardiovascular disease. Int Heart J 2014;55(4):281–6. https://doi.org/10.1536/ihj.14-150.

[85] Choi KA, Kim JH, Ryu K, Kaushik N. Current nanomedicine for targeted vascular disease treatment: trends and perspectives. Int J Mol Sci 2022;23(20):12397. https://doi.org/10.3390/ijms232012397.

[86] Singh S, Behl T, Sharma N, Zahoor I, Chigurupati S, Yadav S, Rachamalla M, Sehgal A, Naved T, Arora S, Bhatia S. Targeting therapeutic approaches and highlighting the potential role of nanotechnology in atopic dermatitis. Environ Sci Pollut Res 2022;29 (22):32605–30. https://doi.org/10.1007/s11356-021-18429-8.

[87] Lombardo D, Kiselev MA, Caccamo MT. Smart nanoparticles for drug delivery application: development of versatile nanocarrier platforms in biotechnology and nanomedicine. J Nanomater 2019;2019. https://doi.org/10.1155/2019/3702518.

[88] Sharma N, Zahoor I, Sachdeva M, Subramaniyan V, Fuloria S, Fuloria NK, Naved T, Bhatia S, Al-Harrasi A, Aleya L, Bungau S. Deciphering the role of nanoparticles for management of bacterial meningitis: an update on recent studies. Environ Sci Pollut Res 2021;28(43):60459–76. https://doi.org/10.1007/s11356-021-16570-y.

[89] Bao R, Tan B, Liang S, Zhang N, Wang W, Liu W. A π-π conjugation-containing soft and conductive injectable polymer hydrogel highly efficiently rebuilds cardiac function after myocardial infarction. Biomaterials 2017;122:63–71. https://doi.org/10.1016/j.biomaterials.2017.01.012.

[90] Garbern JC, Minami E, Stayton PS, Murry CE. Delivery of basic fibroblast growth factor with a pH-responsive, injectable hydrogel to improve angiogenesis in infarcted myocardium. Biomaterials 2011;32(9):2407–16. https://doi.org/10.1016/j.biomaterials.2010.11.075.

[91] Nguyen MM, Carlini AS, Chien MP, Sonnenberg S, Luo C, Braden RL, Osborn KG, Li Y, Gianneschi NC, Christman KL. Enzyme-responsive nanoparticles for targeted accumulation and prolonged retention in heart tissue after myocardial infarction. Adv Mater 2015;27(37):5547–52. https://doi.org/10.1002/adma.201502003.

[92] Zhang C, Hsieh MH, Wu SY, Li SH, Wu J, Liu SM, Wei HJ, Weisel RD, Sung HW, Li RK. A self-doping conductive polymer hydrogel that can restore electrical impulse propagation at myocardial infarct to prevent cardiac arrhythmia and preserve ventricular function. Biomaterials 2020;231, 119672. https://doi.org/10.1016/j.biomaterials.2019.119672.

[93] Perez C, Sanchez A, Putnam D, Ting D, Langer R, Alonso MJ. Poly (lactic acid)-poly (ethylene glycol) nanoparticles as new carriers for the delivery of plasmid DNA. J Control Release 2001;75(1–2):211–24. https://doi.org/10.1016/S0168-3659(01)00397-2.

[94] Chang MY, Yang YJ, Chang CH, Tang AC, Liao WY, Cheng FY, Yeh CS, Lai JJ, Stayton PS, Hsieh PC. Functionalized nanoparticles provide early cardioprotection after acute myocardial infarction. J Control Release 2013;170(2):287–94. https://doi.org/10.1016/j.jconrel.2013.04.022.

[95] Nagaoka K, Matoba T, Mao Y, Nakano Y, Ikeda G, Egusa S, Tokutome M, Nagahama R, Nakano K, Sunagawa K, Egashira K. A new therapeutic modality for acute myocardial infarction: nanoparticle-mediated delivery of pitavastatin induces cardioprotection from ischemia-reperfusion injury via activation of PI3K/Akt pathway and anti-inflammation in a rat model. PLoS One 2015;10(7), e0132451. https://doi.org/10.1371/journal.pone.0132451.

[96] Quadros HC, Santos LD, Meira CS, Khouri MI, Mattei B, Soares MB, De Castro-Borges W, Farias LP, Formiga FR. Development and in vitro characterization of polymeric nanoparticles containing recombinant adrenomedullin-2 intended for therapeutic angiogenesis. Int J Pharm 2020;576, 118997. https://doi.org/10.1016/j.ijpharm.2019.118997.

[97] Levchenko TS, Hartner WC, Torchilin VP. Liposomes in diagnosis and treatment of cardiovascular disorders. Methodist Debakey Cardiovasc J 2012;8(1):36. https://doi.org/10.14797/mdcj-8-1-36.

[98] Verma DD, Levchenko TS, Bernstein EA, Torchilin VP. ATP-loaded liposomes effectively protect mechanical functions of the myocardium from global ischemia in an isolated rat heart model. J Control Release 2005;108(2–3):460–71. https://doi.org/10.1016/j.jconrel.2005.08.029.

[99] Luciani P, Fevre M, Leroux JC. Development and physico-chemical characterization of a liposomal formulation of istaroxime. Eur J Pharm Biopharm 2011;79(2):285–93. https://doi.org/10.1016/j.ejpb.2011.04.013.

[100] Sarter B. Coenzyme Q10 and cardiovascular disease: a review. J Cardiovasc Nurs 2002;16(4):9–20. https://doi.org/10.1097/00005082-200207000-00003.

[101] Allijn IE, Czarny BM, Wang X, Chong SY, Weiler M, Da Silva AE, Metselaar JM, Lam CS, Pastorin G, De Kleijn DP, Storm G. Liposome encapsulated berberine treatment attenuates cardiac dysfunction after myocardial infarction. J Control Release 2017;247:127–33. https://doi.org/10.1016/j.jconrel.2016.12.042.

[102] Ge X, Wei M, He S, Yuan WE. Advances of non-ionic surfactant vesicles (niosomes) and their application in drug delivery. Pharmaceutics 2019;11(2):55. https://doi.org/10.3390/pharmaceutics11020055.

[103] Khatoon M, Shah KU, Din FU, Shah SU, Rehman AU, Dilawar N, Khan AN. Proniosomes derived niosomes: recent advancements in drug delivery and targeting. Drug Deliv 2017;24(2):56–69. https://doi.org/10.1080/10717544.2017.1384520.

[104] Abdelkader H, Alani AW, Alany RG. Recent advances in non-ionic surfactant vesicles (niosomes): self-assembly, fabrication, characterization, drug delivery applications and limitations. Drug Deliv 2014;21(2):87–100. https://doi.org/10.3109/10717544.2013.838077.

[105] Ritwiset A, Krongsuk S, Johns JR. Molecular structure and dynamical properties of niosome bilayers with and without cholesterol incorporation: a molecular dynamics simulation study. Appl Surf Sci 2016;380:23–31. https://doi.org/10.1016/j.apsusc.2016.02.092.

[106] Rentel CO, Bouwstra JA, Naisbett B, Junginger HE. Niosomes as a novel peroral vaccine delivery system. Int J Pharm 1999;186(2):161–7. https://doi.org/10.1016/S0378-5173(99)00167-2.

[107] Behl T, Singh S, Sharma N, Zahoor I, Albarrati A, Albratty M, Meraya AM, Najmi A, Bungau S. Expatiating the pharmacological and nanotechnological aspects of the alkaloidal drug Berberine: current and future trends. Molecules 2022;27(12):3705. https://doi.org/10.3390/molecules27123705.

[108] Wennink JW, Liu Y, Mäkinen PI, Setaro F, de la Escosura A, Bourajjaj M, Lappalainen JP, Holappa LP, van den Dikkenberg JB, Al Fartousi M, Trohopoulos PN. Macrophage selective photodynamic therapy by meta-tetra (hydroxyphenyl) chlorin loaded polymeric micelles: a possible treatment for cardiovascular diseases. Eur J Pharm Sci 2017;107:112–25. https://doi.org/10.1016/j.ejps.2017.06.038.

[109] Peters DT. Targeting atherosclerosis: nanoparticle delivery for diagnosis and treatment. San Diego: University of California; 2009.

[110] Nakashiro S, Matoba T, Umezu R, Koga JI, Tokutome M, Katsuki S, Nakano K, Sunagawa K, Egashira K. Pioglitazone-incorporated nanoparticles prevent plaque destabilization and rupture by regulating monocyte/macrophage differentiation in ApoE−/− mice. Arterioscler Thromb Vasc Biol 2016;36(3):491–500. https://doi.org/10.1161/ATVBAHA.115.307057.

[111] Geelen T, Paulis LE, Coolen BF, Nicolay K, Strijkers GJ. Passive targeting of lipid-based nanoparticles to mouse cardiac ischemia–reperfusion injury. Contrast Media Mol Imaging 2013;8(2):117–26. https://doi.org/10.1002/cmmi.1501.

[112] Lukyanov AN, Hartner WC, Torchilin VP. Increased accumulation of PEG–PE micelles in the area of experimental myocardial infarction in rabbits. J Control Release 2004;94 (1):187–93. https://doi.org/10.1016/j.jconrel.2003.10.008.

[113] Wang J, Seo MJ, Deci MB, Weil BR, Canty JM, Nguyen J. Effect of CCR2 inhibitor-loaded lipid micelles on inflammatory cell migration and cardiac function after myocardial infarction. Int J Nanomedicine 2018;13:6441. https://doi.org/10.2147/IJN.S178650.

[114] Samidurai A, Salloum FN, Durrant D, Chernova OB, Kukreja RC, Das A. Chronic treatment with novel nanoformulated micelles of rapamycin, rapatar, protects diabetic heart against ischaemia/reperfusion injury. Br J Pharmacol 2017;174(24):4771–84. https://doi.org/10.1111/bph.14059.

[115] Ávan der Vlies AJ. Hydrogen sulfide donor micelles protect cardiomyocytes from ischemic cell death. Mol BioSyst 2017;13(9):1705–8. https://doi.org/10.1039/C7MB00191F.

[116] Carlson LJ, Cote B, Alani AW, Rao DA. Polymeric micellar co-delivery of resveratrol and curcumin to mitigate in vitro doxorubicin-induced cardiotoxicity. J Pharm Sci 2014;103(8):2315–22. https://doi.org/10.1002/jps.24042.

[117] Cote B, Carlson LJ, Rao DA, Alani AW. Combinatorial resveratrol and quercetin polymeric micelles mitigate doxorubicin induced cardiotoxicity in vitro and in vivo. J Control Release 2015;213:128–33. https://doi.org/10.1016/j.jconrel.2015.06.040.

[118] Li L, Ni J, Li M, Chen J, Han L, Zhu Y, Kong D, Mao J, Wang Y, Zhang B, Zhu M. Ginsenoside Rg3 micelles mitigate doxorubicin-induced cardiotoxicity and enhance its anticancer efficacy. Drug Deliv 2017;24(1):1617–30. https://doi.org/10.1080/10717544.2017.1391893.

[119] Wang Y, Wang Y, Wang X, Hu P. Tilianin-loaded reactive oxygen species-scavenging nano-micelles protect H9c2 cardiomyocyte against hypoxia/reoxygenation-induced injury. J Cardiovasc Pharmacol 2018;72(1):32–9. https://doi.org/10.1097/FJC.0000000000000587.

[120] Li W, Wu J, Zhang J, Wang J, Xiang D, Luo S, Li J, Liu X. Puerarin-loaded PEG-PE micelles with enhanced anti-apoptotic effect and better pharmacokinetic profile. Drug Deliv 2018;25(1):827–37. https://doi.org/10.1080/10717544.2018.1455763.

[121] Li WQ, Wu JY, Xiang DX, Luo SL, Hu XB, Tang TT, Sun TL, Liu XY. Micelles loaded with puerarin and modified with triphenylphosphonium cation possess mitochondrial targeting and demonstrate enhanced protective effect against isoprenaline-induced H9c2 cells apoptosis. Int J Nanomedicine 2019;14:8345. https://doi.org/10.2147/IJN.S219670.

[122] Chen CC, Lin YP, Wang CW, Tzeng HC, Wu CH, Chen YC, Chen CP, Chen LC, Wu YC. DNA–gold nanorod conjugates for remote control of localized gene expression by near infrared irradiation. J Am Chem Soc 2006;128(11):3709–15. https://doi.org/10.1021/ja0570180.

[123] Chithrani BD, Chan WC. Elucidating the mechanism of cellular uptake and removal of protein-coated gold nanoparticles of different sizes and shapes. Nano Lett 2007;7(6): 1542–50. https://doi.org/10.1021/nl070363y.

[124] Ghann WE, Aras O, Fleiter T, Daniel MC. Synthesis and characterization of lisinopril-coated gold nanoparticles as highly stable targeted CT contrast agents in cardiovascular diseases. Langmuir 2012;28(28):10398–408. https://doi.org/10.1021/la301694q.

[125] Spivak MY, Bubnov RV, Yemets IM, Lazarenko LM, Tymoshok NO, Ulberg ZR. Development and testing of gold nanoparticles for drug delivery and treatment of heart failure:

a theranostic potential for PPP cardiology. EPMA J 2013;4(1):1–23. https://doi.org/
10.1186/1878-5085-4-20.

[126] Sharma N, Zahoor I, Singh S, Behl T, Antil A. Expatiating the pivotal role of dendrimers
as emerging nanocarrier for management of liver disorders. J Integr Sci Technol 2023;11
(2):489.

[127] Zahoor I, Singh S, Sharma N, Behl T, Wani SN. Dendrimers: versatile and revolutionary
nanocarriers for infectious diseases. ECS Trans 2022;107(1):8619.

[128] Singh I, Rehni AK, Kalra R, Joshi G, Kumar M. Dendrimers and their pharmaceutical
applications—a review. Pharmazie 2008;63(7):491–6. https://doi.org/10.1691/
ph.2008.8052.

[129] Lee CC, MacKay JA, Fréchet JM, Szoka FC. Designing dendrimers for biological applications. Nat Biotechnol 2005;23(12):1517–26. https://doi.org/10.1038/nbt1171.

[130] Röglin L, Lempens EH, Meijer EW. A synthetic "tour de force": well-defined multiva-
lent and multimodal dendritic structures for biomedical applications. Angew Chem Int
Ed 2011;50(1):102–12. https://doi.org/10.1002/anie.201003968.

[131] Meel RV, Vehmeijer LJ, Kok RJ, Storm G, van Gaal EV. Ligand-targeted particulate
nanomedicines undergoing clinical evaluation: current status. Intracellular Deliv 2016;
III:163–200. https://doi.org/10.1007/978-3-319-43525-1_7.

[132] Almutairi A, Rossin R, Shokeen M, Hagooly A, Ananth A, Capoccia B, Guillaudeu S,
Abendschein D, Anderson CJ, Welch MJ, Fréchet JM. Biodegradable dendritic positron-
emitting nanoprobes for the noninvasive imaging of angiogenesis. Proc Natl Acad Sci
2009;106(3):685–90. https://doi.org/10.1073/pnas.0811757106.

[133] Lu Y, Sun B, Li C, Schoenfisch MH. Structurally diverse nitric oxide-releasing poly (propylene imine) dendrimers. Chem Mater 2011;23(18):4227–33. https://doi.org/10.1021/
cm201628z.

[134] Tomalia DA. Starburstr dendrimers—nanoscopic supermolecules according to dendritic
rules and principles. In: Macromolecular symposia, vol. 101. Basel: Hüthig&Wepf Ver-
lag; 1996. p. 243–55. https://doi.org/10.1002/masy.19961010128.

[135] Venkatesh S, Li M, Saito T, Tong M, Rashed E, Mareedu S, Zhai P, Bárcena C, López-
Otín C, Yehia G, Sadoshima J. Mitochondrial LonP1 protects cardiomyocytes from
ischemia/reperfusion injury in vivo. J Mol Cell Cardiol 2019;128:38–50. https://doi.
org/10.1016/j.yjmcc.2018.12.017.

[136] Taite LJ, West JL. Poly (ethylene glycol)-lysine dendrimers for targeted delivery of nitric
oxide. J Biomater Sci Polym Ed 2006;17(10):1159–72. https://doi.org/10.1163/
156856206778530696.

[137] Bhadra D, Bhadra S, Jain NK. Pegylated lysine based copolymeric dendritic micelles for
solubilization and delivery of artemether. J Pharm Pharm Sci 2005;8(3):467–82.

[138] Sharma AP, Maheshwari R, Tekade M, Kumar Tekade R. Nanomaterial based
approaches for the diagnosis and therapy of cardiovascular diseases. Curr Pharm Des
2015;21(30):4465–78.

[139] Jayagopal A, Russ PK, Haselton FR. Surface engineering of quantum dots for in vivo vascular imaging. Bioconjug Chem 2007;18(5):1424–33. https://doi.org/10.1021/bc070020r.

[140] Bhatt A, Thekkuveettil A, Ganapathy S, Panniyammakal J, Sivadasanpillai H, Gopi
M. To evaluate the feasibility of cadmium/tellurium (Cd/Te) quantum dots for develop-
ing N-terminal natriuretic peptide (NT-proBNP) in-vitro diagnostics. J Immunoass
Immunochem 2022;(1). https://doi.org/10.1080/15321819.2022.2103430.

[141] Chen Z, Zhang X, Yang R, Zhu Z, Chen Y, Tan W. Single-walled carbon nanotubes as
optical materials for biosensing. Nanoscale 2011;3(5):1949–56. https://doi.org/10.1039/
C0NR01014F.

[142] Strus MC, Chiaramonti AN, Kim YL, Jung YJ, Keller RR. Accelerated reliability testing of highly aligned single-walled carbon nanotube networks subjected to DC electrical stressing. Nanotechnology 2011;22(26), 265713. https://doi.org/10.1088/0957-4484/22/26/265713.

[143] Xu Q, Yang Y, Hou J, Chen T, Fei Y, Wang Q, Zhou Q, Li W, Ren J, Li YG. A carbon nanotubes based in situ multifunctional power assist system for restoring failed heart function. BMC Biomed Eng 2021;3(1):1–3. https://doi.org/10.1186/s42490-021-00051-x.

[144] Albekairi NA, Al-Enazy S, Ali S, Rytting E. Transport of digoxin-loaded polymeric nanoparticles across BeWo cells, an in vitro model of human placental trophoblast. Ther Deliv 2015;6(12):1325–34. https://doi.org/10.4155/tde.15.79.

[145] Pramanik D, Campbell NR, Das S, Gupta S, Chenna V, Bisht S, Sysa-Shah P, Bedja D, Karikari C, Steenbergen C, Gabrielson KL. A composite polymer nanoparticle overcomes multidrug resistance and ameliorates doxorubicin-associated cardiomyopathy. Oncotarget 2012;3(6):640. https://doi.org/10.18632/oncotarget.543.

[146] Sezgin-Bayindir Z, Antep MN, Yuksel N. Development and characterization of mixed niosomes for oral delivery using candesartan cilexetil as a model poorly water-soluble drug. AAPS PharmSciTech 2015;16:108–17. https://doi.org/10.1208/s12249-014-0213-9.

[147] Gunes A, Guler E, Un RN, Demir B, Barlas FB, Yavuz M, Coskunol H, Timur S. Niosomes of *Nerium oleander* extracts: in vitro assessment of bioactive nanovesicular structures. J Drug Delivery Sci Technol 2017;37:158–65. https://doi.org/10.1016/j.jddst.2016.12.013.

[148] Ibrar M, Khan MA, Imran M. Evaluation of Paeonia emodi and its gold nanoparticles for cardioprotective and antihyperlipidemic potentials. J Photochem Photobiol B Biol 2018;189:5–13. https://doi.org/10.1016/j.jphotobiol.2018.09.018.

[149] Park S, Yoon J, Bae S, Park M, Kang C, Ke Q, Lee D, Kang PM. Therapeutic use of H_2O_2-responsive anti-oxidant polymer nanoparticles for doxorubicin-induced cardiomyopathy. Biomaterials 2014;35(22):5944–53. https://doi.org/10.1016/j.biomaterials.2014.03.084.

[150] Satalkar P, Elger BS, Hunziker P, Shaw D. Challenges of clinical translation in nanomedicine: a qualitative study. Nanomedicine 2016;12(4):893–900. https://doi.org/10.1016/j.nano.2015.12.376.

[151] Ho YT, Poinard B, Kah JC. Nanoparticle drug delivery systems and their use in cardiac tissue therapy. Nanomedicine 2016;11(6):693–714. https://doi.org/10.2217/nnm.16.6.

[152] Weis SM. Vascular permeability in cardiovascular disease and cancer. Curr Opin Hematol 2008;15(3):243–9. https://doi.org/10.1097/MOH.0b013e3282f97d86.

[153] Rossello X, Yellon DM. The RISK pathway and beyond. Basic Res Cardiol 2018;113 (1):1–5. https://doi.org/10.1007/s00395-017-0662-x.

[154] Davidson SM, Ferdinandy P, Andreadou I, Bøtker HE, Heusch G, Ibáñez B, Ovize M, Schulz R, Yellon DM, Hausenloy DJ, Garcia-Dorado D. Multitarget strategies to reduce myocardial ischemia/reperfusion injury: JACC review topic of the week. J Am Coll Cardiol 2019;73(1):89–99. https://doi.org/10.1016/j.jacc.2018.09.086.

[155] Mendez-Fernandez A, Cabrera-Fuentes HA, Velmurugan B, Irei J, Boisvert WA, Lu S, Hausenloy DJ. Nanoparticle delivery of cardioprotective therapies. Cond Med 2020;3 (1):18.

[156] Simon-Yarza T, Tamayo E, Benavides C, Lana H, Formiga FR, Grama CN, Ortiz-de-Solorzano C, Kumar MR, Prosper F, Blanco-Prieto MJ. Functional benefits of PLGA particulates carrying VEGF and CoQ10 in an animal of myocardial ischemia. Int J Pharm 2013;454(2):784–90. https://doi.org/10.1016/j.ijpharm.2013.04.015.

[157] Miragoli M, Ceriotti P, Iafisco M, Vacchiano M, Salvarani N, Alogna A, Carullo P, Ramirez-Rodríguez GB, Patrício T, Esposti LD, Rossi F. Inhalation of peptide-loaded nanoparticles improves heart failure. Sci Transl Med 2018;10(424), eaan6205. https://doi.org/10.1126/scitranslmed.aan6205.

[158] Abdelhalim MA. Exposure to gold nanoparticles produces cardiac tissue damage that depends on the size and duration of exposure. Lipids Health Dis 2011;10(1):1–9. https://doi.org/10.1186/1476-511X-10-205.

Expatiating the role of angiogenesis, inflammation, and oxidative stress in angina pectoris: A state-of-the-art on the drug delivery approaches

9

Sukhbir Singh[a], Ishrat Zahoor[a], Priya Dhiman[a], Neelam Sharma[a], Sonam Grewal[b], Tapan Behl[b], and Shahid Nazir Wani[c]
[a]Department of Pharmaceutics, MM College of Pharmacy, Maharishi Markandeshwar (Deemed to be University), Mullana-Ambala, Haryana, India, [b]Amity School of Pharmaceutical Sciences, Amity University, Mohali, Punjab, India, [c]Chitkara College of Pharmacy, Chitkara University, Rajpura, Punjab, India

1 Introduction

Angina pectoris is a form of chest pain that may be caused by a shortage of oxygenated blood reaching the cardiac muscles. The blood arteries that supplies blood to the cardiac muscles of heart are mostly obstructed as a result of cholesterol accumulation. The most frequent symptom of this disorder is discomfort in the chest that is located behind the breastbone and can radiate to the anterior region as well as the left upper limbs [1,2]. Individuals having this disorder have an increased chance of experiencing a heart attack. Patients who have severe angina pectoris may have discomfort even when they are resting, and there is no external stressor present [3]. Angina pectoris may produce episodes of chest discomfort that can last from 5 to 30 min. Patients who have coronary artery disease frequently complain of chest discomfort in the form of angina, that is a common presenting symptom of coronary artery disease [4]. The prevalence of stable angina is difficult to assess as it is a diagnosis based upon history and therefore requires clinical judgement. The prevalence of stable angina increases with age in both sexes. Women aged 45–64 years have a prevalence of 5%–7%, while men in the same age range have a prevalence of 4%–7%. This increases to a prevalence of 10%–12% in women aged 65–84 years and 12%–14% in similarly aged men. The patient-reported Health Survey for England, which asks individuals whether they have ever been given a diagnosis of angina by a health professional, has a similar prevalence with angina having a 3% prevalence in all adults, and the highest burden in those aged over 75 years where the prevalence is 11%. Within the average GP practice in the UK of 8000 patients, around 300 (3.75%) will have a diagnosis of angina. In 2017, there were 1.7 million inpatient episodes related to all cardiovascular diseases,

Targeting Angiogenesis, Inflammation and Oxidative Stress in Chronic Diseases. https://doi.org/10.1016/B978-0-443-13587-3.00001-1

with 4.4% of these for angina pectoris in the UK [5–7]. Angina pectoris is significantly more prevalent in females than in males. Additionally, angina may manifest atypically in females. It is possible that the discomfort is not in the chest and that it has a different nature [8]. To enhance the prognosis, one must alter his or her lifestyle. If patient does not adhere to these modifications, then they will experience a return of their angina symptoms. The greatest preventative steps one can take are adopt an exercise routine, give up smoking, cut down on the alcohol use, avoid stressful situations, and avoid a lot of meals that are high in fat and heavy on the calories [9].

2 Types of angina pectoris

2.1 Stable angina

Clinically, stable angina is a condition that is characterized by pain in the chest, shoulder, jaw, back, or arms. This discomfort is generally induced by physical activity or experiencing high levels of mental stress, and it may be eased by either taking nitroglycerin or taking some rest [10]. Symptoms of discomfort in the epigastric area occur significantly less often. Myocardial infarction is characterized by a number of different symptoms, but the one that is most prevalent is stable angina pectoris. MI develops when the requirement of oxygen of the cardiac muscles is greater than the oxygen that is being supplied to it [11]. There are three main elements that govern the oxygen demand of the myocardium: the heart rate, the contractility of the myocardium, and the intramyocardial wall tension, the latter of which is regarded as the most crucial component. The oxygen demand rises in result of the elevation in heart rate or in after load or preload of the left ventricle [12]. An elevated volume of end-diastolic will increase preload of left ventricles, and a raise in systolic blood pressure and/or stiffness of arteries will enhance after load of left ventricles of heart and, as a result, myocardial oxygen requirement. Plaque accumulation in arteries caused by atherosclerosis and/or coronary artery spasm may reduce the amount of blood that is able to reach the heart [13].

2.2 Unstable angina

In comparison to stable angina, the prevalence of unstable angina is quite low [14]. Autonomic dysfunction plays a significant role in the development of unstable angina pectoris. It has an effect on the systolic function of the coronary arteries, produces coronary spasm, makes myocardial ischemia worse, lowers the threshold for ventricular fibrillation, and raises the frequency of sudden cardiac death in individuals with unstable angina pectoris. It has been suggested that a fundamental pathway that leads to the onset and development of cardiovascular disorders is oxidative stress, which is described as an excessive generation of reactive oxygen species (ROS) that cannot be counteracted by the action of antioxidants. It has been hypothesized that oxidative stress is one of the most important mechanisms that contributes to the start and development of neurodegenerative disorders by causing excitotoxicity, neuronal loss, and axonal degeneration. Although the cause-and-effect relationship between autonomic dysfunction and oxidative stress has not been established, it has been indicated that autonomic dysfunction is one of the most important causes of oxidative stress [15,16]. Inflammation within coronary plaques that are prone to rupture and disintegration may be the source of unstable angina. Symptoms of angina may appear suddenly and sometimes appear while a person is at rest. This might be an indication of a worsening of stable angina; however, some people experience unstable angina for the very first time [14,17]. In unstable angina, the symptoms are more severe. The discomforts are much more common, more intense, last longer, occur while the patient is at rest, and cannot be eased by placing nitroglycerin beneath the tongue. Although unstable angina is not similar as a heart attack, it is serious enough to need an urgent visit to a healthcare professional or the emergency department of a hospital. In order to prevent the patient from having a heart attack, it is necessary that they may need to be hospitalized [18,19].

2.3 Microvascular angina

Microvascular angina is a form of angina that develops when there is a blockage in any of the coronary arteries, the smallest blood vessels that surrounds the heart, provides oxygenated blood to the heart continuously, and is an important component of the heart's blood supply [20,21]. The clinical manifestation of microvascular angina is similar to that of chest pain induced by obstructive epicardial coronary artery disease. It is quite common, and it is believed that it affects almost half of people who have angina but do not have obstructive epicardial coronary artery disease. It has also been shown that individuals suffering from microvascular angina have a slow or poor response to short-acting nitrates [22,23].

2.4 Variant angina

Variant angina is defined by recurring incidents of chest discomfort, followed by transient ST-segment spikes on an electrocardiogram (ECG), and also has a rapid response to treatment with sublingual nitrate when given by sublingual route [24]. The discomfort associated with variant angina may be experienced when the patient is either at rest or engaging in normal activities; it is not triggered on by exercise. Variant angina involves a broad range of clinical symptoms, ranging from short asymptomatic attacks to abrupt cardiac death. Variant angina may be linked to greater ventricular susceptibility and repolarization problems [25,26].

3 Oxidative stress and inflammation in pathophysiology of angina pectoris

The angina pectoris represents the common clinical manifestation of myocardial infarction which occurs when the blood circulation to the coronary arteries falls short of the oxygen requirements of the cardiac muscle. Myocardial cells undergo a change from an aerobic to anaerobic metabolism, which causes a progressive impairment of mechanical, metabolic, and electrical functions [27]. The mechanical and the chemical activation of the sensory afferent nerve terminals in the coronary arteries and myocardium are the root causes of angina. These nerve fibers rise from the 1st to the 4th thoracic spinal nerves, continue to ascend via the spinal cord until they reach the thalamus, and then pass from the thalamus to the cerebral cortex [28,29]. Adenosine is the primary chemical mediator that may be responsible for the feeling of angina pain. Ischemia leads to the breakdown of ATP into adenosine, which then diffuses into the extracellular space and induces arteriolar dilatation as well as anginal discomfort. Angina is caused by adenosine mostly because it stimulates the A1 receptors that are located in the afferent nerve endings of the cardiac muscles [30–33]. The rate of the heart, the myocardial wall tension, and the myocardial inotropic state are the primary factors that determine metabolic function of myocardial muscles, and the oxygen demand of myocardial muscles rises in the heart and myocardial contractile state leads to a rise in the oxygen demand of myocardial muscles [34]. Elevation in both aortic pressure (i.e., after load) and ventricular end-diastolic volume (i.e., preload) resulting in a proportionate increase in pressure in walls of myocardium causes an increased demand for oxygen from the myocardium. The amount of oxygen that is delivered to each organ system is dependent upon the flow of blood and the oxygen extraction during times of high demand; the myocardium can only enhance its oxygen extraction to a certain extent because the saturation of coronary venous oxygen at rest has already been low (approximately 30%) [35,36]. As a consequence of this, an enhancement in the demand of the oxygen of myocardium (such as during exercise) has to be counterbalanced by a comparable enhancement in coronary blood flow. The capacity of coronary arteries to enhance flow of blood in response to increasing demands imposed on the heart by its metabolic processes is known as coronary flow reserve. In individuals who are healthy, the maximum coronary blood flow that occurs following complete dilatation of the coronary arteries is about four to six times higher than the coronary blood flow that occurs at rest [37,38].

It is believed that inflammation plays a crucial part in every stage of the atherosclerotic process, which is also characterized by the proliferation of inflammatory cells and other inflammatory indicators [39,40]. In fact, both inflammation and atherosclerosis are associated with a reduction in vasoreactivity. The atheromatous or fibro-fatty plaque is the primary lesion in atherosclerosis, and it is responsible for narrowing of artery, predisposing to thrombosis, calcifying, and weakening the muscle, which ultimately leads to aneurismal dilation [41,42]. The process of vascular inflammation usually results in the modification of the vascular wall, which is then followed by endothelial dysfunction. As a result, it is crucial in the onset and progression of angina pectoris [43,44]. Oxidative stress has been discovered to have a connection with chronic inflammation. Superoxide anion and other ROS can activate nuclear factor kappa B (NF-κB), a transcription factor linked to a variety of inflammatory diseases. NF-κB is responsible for activating a diverse range of proinflammatory cytokines that contribute to inflammation, including tumor necrosis factor alpha (TNF-α) and interleukin-1 [45,46]. It has been demonstrated that TNF-α targets endothelium cells, which then causes inflammation in the vascular wall via a variety of mechanisms.

TNF-α enhances the expression of cell adhesion molecules by activating NF-κB. In addition to this, it has been discovered that TNF-α can upregulate NADPH oxidase activity in endothelium cells, which results in an increase in the amount of O_2^- in the arterial wall. These activities, when combined, accelerate the progression of endothelial dysfunction and recruit monocytes into the intima, which ultimately results in chronic inflammation within the vascular wall [47,48].

Oxidative stress and endothelial dysfunction are major contributors to the development of a number of vascular and metabolic conditions that affect humans, including angina pectoris. Endothelial activation is a proinflammatory and procoagulant state marked by the expression of endothelium cell-surface binding molecules needed for inflammatory cell recruitment, which is triggered by cytokines released by inflamed tissues [49,50]. Endothelial function can be affected by a number of pathophysiological conditions, such as hyperlipidemia, hyperglycemia, and hypertension. Other factors, such as aging and exposure to certain drugs, can also have an effect on the molecular processes that regulate the bioavailability of nitric oxide (NO) [51]. Endothelial dysfunction has been characterized as a widespread, insidious, and reversible pathological condition of the endothelium, resulting from decreased NO bioavailability and compromised vasodilation, accompanied by proinflammatory and prothrombotic status [52–55]. An elevation in oxidative stress has been related to a reduction in endothelium function, and it is possible that this enhanced oxidative stress is involved in the pathogenesis of the adverse effects of angina pectoris [56,57]. Endothelial NO synthase (eNOS) can become uncoupled for a number of reasons. These reasons include the lack of the NOS cofactor BH4 or the substrate L-arginine, the production of the endogenous NOS inhibitor asymmetrical dimethylarginine (ADMA), and eNOS S-glutathionylation. Among these, an absence of BH4 is regarded as the most important factor in eNOS uncoupling [58,59]. The primary way through which uncoupling eNOS leads to atherogenesis is not only by lowering the amount of NO bioavailability, but also by increasing oxidative stress. A decrease in NO bioavailability can be caused by a number of pathological conditions, like shear stress and the development of oxidative stress. The inactivation of NO can occur during the process of uncoupling eNOS or during the formation of peroxynitrite ($ONOO^-$) from the interaction of NO and superoxide anion (O_2^-), and both of these processes are linked to the progression of endothelial cell dysfunction that is facilitated by atherosclerosis [60].

Atherosclerosis is the leading factor of epicardial coronary artery stenosis and, therefore, angina pectoris. Angina pectoris is a form of chest pain that may be induced by narrowing of the coronary arteries due to plaque accumulation [61]. Patients who have a fixed coronary atherosclerotic lesion of at least 50% experience MI under increasing cardiac metabolic demand as a consequence of a considerable decrease in coronary flow reserve. Angina occurs when a patient's coronary blood flow is not able to rise under stress to meet the higher metabolic demand on the heart muscle. These individuals are unable to improve the blood flow to their coronary arteries at times of stress to meet the enhanced metabolic demand exerted on the myocardium, and as a result, they suffer from angina [62–65]. The sequential stepwise pathophysiology of angina pectoris is described in Fig. 1.

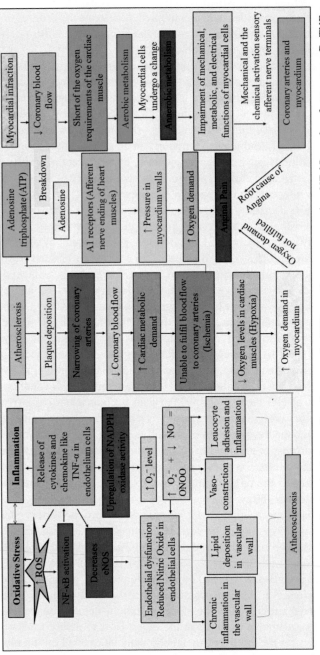

Fig. 1 Schematic representation of pathophysiology of angina pectoris. *ROS*, reactive oxygen species; *NF-kB*, nuclear factor κappa B; *TNF-α*, tumor necrosis factor-α; *NADPH*, nicotinamide adenine dinucleotide phosphate oxidase.

4   Role of angiogenesis in angina pectoris

Angiogenesis plays a crucial role in the pathogenesis and treatment of cardiovascular disease and has become a major topic of discussion in recent decades. Angiogenesis is a complicated process of blood vessel development that involves the interaction of several angiogenic growth factors [1]. Vascular endothelial growth factor (VEGF), angiopoietin-1 (Ang-1), and angiopoietin-2 (Ang-2) play crucial roles in developmental blood vessel formation and regulation of hypoxia-induced tissue angiogenesis. In addition, some other angiogenic factors and protein are also involved in angiogenesis, such as angiogenin, angiostatin, basic fibroblast growth factor (bFGF), and platelet-derived growth factor-BB (PDGF-BB) (Fig. 2) [66–68]. Angiogenesis is subdivided into two categories (physiological and pathological), with pathological angiogenesis always resulting in numerous diseases, including cardiovascular conditions such as angina pectoris, tumors, and inflammation [69,70].

5   Treatment for angina pectoris

The objectives of the therapy are to provide the patient with as much symptomatic relief as is clinically feasible while also preventing any long-term potential risks,

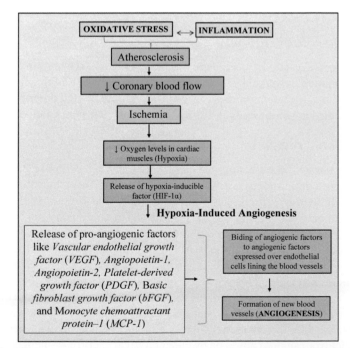

Fig. 2 Schematic representation of role of angiogenesis in management of angina pectoris in response to oxidative stress and inflammation.

including as myocardial infarction, arrhythmias, and left ventricular failure that might potentially be fatal. The most common approaches that are all used in the treatment of angina are as following.

5.1 Lifestyle modifications

It is highly recommended that people quit smoking and stay away from second-hand smoke. The therapy of nicotine replacement is one method that may be used to help people quit the smoking habit. Individuals having stable angina pectoris should also keep a regular level of exercise [71,72]. The American College of Cardiology Foundation and the American Heart Association both suggest that adults who are in a healthy enough condition actively engage in 7 days per week of exercise of a moderate intensity for 30–60 min, like as brisk walking (5 days at least). It is appropriate to include strength training in the treatment plan if the individual's health allows for it. A cardiac rehabilitation program that is medically supervised is suggested for individuals who are considered to be at a high risk, including those who have just had revascularization, had a myocardial infarction, or who have heart failure. Maintaining a healthy weight is highly recommended. It is essential to have a diet that is abundant in fresh fruits, vegetables, and dairy products that are low in fat. Consumption of alcohol and salt should be kept to a minimum. The amount of cholesterol that should be taken in should not exceed 200 mg/day, and the amount of saturated fat that should be ingested should not exceed 7% of the total number of calories [73–75].

5.2 Pharmacologic therapy

Drugs that lessen exercise-induced increases in demand of oxygen of myocardial tissues and/or improve oxygen supply of myocardial tissues have traditionally been used to avoid exertional symptoms of angina [76]. The commonly used drugs that are helpful in preventing the various signs and symptoms of angina pectoris include following:

5.2.1 Organic nitrates

Both organic nitrates, short-acting and long-acting that are currently present in the market for therapeutic usage, are very successful in the treatment of angina once administered properly [77]. Nitrates have an efficacy that is comparable to that of β-blockers and calcium channel blockers (CCBs). These are powerful vasodilators, which mean that they lower the amount of blood that returns from the veins to the heart. This lowers the preload on the heart by lowering the volume of the ventricles, which in turn lowers the demand of oxygen of the myocardial tissues. This impact of nitrates is more pronounced while seated or standing and after the first dose [78]. Nitroglycerin tablets that are taken sublingually and nitroglycerin sprays that are given transorally are quickly absorbed and, when used as a prophylactic measure, increase exercise tolerance and lower the risk of myocardial ischemia. Their use often provides relief from angina caused on by either physical exertion or mental stress [79]. It has been shown that taking sublingual isosorbide dinitrate may enhance angina-free walking; however, its effects are delayed, and it is not recommended that this drug be used as a prophylactic measure. Nitroglycerin patches are equally efficient in improving exercise tolerance; but, since they are used constantly, users quickly build up a tolerance to their effects [80,81].

5.2.2   Calcium channel blockers

The CCBs restrict calcium from reaching the cardiac muscle cells, as well as the smooth muscles which induce blood vessels to contract. CCBs "relax" the affected muscle cells by lowering the amount of calcium that enters the cells. As a result of this relaxing impact, blood vessels dilate, and the force with which the heart muscle contracts reduces. Some CCBs also have the additional effect of slowing down the sinus node and the AV node's transmission of electrical impulses to the rest of the heart. As a consequence of these actions, CCBs are beneficial in the treatment of certain arrhythmias [82,83].

All of the actions of CCBs like dilatation of blood vessel, a decrease in contraction of heart muscles, and a slower heart rate have the cumulative effect of lowering the volume of oxygen that the heart muscle needs. Also, when the blood flow of coronary arteries is blocked partially because of the plaque development during atherosclerosis, the heart is able to continue functioning for longer without suffering ischemia when the quantity of oxygen that is required for the heart is decreased [84]. In individuals having stable angina, CCBs often improve the duration of physical exercise that may be performed before the occurrence of angina [85]. CCBs are particularly effective for people suffering from Prinzmetal's angina because they directly inhibit spasm of coronary artery. Diltiazem and verapamil are both nondihydropyridine calcium channel blockers that may be utilized as first-line therapeutic alternatives in situations when β-blockers are contraindicated. When there is inadequate response to the β-blockers therapy, it may be necessary to add CCBs [86]. All CCBs were employed for the therapeutic purpose of angina. All calcium blockers have been employed in the treatment of angina. However, the longer-acting calcium blockers such as diltiazem and verapamil, felodipine or amlodipine are the ones that are utilized most frequently. In general, patients with angina should avoid taking nifedipine, and particularly short-acting forms of the drug, because it causes marked dilation of blood vessel, which can significantly enhance adrenaline, resulting in faster heart rate and therefore, an upsurge in the demand of oxygen of cardiac muscles (which can raise the risk of developing MI). Despite CCBs may be effective in treating angina, in comparison to β-blockers, they are generally regarded as being less effective [87,88].

5.2.3   β-Blockers

The β-blockers have been demonstrated to be an effective therapy for patients who have stable angina. These drugs reduce myocardial workload and enhance exercise tolerance. This advantage is primarily mediated by competitive inhibition of ß-adrenergic receptors, which decreases heart rate, increase in blood pressure caused by exercise, and the ability of cardiac muscles to contract and as a result, demand of oxygen of myocardium [89]. The diastolic perfusion time is increased by β-Blockers, which in turn increases coronary perfusion. When treating individuals who have coronary artery disorder, the doses that are often advised are those that maintain a heart rate of less than or equal to 60 beats per minute, or even less in the case of severe angina [90]. The exercise test may be utilized to modify the exercise heart rate for therapy that is below the threshold of pain. The β-Blockers are more efficacious than CCBs and nitrates in reducing the number of silent ischemic episodes and early morning ischemia, as well as improving post-MI mortality. Individuals having normal left ventricular function should continue β-Blockers treatment for 3 years [91]. The β-Blockers are the ideal first treatment for individuals having stable angina and decreased ejection fraction of the left ventricles that is less than 40%, as well as post-infarct angina. The β-blockers are also ideal treatment option for the individuals having stable angina with elevated blood pressure and fast heart rates, a history of supraventricular tachycardia such as atrial fibrillation, as well as individuals having hypertrophic cardiomyopathy with exertional angina [72,92]. The β-Blockers therapy with metoprolol, carvedilol, or bisoprolol should be considered in patients who have left ventricular dysfunction. Long-term therapy of β-Blockers may be explored for every individual who has coronary as well as other vascular disorder. Carvedilol is an effective treatment option for individuals who suffer from congestive heart failure because it produces peripheral vasodilation by blocking α1-adrenergic receptor [93]. Nebivolol is superior to metoprolol, bisoprolol, and carvedilol in terms of the selectivity it has for the β1-receptor. In addition to this, nebivolol stimulates the release of nitric oxide, which results in peripheral vasodilation [17].

5.2.4 Other approved therapies

In the last several years, numerous novel drugs have been given approval for use in the treatment of stable angina. These compounds are distinct from others because they will not have the effect of lowering blood pressure or the heart rate, nor can they have any impact on the degree to which cardiac muscles are able to contract.

Nicorandil: The antianginal drug nicorandil has two different mechanisms of action and is a nicotinamide derivative containing a nitrate moiety. Initially, it improves the permeability of potassium ions by opening potassium channels that are sensitive to adenosine triphosphate, which then stimulates the guanylate cyclase enzyme. Second, nicorandil is similar to nitrates in that it has the ability to relax smooth muscle and cause vasodilation, which in turn reduces preload by increasing vasodilation [94]. In addition to these effects, the drug enhances the production of endothelial nitric oxide synthase and decreases after load. The use of this compound is related to enhanced myocardial function throughout ischemia-reperfusion, prevention of myocardium mostly during ischemia, reduced action potential duration, and reduction of intracellular calcium toxicity, all of which are important in controlling ischemic cell damage and mortality [95]. Nicorandil has been given approval in Japan, France, Iran, Lebanon, and the United Kingdom as a therapy for angina [86].

Ranolazine: In January 2006, ranolazine was approved for the therapy of chronic angina [17]. Ranolazine treatment is recommended as alternative of β-Blockers therapy in the most recent guidelines if β-Blockers therapy is either contraindicated or intolerable in patients. When BB monotherapy is inadequate, ranolazine might be added to the treatment as an adjunctive therapy [96]. It is an antiangina drug and does not function via the typical hemodynamic approaches to either increase the supply of oxygen or decrease the demand for oxygen [97]. It does this by inhibiting the late sodium current during the later stage of the action potential, which in turn lowers the amount of sodium accumulation within the cell. Subsequently, it does this by lowering the amount of sodium that is exchanged for calcium, which has the end result of lowering the amount of calcium overload that is present in ischemic cardiomyocytes. Since there is a reduction in overload of calcium, the cells of myocardium are able to relax more effectively during the diastole, which results in an increase in microvascular perfusion [98].

Ivabradine: The ivabradine is an inhibitor of sinus node which reduces the heart rate without having any impact on the degree to which cardiac muscles are able to contract or having a significant influence on blood pressure [99]. The hyperpolarization-activated mixed sodium-potassium inward I_f current may be inhibited by ivabradine, which is a selective inhibitor. This slows down the rate at which the cells of pacemaker in the sinus node depolarize during diastole. This, in turn, makes the heart rate less responsive, both during the exercise and at the rest. Ivabradine induces a decrease in heart rate that is on average around ten beats slower than normal per minute on average [100].

Trimetazidine: This antianginal drug has a beneficial effect by decreasing the rate at which fatty acids are oxidized and shifting metabolism of energy toward the consumption of glucose. The oxygen demand of the pathways of glucose in the myocardium is much lower as compared to the oxygen demand of pathway of free fatty acids. Ischemia is believed to cause an increase in the amounts of oxidized free fatty acids inside the cardiomyocyte, and this reduces the effectiveness of the pathway of glucose [101,102]. Trimetazidine inhibits Acyl CoA β from being oxidized to acetyl-CoA, which in turn promotes the breakdown of glucose into pyruvate, which ultimately results in the formation of acetyl-CoA [43]. The drug increases the myocardium's utilization of glucose, inhibits a decline in ATP and phosphocreatine concentrations in response to hypoxia or ischemia, conserves the function of ionic pumps, reduces the formation of free radicals, and serves as a protective against intracellular calcium overload and acidosis. It enhances exercise performance, decreases the frequency of episodes in angina, increases coronary flow reserve, and preserves the utilization of nitrates without causing fluctuations in heart rate, negative inotropic effects, or vasodilator effects [103,104].

6 Nanoparticles and their role in angina pectoris

The introduction of nanotechnology has paved the way for the discovery of a novel approach to the use of nanoparticles in the area of medicine. Nanotherapies based on nanoparticles are very promising for the treatment of cardiovascular diseases. Nanomedicine is an advanced version of a conventional drug that includes the development and application of nanomaterials and nanotechnologies. Nanomedicine is highly specific for targeted drug delivery, improves drug bioavailability, minimizes associated side effects, and reduces costs. Nanoparticles have the potential to increase the production of vascular endothelial growth factor, dilatation coronary arteries, enhance cardiovascular mechanism, and lower the frequency at which patients experience angina pectoris recurrences. Nanoparticles offer a novel tool for biomedical research because of their huge specific surface area, greater surface reactivity, excellent catalytic effectiveness, and excellent adsorption potential [105]. Various nanoparticles are commonly used as carriers for various drugs that are used for the therapy of angina pectoris in order to improve certain properties of these drugs which enhance their therapeutic effectiveness and are summarized in Table 1.

6.1   Polymeric nanoparticles

In recent years, polymeric nanoparticles have gained a significant amount of attention due to the characteristics that result from their extremely small dimension. The use of polymeric nanoparticles as drug carriers has a number of potential benefits. These benefits include the ability to increase bioavailability and therapeutic index, the ability to protect drugs and other biologically active compounds from the environment, and the potential use of polymeric NPs for controlled release [119,120]. The term "nanoparticle" refers to both nanospheres and nanocapsules, which have distinct morphologies. Nanocapsules are made up of a viscous core, in which the drug is typically dispersed, encapsulated in a polymeric shell that regulates the rate at which the drug is released from the core. The nanospheres are comprised of a continuous polymeric network that allows the substance to be retained within or adsorb onto their surface. These two types of polymeric nanoparticles are known as reservoir (nanocapsule) and matrix systems (nanosphere) [121,122].

6.2   Nanosuspension

Nanosuspension is made up of the pure drug that does not easily dissolved in water and is suspended in dispersion. Nanosuspensions are colloidal submicron dispersions of drug particles that are nanoscale in size and are stabilized with surfactants. The preparation of nanosuspensions is a simple process that can be applied to any drug that is insoluble in water. The use of a nanosuspension not only addresses the issues of inadequate bioavailability and solubility, but also modifies the pharmacokinetics of the drug, which ultimately results in an improvement in both the drug's safety and its effectiveness [123,124].

6.3   Nanocrystals

Nanocrystals have gained considerable interest in the field of drug delivery for the efficient treatment of a wide variety of diseases due to the high drug loading characteristics and the low particulate size that they possess. Nanocrystals, which are described as crystalline fragments with a dimension varying from 1 to 1000 nm, have

Table 1 An outline of nanoparticles explored for therapeutics of angina pectoris.

Drug (technique)	Excipients	Dosage form	Outcome and significance	References
Ivabradine hydrochloride (Double emulsion method)	Polylactic-*co*-glycolic acid, Span 80, D-α-tocopherol polyethylene glycol 1000 succinate	Polymeric nanoparticle	Optimized nanoparticles exhibited entrapment efficiency of $60 \pm 4.8\%$ and zeta potential of $-43.75\,mv$ respectively. In comparison to ivabradine hydrochloride tablets, they showed biphasic release and an enhancement in intestinal permeability that was 1.85 times greater than what was seen with the tablets.	[106]
Lercanidipine hydrochloride (Evaporative antisolvent precipitation method)	Sodium alginate, D-α-tocopherol polyethylene glycol 1000 succinate, polyvinyl alcohol, hypromellose E15, methyl cellulose	Nanoparticles (Nanosuspension)	Increased in vitro dissolution and 4.5 times increase in ex vivo permeation via porcine buccal mucosa.	[107]
Felodipine (Solvent-antisolvent sonoprecipitation method)	Polyvinylpyrrolidone K30	Nanoparticles (nanosuspension)	Optimized F5 batch exhibited $82 \pm 0.47\%$ release within 4 min and follows nonfickian diffusion-controlled release pattern.	[108]
Verapamil (high-shear homogenization technique)	Poloxamer 188, Compritol ATO 888, Tween 80, dextran sulfate, oleic acid	Hybrid nanostructured lipid carriers	Optimized (VER-9) formula showed E.E% of $93.26 \pm 2.66\%$ and the addition of dextran sulfate to the preparation prolonged verapamil release($\sim 85\%$ in 48 h) in both stimulated gastric fluid (pH 1.2) and simulated intestinal fluid (pH 6.8) and simulated gastric fluid (pH 1.2). The study of cellular uptake utilizing the Caco-2 cell line revealed that verapamil uptake from these nanoparticles was greater than that of verapamil-dextran complex and verapamil solution.	[109]

Continued

Table 1 Continued

Drug (technique)	Excipients	Dosage form	Outcome and significance	References
Nifedipine (Solvent/ anti-solvent precipitation method)	Stearic acid	Nanoparticle agglomerate	Nanoparticles and nanoparticle agglomerates showed increased drug dissolution in comparison to stock drug	[110]
Isradipine (Solvent/ antisolvent precipitation method)	Soluplus, Poloxamer 188, PVA	Nanoparticles	Nanoparticles of F9 formula containing Soluplus as a stabilizer exhibited 10-fold increase in the solubility of drug as well as particle size ranging from 77.34–1582.56 nm and the reduced particle size leads to higher in vitro release and dissolution compared to pure drug	[111]
Felodipine (Nanoprecipitation technique)	PluronicF-68, Poly (lactide-*co*-glycolide) (PLGA)	PLGA nanoparticles	PLGA nanoparticles showed entrapment efficiency and particle size of more than 90% and less than 200 nm respectively, as well as the results of in vitro and ex vivo release studies from the felodipine loaded PLGA nanoparticles via stomach and intestine exhibited sustained drug release.	[112]
Isradipine (Solvent evaporation technique)	Poly(methyl methacrylate) (PMMA), Poloxamer 40	Nanoparticles	A drop in blood pressure was delayed by the PMMA isradipine nanoparticles and reached 15,272 mmHg at 1 h as well as the nanoparticles showed superior bioavailability to that of the solution form.	[113]
Nicorandil (Electrospinning method)	PVA, hyaluronic acid	Polymeric electrospun nanofiber	The developed nanofibers showed diameter ranging from 200 to 450 nm and demonstrated controlled release behavior in an in vitro drug release study. No mucosal ulceration was found at the application site, as determined by histopathology.	[114]

Drug (method)	Nanoparticle type	Materials	Description	Ref.
Carvedilol (Emulsion solvent evaporation technique)	Polymeric nanoparticles	Carbon tetrachloride, poly(ethylene-co-vinyl acetate), polyethylene glycol 400, chitosan	The enhancement in chitosan concentration causes a reduction in the in vitro drug release from coated nanoparticles. Formulation based on mannitol had a low density, improved flow ability, a reduced aerodynamic diameter, and greater fine powder fraction.	[115]
Propranolol (Ionic gelation method)	Chitosan nanoparticles	Tripolyphosphate, chitosan	All the developed nanoparticles exhibited 70% of entrapment efficiency and the gel formulation containing these nanoparticles exhibited thixotropic behavior and prolonged drug release properties.	[116]
Atenolol (Ionotropic gelation)	Biopolymeric nanoparticles	Barium chloride, gellan gum	Increased release rates were found with lower concentrations of both the cross linker and the gellan gum, and around 88% of the drug was released (in 48 h) in an aqueous environment with a pH (6.8–7.0).	[117]
Nifedipine (High-pressure homogenization method)	Nanocrystals	Hydroxy-propylmethyl cellulose	Following particle size reduction, it has been demonstrated that the initial crystalline state is preserved, and it has also been demonstrated that the dissolution properties of nifedipine nanoparticles are greatly improved in comparison to the commercial product.	[118]

discovered a wide variety of applications in the fields of science and engineering. Some examples of these applications include optical devices, drug delivery, chemical catalysis, and biological sensors [125,126].

7 Conclusion and future perspective

Angina pectoris remains to be connected with considerable morbidity and serious cardiovascular serious complications. The objective of treatment should not only to address the potential risks of atherosclerotic disease, most of which have been found to minimize significant adverse cardiac events, but treatment should also be aimed toward minimizing or eliminating angina. The common therapies for angina, such as nitrates, calcium blockers, β-blockers, and ranolazine, are effective however they may not always eliminate the condition completely. Nanoparticles offer a novel tool for biomedical research because of their huge specific surface area, greater surface reactivity, excellent catalytic effectiveness, and excellent adsorption potential. Various nanoparticles are commonly used as carriers for various drugs that are used for the management and therapy of angina pectoris in order to improve certain properties of these drugs which enhance their therapeutic effectiveness. Nanoparticles have the potential to increase the production of vascular endothelial growth factor, dilatation coronary arteries, enhance cardiovascular mechanism, and lower the frequency at which patients experience angina pectoris recurrences.

Acknowledgments

The authors would like to thank Department of Pharmaceutics, MM College of Pharmacy, Maharishi Markandeshwar (Deemed to be University), Mullana-Ambala, Haryana, India 133207 and School of Health Science, University of Petroleum and Energy Studies, Dehradun, Uttarakhand, India, for providing facilities for the completion of this review.

References

[1] Geng N, Su G, Wang S, Zou D, Pang W, Sun Y. High red blood cell distribution width is closely associated with in-stent restenosis in patients with unstable angina pectoris. BMC Cardiovasc Disord 2019;19(1):1–7. https://doi.org/10.1186/s12872-019-1159-3.

[2] Shi J, Wang J, Wang Y, Liu K, Fu Y, Sun JH, Zhao JP, Shao XM, Feng SF, Yang YW, Li J. Correlation between referred pain region and sensitized acupoints in patients with stable angina pectoris and distribution of sensitized spots in rats with myocardial ischemia. Zhen ci yanjiu Acupunct Res 2018;43(5):277–84. https://doi.org/10.13702/j.1000-0607.180123.

[3] Al-Janabi F, Mammen R, Karamasis G, Davies J, Keeble T. In-flight angina pectoris; an unusual presentation. BMC Cardiovasc Disord 2018;18(1):1–4. https://doi.org/10.1186/s12872-018-0797-1.

[4] Wang JS, Yu XD, Deng S, Yuan HW, Li HS. Acupuncture on treating angina pectoris: a systematic review. Medicine 2020;99(2). https://doi.org/10.1097/MD.0000000000018548.

[5] Montalescot G, Sechtem U, Achenbach S, Andreotti F, Arden C, Budaj A, Bugiardini R, Crea F, Cuisset T, Di Mario C, Ferreira JR. ESC Committee for Practice Guidelines. 2013 ESC Guidelines on the management of stable coronary artery disease: the Task Force on the management of stable coronary artery disease of the European Society of Cardiology. Eur Heart J 2013;34(38):2949–3003. https://doi.org/10.1093/eurheartj/eht296.

[6] Daly CA, Clemens F, Sendon JL, Tavazzi L, Boersma E, Danchin N, Delahaye F, Gitt A, Julian D, Mulcahy D, Ruzyllo W. The initial management of stable angina in Europe, from the Euro Heart Survey: a description of pharmacological management and revascularization strategies initiated within the first month of presentation to a cardiologist in the Euro Heart Survey of Stable Angina. Eur Heart J 2005;26(10):1011–22. https://doi.org/10.1093/eurheartj/ehi109.

[7] Stewart S, Murphy N, Walker A, McGuire A, McMurray JJ. The current cost of angina pectoris to the National Health Service in the UK. Heart 2003;89(8):848–53. https://doi.org/10.1136/heart.89.8.848.

[8] Bhowmik D, Das BC, Dutta AS. Angina pectories-a comprehensive review of clinical features, differential diagnosis, and remedies. Elixir Pharmacy 2011;40:5125–30.

[9] Dose N, Michelsen MM, Mygind ND, Pena A, Ellervik C, Hansen PR, Kanters JK, Prescott E, Kastrup J, Gustafsson I, Hansen HS. Ventricular repolarization alterations in women with angina pectoris and suspected coronary microvascular dysfunction. J Electrocardiol 2018;51(1):15–20. https://doi.org/10.1016/j.jelectrocard.2017.08.017.

[10] Zhang Z, Chen M, Zhang L, Zhang Z, Wu W, Liu J, Yan J, Yang G. Meta-analysis of acupuncture therapy for the treatment of stable angina pectoris. Int J Clin Exp Med 2015;8(4):5112.

[11] Skalidis EI, Vardas PE. Guidelines on the management of stable angina pectoris. Eur Heart J 2006;27(21):2606–7. https://doi.org/10.1093/eurheartj/ehl258.

[12] Albrecht S. The pathophysiology and treatment of stable angina pectoris. US Pharm 2013;38(2):43–60.

[13] Kaski JC, Arrebola-Moreno A, Dungu J. Treatment strategies for chronic stable angina. Expert Opin Pharmacother 2011;12(18):2833–44. https://doi.org/10.1517/14656566.2011.634799.

[14] Wang Q, Wu T, Chen XY, Duan X, Zheng J, Qiao J, Zhou L, Wei J, Ni J. Puerarin injection for unstable angina pectoris. Cochrane Database Syst Rev 2006;3. https://doi.org/10.1002/14651858.CD004196.pub2.

[15] Wang YC, Ma DF, Jiang P, Yang JL, Zhang YM, Li X. Serum levels of homocysteine and circulating antioxidants associated with heart rate variability in patients with unstable angina pectoris. Chin Med J (Engl) 2019;132(01):96–9.

[16] He F, Zuo L. Redox roles of reactive oxygen species in cardiovascular diseases. Int J Mol Sci 2015;16(11):27770–80. https://doi.org/10.3390/ijms161126059.

[17] Buffon A, Biasucci LM, Liuzzo G, D'Onofrio G, Crea F, Maseri A. Widespread coronary inflammation in unstable angina. N Engl J Med 2002;347(1):5–12.

[18] Ambrose JA, Dangas G. Unstable angina: current concepts of pathogenesis and treatment. Arch Intern Med 2000;160(1):25–37. https://doi.org/10.1001/archinte.160.1.25.

[19] Newby DE, Fox KA. Unstable angina: the first 48 hours and later in-hospital management. Br Med Bull 2001;59(1):69–87. https://doi.org/10.1093/bmb/59.1.69.

[20] Marinescu MA, Löffler AI, Ouellette M, Smith L, Kramer CM, Bourque JM. Coronary microvascular dysfunction, microvascular angina, and treatment strategies. JACC Cardiovasc Imaging 2015;8(2):210–20.

[21] Soleymani M, Masoudkabir F, Shabani M, Vasheghani-Farahani A, Behnoush AH, Khalaji A. Updates on pharmacologic management of microvascular angina. Cardiovasc Ther 2022;2022. https://doi.org/10.1155/2022/6080258.

[22] Lanza GA, Parrinello R, Figliozzi S. Management of microvascular angina pectoris. Am J Cardiovasc Drugs 2014;14(1):31–40. https://doi.org/10.1007/s40256-013-0052-1.

[23] Pelletier-Galarneau M, Dilsizian V. Microvascular angina diagnosed by absolute PET myocardial blood flow quantification. Curr Cardiol Rep 2020;22(2):1–9. https://doi.org/10.1007/s11886-020-1261-2.

[24] Kusama Y, Kodani E, Nakagomi A, Otsuka T, Atarashi H, Kishida H, Mizuno K. Variant angina and coronary artery spasm: the clinical spectrum, pathophysiology, and management. J Nippon Med Sch 2011;78(1):4–12. https://doi.org/10.1272/jnms.78.4.

[25] Kundu A, Vaze A, Sardar P, Nagy A, Aronow WS, Botkin NF. Variant angina and aborted sudden cardiac death. Curr Cardiol Rep 2018;20(4):1–6. https://doi.org/10.1007/s11886-018-0963-1.

[26] Kundu A, Vaze A, Sardar P, Nagy A, Aronow WS, Botkin NF. Variant angina and aborted sudden cardiac death. Curr Cardiol Rep 2018;20(4):1–6. https://doi.org/10.1007/s11886-018-0963-1.

[27] Pepine CJ, Nichols WW. The pathophysiology of chronic ischemic heart disease. Clin Cardiol 2007;30(S1):1–4. https://doi.org/10.1002/clc.20048.

[28] Rosen SD. From heart to brain: the genesis and processing of cardiac pain. Can J Cardiol 2012;28(2):S7–19. https://doi.org/10.1016/j.cjca.2011.09.010.

[29] Eckert S, Horstkotte D. Management of angina pectoris. Am J Cardiovasc Drugs 2009;9(1):16–28. https://doi.org/10.1007/BF03256592.

[30] Henning RJ. Therapeutic angiogenesis: angiogenic growth factors for ischemic heart disease. Future Cardiol 2016;12(5):585–99. https://doi.org/10.2217/fca-2016-0006.

[31] Stanley WC. Changes in cardiac metabolism: a critical step from stable angina to ischaemic cardiomyopathy. Eur Heart J Suppl 2001;3(suppl_O):2–7. https://doi.org/10.1016/S1520-765X(01)90147-6.

[32] Barron HV, Sciammarella MG, Lenihan K, Michaels AD, Botvinick EH. Effects of the repeated administration of adenosine and heparin on myocardial perfusion in patients with chronic stable angina pectoris. Am J Cardiol 2000;85(1):1–7. https://doi.org/10.1016/S0002-9149(99)00596-2.

[33] Yu L, Lu X, Xu C, Li T, Wang Y, Liu A, Wang Y, Chen L, Xu H. Overview of microvascular angina pectoris and discussion of traditional Chinese medicine intervention. Evid Based Complement Alternat Med 2022;2022. https://doi.org/10.1155/2022/1497722.

[34] Levy BI, Heusch G, Camici PG. The many faces of myocardial ischaemia and angina. Cardiovasc Res 2019;115(10):1460–70. https://doi.org/10.1093/cvr/cvz160.

[35] Tousoulis D, Androulakis E, Kontogeorgou A, Papageorgiou N, Charakida M, Siama K, Latsios G, Siasos G, Kampoli AM, Tourikis P, Tsioufis K. Insight to the pathophysiology of stable angina pectoris. Curr Pharm Des 2013;19(9):1593–600.

[36] Yilmaz A, Sechtem U. Angina pectoris in patients with normal coronary angiograms: current pathophysiological concepts and therapeutic options. Heart 2012;98(13):1020–9. https://doi.org/10.1136/heartjnl-2011-301352.

[37] Heusch G. Myocardial ischemia: lack of coronary blood flow or myocardial oxygen supply/demand imbalance? Circ Res 2016;119(2):194–6. https://doi.org/10.1161/CIRCRESAHA.116.308925.

[38] Kloner RA, Chaitman B. Angina and its management. J Cardiovasc Pharmacol Ther 2017;22(3):199–209. https://doi.org/10.1177/1074248416679733.

[39] Pi X, Xie L, Patterson C. Emerging roles of vascular endothelium in metabolic homeostasis. Circ Res 2018;123(4):477–94. https://doi.org/10.1161/CIRCRESAHA.118.313237.

[40] Ali L, Schnitzler JG, Kroon J. Metabolism: the road to inflammation and atherosclerosis. Curr Opin Lipidol 2018;29(6):474–80. https://doi.org/10.1097/MOL.0000000000000550.

[41] Conti P, Shaik-Dasthagirisaeb Y. Atherosclerosis: a chronic inflammatory disease mediated by mast cells. Central Eur J Immunol 2015;40(3):380–6. https://doi.org/10.5114/ceji.2015.54603.

[42] Spagnoli LG, Bonanno E, Sangiorgi G, Mauriello A. Role of inflammation in atherosclerosis. J Nucl Med 2007;48(11):1800–15. https://doi.org/10.2967/jnumed.107.038661.

[43] Kulkarni NM, Muley MM, Jaji MS, Vijaykanth G, Raghul J, Reddy NK, Vishwakarma SL, Rajesh NB, Mookkan J, Krishnan UM, Narayanan S. Topical atorvastatin ameliorates 12-O-tetradecanoylphorbol-13-acetate induced skin inflammation by reducing cutaneous cytokine levels and NF-κB activation. Arch Pharm Res 2015;38:1238–47. https://doi.org/10.1007/s12272-014-0496-0.

[44] Gimbrone Jr MA, García-Cardeña G. Endothelial cell dysfunction and the pathobiology of atherosclerosis. Circ Res 2016;118(4):620–36. https://doi.org/10.1161/CIRCRESAHA.115.306301.

[45] Vara D, Watt JM, Fortunato TM, Mellor H, Burgess M, Wicks K, Mace K, Reeksting S, Lubben A, Wheeler-Jones CP, Pula G. Direct activation of NADPH oxidase 2 by 2-deoxyribose-1-phosphate triggers nuclear factor kappa B-dependent angiogenesis. Antioxid Redox Signal 2018;28(2):110–30. https://doi.org/10.1089/ars.2016.6869.

[46] Binesh A, Devaraj SN, Halagowder D. Molecular interaction of NFκB and NICD in monocyte–macrophage differentiation is a target for intervention in atherosclerosis. J Cell Physiol 2019;234(5):7040–50. https://doi.org/10.1002/jcp.27458.

[47] Heo KS, Le NT, Cushman HJ, Giancursio CJ, Chang E, Woo CH, Sullivan MA, Taunton J, Yeh ET, Fujiwara K, Abe JI. Disturbed flow-activated p90RSK kinase accelerates atherosclerosis by inhibiting SENP2 function. J Clin Invest 2015;125(3):1299–310. https://doi.org/10.1172/JCI76453.

[48] Kleinbongard P, Heusch G, Schulz R. TNFα in atherosclerosis, myocardial ischemia/reperfusion and heart failure. Pharmacol Ther 2010;127(3):295–314. https://doi.org/10.1016/j.pharmthera.2010.05.002.

[49] Incalza MA, D'Oria R, Natalicchio A, Perrini S, Laviola L, Giorgino F. Oxidative stress and reactive oxygen species in endothelial dysfunction associated with cardiovascular and metabolic diseases. Vascul Pharmacol 2018;100:1–9. https://doi.org/10.1016/j.vph.2017.05.005.

[50] Rehan R, Weaver J, Yong A. Coronary vasospastic angina: a review of the pathogenesis, diagnosis, and management. Life 2022;12(8):1124. https://doi.org/10.3390/life12081124.

[51] Sena CM, Pereira AM, Seiça R. Endothelial dysfunction—a major mediator of diabetic vascular disease. Biochim Biophys Acta (BBA) Mol Basis Dis 2013;1832(12):2216–31. https://doi.org/10.1016/j.bbadis.2013.08.006.

[52] Sitia S, Tomasoni L, Atzeni F, Ambrosio G, Cordiano C, Catapano A, Tramontana S, Perticone F, Naccarato P, Camici P, Picano E. From endothelial dysfunction to atherosclerosis. Autoimmun Rev 2010;9(12):830–4. https://doi.org/10.1016/j.autrev.2010.07.016.

[53] Jamwal S, Sharma S. Vascular endothelium dysfunction: a conservative target in metabolic disorders. Inflamm Res 2018;67:391–405. https://doi.org/10.1007/s00011-018-1129-8.

[54] Hadi HA, Carr CS, Al SJ. Endothelial dysfunction: cardiovascular risk factors, therapy, and outcome. Vasc Health Risk Manag 2005;1(3):183–98.

[55] Flammer AJ, Anderson T, Celermajer DS, Creager MA, Deanfield J, Ganz P, Hamburg NM, Lüscher TF, Shechter M, Taddei S, Vita JA. The assessment of endothelial function: from research into clinical practice. Circulation 2012;126(6):753–67. https://doi.org/10.1161/CIRCULATIONAHA.112.093245.

[56] Heitzer T, Schlinzig T, Krohn K, Meinertz T, Munzel T. Endothelial dysfunction, oxidative stress, and risk of cardiovascular events in patients with coronary artery disease. Circulation 2001;104(22):2673–8. https://doi.org/10.1161/hc4601.099485.

[57] Scioli MG, Storti G, D'Amico F, Rodríguez Guzmán R, Centofanti F, Doldo E, Céspedes Miranda EM, Orlandi A. Oxidative stress and new pathogenetic mechanisms in endothelial dysfunction: potential diagnostic biomarkers and therapeutic targets. J Clin Med 2020;9(6):1995. https://doi.org/10.3390/jcm9061995.

[58] Cai H, Harrison DG. Endothelial dysfunction in cardiovascular diseases: the role of oxidant stress. Circ Res 2000;87(10):840–4. https://doi.org/10.1161/01.RES.87.10.840.

[59] Bai J, Zhang N, Hua Y, Wang B, Ling L, Ferro A, Xu B. Metformin inhibits angiotensin II-induced differentiation of cardiac fibroblasts into myofibroblasts. PloS One 2013;8(9), e72120. https://doi.org/10.1371/journal.pone.0072120.

[60] Heo KS, Lee H, Nigro P, Thomas T, Le NT, Chang E, McClain C, Reinhart-King CA, King MR, Berk BC, Fujiwara K. PKCζ mediates disturbed flow-induced endothelial apoptosis via p53 SUMOylation. J Cell Biol 2011;193(5):867–84. https://doi.org/10.1083/jcb.201010051.

[61] Paswan S, Sharma RK, Gaur AR, Sachan A, Yadav MS, Sharma P, Gautam M. Angina pectoris epidemic in India: a comprehensive review of clinical features, differential diagnosis, and remedies. Ind J Res Pharm Biotech 2013;1(3):339–45.

[62] WelenSchef K, Hagstrom E, Ravn-Fischer A, Soderberg S, Yndigegn T, Tornvall P, Jernberg T. Prevalence and risk factors of angina pectoris and its association with coronary atherosclerosis in a general population, a cross-sectional study. Eur Heart J 2022;43(Suppl 2). https://doi.org/10.1093/eurheartj/ehac544.1155. ehac544-1155.

[63] Bentzon JF, Otsuka F, Virmani R, Falk E. Mechanisms of plaque formation and rupture. Circ Res 2014;114(12):1852–66. https://doi.org/10.1161/CIRCRESAHA.114.302721.

[64] Verhagen SN, Rutten A, Meijs MF, Isgum I, Cramer MJ, van der Graaf Y, Visseren FL. Relationship between myocardial bridges and reduced coronary atherosclerosis in patients with angina pectoris. Int J Cardiol 2013;167(3):883–8. https://doi.org/10.1016/j.ijcard.2012.01.091.

[65] Ferrari R, Pavasini R, Balla C. The multifaceted angina. Eur Heart J Suppl 2019;21(Suppl C):1–5. https://doi.org/10.1093/eurheartj/suz035.

[66] Krock BL, Skuli N, Simon MC. Hypoxia-induced angiogenesis: good and evil. Genes Cancer 2011;2(12):1117–33. https://doi.org/10.1177/1947601911142365.

[67] Braile M, Marcella S, Cristinziano L, Galdiero MR, Modestino L, Ferrara AL, Varricchi G, Marone G, Loffredo S. VEGF-A in cardiomyocytes and heart diseases. Int J Mol Sci 2020;21(15):5294. https://doi.org/10.3390/ijms21155294.

[68] Gui C, Li SK, Nong QL, Du F, Zhu LG, Zeng ZY. Changes of serum angiogenic factors concentrations in patients with diabetes and unstable angina pectoris. Cardiovasc Diabetol 2013;12(1):1–8.

[69] Guo M, Shi JH, Wang PL, Shi DZ. Angiogenic growth factors for coronary artery disease: current status and prospects. J Cardiovasc Pharmacol Ther 2018;23(2):130–41. https://doi.org/10.1177/1074248417735399.

[70] Liao YY, Chen ZY, Wang YX, Lin Y, Yang F, Zhou QL. New progress in angiogenesis therapy of cardiovascular disease by ultrasound targeted microbubble destruction. Biomed Res Int 2014;2014. https://doi.org/10.1155/2014/872984.

[71] Critchley JA, Unal B. Is smokeless tobacco a risk factor for coronary heart disease? A systematic review of epidemiological studies. Eur J Cardiovasc Prev Rehabil 2004;11 (2):101–12. https://doi.org/10.1097/01.hjr.0000114971.39211.d7.

[72] Rousan TA, Mathew ST, Thadani U. Drug therapy for stable angina pectoris. Drugs 2017;77(3):265–84. https://doi.org/10.1007/s40265-017-0691-7.

[73] Levine GN, Bates ER, Bittl JA, Brindis RG, Fihn SD, Fleisher LA, Granger CB, et al. 2016 ACC/AHA guideline focused update on duration of dual antiplatelet therapy in patients with coronary artery disease: a report of the American College of Cardiology/ American Heart Association Task Force on Clinical Practice Guidelines: an update of the 2011 ACCF/AHA/SCAI guideline for percutaneous coronary intervention, 2011 ACCF/AHA guideline for coronary artery bypass graft surgery, 2012 ACC/AHA/ ACP/AATS/PCNA/SCAI/STS guideline for the diagnosis and management of patients with stable ischemic heart heart disease, 2013 ACCF/AHA guideline for the management of ST-elevation myocardial infarction, 2014 AHA/ACC guideline for the management of patients with non–ST-elevation acute coronary syndromes, and 2014 ACC/AHA guide-line on perioperative cardiovascular evaluation and management of patients undergoing noncardiac surgery. Circulation 2016;134(10):e123–55. https://doi.org/10.1161/ CIR.0000000000000404.

[74] Fihn SD, Gardin JM, Abrams J, Berra K, Blankenship JC, Dallas AP, Douglas PS, et al. 2012 ACCF/AHA/ACP/AATS/PCNA/SCAI/STS guideline for the diagnosis and man-agement of patients with stable ischemic heart disease: a report of the American College of Cardiology Foundation/American Heart Association task force on practice guidelines, and the American College of Physicians, American Association for Thoracic Surgery, Preventive Cardiovascular Nurses Association, Society for Cardiovascular Angiography and Interventions, and Society of Thoracic Surgeons. Circulation 2012;126(25):e354–471. https://doi.org/10.1161/CIR.0b013e318277d6a0.

[75] Hambrecht R, Walther C, Mobius-Winkler S, Gielen S, Linke A, Conradi K, Erbs S, Kluge R, Kendziorra K, Sabri O, Sick P. Percutaneous coronary angioplasty compared with exercise training in patients with stable coronary artery disease: a randomized trial. Circulation 2004;109(11):1371–8. https://doi.org/10.1161/01.CIR.0000121360. 31954.1F.

[76] Ong P, Aziz A, Hansen HS, Prescott E, Athanasiadis A, Sechtem U. Structural and func-tional coronary artery abnormalities in patients with vasospastic angina pectoris. Circ J 2015;79(7):1431–8. https://doi.org/10.1253/circj.CJ-15-0520.

[77] Kosmicki MA. Long-term use of short-and long-acting nitrates in stable angina pectoris. Curr Clin Pharmacol 2009;4(2):132–41. https://doi.org/10.2174/157488409788185016.

[78] Pascual I, Moris C, Avanzas P. Beta-blockers and calcium channel blockers: first line agents. Cardiovasc Drugs Ther 2016;30(4):357–65. https://doi.org/10.1007/s10557-016-6682-1.

[79] Boden WE, Padala SK, Cabral KP, Buschmann IR, Sidhu MS. Role of short-acting nitro-glycerin in the management of ischemic heart disease. Drug Des Devel Ther 2015;9:4793. https://doi.org/10.2147/DDDT.S79116.

[80] Fathi M, Alami-Milani M, Salatin S, Sattari S, Montazam H, Fekrat F, Jelvehgari M. Fast dissolving sublingual strips: a novel approach for the delivery of isosorbide dinitrate. Pharm Sci 2019;25(4):311–8. https://doi.org/10.15171/PS.2019.34.

[81] Cao HY, Song ZK, Tang ML, Yang S, Liu Y, Qin L. Higher than recommend dosage of sublingual isosorbide dinitrate for treating angina pectoris: a case report and review of the literature. Pan Afr Med J 2021;39(1). https://doi.org/10.11604/pamj.2021.39.28. 22180.

[82] Parker JD, Parker JO. Stable angina pectoris: the medical management of symptomatic myocardial ischemia. Can J Cardiol 2012;28(2):S70–80. https://doi.org/10.1016/j. cjca.2011.11.002.

[83] Husted SE, Ohman EM. Pharmacological and emerging therapies in the treatment of chronic angina. Lancet 2015;386(9994):691–701. https://doi.org/10.1016/S0140-6736 (15)61283-1.

[84] Elliott WJ, Ram CV. Calcium channel blockers. J Clin Hypertens 2011;13 (9):687. https://doi.org/10.1111/j.1751-7176.2011.00513.x.

[85] Shu DF, Dong BR, Lin XF, Wu TX, Liu GJ. Long-term beta blockers for stable angina: systematic review and meta-analysis. Eur J Prev Cardiol 2012;19(3):330–41. https://doi. org/10.1177/1741826711409325.

[86] Balla C, Pavasini R, Ferrari R. Treatment of angina: where are we? Cardiology 2018;140 (1):52–67. https://doi.org/10.1159/000487936.

[87] Thadani U. Management of stable angina–current guidelines: a critical appraisal. Cardi-ovasc Drugs Ther 2016;30(4):419–26. https://doi.org/10.1007/s10557-016-6681-2.

[88] Gayet JL, Paganelli F, Cohen-Solal A. Update on the medical treatment of stable angina. Arch Cardiovasc Dis 2011;104(10):536–44. https://doi.org/10.1016/j.acvd.2011.08.001.

[89] Kallistratos MS, Poulimenos LE, Manolis AJ. Beta blockers, calcium channel blockers, and long-acting nitrates for patients with stable angina and low blood pressure levels: should this recommendation be reconsidered? Eur Heart J 2020;41(3):479. https://doi. org/10.1093/eurheartj/ehz900.

[90] Huang HL, Fox KA. The impact of beta-blockers on mortality in stable angina: a meta-analysis. Scott Med J 2012;57(2):69–75. https://doi.org/10.1258/smj.2011.011274.

[91] Bangalore S, Steg G, Deedwania P, Crowley K, Eagle KA, Goto S, Ohman EM, Cannon CP, Smith SC, Zeymer U, Hoffman EB. β-Blocker use and clinical outcomes in stable outpatients with and without coronary artery disease. JAMA 2012;308(13):1340–9. https://doi.org/10.1001/jama.2012.12559.

[92] Manolis AJ, Poulimenos LE, Ambrosio G, Kallistratos MS, Lopez-Sendon J, Dechend R, Mancia G, Camm AJ. Medical treatment of stable angina: a tailored therapeutic approach. Int J Cardiol 2016;220:445–53. https://doi.org/10.1016/j.ijcard.2016.06.150.

[93] Kones R. Recent advances in the management of chronic stable angina II. Anti-ischemic therapy, options for refractory angina, risk factor reduction, and revascularization. Vasc Health Risk Manag 2010;6:749. https://doi.org/10.2147/vhrm.s11100.

[94] Ahmed LA. Nicorandil: a drug with ongoing benefits and different mechanisms in var-ious diseased conditions. Indian J Pharmacol 2019;51(5):296. https://doi.org/10.4103/ ijp.IJP_298_19.

[95] Tarkin JM, Kaski JC. Nicorandil and long-acting nitrates: vasodilator therapies for the management of chronic stable angina pectoris. Eur Cardiol Rev 2018;13 (1):23. https://doi.org/10.15420/ecr.2018.9.2.

[96] Chaitman BR, Pepine CJ, Parker JO, Skopal J, Chumakova G, Kuch J, Wang W, Skettino SL, Wolff AA. Combination Assessment of Ranolazine In Stable Angina (CARISA) Investigators, Combination Assessment of Ranolazine In Stable Angina (CARISA) Investigators. Effects of ranolazine with atenolol, amlodipine, or diltiazem on exercise tolerance and angina frequency in patients with severe chronic angina: a ran-domized controlled trial. JAMA 2004;291(3):309–16. https://doi.org/10.1001/ jama.291.3.309.

[97] Dobesh PP, Trujillo TC. Ranolazine: a new option in the management of chronic stable angina. Pharmacotherapy 2007;27(12):1659–76. https://doi.org/10.1592/phco.27.12. 1659.

[98] Tamargo J, Lopez-Sendon J. Ranolazine: a better understanding of its pathophysiology and patient profile to guide treatment of chronic stable angina. Future Cardiol 2021;18 (3):235–51. https://doi.org/10.2217/fca-2021-0058.

[99] Thadani U. Management of stable angina–current guidelines: a critical appraisal. Cardiovasc Drugs Ther 2016;30(4):419–26. https://doi.org/10.1007/s10557-016-6681-2.

[100] Borer JS, Fox K, Jaillon P, Lerebours G. Antianginal and antiischemic effects of ivabradine, an if inhibitor, instable angina: a randomized, double-blind, multicentered, placebo-controlled trial. Circulation 2003;107(6):817–23. https://doi.org/10.1161/01.CIR.0000048143.25023.87.

[101] Stanley WC, Marzilli M. Metabolic therapy in the treatment of ischaemic heart disease: the pharmacology of trimetazidine. Fundam Clin Pharmacol 2003;17(2):133–45. https://doi.org/10.1046/j.1472-8206.2003.00154.x.

[102] Ciapponi A, Pizarro R, Harrison J. Trimetazidine for stable angina. Cochrane Database Syst Rev 2005;4. https://doi.org/10.1002/14651858.CD003614.pub2.

[103] Chazov EI, Lepakchin VK, Zharova EA, Fitilev SB, Levin AM, Rumiantzeva EG, Fitileva TB. Trimetazidine in Angina Combination Therapy-the TACT study: trimetazidine versus conventional treatment in patients with stable angina pectoris in a randomized, placebo-controlled, multicenter study. Am J Ther 2005;12(1):35–42.

[104] Kones R. Recent advances in the management of chronic stable angina II. Anti-ischemic therapy, options for refractory angina, risk factor reduction, and revascularization. Vasc Health Risk Manag 2010;6:749. https://doi.org/10.2147/vhrm.s11100.

[105] Zang L. Nanoparticles in the diagnosis and treatment of coronary artery diseases under sports rehabilitation intervention. Ferroelectrics 2021;580(1):283–97. https://doi.org/10.1080/00150193.2021.1902781.

[106] Sharma V, Dewangan HK, Maurya L, Vats K, Verma H. Rational design and in-vivo estimation of Ivabradine hydrochloride loaded nanoparticles for management of stable angina. J Drug Deliv Sci Technol 2019;54, 101337. https://doi.org/10.1016/j.jddst.2019.101337.

[107] Chonkar AD, Rao JV, Managuli RS, Mutalik S, Dengale S, Jain P, Udupa N. Development of fast dissolving oral films containing lercanidipine HCl nanoparticles in semicrystalline polymeric matrix for enhanced dissolution and ex vivo permeation. Eur J Pharm Biopharm 2016;103:179–91. https://doi.org/10.1016/j.ejpb.2016.04.001.

[108] Chavan DU, Marques SM, Bhide PJ, Kumar L, Shirodkar RK. Rapidly dissolving felodipine nanoparticle strips-formulation using design of experiment and characterisation. J Drug Deliv Sci Technol 2020;60, 102053. https://doi.org/10.1016/j.jddst.2020.102053.

[109] Khan AA, Abdulbaqi IM, AbouAssi R, Murugaiyah V, Darwis Y. Lyophilized hybrid nanostructured lipid carriers to enhance the cellular uptake of verapamil: statistical optimization and in vitro evaluation. Nanoscale Res Lett 2018;13(1):1–6. https://doi.org/10.1186/s11671-018-2744-6.

[110] Plumley C, Gorman EM, El-Gendy N, Bybee CR, Munson EJ, Berkland C. Nifedipine nanoparticle agglomeration as a dry powder aerosol formulation strategy. Int J Pharm 2009;369(1–2):136–43. https://doi.org/10.1016/j.ijpharm.2008.10.016.

[111] Hussien RM, Ghareeb MM. Formulation and characterization of isradipine nanoparticle for dissolution enhancement. Iraqi J Pharm Sci 2021;30(1):218–25. https://doi.org/10.31351/vol30iss1pp218-225.

[112] Shah U, Joshi G, Sawant K. Improvement in antihypertensive and antianginal effects of felodipine by enhanced absorption from PLGA nanoparticles optimized by factorial design. Mater Sci Eng C 2014;35:153–63. https://doi.org/10.1016/j.msec.2013.10.038.

[113] Venugopal V, Kumar KJ, Muralidharan S, Parasuraman S, Raj PV, Kumar KV. Optimization and in-vivo evaluation of isradipine nanoparticles using Box-Behnken design surface response methodology. OpenNano 2016;1:1–5. https://doi.org/10.1016/j.onano.2016.03.002.

[114] Singh B, Garg T, Goyal AK, Rath G. Development, optimization, and characterization of polymeric electrospun nanofiber: a new attempt in sublingual delivery of nicorandil for the management of angina pectoris. Artif Cells Nanomed Biotechnol 2016;44(6):1498–507. https://doi.org/10.3109/21691401.2015.1052472.

[115] Varshosaz J, Taymouri S, Hamishehkar H. Fabrication of polymeric nanoparticles of poly (ethylene-co-vinyl acetate) coated with chitosan for pulmonary delivery of carvedilol. J Appl Polym Sci 2014;131(1). https://doi.org/10.1002/app.39694.

[116] Al-Kassas R, Wen J, Cheng AE, Kim AM, Liu SS, Yu J. Transdermal delivery of propranolol hydrochloride through chitosan nanoparticles dispersed in mucoadhesive gel. Carbohydr Polym 2016;153:176–86. https://doi.org/10.1016/j.carbpol.2016.06.096.

[117] Sharma R, Sharma U. Formulation and characterization of atenolol-loaded gellan gum nanoparticles. Indian J Pharm Sci 2021;83(1):60–5.

[118] Hecq J, Deleers M, Fanara D, Vranckx H, Amighi K. Preparation and characterization of nanocrystals for solubility and dissolution rate enhancement of nifedipine. Int J Pharm 2005;299(1–2):167–77. https://doi.org/10.1016/j.ijpharm.2005.05.014.

[119] Owens III DE, Peppas NA. Opsonization, biodistribution, and pharmacokinetics of polymeric nanoparticles. Int J Pharm 2006;307(1):93–102. https://doi.org/10.1016/j.ijpharm.2005.10.010.

[120] Jawahar N, Meyyanathan SN. Polymeric nanoparticles for drug delivery and targeting: a comprehensive review. Int J Health Allied Sci 2012;1(4):217.

[121] Schaffazick SR, Pohlmann AR, Dalla-Costa T, Guterres SS. Freeze-drying polymeric colloidal suspensions: nanocapsules, nanospheres and nanodispersion. A comparative study. Eur J Pharm Biopharm 2003;56(3):501–5. https://doi.org/10.1016/S0939-6411(03)00139-5.

[122] Guterres SS, Alves MP, Pohlmann AR. Polymeric nanoparticles, nanospheres and nanocapsules, for cutaneous applications. Drug Target Insights 2007;2, 117739280700200002. https://doi.org/10.1177/117739280700200002.

[123] Patel VR, Agrawal YK. Nanosuspension: an approach to enhance solubility of drugs. J Adv Pharm Technol Res 2011;2(2):81. https://doi.org/10.4103/2231-4040.82950.

[124] Rabinow BE. Nanosuspensions in drug delivery. Nat Rev Drug Discov 2004;3(9):785–96. https://doi.org/10.1038/nrd1494.

[125] Boukouvala C, Daniel J, Ringe E. Approaches to modelling the shape of nanocrystals. Nano Converg 2021;8(1):1–5. https://doi.org/10.1186/s40580-021-00275-6.

[126] Kambhampati P. Nanoparticles, nanocrystals, and quantum dots: what are the implications of size in colloidal nanoscale materials? J Phys Chem Lett 2021;12(20):4769–79. https://doi.org/10.1021/acs.jpclett.1c00754.

Targeting angiogenesis, inflammation, and oxidative stress in Alzheimer's diseases

Manorama Bhandari[a], Raj Kumar Tiwari[b], Silpi Chanda[c,d], and Gunjan Vasant Bonde[a]

[a]School of Health Sciences & Technology, University of Petroleum and Energy Studies, Dehradun, Uttarakhand, India, [b]Department of Pharmacognosy, Era College of Pharmacy, Era University, Lucknow, UP, India, [c]Department of Pharmacognosy, Amity Institute of Pharmacy, Lucknow, UP, India, [d]Amity University, Noida, UP, India

1 Introduction

Alzheimer's disease is a neurological condition that affects older people and is defined by impaired memory and brain function. Alois Alzheimer, a German neuropathologist and psychiatrist, discovered AD in 1906. AD is characterized by memory loss, speech impairment, impaired performance disorders, disorientation, behavioral disturbances, irregular gait, and slow thinking [1]. Alzheimer's disease is declared as a "global public health priority" by the WHO because there is no long-term treatment for AD [2]. Most commonly, AD is to cause dementia among those who are 60 years of age or older. According to research, 50%–75% of people diagnosed with dementia have AD.

According to statistical data collected around the world, women are more prone to male and the risk increases with age [3]. Dementia is most frequently seen in the form of AD. It affects more than 50 million people worldwide, and if a treatment or preventative measures are not discovered, that number will significantly increase to 152 million by the year 2050 [4]. The prevalence of AD quickly rises from around 2%–3% among the people with age of 70–75 years to 20%–25% for the people of 85 years or more of age. Over this, there's not adequate information to affirm whether AD pervasiveness continues expanding or balances out. Especially in old age, ladies are more likely to be afflicted with AD than men, fundamentally because of age-adjusted increased risk of AD. A few investigations have demonstrated that the general pervasiveness of AD differs broadly among nations, being impacted by social and financial components [5].

Over 50 million individuals globally have dementia, according to a 2018 estimate from Alzheimer's Disease International, and that number is anticipated to quadruple by 2050, with two-thirds of them residing in low- and middle-income nations. According to the latest data estimates, prevalence of dementia in Europe will double by 2050. There is significant evidence that dementia is becoming less common in high-income nations [6], but the evidence to suggest that prevalence is declining is not convincing [7]. AD is the most common form of neurodegenerative dementia, especially in the elderly. Age, genetics, gender, trauma, and air pollution are some of the relevant risk factors associated with AD, although the actual reason is still unknown. Extracellular neurofibrillary tangles and amyloid plaques in neurons are two characteristics of AD. Although AD cannot be cured, early disease stages may benefit more from some therapy options. According to recent finding, the pathophysiology of AD is influence by inflammation, angiogenesis, and oxidative stress. OS continues to be present at all phases of the disease, including its onset, development, and aggravation and has been related to AD for a long time. Additionally, excessive reactive oxidative species devastate and deplete the skin's antioxidant defense, which eventually promotes and exacerbates AD. Angiogenesis, a feature of chronic inflammatory illnesses, is also implicated in AD in addition to OS [8]. Alzheimer's patients have high levels of VEGF in their brains, and angiogenesis has been demonstrated to be dysregulated in patients or AD model type. Meanwhile, Chen et al. found that vascular VEGF-A mRNA gradually raised in an Alzheimer's mouse model's skin [9]. The VEGF pathway has been acknowledged as a potential connection between oxidative stress and angiogenesis in AD. Overproduction of reactive oxygen species in the brain has been shown to contribute to AD pathogenesis, while VEGF is a potent angiogenic factor that plays a critical role in the growth of new blood vessels. Studies have suggested that VEGF may also have a protective effect against oxidative stress in the brain. Therefore, targeting the VEGF pathway may represent a potential therapeutic strategy for AD, particularly in reducing oxidative stress and promoting angiogenesis. Another potential target for AD treatment is for Tie-2 receptor. Tie-2 is a receptor that is stated on endothelial cells, and it is essential for vascular homeostasis and angiogenesis. Inhibition of Tie-2 has been observed to decrease amyloid-beta accumulation and enhancement of cognitive function in animal models of AD. On the other hand, the immunological and neurological systems have an enduring connection since they regularly interact and share regulation mechanisms [10].

Neurotransmitters are important substances in neuronal communication because they regulate the production of cytokines and the inflammatory response. A large and diverse family of proteins, including growth factors and enzymes, interleukins, interferons, tumor necrosis factors, chemokines, and other inflammatory substances, play crucial roles in cellular communication and immunological control. While the function of inflammation in the etiology of AD is not properly known, but some inflammatory molecules, as well as IL-1, IL-6, TNF-α, have been related to the development of AD. Immune molecules and their imbalance may contribute to neurodegeneration, even if the immune system defends the brain. It was hypothesized that levels of cytokines and other molecules fluctuate in dementia; as a result, examining changes in their levels in AD patients may help distinguish between AD and cognitive decline [11].

2   Molecular pathways of pathogenesis

Alzheimer's disease (AD) has been fraught with debate about cause and effect ever since the disorder was first identified. According to Alois Alzheimer, the histopathological features indicated an upstream process rather than a primary cause of disease.

Inhibition of cytochrome oxidase activity may therefore represent a potential treatment method for AD. By suppressing the activity of this enzyme, it may be possible to shift amyloid precursor protein metabolism away from production of amyloidogenic and nonamyloidogenic by-products. This may prevent the development of amyloid plaques in the brain and halt the disease's progression. However, further research is needed to fully understand the role of cytochrome oxidase in AD [12]. Beta secretase (BACE) activity is activated by oxidative stress, which is a necessary step in the conversion of APP to Aβ [13]. This kind of evidence suggests that amyloidosis in AD occurs later in the disease's progression. In contrast, it's not hard to see how Alzheimer's disease may be misdiagnosed as primary amyloidosis. In 1991, a mutation in the APP gene was found to induce a form of AD with early onset and autosomal dominant inheritance. Further findings indicated that there are two additional presenilin genes associated with the autosomal dominant, early-onset variant of AD, presenilin 1, and presenilin 2. Functional investigations showed that these mutations affect how the APP is processed. A42 to A40 ratios rise in each of these situations. The argument over AD etiology centers on whether the majority of AD is caused by primary or secondary amyloidosis [14].

According to the amyloid cascade hypothesis, the pathophysiology of AD is thought to begin with overproduction and extracellular deposition of amyloid-β peptide (Aβ) and intracellular deposition of neurofibrillary tangles, as proposed by the amyloid cascade theory (NFT) (Fig. 1). These accumulations serve as triggers for a wide variety of neurotoxic pathways, such as excitotoxicity, Ca2β homeostasis disturbance, free radical generation, and neuroinflammation. The cholinergic theory postulates that declines in acetyl choline neurotransmitter and cholinergic indicators like cognitive and memory problems are brought on by choline acetyltransferase (ChAT) and acetyl cholinesterase (AChE).

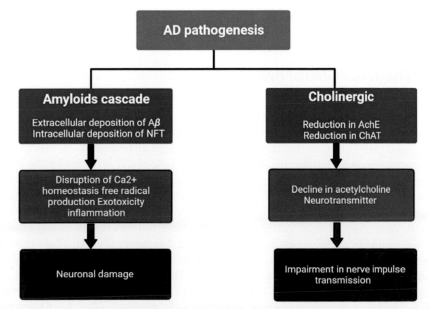

Fig. 1 Schematic representation of major hypothesis in Alzheimer's disease (AD).

2.1 The cholinergic hypothesis

When it comes to molecular theories that attempt to explain the pathophysiology of AD, the cholinergic hypothesis is the oldest and most thoroughly studied. Primarily, a degenerative process was initially recognized as such over 30 years ago. Some changes cause a downregulation of cholinergic indicators like acetyltransferase and acetylcholinesterase, which are associated with the development of cognitive impairment [15,16]. This downregulation is related to the density of neurofibrillary tangles and the severity of injury. Certain M1 muscarinic agonists are one of the most crucial links between the major AD symptoms. Immunohistochemical, neuroimaging, and other analyses corroborated that this theory showed a reduction in nicotinic receptor number and density in AD patients (primarily 42 subtype) [17]. The cholinergic theory is confirmed by the fact that there is a substantial interaction between the glutamatergic and cholinergic systems during changes in glutamatergic neurotransmission. Since glutamate is involved in mediating neuronal excitability, acetylcholine, and its receptors, they are thought to be neuroprotective through regulation [18,19]. The entorhinal cortex (EC) is the first brain region to experience alterations in glutamatergic neurotransmission due to Alzheimer's disease [20]. The long-term potentiation mediated by glutamatergic neurotransmission in the hippocampus is important for memory retention and learning [19]. NMDA glutamate receptors' hyperactivation has been connected to neurodegeneration, cell death, and an increase in chloride, sodium, and calcium ions in the postsynaptic membrane [21]. Some of the possible mechanism involved in the pathophysiology of AD. Specifically, distinct neuronal depolarization, activation of nitric oxide generation by amyloid, and blockage of presynaptic and glial glutamate transport are thought to contribute to the cognitive and neuronal dysfunction seen in AD [22,23]. This results in continued receptor activation and excitotoxicity [24]. Treatment strategies and medication development for Alzheimer's disease are heavily grounded in the cholinergic hypothesis.

2.2 The mitochondrial hypothesis

The mitochondrial cascade theory aims to provide a comprehensive explanation for the histological, biochemical, and clinical characteristics of AD [25]. Numerous conceptual leaps are made by the mitochondrial cascade theory. It's assumed that brain aging, and AD is caused by comparable physiological processes. It asserts that because mitochondrial failure in AD affects the whole body, it cannot be the result of neurodegeneration alone. It suggests that amyloidosis, tau phosphorylation, and cell cycle re-entry are caused by mitochondrial malfunction in AD brains. According to the free radical hypothesis of aging, oxidative byproducts cause structural damage to cells over time. Recent findings indicating that mtDNA mutation acquisition increases aging in experimental animals to support this theory [26,27]. There is mitochondrial dysfunction in many AD tissues. At the very least, mitochondria in fibroblasts, platelets, and brain cells are implicated. Three mitochondrial enzymes have been identified as being broken. Decreased activity of cytochrome oxidase, pyruvate, and α-ketoglutarate dehydrogenase complexes is required [28,29]. Exogenous

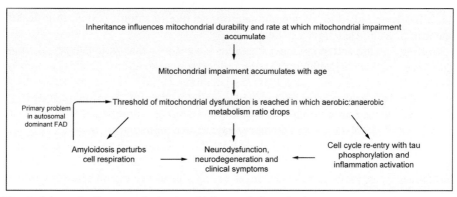

Fig. 2 Schematic diagram of mitochondrial cascade hypothesis.

mitochondrial DNA is transferred to cultivated cells that are devoid of indigenous mtDNA for cybrid research. An alternative name for mitochondrial DNA is "DNA." Therefore, these mtDNA-depleted cells are also known as "0 cells." They lack cytochrome oxidase function and do not generate proteins that are encoded by mitochondrial DNA. Transferred mitochondrial DNA reproduces and replaces the mtDNA in recipient cells. This permits expression of the ETC subunits encoded by the mtDNA and the recovery of cytochrome oxidase activity. Cybrid lines with mitochondrial DNA from AD patient platelets had lower levels of cytochrome oxidase activity than age-matched controls without AD, according to comparisons made between the two types of cybrid lines (Fig. 2) [30–33].

2.3 The amyloid hypothesis

The amyloid cascade principle states that Alzheimer's aliment is because of an imbalance among amyloid synthesis and clearance [34], the imbalance being the fundamental cause and the event that initiates subsequent abnormalities identified in AD. High resistance to proteolytic degradation characterizes the peptide known as amyloid [35]. The 142 amyloid peptide isoform is hydrophobic and is considered to be the most toxic isoform of amyloid peptide. It takes the shape of an β-pleated sheet [36] due to its physical characteristics and has a stronger propensity to accumulate and form amyloid plaque [37]. Amyloid neuritic plaques primarily consist of this isoform [38,39]. The amyloid peptide formation is caused processing of amyloid precursor protein (APP) across the plasma membrane. The precise function of the type I transmembrane glycoprotein known as APP is unknown, but it is known that cellular stress-related phenomena cause it to express itself more. First, APP is damaged by enzymes with α-secretase activity; the ADAM (disintegrin and metalloprotease) family of enzymes, such as ADAM19, ADAM10, and ADAM17, is usually the largest [40]. ADAM activity may be altered by a variety of factors, including receptor activation, growth factors, cytokines, and hormones. When APP is cut, a fragment of peptide of APP and the carboxyl terminal part are formed and released [41]. The APPs, which has been linked to trophic and neuroprotective functions, is present in lesser concentrations in Alzheimer's patients.

Multiple in vitro and in vivo investigations have shown that various types of amyloid are neurotoxic [42], although the exact processes are complicated and not entirely understood. The main structure between positions 25 and 35 has an amino acid sequence with increased neurotoxicity and is well documented. Amyloid peptide concentrations are managed through the equilibrium among synthesis and clearance; in AD patients, an imbalance in clearance ends in accumulation of amyloid in brain [43]. Peptides are transported to the blood from the brain through LRP-1 receptor contact and p-glycoprotein activity. Receptors in the brain allow amyloid to cross into the bloodstream. Age increased amyloid levels, and reduced expression of LRP-1 receptors all contribute to amyloid peptide buildup in the CNS [44].

2.4 The infectious hypothesis

A long-time hypothesized theory which states that underlying infections could be a cause of Alzheimer's which is still in interest and has been rekindled by a number of recent research. A study found the herpes virus in the brains of patients diagnosed with AD as early as 1991 [45]. In the preclinical studies, researchers have additionally discovered that the brains of individuals with AD contain a group of network driver genes that are related to vulnerability to viruses. Human herpesvirus (HHV)-6 and Herpes simplex virus (HSV)-1 DNA were detected in three different cohorts of AD brain specimens after further analysis, and interactions of numerous putative viralhost that controls the gene network relevant to biology of AD, such as APP processing and innate immunity, were also discovered [46].

3 Role of angiogenesis in AD

There has not been much research done on AD's potential relationship with angiogenesis. However, evidence is growing that suggests angiogenesis-related elements and processes that exist in the AD brain. Genome-wide analysis of gene expression has revealed that angiogenesis-related genes are significantly elevated in the AD brain [47]. One of the main clinical manifestations of AD is cerebral hypoperfusion, which probably contributes significantly to its etiology [48]. As well as influencing AD's clinical and pathological symptoms, hypoxia is known to induce angiogenesis [49,50]. VEGF and various hypoxia-inducible genes are upregulated in response to hypoxia [50,51]. In the AD brain, VEGF, a powerful modulator of angiogenesis, is found in clusters of reactive astrocytes, diffuse perivascular deposits, and the walls of intraparenchymal arteries [51]. Additionally, the clinical severity of AD and the intrathecal levels of A are related to intrathecal levels of VEGF [52]. The VEGF gene's promoter region may have polymorphisms that increase the risk of AD, according to growing evidence [52].

Endothelial activation is the first identifiable stage in the complicated angiogenic process. Numerous proangiogenic mediators are produced and secreted by the active endothelium [53,54]. During normal angiogenesis, the activation of endothelial cells is temporary and self-contained. One of the most prominent clinical hallmarks of AD is cerebral hypoperfusion, which triggers activation of brain endothelial cells in response to a continuous or intermittent stimulation. Activated endothelial cells are extraordinarily synthetic, releasing a wide range of substances that may affect the survival of nearby neurons and neighboring cells activation (including astrocytes and microglia). No new vessel development occurs despite the stimulus still being present. Multiple factors may be perform in the inhibition of vascular growth, including A's antiangiogenic activity, faulty homeobox signaling, and combinations of these and other mechanisms, [55–58]. Endothelial dysfunction that is irreversible in AD results from reversible endothelial activation. A permanently dysfunctional endothelium may produce vascular byproducts that harm neurons.

Brain endothelial cells are activated in response to continuous or intermittent stimuli, such as cerebral hypoxia. The activation of neighboring cells can be influenced by substances released by activated endothelial cells, which are themselves highly synthetic. No new blood vessels form even if the stimulation is still present. Without the formation of new blood arteries, vascular activation endothelial cells are not inhibited by feedback signals, as they are during physiological angiogenesis. Endothelial activation is normally reversible, but in Alzheimer's disease it causes permanent malfunction. Neuronal injury/death may be caused by the circulatory products of a persistently defective endothelium either directly or by activating astrocytes and/or microglia [59].

4   Role of inflammation in AD

Histologic samples of Auguste D's brain were analyzed by Alois Alzheimer over a century ago, and he found not just the neurofibrillary tangles and amyloid plaques that are today described as indicators of AD, but also glial cells collected around the plaques [60]. Microglia and astrocytes, two types of innate immune cells, are expected to serve a major component withinside the improvement of AD, and they will offer new healing goals which might be simply as critical because the proteins tau and amyloid that produce plaques and tangles [59].

It is becoming clearer neuronal dysfunction is not the only cause of AD development and progression, but glia-dependent neuroinflammatory pathways also play a crucial role. Many innate immune responses, including as astrocyte, microglial synthesis, and release of inflammatory mediators, are triggered in response to disease stresses. Neuroinflammatory signaling additionally includes both peripheral myeloid molecular and perivascular macrophages populations which might be capable of migrating to the right into a dysfunctional brain (Fig. 3).

The protein known as YKL-40, and also known as chitinase 3-like protein 1, is made by astrocytes and is expressed in the cerebrospinal fluid, making it a viable indicator of AD neuroinflammation. CSF YKL-40 shows a nonlinear correlation with cortical thickness in early AD patients, a pattern that is different due to that shown to be associated with p-tau-induced neurodegeneration. These findings lend credence to the idea that neuroinflammatory and neurodegenerative processes occur simultaneously during the early clinical stages of AD and imply that YKL-40 may be beneficial for monitoring inflammatory processes associated with neurodegeneration [61–63].

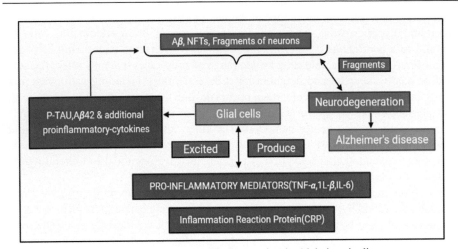

Fig. 3 Schematic representation of role of inflammation in Alzheimer's disease.

5 Role of oxidative stress in AD

Aging brains are greater vulnerable to oxidative stress because of a growth in redox-associated damage, which include the formation of ROS or the breakdown antioxidant system [63,64]. The chain of electron transport of mitochondrial uses almost all of the oxygen (molecular); it takes in the cytochrome oxidase complex, converting the rest into hydrogen peroxide and superoxide radicals. Superoxide ($O_2\cdot$), a radical form of oxygen, is produced during normal metabolism and a wide variety of activities, and the other nonradical oxidants such as hydrogen peroxide (H_2O_2) [65]. Extremely hydroxyl radical ($OH\cdot$) and different oxidant substances are shaped with in the presence of catalytic iron or copper ions at some point during the overproduction of oxygen radicals and hydrogen peroxide, which might also additionally motivate tissue damage [66]. Under stress and in old age, the mitochondrial electron transport system produces large amounts of ROS, putting the cell at danger of developing AD if an adequate antioxidant mechanism either is not present or not efficiently working. Both the source and the target of ROS damage are mitochondria; this is because oxidative stress as well as mitochondrial dysfunction plays crucial role in neurodegenerative diseases and aging [67]. Fig. 4 represents the role of ROS in AD progression.

6 Natural approach in the management of AD

There are multiple herbs, and their bioactive compounds those offer some options for changing the progress and alleviate the symptoms of AD. The standardized plant extract and their fraction promisingly show the efficacy and safety for the treatment of ailments. Ayurveda is one of the ancient systems of medicine practiced mainly in

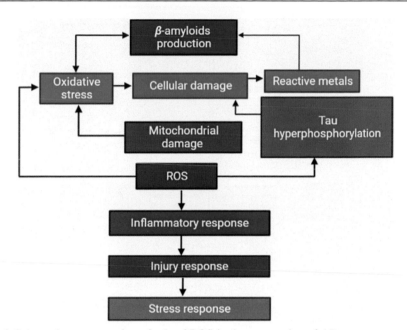

Fig. 4 Schematic representation of role of ROS in the progression of AD.

Asia and documented several herbs for the management of AD. According to Ayurveda, AD is imbalance in Vata and Kapha doshas. Many secondary metabolites have important roles in the treatment of the disease such as lignans, polyphenols, tannic acid flavonoids, triterpenes, sterols, alkaloids, etc. (Fig. 5) [68–72].

Diet for AD is not surprisingly different from a general healthy diet. A diet rich in saturated fatty acids, cholesterol, sugar, salt, and branched-chain of amino acids promotes the progression of the disease. In AD, conventional medications can cause constipation; thus, a diet high in fiber is always recommended. According to research, frequent dietary changes alter the gut flora, which may hasten the onset of AD. Additionally, consumption of alcohol, a nervous system depressant can interfere vitamins from being absorbed. There should be a balance between sugar need and cravings. In aspect of management of blood sugar, the body should not be deprived of carbohydrate which results in a low serotonin level. Supplementing with omega-3 fatty acids is good for the improvement of brain function impairment. Cigarettes smoking and caffeine consumption must be stopped. Corollary to that research supports that low risk of may be associated with moderate coffee consumption in neurodegenerative disorders including AD [73]. Regular exercise always acts as a source of body fuel and can control norepinephrine.

Some other medicinal plants and its part use in AD treatment are *Magnolia officinalis* (Magnoliaceae), *Glycyrrhiza glabra* (Fabaceae), *Tinospora cordifolia* (Menispermaceae), *Allium sativum* (Amaryllidaceae), *Collinsonia canadensis* (Lamiaceae), *Bertholettia excelsa* (Lecythidaceae), *Zingiber officinale* (Zingiberaceae), *Matricaria recutita* (Asteraceae), *Angelica archangelica* (Umbelliferae), *Rosmarinus officinalis*

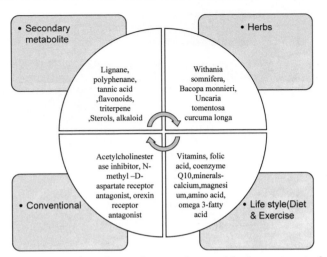

Fig. 5 Schematic representation of natural approaches used in the treatment of AD.

(Lamiaceae), *Melissa officinalis* (Lamiaceae), *C. canadensis* (Lamiaceae), *Bertholettia excelsa* (Lecythidaceae), *Camellia sinensis* (Theaceae), *Urtica dioica* (Clusiacea), *Lepidium meyenii* (Brassicaceae), etc. Details of the plants used for the management of AD are shown in Table 1.

7   Synthetic approach in the management of AD

Alzheimer's is characterized as the most prevalent neurodegenerative diseases which leads to occurrence of dementia. Numerous studies have indicated that elevated ROS levels and oxidative stress are important factors in the progression of AD and that the advancement of the disease may be reduced by using antioxidants to prevent the production of free radicals. Since AD mostly affects elderly individuals, there is currently no effective therapy. There are several herbal treatments available for the disease's palliative care. To make a significant advancement in order to cure AD, researchers are now concentrating on synthetic substances.

    Among the medications for AD that are now marketed are rivastigmine, donepezil, memantine, and galantamine. Without addressing the aforementioned crucial fundamental elements responsible for the neurodegeneration, the most of the currently available medications can provide is symptomatic alleviation for the first 1–2 years of the illness. The majority of recently available medications used to treat AD either improve ACh-mediated transmission or make up for ACh deficit [103]. This is accomplished by inhibiting cholinesterases (ChEs), including the acetylcholinesterase and butyrylcholinesterase that degrades ACh [79]. Although none of the ChE inhibitors (ChEIs) galantamine, donepezil, tacrine, and rivastigmine have been shown to be useful in slowing the course of AD, they are all capable of improving functional, cognitive, and behavioral functions [47].

Table 1 An outline of commonly used herbs explored for management of Alzheimer's diseases.

Plant name	Family	Plant parts	Formulation/Bioactives	Predicted mechanism	References
Ashwagandha (*Withania somnifera*)	Solanaceae	Root	Steroidal lactone: e steroidal lactones, such as dehydro-withanolide-R, withanolides A-Y, withanamides A and C, and withasomniferin-A	Antioxidant, free radical scavengers, rejuvenate immune system	[74,75]
Brahmi (*Bacopa monnieri*)	Plantaginaceae	Whole plant	Bacopasides III, IV, V, saponins, bacosaponins A, B, C, D, E, F and bacosides A and B	Antioxidant, free radical scavengers, neuroprotective	[76]
Cat's claw (*Uncaria tomentosa*)	Rubiaceae	Root/bark	Spiroindole alkaloids isopteropodine and rynchophylline	Neuroprotective, inhibit the hyperphosphorylation of tau protein, enhance activities of antioxidant enzymes, lower the increased levels of IL-6, IL-1β, and TNF-α	[77]
Ginkgo (*Ginkgo Biloba*)	Ginkgoaceae	Leaf	Terpene lactones: ginkgolides and diterpenes and ginkgo flavone. Ginkgolides A, B, C, J, M, K, L sesquiterpene–bilobalide glycosides: ginkgetin, bilobetin, and sciadopitysin	Antioxidant, inhibition of lipid peroxidation, uptake of serotonin, dopamine, norepinephrine, platelet aggregation	[78,79]

Continued

Table 1 Continued

Plant name	Family	Plant parts	Formulation/Bioactives	Predicted mechanism	References
Huperzia serrata/ Lycopodium serratum	Lycopodiaceae	Leaf and stem	Quinolizidine alkaloid: Huperzine A	Neuroprotective, inhibition of AChE	[80]
Ginseng (*Panax ginseng*)	Araliaceae	Root	Saponin glycosides: Ginsenosides-protopanaxadiol and protopanaxatriol	Reduced Aβ deposition and tau protein hyper phosphorylation, autophagy stimulation, neurogenesis, antioxidative stress, antiapoptosis effects, and antiinflammatory activities	[81]
Sage (*Salvia officinalis*)	Lamiaceae	Leaf	Essential oils: (*E*)-β-caryophyllene (7.47%), α-thujone (26.68%)	Amyloid-β activation, cholinergic activity, neurotrophins, suppression of oxidative stress, inflammation, antioxidant, antidepressant	[82,83]
Turmeric (*Curcuma longa*)	Zingiberaceae	Rhizome	Sesquiterpenoids and monoterpenoids Curcuminoids: curcumin, demethoxycurcumin, and bisdemethoxycurcumin	Inhibition of inflammation and oxidative stress, activation of amyloid-β Delayed in beta-amyloid plaques, degradation of neurons	[84]

Plant	Family	Part	Phytochemicals	Activity	Ref.
Shankhpushpi (*Convolvulus pluricaulis, Convolvulus microphyllus, Clitoria ternatea,* and *Evolvulus alsinoides*)	Gentianaceae	Whole plant	Flavanol glycosides, triterpenoids, steroids, and anthocyanins	Anxiolytic, memory enhancement, antidepressant, tranquilizing, neurodegenerative, antistress, antioxidant, and antiinflammatory	[85]
Centella (*Centella asiatica*)	Apiaceae	Whole plant	Pentacyclic triterpenoids, such as brahmoside, asiaticoside, brahmic acid, and asiatic acid	Reducing oxidative stress, improving mitochondrial function, and increasing synaptic density	[86,87]
Black oil plant (*Celastrus paniculatus*)	Celastraceae	Seed	Sterols, alkaloids, and bright coloring substance, celapanigine, celapanine, celastrine, celapagine, and paniculatine	Antioxidant, inhibition of ROS	[88]
Jatamansi (*Nardostachys jatamansi*)	Valerianaceae	Rhizome	Pyranocoumarins seselin, jatamansinol, jatamansine, jatamansinone, and dihydrojatamansin Sesquiterpenes terpenoids include nardostachysin, spirojatamol, calarenol, and jatamols A and B	Acetylcholinesterase, butyrylcholinesterase inhibitors	[89]
Guggul (*Commiphora mukul, C. abyssinica, C. molmol, C. whighitii,* and *C. Burseraceae*)	Burseraceae	Gum	Ferulic acids, phenols, and other nonphenolic aromatic acids	Antioxidant, anticholinesterase activity	[90]

Continued

Table 1 Continued

Plant name	Family	Plant parts	Formulation/Bioactives	Predicted mechanism	References
Panax notoginseng	Araliaceae	Root	Ginsenoside Rg1	Secretase activity	[91]
Dipsacus asper	Caprifoliaceae	Root	Akebia saponin D	Aβ toxicity	[92]
Paeonia suffruticosa	Paeoniaceae	Root bark	1,2,3,4,6-penta-O-galloyleta D-glucopyranose	Aβ fibril formation, stability, and impairment of in vivo long-term memory	[93]
Polygala tenuifolia	Polygalaceae	Leaves	Tenuifolin (extract)	Secretase activity; morphological plasticity	[94]
Radix salviae miltiorrhizae	Lamiaceae	Seed	Triterpenoids; Tanshinone	Activity of AChE, in vivo and in vitro toxicity of Aβ, and NOS protein	[95]
Murraya koenigii	Rutaceae	Leaves	Scoponin, carbazole alkaloids	Antimemory effect and decrease ChE activity	[96]
Cassia obtusafolia	Fabaceae	Seeds	Obtusifolin	AChE inhibition	[97]
Centella asiatica	Apiaceae	Whole plant	Triterpene glycosides, saponies	Reducing Aβ in vivo, neuroceutical, Aβ levels, and apoptosis, improves mood and memory, promotes dendritic growth and mitochondrial health	[79]
Desmodium gangeticum	Fabaceae	Whole plant	Aminoglucosyl-glycerolipids, cerebrosides	Reserved amnesia, AChE inhibition	[98,99]
Ginkgo biloba	Ginkgoaceae	Leaf	Flavonoids, glycosides, and ginkgolides	Stimulates cerebral blood flow, improves mitochondrial function, block neural cell death, antioxidant	[100]
Lion's mane (Hericium erinaceus)	Hericiaceae	Fruiting bodies	Isoindolinones, myconutrients, β-glucan	Improves cognition, has antiinflammatory properties,	[78]

			polysaccharides, hericenones, and erinacine terpenoids	block Aβ production, stimulates neurotransmission and neurite outgrowth, neuroprotective	
Saffron (*Crocus sativus*)	Iridaceae	Stigma	Safranal, crocin, crocetin, picrocrocin, kaempferol, and flavonoid	Antioxidant, antiamyloidogenic, antiinflammatory, immunomodulation, neuroprotection	[101]
Triphala (*Emblica officinalis, Terminalia bellerica*, and *Terminalia chebula*)	Phyllanthaceae	fruit	Punicafolin, flavonoids, kaempferol, ellagic acid, and gallic acid	Antimicrobial, antiinflammatory, immunostimulating, and prevents dental caries, antioxidant	[102]

The first ChEI licensed for treating AD was tacrine (9-amino-1,2,3,4-tetra-hydroacridine, THA). But hepatotoxicity forced the withdrawal of tacrine. AD and Parkinson's disease (PD) are both treated with aminoadamantanes (amantadine and memantine). A recent anti-PD and antiviral medication is amantadine (1-aminoadamantane, adamantylamine). Amantadine, a mild NMDA receptor antagonist, is thought to be helpful for patients with AD and PD who need an enhance dopaminergic transmission to make up for the imbalance between glutamate and dopamine. Memantine (1-amino-3,5-dimethyladamantane) can act through many mechanisms of action (MOA) for its neuroprotective consequences for the remedy of AD. For mild to excessive AD, memantine, a nonaggressive NMDA glutamate receptor antagonist, is authorized [104].

The tacrine can be modify by modification in its chemical structure. A five-membered heterocyclic ring was also generated by converting an amino group into a 1,2-dithiolane derivative. This was done by replacing 1,2-dithiolane for the amino group. The tacrine-lipoic acid dimer was the most prevalent derivative that was produced. A carbon chain of varying length has been used to combine the universal antioxidant lipoic acid with a tacrine moiety. The most effective derivative had a linker composed of three methylene groups, and it showed good antioxidant activity [105].

Imidazole-substituted derivatives include the replacement of the amino group in the 4-aminopyridine ring of tacrine with a substituted imidazole ring. Along with the inhibition of acetylcholinesterase, this kind of alteration has other abilities like the suppression of N-methyltransferase (NMT), histamine metabolism, and histamine H3 receptor antagonism, all of which may ameliorate impaired cognitive skills [106].

Recent research has shown that a serine-threonine kinase known as Rho-related protein kinase 2 (additionally called ROCK2) acts as a bad regulator of Parkin-based mitophagy. Inhibition of ROCK2 has been discovered to have defensive advantages in more than a few approach of neurodegenerative diseases (NDDs), despite the fact that the mechanisms that underlie ROCK2-mediated control of mitophagy remain unclear. The ROCK2-dependent pathway is traditionally thought to be predominantly responsible for abnormal brain processes. It is also thought to be involved in crucially influencing the inflammatory response of microglial cells [107]. In spite of this, ROCK2's function as a negative regulator of mitophagy underlines its significance as a possible target for the therapy of AD. This would be accomplished by inducing mitophagy of damaged mitochondria [108].

Isoxazole is a crucial nucleus that contains a five-membered heterocycle ring. The atoms of nitrogen and oxygen are situated next to one another in this structure. The significance of this nucleus may be attributed to the fact that it engages in a wide variety of biological and therapeutic activities. Antimicrobial, anticancer, anti-Parkinson, anti-Alzheimer, anticonvulsant, and antiviral activities are attributed to isoxazole nucleus derivatives. However, it has been found to have a low level of cytotoxicity. These derivatives have also shown promise in terms of their multiple biological activities [109]. Isoxazolones contain an isoxazole nucleus, and it has been discovered that pretreatment with isoxazolone derivatives improves the cholinergic activities of the brain, which in turn enhances the activity of the brain's intrinsic cognitive processes. In addition to this, it has been observed that the derivatives in question may feature superior anxiolytic activities [110].

Choubey et al. [85] designed, synthesized, and clinically evaluated molecular multitarget hybrids of N-benzyl pyrrolidine derivatives with the aim of treating AD. Effect of in silico molecular docking and lead dynamics simulations confirmed consensus binding affinities similar to those of PAS-AchE and BACE-1 [111]. By inserting multiple structurally unique amino acetamide a group at 6-position of the deoxyvasicinone group, a series of multitarget ligands were created with the goal of developing new multifunctional by employing deoxyvasicinone as the substrate. This was done with the aim of securing innovative multifunctional derivatives, specifically targeting AD. In vitro tests found out that the majority of the synthesized derivatives had IC50 values in the low nanomolar range for inhibition of human recombinant acetylcholinesterase (hAChE) and human serum butyrylcholinesterase and the compounds showed moderate suppression of A142 self-aggregation (hBChE). Inhibition of human serum butyrylcholinesterase via compound becomes additionally observed (hBChE) [111].

The search for BACE1 inhibitors has rapidly risen to the forefront of medicinal approaches to AD. Peptidomimetic BACE1 inhibitors such as hydroxyethylamine, hydroxyethylene, and others similar do not have sufficient metabolic stability or pharmacokinetic properties. Inhibitors' nonpeptides are being created because of their enhanced metabolic stability, reduced size, and increased capacity to penetrate the BBB. A wide variety of scaffolds, including aminoimidazole, acyl guanidine, aminothiazoline, amino/iminohydantoin, 2-aminopyridine, and aminooxazoline, have been produced over the course of the last several decades. The effectiveness of these small compounds suggests that BACE1 inhibitors could slow down AD's development [112].

8   Strategic approaches for AD treatment

The use of polymeric NPs is one of the most innovative and promising approaches in delivering therapeutic medications in a targeted way directly to the brains. Being Innovative and nonhazardous, these nanosystems may enhance the pharmacokinetics and biodistribution of parenterally delivered drugs, hence boosting their therapeutic effectiveness [113]. Block copolymers comprising simple monomers of either synthetic or natural origin make up polymeric NPs, or natural origin, like polysaccharides and albumin, and are therefore typically biocompatible, biodegradable, and quickly excretable [114]. Albumin, chitosan, polylactic acid (PLA), poly(butylcyanoacrylate), polyethylenimine, and polyethylene glycol (PEG), are only a few of the polymers that are frequently employed in the production of these NPs. The NPs improve AMT delivery of macromolecular drug-encapsulating into the brain in AD, their surfaces can be conjugated with protein, cationic oligos or basic cell-penetrating peptides. Further, capillary endothelial cells have a negative charge and can attract positively charged NPs [115,116]. When it comes to selective conjugation between complementary functional groups without generating unwanted byproducts, orthogonal chemistry is the gold standard. As the nearest approximation to biological tissues and materials, NPs are designed by functionalizing polymer backbones with particular moieties to allow for the biomolecules grafting and the production of diverse carbon-based systems [117,118].

8.1 Nanotechnology approach

The goal of nanotechnology is to fabricate and use nanomaterials or nanocarriers, which are substances with at least one dimension between 1 and 100 nm. Superparamagnetism or surface plasmon resonance, two bulk-independent phenomena of interest to the medical community, are frequently seen in materials at this scale [119]. Additionally, protein and nucleic acids are well suited for interacting with those biomolecules and with cells because they are in the same size range as nanomaterial, particularly nanoparticles. Similarly, a high surface-to-volume ratio associated with nanometric size is advantageous in biological application, notably in sensing. These nanocarriers had received significant attention and evaluation for therapeutic application [120–122]. Over the past few years, research on the further employment of nanomaterial in precision medicine had also increased [123]. The individual treatment for AD now available on the market is on symptomatic relief because the medication is unable to cross over the BBB. Nanotechnology-based therapy may be able to get over this restriction because of its many advantages [124]. The FDA had approved the use of a wide range of nanocarriers, ranging in size from nanometers to micrometers, for the delivery of commercially available drugs. These nanocarriers are utilized to treat neurological disorders like Alzheimer's disease and brain cancer [125]. Several nanocarriers loaded with distinct nanomedicines were fabricated and evaluated. A substantial amount of research is being done on the possibility of using nanomaterials for the treatment of AD. The following section throws light on some such approaches. Most of them fall into one of three categories: lipid-based nanoparticles, organic nanostructures, or metallic nanoparticles (NPs).

8.1.1 Liposomes

The most likely solution to the issue of delivering the drugs across the BBB is the phospholipid bilayer structure of liposomes. To improve the transport across the BBB, numerous surface modifications of the surface of liposomes were evaluated [126]. The surface of the BBB is characterized by the presence of large number of proteins, peptides, antibodies, and other ligand receptors. Surface-active ligands targeted to these receptors can be used for functionalization of liposomes and targeting the BBB for speeding up the transcytosis of drugs or nanocarriers. The BBB is concurrently penetrated by cationic liposomes and transcytosis. To facilitate their movement through the body, liposomes are frequently coated with nutrients like glucose. The passive diffusion mechanism can start working after the liposomes have reached the brain. The brain's passive efflux initiates this process [127]. Curcumin-loaded liposomes can significantly enhance medication transport to the CNS through related receptors on BBB cells [128]. The delivery of apolipoprotein E (ApoE2) in the AD-damaged brain had been accomplished using the liposome carrier system modified with a surface containing mannose ligand and cell-penetrating peptides (CPPs). The results demonstrated that functionalized liposomes can safely and successfully transfer a significant concentration of drugs to the target tissues for the treatment of AD [129]. Because of its preventative effects on hippocampal neurons and anti-

A properties, osthole (Ost) is an anti-AD substance. An Ost-liposomes carrier system was created to address bioavailability and exposure to target regions in the AD mouse brain [130].

8.1.2 Polymeric nanoparticles

The polymeric biodegradable NPs can be functionalized with an antibody and polyethylene glycol (PEG) and can be administered in transgenic AD mice. Recent studies suggested that exposure to PEGylated NPs could significantly improve memory impairments and considerably reduce A-soluble peptide levels. The fabricated formulation could therefore be used to treat AD illness [131]. The double emulsion approach has been studied for the synthesis of biodegradable polymeric NPs. The aim of this study is to investigate whether loading of memantine in these NPs will improve its ability to cure AD. When administered to AD brain tissue, memantine-loaded NPs demonstrated drastically diminished A plaques and AD-related inflammatory processes [132]. Targeting the brain with zinc-loaded polymeric NPs in AD mice can decrease the size of amyloid plaques and assist in treating other neuronal deficits [133]. Mucoadhesive and target poly lactic-*co*-glycolic acid nanoparticles (PLGA-NPs) with surfaces modified with lactoferrin-conjugated *N*-trimethylated chitosan had been employed to transport the acetylcholinesterase inhibitor known as Huperzine A. Both the formulation's sustained-release activity and its capacity to address AD pathology had demonstrated promising outcomes [134]. The bioactive substance thymoquinone (TQ), which was present in *Nigella sativa* seed essential oil, has been shown to have a number of therapeutic uses [135].

Numerous preliminary pharmacological studies on the therapeutic use of TQ have been conducted. Recent research had shown that TQ might be a possible AD therapy [136]. Polysorbate-80 (P-80)-coated TQ-containing NPs may be a reliable and practical solution to administer nanocarriers to the brain for crossing BBB effectively [137]. PLGA is a biodegradable polymer. Although hydrophobic PLGA is frequently used for CNS-targeted medication administration, it has a tendency to opsonize and get removed by the reticuloendothelial system (RES). The nontoxic, nonionic, biodegradable, and hydrophilic surfactant P-80 that was coated on the PLGA-NPs prevents them from being opsonized and getting cleared from systemic circulation [138]. The autocatalytic hydrolytic degradation of PLGA to lactic and glycolic acid significantly contributes the high porosity of NPs that leads to high drug release from the matrix. The hydrophilicity of the P-80 covering allowed for the straightforward derivation of TQ from P-80-TQN [139,140]. The development of A inhibitors may aid in the treatment of AD.

The fibrils can be destabilized by a variety of medicines in vitro, preventing A buildup and neurotoxic consequences. The inhibitory effects of NPs have been the subject of extensive study. The effective prevention of protein aggregation may be achieved by the functional NPs. Preformed fibrils might dissolve when exposed to light-activated gold nanoparticles (AuNPs) that contain peptides. When the surface is properly functionalized, harmful ions are prevented from leaving the nanocarrier [141]. A naturally occurring phytochemical called curcumin is an antioxidant with minimal toxicity and a free radical scavenger [142,143]. Curcumin, that inhibits tau protein buildup and exhibits antiamyloid properties at doses in the micromolar range, is the most promising therapy for treating AD [144]. When delivered to the brain, curcumin had shown less stability and lower brain absorption resulting in lower bioavailability in brain. In order to solve these issues, the use of nanocarriers was encouraged due to their enhanced and sustained brain exposure as well as the related safety advantages. By using a nanoemulsion of PLGA particles coated with red blood cell membranes and loaded with T807 molecules, a stable and sustained release of curcumin could be achieved across the BBB. The interactions between these proteins demonstrated the potent inhibitory effects of T807, 807/RPCNP's on tau-associated pathogenesis [145]. Additionally, curcumin loaded with chitosan and bovine serum albumin NPs were utilized to reduce AD symptoms in order to speed up the phagocytosis of the A peptide and improve medication permeability [146]. Curcuminoids regulate the growth of neural stem cells in the A-induced rat model through different kinase pathways [124]. Curcumin that had been loaded with chitosan and bovine serum albumin NPs had been demonstrated to increase the drug penetration and expedite the phagocytosis of the A peptide, hence reducing the symptoms of AD. Activating the transcription factor Nrf2, known as a master regulator of the antioxidant response, is one of the biological advantages of curcumin-based nanomedicines for treating brain illnesses. This ability to protect neurons against dopaminergic toxicity is one of these biological advantages [146–148].

8.1.3 Dendrimers

Dendrimers are thought to be a potentially beneficial substance for treating AD [149]. A finding was made by combining low-generation dendrimers and lactoferrin to deliver memantine to particular brain areas in AD-induced mice. Recent research found that the target mice's memory had a significant impact [150]. To increase the effectiveness of drugs used to treat drug-related CNS disorders like AD and polyamidoamine, dendrimers with an ethylenediamine core (generation 4.0 and 4.5) were frequently used. These dendrimers increase the drug's solubility and bioavailability for greater permeation across the BBB and can target the damaged parts of the brain (PAMAM). This was done to improve the effectiveness of CNS diseases caused by drugs, such as AD and PAMAM [151,152]. The development of dendrimers with a poly (propylene imine) core and a maltose-histidine shell (G4HisMal) may significantly lessen the symptoms of AD, such as memory impairment. Tacrine had also been used in nanocomposites with generation 4.0 and PAMAM dendrimers to increase biocompatibility and minimize the harmful effects of drugs used to treat AD [153,154]

8.1.4 Nanogel

It had been established that administering free medications is less successful than using nanogels to deliver pharmaceuticals. Nanogels were prepared in light of the improvement of the efficacy of drugs to target BBB, which was attributed to the improved drug cellular absorption, reduced drug toxicity, increased drug loading, and controlled release of the loaded drug at the desired spot [155]. Because of their capacity to bind active substances, macromolecules, and medications, nanogels are appealing drug delivery systems that had been employed to address a variety of issues connected to different diseases, such as AD [156]. One of the most successful therapies for AD, according to a recent study, included administering deferoxamine as nanogels using the chitosan and tripolyphosphate method [157]. By reducing the formation of A amyloids, modified polysaccharide pullulan backbones containing cholesterol moieties function as artificial chaperones that had been shown to reduce AD pathogenesis [158]. In a preclinical study using mice, it was determined that using nanogels as a carrier increased the amount of insulin that was delivered from the nose to the brain [159]. The NPs have several advantages when mixed with polysaccharides, including being nontoxic, extremely stable, hydrophilic, and biodegradable [134].

8.1.5 Selenium nanoparticles

Lowering the amount of ROS in the brain is one of the key treatments for AD. Numerous distinct trace elements, such as selenium (II), sodium selenite (VI), and sodium selenite, have active ROS inhibitors (IV). Nanoparticles containing selenium and selenite had been demonstrated to reduce oxidative stress and prevent cell cytotoxicity since selenium and selenite are vital micronutrients for human health and have the potential to be used in biomedical applications of selenium nanoformulation. They might therefore be applied to the treatment of neurodegenerative illnesses like AD [160]. Modified selenium NPs with sialic acid have been found to be permeable to the BBB, and exposure to them has been proven to prevent the A accumulation reactions [161]. Using sialic acid-modified selenium nanoparticles coated in high BBB permeability peptide-B6 and epigallocatechin-3-gallate could also prevent A aggregation (EGCG) [162]. A unique modified nanoformulation of selenium NPs enclosed in PLGA nanospheres with curcumin, showing substantial inhibitory benefits against A buildup in a transgenic AD mouse model, was one promising delivery strategy for AD treatment [163].

8.2 Target-based approaches

8.2.1 Immunotherapy approach

As a strategy to improve removal, immunotherapy strategies have been investigated. The A-based vaccination AN1792 (A connected to an adjuvant) was the subject of the most in-depth study. The vaccination was well tolerated in mice after administration, and A burdens were decreased while cognitive function was preserved. A phase II study was started after phase I human trials that looked for evident toxicity and did not show effective results. The ensuing encephalitis, a potentially fatal brain inflammation, was experienced by a significant portion of individuals producing a strong immune response to the vaccination, prompting the early termination of this research. A strong immunological response was elicited by over 40 research participants who received the vaccine but did not experience encephalitis. Most of these participants, along with the placebo group in the clinical trial, had continued clinical monitoring. Neuropsychological information on these participants that was collected 1 year after immunization has been published. Significant statistical benefits could not be observed as per previous clinical studies.

Both the placebo and therapy groups included declining subjects. AN1792 study-related autopsy of vaccinated participants who later died for a variety of causes provided further data. It was obvious from the brain histopathology of these dead persons' brains that the brain parenchyma A had decreased significantly. According to the available clinical data, engaging the immune system to eliminate, during the course of a year, A does not have a major impact on cognition, which sums up the AN1792 experience from an efficacy standpoint [72].

8.2.2   Anti-NF-κB strategies

At first glance, it may seem that suppression of proinflammatory NF-κB complex might offer a unique and effective treatment method, since all six proinflammatory dimeric transcription factors of NF-κB complex, miRNAs are abundantly increased and activated. Neurological researchers and clinicians can choose from a large pool of 1000 NF-κB-inhibitors (including IkB-phosphorylation, IkB-activation and NF-κB-translocation inhibitors, etc.) in the form of small RNA/DNA, natural and synthetic antioxidants, peptides, engineered dominant-negative or constitutively active polypeptides, and antibodies. Many herbal compounds had been proven to have therapeutic, preventive, and useful antiinflammatory properties [164]. This may be attributed, partly, based on their ability to prevent NF-κB signaling. However, significant barriers prevent, broad application of anti-NF-κB medicinal approaches. These include "(i) the hepatotoxicity, nephrotoxicity, and neurotoxicity of many anti-NF-kB compounds, especially in elderly humans; (ii) challenges in drug-delivery and the selective targeting of NF-kB signaling-pathways within CNS compartments; and (iii) significant off-target effects for NF- This is because NF-kB is a critical and pluripotent transcription factor that orchestrates a wide variety of physiological and pathological processes and responses. Multiple, well-illustrated peer-reviewed articles have demonstrated the unique and significant role of NF-kB in the production of potentially harmful miRNA in cancer and neurological illness" [6,165].

8.2.3   Anti-microRNA (AM) strategies

Anti-miRNAs (AMs) have a lot of potential for effective therapeutic use as AM-based treatments for human central nervous system aliments due to their many advantageous intrinsic qualities and advantage in their molecular structure and function. They consist of the following: "(i) At physiologically relevant concentrations, both AMs and miRNAs are nontoxic to human brain cells in primary culture; (ii) miRNAs (18-22 nucleotide) are likely too small to elicit a robust immune response in the host because they are less than $1/375^{th}$ the size of the smallest known single-stranded RNA virus (poliovirus, 7500 nucleotide); and (iii) there is a great deal of leeway in the management of their stabilities and, by extension, functional lifetime in the cell, despite the fact that many miRNAs and AMs have a short half-life in the brain and retina. Their structure can be chemically and/or physically modified to significantly increase their stability *in vivo*, allowing them to be long acting and effective throughout the treatment regimen; miRNAs in the brain and retina can also be relatively unstable entities, and this may be useful in limiting the efficacy of the AMA itself; just like, (iv) like antibody-protein recognition, miRNAs are highly selective with minimal toxicity and negligible off-target effects; (v) AMs can be designed so that their complementary anti-miRNA ribonucleotide sequence is either fully or only partly complementary to modulate the full AM effect; (vi) miRNAs are well tolerated by the aged AD patient and they often compromised physiological condition; and (c) miRNAs have an intrinsic capacity to neutralize multiple disease target; (viii) depending on complementarity between miRNA & AM can restricted or totally negated the abundance and actions of a particular miRNA; (ix) a pharmacological treatment that focus on at least one crucial molecular-genetic component upstream from the branching of a crucial pathological pathway relevant to the AD process could be defined by a strategic AM targeting a particular miRNA; (x) high selectivity based on complementary AM-miRNA pairing, one of the most specific and selective links in human neurobiology; (xi) delivery of AMs via exosomes and viral vectors; and (xi) a strategic approach of an AM targeting a specific miRNA could define a pharmacological treatment that focuses on at least one critical molecular-genetic component lying upstream from the branching of a critical pathological pathway relevant to the AD process. Many sncRNAs and AM stabilities can be kept short, (xii) especially in the brain and retina, suggesting that the therapeutic effect of miRNAs or sncRNAs can be tailored to be self-limiting; and (xiii) if necessary, anti-NF-κB and AM strategies can be combined to achieve the desired pharmacotherapeutic effect" [166].

8.2.4 Targeting tau kinases

Tau kinases such as CDK5, GSK-3β, MAPK, and MARK are still being investigated as possible therapeutic targets for AD since several evidence indicated that aberrant phosphorylation that plays a role in tau pathogenicity [167]. The fact that each of these kinase has several substrates beside tau, however, raises questions regarding the safety of each of their inhibitors, and it seems that no tau kinase inhibitors has progressed to ADs late-phase clinical trial. Lowering the total amount of tau might be an option to modify tau phosphorylation. Current research shows that lowering the levels of overall tau is safe and effective, even in both mature and elderly animals with diseases related to AD. Therefore, it could be beneficial to lower the levels of tau by focusing on tau itself or molecule that regulate tau expression or clearance.

Although tau has been the subject of far fewer pharmacological trials, interest in pharmaceuticals targeting the tau protein has progressively increased in recent year [90]. Once it was believed the methylene blue impedes tau-tau interactions. However, in a one-year phase II clinical study, phenothiazine methylene blue demonstrated moderate evidence for delaying disease development in adult AD patient. Phase III experiments with methylene blue, a more current formulation, are anticipated. Uncertainty surrounds whether tau assembly or conformation is the root of tau-dependent neuronal degeneration and dysfunction, as well as if any of the tau aggregation inhibitors available in the market may diminish the abundance of this structure. In fact, a few inhibitors of tau aggregation promote the production of potentially harmful tau oligomers [168].

8.2.5 Targeting Aβ

On the basis of ground-breaking findings made over the last 20 years, a number of therapies have been created to reduce Aβ production or improve Aβ clearance. The former group of medications aims to block the enzymes known as g- or β-secretase, which liberate Aβ from their precursor [169]. It has been challenging to develop pharmaceuticals that can cross the barrier of blood-brain and specifically inhibit the APP cleavage without affecting the cleavage of substitute substrates like Notch and voltage-gated sodium channel subunits. It has been a difficult task, despite the fact that the identities of these enzymes have been known for a while [170].

The Aβ42/43 formation can be reduced in favor of the short acting species of Aβ, that could be less hazardous than longer species, by another class of medicines that control g-secretase cleavage of APP [171]. It has not yet been completely determined if the shorter species of Aβ also contributes to the neuronal dysfunction in vivo, and this might have an impact on the effectiveness and safety of g-secretase modulators.

8.2.6 Combination therapies

It has now been hypothesized that combinatorial treatment for AD would more likely provide positive clinical results in comparison to monotherapy, according to treatment paradigms for other chronic illnesses like congestive heart failure, hypertension, HIV, epilepsy, and others [172]. Many of the disorders already mentioned can only be adequately treated with a combination of drugs from various pharmacological classes. Given the underlying complexity of AD pathobiology, it is irrational to assume that if multiple pathologies of various types are present, a single treatment targeting a single pathogenic mechanism will be particularly successful in slowing disease development. In reality, clinical trial design has started to use combination tactics. Donanemab, a monoclonal antibody against pyroglutamate, was originally investigated in conjunction with the BACE1 inhibitor LY3202626 in the Lilly TRAIL-BLAZER phase II study. Unfortunately, LY3202626 was shown to be linked to a small impairment of cognition in focused phase II studies, leading to the termination of the combined therapy arm. Nevertheless, it is important to support combination methods in clinical trials, particularly those involving several medication classes. Ideally, preclinical, or preventive AD studies would also include this kind of therapy strategy [72,173].

9 Conclusion and future prospective

In summary, the pathophysiology of AD is characterized by disruptions and imbalances that may arise from a wide number of processes. It is surprising that despite the vast amount of information that is presently accessible on AD, there are currently just a few alternatives available for the treatment of the condition. The progression of the illness is also complicated in its own unique ways. The treatment of symptoms is the most effective aspect of the management process at the moment; nonetheless, tremendous, and exciting strides have been made in the development of disease-modifying techniques. The goal of both pharmacologic and non-pharmacologic therapies, such as FDA-approved AD drugs (ChEIs and memantine), is to lessen the debilitating consequences of cognitive and functional loss as well as BPSD. Given that neuroinflammation, angiogenesis, and oxidative stress all of which contribute to the etiology of AD, additional research into treatment techniques that target these pathways should be discussed. In order to provide compassionate care for people living with Alzheimer's disease and the people who take care of them, clinicians should develop proactive and adaptable patient- and caregiver-centered approaches. The establishment and preservation of a robust healing alliance include appropriate medication, behavioral and environmental approaches, psychoeducation, and is holistic and pragmatic, preparing for present and future healthcare requirements, promoting brain health, as well as the psychosocial wellbeing of the patient-caregiver dyad, is vital to supplying equipped and ethical care for patients with AD. Intense research efforts are still being put forward to create diagnostic methods and treatments for Alzheimer's disease that are more accurate.

Acknowledgments

We would like to thank Department of Pharmacognosy, Era College of Pharmacy, Era University, Tondan Marg, Sarfarazganj, Lucknow, Uttar Pradesh 226003, India, Amity Institute of Pharmacy, Lucknow, Amity University Uttar Pradesh, Noida, UP, India and School of Health Sciences & Technology, University of Petroleum & Energy Studies, Energy Acres, Bidholi, Dehradun, Uttarakhand, India, for providing the facilities for the completion of this review.

References

[1] Pratap GKAS, Shantaram M. Alzheimer's disease: a challenge in its management with certain medicinal plants—a review. Int J Pharmaceut Sci Res 2017;8(12):4960–72.
[2] McKeown A, Turner A, Angehrn Z, Gove D, Ly A, Nordon C, et al. Health outcome prioritization in Alzheimer's disease: understanding the ethical landscape. J Alzheimers Dis 2020;77(1):339–53.
[3] Fish PV, Steadman D, Bayle ED, Whiting P. New approaches for the treatment of Alzheimer's disease. Bioorg Med Chem Lett 2019;29(2):125–33.
[4] Livingston G, Huntley J, Sommerlad A, Ames D, Ballard C, Banerjee S, et al. Dementia prevention, intervention, and care: 2020 report of the lancet commission. Lancet 2020;396(10248):413–46.
[5] Rizzi L, Rosset I, Roriz-Cruz M. Global epidemiology of dementia: Alzheimer's and vascular types. Biomed Res Int 2014;2014, 908915.
[6] Wu YT, Beiser AS, Breteler MMB, Fratiglioni L, Helmer C, Hendrie HC, et al. The changing prevalence and incidence of dementia over time—current evidence. Nat Rev Neurol 2017;13(6):327–39.

[7] Prince M, Ali GC, Guerchet M, Prina AM, Albanese E, Wu YT. Recent global trends in the prevalence and incidence of dementia, and survival with dementia. Alzheimers Res Ther 2016;8(1):23.

[8] Adini A, Adini I, Chi ZL, Derda R, Birsner AE, Matthews BD, et al. A novel strategy to enhance angiogenesis in vivo using the small VEGF-binding peptide PR1P. Angiogenesis 2017;20(3):399–408.

[9] Richarz N, Boada A, Carrascosa JM. Angiogenesis in dermatology—insights of molecular mechanisms and latest developments. Actas Dermo-Sifiliográficas (English Edition) 2017;108:515–23.

[10] Touyz RM, Briones AM. Reactive oxygen species and vascular biology: implications in human hypertension. Hypertens Res 2011;34(1):5–14.

[11] Heneka MT, O'Banion MK, Terwel D, Kummer MP. Neuroinflammatory processes in Alzheimer's disease. J Neural Transm (Vienna) 2010;117(8):919–47.

[12] Gabuzda D, Busciglio J, Chen LB, Matsudaira P, Yankner BA. Inhibition of energy metabolism alters the processing of amyloid precursor protein and induces a potentially amyloidogenic derivative. J Biol Chem 1994;269(18):13623–8.

[13] Tamagno E, Bardini P, Obbili A, Vitali A, Borghi R, Zaccheo D, et al. Oxidative stress increases expression and activity of BACE in NT2 neurons. Neurobiol Dis 2002;10(3):279–88.

[14] Scheuner D, Eckman C, Jensen M, Song X, Citron M, Suzuki N, et al. Secreted amyloid beta-protein similar to that in the senile plaques of Alzheimer's disease is increased in vivo by the presenilin 1 and 2 and APP mutations linked to familial Alzheimer's disease. Nat Med 1996;2(8):864–70.

[15] Schaeffer EL, Gattaz WF. Cholinergic and glutamatergic alterations beginning at the early stages of Alzheimer disease: participation of the phospholipase A2 enzyme. Psychopharmacology (Berl) 2008;198(1):1–27.

[16] Watanabe S, Kato I, Koizuka I. Retrograde-labeling of pretecto-vestibular pathways in cats. Auris Nasus Larynx 2003;30(Suppl):S35–40.

[17] Saraf MK, Anand A, Prabhakar S. Scopolamine induced amnesia is reversed by Bacopa monniera through participation of kinase-CREB pathway. Neurochem Res 2010;35(2):279–87.

[18] Caccamo A, Oddo S, Billings LM, Green KN, Martinez-Coria H, Fisher A, et al. M1 receptors play a central role in modulating AD-like pathology in transgenic mice. Neuron 2006;49(5):671–82.

[19] Fisher A. M1 muscarinic agonists target major hallmarks of Alzheimer's disease—the pivotal role of brain M1 receptors. Neurodegen Dis 2008;5(3–4):237–40.

[20] Fisher A, Medeiros R, Barner N, Natan N, Brandeis R, Elkon H, et al. M1 muscarinic agonists and a multipotent activator of sigma1/M1 muscarinic receptors: future therapeutics of Alzheimer's disease. Alzheimers Dement 2014;10:P123.

[21] Ellis JR, Ellis KA, Bartholomeusz CF, Harrison BJ, Wesnes KA, Erskine FF, et al. Muscarinic and nicotinic receptors synergistically modulate working memory and attention in humans. Int J Neuropsychopharmacol 2006;9(2):175–89.

[22] Wu J, Ishikawa M, Zhang J, Hashimoto K. Brain imaging of nicotinic receptors in Alzheimer's disease. Int J Alzheimers Dis 2010;2010, 548913.

[23] Wenk GL. Neuropathologic changes in Alzheimer's disease: potential targets for treatment. J Clin Psychiatry 2006;67(Suppl 3):3–7 [quiz 23].

[24] Mesulam M. The cholinergic lesion of Alzheimer's disease: pivotal factor or side show? Learn Mem 2004;11(1):43–9.

[25] Shen J, Wu J. Nicotinic cholinergic mechanisms in Alzheimer's disease. Int Rev Neurobiol 2015;124:275–92.

[26] Ni R, Marutle A, Nordberg A. Modulation of α7 nicotinic acetylcholine receptor and fibrillar amyloid-β interactions in Alzheimer's disease brain. J Alzheimers Dis 2013;33(3):841–51.

[27] Lin H, Vicini S, Hsu FC, Doshi S, Takano H, Coulter DA, et al. Axonal α7 nicotinic ACh receptors modulate presynaptic NMDA receptor expression and structural plasticity of glutamatergic presynaptic boutons. Proc Natl Acad Sci U S A 2010;107(38):16661–6.

[28] Dong X-x, Wang Y, Qin Z-h. Molecular mechanisms of excitotoxicity and their relevance to pathogenesis of neurodegenerative diseases. Acta Pharmacol Sin 2009;30 (4):379–87.

[29] Peña F, Gutiérrez-Lerma A, Quiroz-Baez R, Arias C. The role of beta-amyloid protein in synaptic function: implications for Alzheimer's disease therapy. Curr Neuropharmacol 2006;4(2):149–63.

[30] Doggrell SA, Evans S. Treatment of dementia with neurotransmission modulation. Expert Opin Investig Drugs 2003;12(10):1633–54.

[31] Swerdlow RH, Khan SM. A "mitochondrial cascade hypothesis" for sporadic Alzheimer's disease. Med Hypotheses 2004;63(1):8–20.

[32] Kujoth GC, Hiona A, Pugh TD, Someya S, Panzer K, Wohlgemuth SE, et al. Mitochondrial DNA mutations, oxidative stress, and apoptosis in mammalian aging. Science 2005;309(5733):481–4.

[33] Glabe CG, Kayed R. Common structure and toxic function of amyloid oligomers implies a common mechanism of pathogenesis. Neurology 2006;66(2 Suppl 1):S74–8.

[34] Galimberti D, Scarpini E. Disease-modifying treatments for Alzheimer's disease. Ther Adv Neurol Disord 2011;4(4):203–16.

[35] Deane R, Bell RD, Sagare A, Zlokovic BV. Clearance of amyloid-beta peptide across the blood-brain barrier: implication for therapies in Alzheimer's disease. CNS Neurol Disord Drug Targets 2009;8(1):16–30.

[36] Jucker M, Walker LC. Neurodegeneration: amyloid-β pathology induced in humans. Nature 2015;525(7568):193–4.

[37] McGowan E, Pickford F, Kim J, Onstead L, Eriksen J, Yu C, et al. Abeta42 is essential for parenchymal and vascular amyloid deposition in mice. Neuron 2005;47(2):191–9.

[38] Bertram L, Tanzi RE. Thirty years of Alzheimer's disease genetics: the implications of systematic meta-analyses. Nat Rev Neurosci 2008;9(10):768–78.

[39] Wu LG, Saggau P. Presynaptic inhibition of elicited neurotransmitter release. Trends Neurosci 1997;20(5):204–12.

[40] Leão RN, Colom LV, Borgius L, Kiehn O, Fisahn A. Medial septal dysfunction by Aβ-induced KCNQ channel-block in glutamatergic neurons. Neurobiol Aging 2012;33 (9):2046–61.

[41] Tanzi RE, Bertram L. Twenty years of the Alzheimer's disease amyloid hypothesis: a genetic perspective. Cell 2005;120(4):545–55.

[42] Sennvik K, Fastbom J, Blomberg M, Wahlund LO, Winblad B, Benedikz E. Levels of alpha- and beta-secretase cleaved amyloid precursor protein in the cerebrospinal fluid of Alzheimer's disease patients. Neurosci Lett 2000;278(3):169–72.

[43] Wen Y, Onyewuchi O, Yang S, Liu R, Simpkins JW. Increased beta-secretase activity and expression in rats following transient cerebral ischemia. Brain Res 2004;1009 (1–2):1–8.

[44] Wang DS, Dickson DW, Malter JS. beta-Amyloid degradation and Alzheimer's disease. J Biomed Biotechnol 2006;2006(3):58406.

[45] Carbone I, Lazzarotto T, Ianni M, Porcellini E, Forti P, Masliah E, et al. Herpes virus in Alzheimer's disease: relation to progression of the disease. Neurobiol Aging 2014;35 (1):122–9.

[46] Readhead B, Haure-Mirande JV, Funk CC, Richards MA, Shannon P, Haroutunian V, et al. Multiscale analysis of independent Alzheimer's cohorts finds disruption of molecular, genetic, and clinical networks by human herpesvirus. Neuron 2018;99(1):64–82.e7.

[47] Yiannopoulou KG, Papageorgiou SG. Current and future treatments for Alzheimer's disease. Ther Adv Neurol Disord 2013;6(1):19–33.

[48] Sgambato A, Cittadini A. Inflammation and cancer: a multifaceted link. Eur Rev Med Pharmacol Sci 2010;14(4):263–8.

[49] Pogue AI, Lukiw WJ. Angiogenic signaling in Alzheimer's disease. Neuroreport 2004;15 (9):1507–10.

[50] Miklossy J. Cerebral hypoperfusion induces cortical watershed microinfarcts which may further aggravate cognitive decline in Alzheimer's disease. Neurol Res 2003;25 (6):605–10.

[51] Pugh CW, Ratcliffe PJ. Regulation of angiogenesis by hypoxia: role of the HIF system. Nat Med 2003;9(6):677–84.

[52] Yamakawa M, Liu LX, Date T, Belanger AJ, Vincent KA, Akita GY, et al. Hypoxia-inducible factor-1 mediates activation of cultured vascular endothelial cells by inducing multiple angiogenic factors. Circ Res 2003;93(7):664–73.

[53] Tarkowski E, Issa R, Sjögren M, Wallin A, Blennow K, Tarkowski A, et al. Increased intrathecal levels of the angiogenic factors VEGF and TGF-beta in Alzheimer's disease and vascular dementia. Neurobiol Aging 2002;23(2):237–43.

[54] Del Bo R, Ghezzi S, Scarpini E, Bresolin N, Comi GP. VEGF genetic variability is associated with increased risk of developing Alzheimer's disease. J Neurol Sci 2009;283 (1–2):66–8.

[55] Yin X, Wright J, Wall T, Grammas P. Brain endothelial cells synthesize neurotoxic thrombin in Alzheimer's disease. Am J Pathol 2010;176(4):1600–6.

[56] Lukiw WJ, Ottlecz A, Lambrou G, Grueninger M, Finley J, Thompson HW, et al. Coordinate activation of HIF-1 and NF-kappaB DNA binding and COX-2 and VEGF expression in retinal cells by hypoxia. Invest Ophthalmol Vis Sci 2003;44(10):4163–70.

[57] Jellinger KA. Alzheimer disease and cerebrovascular pathology: an update. J Neural Transm (Vienna) 2002;109(5–6):813–36.

[58] Buée L, Hof PR, Delacourte A. Brain microvascular changes in Alzheimer's disease and other dementias. Ann N Y Acad Sci 1997;826:7–24.

[59] Paris D, Ait-Ghezala G, Mathura VS, Patel N, Quadros A, Laporte V, et al. Anti-angiogenic activity of the mutant Dutch A(beta) peptide on human brain microvascular endothelial cells. Brain Res Mol Brain Res 2005;136(1–2):212–30.

[60] Paris D, Patel N, DelleDonne A, Quadros A, Smeed R, Mullan M. Impaired angiogenesis in a transgenic mouse model of cerebral amyloidosis. Neurosci Lett 2004;366(1):80–5.

[61] Monro OR, Mackic JB, Yamada S, Segal MB, Ghiso J, Maurer C, et al. Substitution at codon 22 reduces clearance of Alzheimer's amyloid-beta peptide from the cerebrospinal fluid and prevents its transport from the central nervous system into blood. Neurobiol Aging 2002;23(3):405–12.

[62] Alcolea D, Vilaplana E, Pegueroles J, Montal V, Sánchez-Juan P, González-Suárez A, et al. Relationship between cortical thickness and cerebrospinal fluid YKL-40 in predementia stages of Alzheimer's disease. Neurobiol Aging 2015;36(6):2018–23.

[63] Gispert JD, Monté GC, Falcon C, Tucholka A, Rojas S, Sánchez-Valle R, et al. CSF YKL-40 and pTau181 are related to different cerebral morphometric patterns in early AD. Neurobiol Aging 2016;38:47–55.

[64] Snyder HM, Carrillo MC, Grodstein F, Henriksen K, Jeromin A, Lovestone S, et al. Developing novel blood-based biomarkers for Alzheimer's disease. Alzheimers Dement 2014;10(1):109–14.

[65] Kiddle SJ, Sattlecker M, Proitsi P, Simmons A, Westman E, Bazenet C, et al. Candidate blood proteome markers of Alzheimer's disease onset and progression: a systematic review and replication study. J Alzheimers Dis 2014;38(3):515–31.

[66] Andreyev AY, Kushnareva YE, Starkov AA. Mitochondrial metabolism of reactive oxygen species. Biochemistry (Mosc) 2005;70(2):200–14.

[67] Leeuwenburgh C, Heinecke JW. Oxidative stress and antioxidants in exercise. Curr Med Chem 2001;8(7):829–38.

[68] Jeong EJ, Lee HK, Lee KY, Jeon BJ, Kim DH, Park JH, et al. The effects of lignan-riched extract of Shisandra chinensis on amyloid-β-induced cognitive impairment and neurotoxicity in the cortex and hippocampus of mouse. J Ethnopharmacol 2013;146(1):347–54.

[69] Gerzson MFB, Bona NP, Soares MSP, Teixeira FC, Rahmeier FL, Carvalho FB, et al. Tannic acid ameliorates STZ-induced Alzheimer's disease-like impairment of memory, neuroinflammation, neuronal death and modulates Akt expression. Neurotox Res 2020;37(4):1009–17.

[70] Burg VK, Grimm HS, Rothhaar TL, Grösgen S, Hundsdörfer B, Haupenthal VJ, et al. Plant sterols the better cholesterol in Alzheimer's disease? A mechanistical study. J Neurosci 2013;33(41):16072–87.

[71] Hussain G, Zhang L, Rasul A, Anwar H, Sohail MU, Razzaq A, et al. Role of plant-derived flavonoids and their mechanism in attenuation of Alzheimer's and Parkinson's diseases: an update of recent data. Molecules 2018;23(4).

[72] Ji S, Li S, Zhao X, Kang N, Cao K, Zhu Y, et al. Protective role of phenylethanoid glycosides, Torenoside B and Savatiside A, in Alzheimer's disease. Exp Ther Med 2019;17 (5):3755–67.

[73] Martins RN, Villemagne V, Sohrabi HR, Chatterjee P, Shah TM, Verdile G, et al. Alzheimer's disease: a journey from amyloid peptides and oxidative stress, to biomarker technologies and disease prevention strategies-gains from AIBL and DIAN cohort studies. J Alzheimers Dis 2018;62(3):965–92.

[74] Howes MJ, Houghton PJ. Plants used in Chinese and Indian traditional medicine for improvement of memory and cognitive function. Pharmacol Biochem Behav 2003;75 (3):513–27.

[75] Namdeo AG, Ingawale DK. Ashwagandha: advances in plant biotechnological approaches for propagation and production of bioactive compounds. J Ethnopharmacol 2021;271, 113709.

[76] Dubey T, Chinnathambi S. Photodynamic treatment modulates various GTPase and cellular signalling pathways in tauopathy. Small GTPases 2022;13(1):183–95.

[77] Xu Q-Q, Shaw PC, Hu Z, Yang W, Ip S-P, Xian Y-F, et al. Comparison of the chemical constituents and anti-Alzheimer's disease effects of Uncaria rhynchophylla and Uncaria tomentosa. Chin Med 2021;16(1):110.

[78] Yasuno F, Tanimukai S, Sasaki M, Ikejima C, Yamashita F, Kodama C, et al. Combination of antioxidant supplements improved cognitive function in the elderly. J Alzheimers Dis 2012;32(4):895–903.

[79] Gregory J, Vengalasetti YV, Bredesen DE, Rao RV. Neuroprotective herbs for the management of Alzheimer's disease. Biomolecules 2021;11(4):543.

[80] Tsai SJ. Huperzine-A, a versatile herb, for the treatment of Alzheimer's disease. J Chin Med Assoc 2019;82(10):750–1.

[81] Choi SH, Lee R, Nam SM, Kim DG, Cho IH, Kim HC, et al. Ginseng gintonin, aging societies, and geriatric brain diseases. Integr Med Res 2021;10(1), 100450.

[82] Lopresti AL. Salvia (sage): a review of its potential cognitive-enhancing and protective effects. Drugs R D 2017;17(1):53–64.

[83] Sanjana D, Shailendra P. Evaluation of traditional herb extract Salvia officinalis in treatment of Alzheimers disease. Pharm J 2020;12(1).

[84] Voulgaropoulou SD, van Amelsvoort T, Prickaerts J, Vingerhoets C. The effect of curcumin on cognition in Alzheimer's disease and healthy aging: a systematic review of preclinical and clinical studies. Brain Res 2019;1725, 146476.

[85] Choubey PK, Tripathi A, Sharma P, Shrivastava SK. Design, synthesis, and multitargeted profiling of N-benzylpyrrolidine derivatives for the treatment of Alzheimer's disease. Bioorg Med Chem 2020;28(22), 115721.

[86] Gray NE, Harris CJ, Quinn JF, Soumyanath A. Centella asiatica modulates antioxidant and mitochondrial pathways and improves cognitive function in mice. J Ethnopharmacol 2016;180:78–86.

[87] Gray NE, Sampath H, Zweig JA, Quinn JF, Soumyanath A. Centella asiatica attenuates amyloid-β-induced oxidative stress and mitochondrial dysfunction. J Alzheimers Dis 2015;45(3):933–46.

[88] Kumar MH, Gupta YK. Antioxidant property of Celastrus paniculatus willd.: a possible mechanism in enhancing cognition. Phytomedicine 2002;9(4):302–11.

[89] Anupama KP, Shilpa O, Antony A, Raghu SV, Gurushankara HP. Jatamansinol from Nardostachys jatamansi (D.Don) DC. protects Aβ(42)-induced neurotoxicity in Alzheimer's disease Drosophila model. Neurotoxicology 2022;90:62–78.

[90] Saxena G, Singh SP, Pal R, Singh S, Pratap R, Gugulipid NC. An extract of Commiphora whighitii with lipid-lowering properties, has protective effects against streptozotocin-induced memory deficits in mice. Pharmacol Biochem Behav 2007;86(4):797–805.

[91] Wang YH, Du GH. Ginsenoside Rg1 inhibits beta-secretase activity in vitro and protects against Abeta-induced cytotoxicity in PC12 cells. J Asian Nat Prod Res 2009;11 (7):604–12.

[92] Zhou YQ, Yang ZL, Xu L, Li P, Hu YZ. Akebia saponin D, a saponin component from Dipsacus asper Wall, protects PC 12 cells against amyloid-beta induced cytotoxicity. Cell Biol Int 2009;33(10):1102–10.

[93] Fujiwara H, Tabuchi M, Yamaguchi T, Iwasaki K, Furukawa K, Sekiguchi K, et al. A traditional medicinal herb Paeonia suffruticosa and its active constituent 1,2,3,4,6-penta-O-galloyl-beta-D-glucopyranose have potent anti-aggregation effects on Alzheimer's amyloid beta proteins in vitro and in vivo. J Neurochem 2009;109(6):1648–57.

[94] Lv J, Jia H, Jiang Y, Ruan Y, Liu Z, Yue W, et al. Tenuifolin, an extract derived from tenuigenin, inhibits amyloid-beta secretion in vitro. Acta Physiol (Oxf) 2009;196 (4):419–25.

[95] Yin Y, Huang L, Liu Y, Huang S, Zhuang J, Chen X, et al. Effect of tanshinone on the levels of nitric oxide synthase and acetylcholinesterase in the brain of Alzheimer's disease rat model. Clin Invest Med 2008;31(5):E248–57.

[96] Tan MA, Sharma N, An SSA. Multi-target approach of Murraya koenigii leaves in treating neurodegenerative diseases. Pharmaceuticals (Basel) 2022;15(2).

[97] Kim DH, Hyun SK, Yoon BH, Seo JH, Lee KT, Cheong JH, et al. Gluco-obtusifolin and its aglycon, obtusifolin, attenuate scopolamine-induced memory impairment. J Pharmacol Sci 2009;111(2):110–6.

[98] Cacabelos R, Cacabelos P, Torrellas C, Tellado I, Carril JC. Pharmacogenomics of Alzheimer's disease: novel therapeutic strategies for drug development. In: Pharmacogenomics in Drug Discovery and Development; 2014. p. 323–556.

[99] Joshi H, Parle M. Antiamnesic effects of Desmodium gangeticum in mice. Yakugaku Zasshi 2006;126(9):795–804.

[100] Nowak A, Kojder K, Zielonka-Brzezicka J, Wróbel J, Bosiacki M, Fabiańska M, et al. The use of Ginkgo biloba L. as a neuroprotective agent in the Alzheimer's disease. Front Pharmacol 2021;12, 775034.

[101] D'Onofrio G, Nabavi SM, Sancarlo D, Greco A, Crocus PS, Sativus L. (Saffron) in Alzheimer's disease treatment: bioactive effects on cognitive impairment. Curr Neuropharmacol 2021;19(9):1606–16.

[102] Adalier N, Parker H, Vitamin E. Turmeric and saffron in treatment of Alzheimer's disease. Antioxidants (Basel) 2016;5(4).

[103] Hansen RA, Gartlehner G, Webb AP, Morgan LC, Moore CG, Jonas DE. Efficacy and safety of donepezil, galantamine, and rivastigmine for the treatment of Alzheimer's disease: a systematic review and meta-analysis. Clin Interv Aging 2008;3(2):211–25.

[104] Peeters M, Page G, Maloteaux JM, Hermans E. Hypersensitivity of dopamine transmission in the rat striatum after treatment with the NMDA receptor antagonist amantadine. Brain Res 2002;949(1–2):32–41.

[105] Manning SM, Boll G, Fitzgerald E, Selip DB, Volpe JJ, Jensen FE. The clinically available NMDA receptor antagonist, memantine, exhibits relative safety in the developing rat brain. Int J Dev Neurosci 2011;29(7):767–73.

[106] Pang YP, Quiram P, Jelacic T, Hong F, Brimijoin S. Highly potent, selective, and low cost bis-tetrahydroaminacrine inhibitors of acetylcholinesterase. Steps toward novel drugs for treating Alzheimer's disease. J Biol Chem 1996;271(39):23646–9.

[107] Simoni E, Daniele S, Bottegoni G, Pizzirani D, Trincavelli ML, Goldoni L, et al. Combining galantamine and memantine in multitargeted, new chemical entities potentially useful in Alzheimer's disease. J Med Chem 2012;55(22):9708–21.

[108] Quinn PMJ, Moreira PI, Ambrósio AF, Alves CH. PINK1/PARKIN signalling in neurodegeneration and neuroinflammation. Acta Neuropathol Commun [Internet] 2020;8 (1):189. https://doi.org/10.1186/s40478-020-01062-w. Available from: http://europepmc.org/abstract/MED/33168089; https://europepmc.org/articles/PMC7654589; https://europepmc.org/articles/PMC7654589?pdf=render.

[109] Agrawal N, Mishra P. The synthetic and therapeutic expedition of isoxazole and its analogs. Med Chem Res 2018;27(5):1309–44.

[110] Agrawal N, Mishra P. Synthesis, monoamine oxidase inhibitory activity and computational study of novel isoxazole derivatives as potential antiparkinson agents. Comput Biol Chem 2019;79:63–72.

[111] Agrawal N, Mishra P. Novel isoxazole derivatives as potential antiparkinson agents: synthesis, evaluation of monoamine oxidase inhibitory activity and docking studies. Med Chem Res 2019;28(9):1488–501.

[112] Ma F, Du H. Novel deoxyvasicinone derivatives as potent multitarget-directed ligands for the treatment of Alzheimer's disease: design, synthesis, and biological evaluation. Eur J Med Chem 2017;140:118–27.

[113] Clarke JR, Lyra ESNM, Figueiredo CP, Frozza RL, Ledo JH, Beckman D, et al. Alzheimer-associated Aβ oligomers impact the central nervous system to induce peripheral metabolic deregulation. EMBO Mol Med 2015;7(2):190–210.

[114] Sayre LM, Smith MA, Perry G. Chemistry and biochemistry of oxidative stress in neurodegenerative disease. Curr Med Chem 2001;8(7):721–38.

[115] Tiwari G, Tiwari R, Sriwastawa B, Bhati L, Pandey S, Pandey P, et al. Drug delivery systems: an updated review. Int J Pharm Investig 2012;2(1):2–11.

[116] Calzoni E, Cesaretti A, Polchi A, Di Michele A, Tancini B, Emiliani C. Biocompatible polymer nanoparticles for drug delivery applications in cancer and neurodegenerative disorder therapies. J Funct Biomater 2019;10(1).

[117] Mukherjee S, Madamsetty VS, Bhattacharya D, Paul M, Mukherjee A. Recent advancements of nanomedicine in neurodegenerative disorders theranostics. Adv Funct Mater 2020;30, 2003054.

[118] Lu W. Adsorptive-mediated brain delivery systems. Curr Pharm Biotechnol 2012;13 (12):2340–8.

[119] Khan I, Saeed K, Khan I. Nanoparticles: properties, applications and toxicities. Arabian J Chem 2019;12(7):908–31.

[120] Auría-Soro C, Nesma T, Juanes-Velasco P, Landeira-Viñuela A, Fidalgo-Gomez H, Acebes-Fernandez V, et al. Interactions of nanoparticles and biosystems: microenvironment of nanoparticles and biomolecules in nanomedicine. Nanomaterials (Basel) 2019;9(10).

[121] Li Y, Yao CF, Xu FJ, Qu YY, Li JT, Lin Y, et al. APC/C(CDH1) synchronizes ribose-5-phosphate levels and DNA synthesis to cell cycle progression. Nat Commun 2019;10 (1):2502.

[122] He X, Zhu Y, Yang L, Wang Z, Wang Z, Feng J, et al. Embryonic stem cell pluripotency: MgFe-LDH nanoparticles: a promising leukemia inhibitory factor replacement for self-renewal and pluripotency maintenance in cultured mouse embryonic stem cells (Adv. Sci. 9/2021). Adv Sci (Weinh) 2021;8(9), 2170049. https://doi.org/10.1002/advs.202170049 [eCollection 2021 May].

[123] Mura S, Couvreur P. Nanotheranostics for personalized medicine. Adv Drug Deliv Rev 2012;64(13):1394–416.

[124] Ling TS, Chandrasegaran S, Xuan LZ, Suan TL, Elaine E, Nathan DV, et al. The potential benefits of nanotechnology in treating Alzheimer's disease. Biomed Res Int 2021;2021.

[125] Patra JK, Das G, Fraceto LF, Campos EVR, Rodriguez-Torres MP, Acosta-Torres LS, et al. Nano based drug delivery systems: recent developments and future prospects. J Nanobiotechnol 2018;16(1):1–33.

[126] Spuch C, Navarro C. Liposomes for targeted delivery of active agents against neurodegenerative diseases (Alzheimer's disease and Parkinson's disease). J Drug Deliv 2011;2011.

[127] Noble GT, Stefanick JF, Ashley JD, Kiziltepe T, Bilgicer B. Ligand-targeted liposome design: challenges and fundamental considerations. Trends Biotechnol 2014;32 (1):32–45.

[128] Lajoie JM, Shusta EV. Targeting receptor-mediated transport for delivery of biologics across the blood-brain barrier. Annu Rev Pharmacol Toxicol 2015;55:613–31.

[129] Arora S, Layek B, Singh J. Design and validation of liposomal ApoE2 gene delivery system to evade blood–brain barrier for effective treatment of Alzheimer's disease. Mol Pharm 2020;18(2):714–25.

[130] Kong L, Li XT, Ni YN, Xiao HH, Yao YJ, Wang YY, et al. Transferrin-modified osthole PEGylated liposomes travel the blood-brain barrier and mitigate Alzheimer's disease-related pathology in APP/PS-1 mice. Int J Nanomedicine 2020;15:2841–58.

[131] Carradori D, Balducci C, Re F, Brambilla D, Le Droumaguet B, Flores O, et al. Antibody-functionalized polymer nanoparticle leading to memory recovery in Alzheimer's disease-like transgenic mouse model. Nanomed Nanotechnol Biol Med 2018;14 (2):609–18.

[132] Sánchez-López E, Ettcheto M, Egea MA, Espina M, Cano A, Calpena AC, et al. Memantine loaded PLGA PEGylated nanoparticles for Alzheimer's disease: in vitro and in vivo characterization. J Nanobiotechnol 2018;16:1–16.

[133] Vilella A, Belletti D, Sauer AK, Hagmeyer S, Sarowar T, Masoni M, et al. Reduced plaque size and inflammation in the APP23 mouse model for Alzheimer's disease after chronic application of polymeric nanoparticles for CNS targeted zinc delivery. J Trace Elem Med Biol 2018;49:210–21.

[134] Meng Q, Wang A, Hua H, Jiang Y, Wang Y, Mu H, et al. Intranasal delivery of Huperzine A to the brain using lactoferrin-conjugated N-trimethylated chitosan surface-modified PLGA nanoparticles for treatment of Alzheimer's disease. Int J Nanomedicine 2018;13:705.

[135] Javidi S, Razavi BM, Hosseinzadeh H. A review of neuropharmacology effects of Nigella sativa and its main component, thymoquinone. Phytother Res 2016;30 (8):1219–29.

[136] Abulfadl Y, El-Maraghy N, Ahmed AE, Nofal S, Abdel-Mottaleb Y, Badary OA. Thymoquinone alleviates the experimentally induced Alzheimer's disease inflammation by modulation of TLRs signaling. Hum Exp Toxicol 2018;37(10):1092–104.

[137] Yusuf M, Khan M, Alrobaian MM, Alghamdi SA, Warsi MH, Sultana S, et al. Brain targeted Polysorbate-80 coated PLGA thymoquinone nanoparticles for the treatment of Alzheimer's disease, with biomechanistic insights. J Drug Deliv Sci Technol 2021;61, 102214.

[138] Sempf K, Arrey T, Gelperina S, Schorge T, Meyer B, Karas M, et al. Adsorption of plasma proteins on uncoated PLGA nanoparticles. Eur J Pharm Biopharm 2013;85 (1):53–60.

[139] Zeng Q, Bie B, Guo Q, Yuan Y, Han Q, Han X, et al. Hyperpolarized Xe NMR signal advancement by metal-organic framework entrapment in aqueous solution. Proc Natl Acad Sci 2020;117(30):17558–63.

[140] Tığlı Aydın RS, Kaynak G, Gümüşderelioğlu M. Salinomycin encapsulated nanoparticles as a targeting vehicle for glioblastoma cells. J Biomed Mater Res A 2016;104(2):455–64.

[141] Meenambal R, Bharath MS. Nanocarriers for effective nutraceutical delivery to the brain. Neurochem Int 2020;140, 104851.

[142] Sharifi-Rad J, Rayess YE, Rizk AA, Sadaka C, Zgheib R, Zam W, et al. Turmeric and its major compound curcumin on health: bioactive effects and safety profiles for food, pharmaceutical, biotechnological and medicinal applications. Front Pharmacol 2020;11:01021.

[143] Azizi M, Pasbakhsh P, Nadji SA, Pourabdollah M, Mokhtari T, Sadr M, et al. Therapeutic effect of perinatal exogenous melatonin on behavioral and histopathological changes and antioxidative enzymes in neonate mouse model of cortical malformation. Int J Dev Neurosci 2018;68:1–9.

[144] Yang H, Zeng F, Luo Y, Zheng C, Ran C, Yang J. Curcumin scaffold as a multifunctional tool for Alzheimer's disease research. Molecules 2022;27(12):3879.

[145] Gao J, Chen X, Ma T, He B, Li P, Zhao Y, et al. PEG-ceramide nanomicelles induce autophagy and degrade tau proteins in N2a cells. Int J Nanomedicine 2020;15:6779–89.

[146] Yang R, Zheng Y, Wang Q, Zhao L. Curcumin-loaded chitosan–bovine serum albumin nanoparticles potentially enhanced Aβ 42 phagocytosis and modulated macrophage polarization in Alzheimer's disease. Nanoscale Res Lett 2018;13:1–9.

[147] Chen L, Huang Y, Yu X, Lu J, Jia W, Song J, et al. Corynoxine protects dopaminergic neurons through inducing autophagy and diminishing neuroinflammation in rotenone-induced animal models of Parkinson's disease. Front Pharmacol 2021;12, 642900.

[148] Szwed A, Miłowska K. The role of proteins in neurodegenerative disease. Adv Hyg Exp Med 2012;66:187–95.

[149] Aliev G, Ashraf GM, Tarasov VV, Chubarev VN, Leszek J, Gasiorowski K, et al. Alzheimer's disease—future therapy based on dendrimers. Curr Neuropharmacol 2019;17(3):288–94.

[150] Gothwal A, Kumar H, Nakhate KT, Ajazuddin DA, Borah A, et al. Lactoferrin coupled lower generation PAMAM dendrimers for brain targeted delivery of memantine in

aluminum-chloride-induced Alzheimer's disease in mice. Bioconjug Chem 2019;30 (10):2573–83.

[151] Igartúa DE, Martinez CS, Temprana CF, Alonso SV, Prieto MJ. PAMAM dendrimers as a carbamazepine delivery system for neurodegenerative diseases: a biophysical and nanotoxicological characterization. Int J Pharm 2018;544(1):191–202.

[152] Yang W, Liu W, Li X, Yan J, He W. Turning chiral peptides into a racemic supraparticle to induce the self-degradation of MDM2. J Adv Res 2023;45:59–71.

[153] Jin H-Y, Wang Z-A. Boundedness, blowup and critical mass phenomenon in competing chemotaxis. J Differ Equ 2016;260(1):162–96.

[154] Igartúa DE, Martinez CS, del V Alonso S, Prieto MJ. Combined therapy for alzheimer's disease: tacrine and PAMAM dendrimers co-administration reduces the side effects of the drug without modifying its activity. AAPS PharmSciTech 2020;21:1–14.

[155] Neamtu I, Rusu AG, Diaconu A, Nita LE, Chiriac AP. Basic concepts and recent advances in nanogels as carriers for medical applications. Drug Deliv 2017;24 (1):539–57.

[156] Aderibigbe BA, Naki T. Design and efficacy of nanogels formulations for intranasal administration. Molecules 2018;23(6):1241.

[157] Ashrafi H, Azadi A, Mohammadi-Samani S, Hamidi M. New candidate delivery system for Alzheimer's disease: deferoxamine nanogels. Biointerface Res Appl Chem 2020;10 (6):7106–19.

[158] Ikeda K, Okada T, Sawada S-i, Akiyoshi K, Matsuzaki K. Inhibition of the formation of amyloid β-protein fibrils using biocompatible nanogels as artificial chaperones. FEBS Lett 2006;580(28–29):6587–95.

[159] Picone P, Sabatino MA, Ditta LA, Amato A, San Biagio PL, Mulè F, et al. Nose-to-brain delivery of insulin enhanced by a nanogel carrier. J Control Release 2018;270:23–36.

[160] Rajeshkumar S, Ganesh L, Santhoshkumar J. Selenium nanoparticles as therapeutic agents in neurodegenerative diseases. Nanobiotechnol Neurodegen Dis 2019;209–24.

[161] Yin T, Yang L, Liu Y, Zhou X, Sun J, Liu J. Sialic acid (SA)-modified selenium nanoparticles coated with a high blood–brain barrier permeability peptide-B6 peptide for potential use in Alzheimer's disease. Acta Biomater 2015;25:172–83.

[162] Zhang L, Yang S, Wong LR, Xie H, Ho PC-L. In vitro and in vivo comparison of curcumin-encapsulated chitosan-coated poly (lactic-co-glycolic acid) nanoparticles and curcumin/hydroxypropyl-β-cyclodextrin inclusion complexes administered intranasally as therapeutic strategies for Alzheimer's disease. Mol Pharm 2020;17(11):4256–69.

[163] Huo X, Zhang Y, Jin X, Li Y, Zhang L. A novel synthesis of selenium nanoparticles encapsulated PLGA nanospheres with curcumin molecules for the inhibition of amyloid β aggregation in Alzheimer's disease. J Photochem Photobiol B Biol 2019;190:98–102.

[164] Gilmore TD, Herscovitch M. Inhibitors of NF-kappaB signaling: 785 and counting. Oncogene 2006;25(51):6887–99.

[165] Je W, Ding J, Yang J, Guo X, Zheng Y. MicroRNA roles in the nuclear factor kappa B signaling pathway in cancer. Front Immunol 2018;9.

[166] Mendiola AS, Cardona AE. The IL-1β phenomena in neuroinflammatory diseases. J Neural Transm (Vienna) 2018;125(5):781–95.

[167] Mosconi L, Rahman A, Diaz I, Wu X, Scheyer O, Hristov HW, et al. Increased Alzheimer's risk during the menopause transition: a 3-year longitudinal brain imaging study. PloS One 2018;13(12), e0207885.

[168] Zempel H, Mandelkow E. Lost after translation: missorting of tau protein and consequences for Alzheimer disease. Trends Neurosci 2014;37(12):721–32.

[169] Taniguchi S, Suzuki N, Masuda M, Hisanaga S, Iwatsubo T, Goedert M, et al. Inhibition of heparin-induced tau filament formation by phenothiazines, polyphenols, and porphyrins. J Biol Chem 2005;280(9):7614–23.

[170] Golde TE, Schneider LS, Koo EH. Anti-Aβ therapeutics in Alzheimer's disease: the need for a paradigm shift. Neuron 2011;69(2):203–13.

[171] De Strooper B, Vassar R, Golde T. The secretases: enzymes with therapeutic potential in Alzheimer disease. Nat Rev Neurol 2010;6(2):99–107.

[172] Karran E, Mercken M, De Strooper B. The amyloid cascade hypothesis for Alzheimer's disease: an appraisal for the development of therapeutics. Nat Rev Drug Discov 2011;10 (9):698–712.

[173] Stephenson D, Perry D, Bens C, Bain LJ, Berry D, Krams M, et al. Charting a path toward combination therapy for Alzheimer's disease. Expert Rev Neurother 2015;15(1):107–13.

Advancement in herbal drugs for the treatment of Parkinson's disease

11

Ankit Shokeen, Bhavya Dhawan, Maryam Sarwat, and Sangeetha Gupta

Amity Institute of Pharmacy (AIP), Amity University Uttar Pradesh, Noida, Uttar Pradesh, India

1 Introduction

James Parkinson, the scientist in 1817, described Parkinson's disease (PD) for the first time as a syndrome comprising various physical signs which include rigidity, bradykinesia/akinesia, postural disturbances, and tremors [1]. It is a neurodegenerative disorder denoted by motor and nonmotor symptoms. There are various studies that evidence the onset of PD shows gender biases as it occurs 2 years earlier in men as compared to women. According to Hayes [2], classically the disease is referred to show motor symptoms, as evidenced in 187 patients suffering from PD with rigidity, postural instability, bradykinesia, and resting tremors. With the passage of time and ongoing studies, it came out as PD shows nonmotor symptoms as well which involve, cognitive decline, anxiety, depression, dysautonomia, and sleep disturbances [2].

Levodopa is considered the standard and most efficacious therapy in the management of PD thereby delaying the progression of the disease. It is the rapid precursor to dopamine, having the ability to cross the blood-brain barrier. This drug is given in combination with carbidopa to inhibit the peripheral degradation of levodopa by dopa-decarboxylase resulting to increased bioavailability in the brain and decreased peripheral side effects, particularly nausea. Other possible side effects associated with levodopa include dyskinesias, hallucinations, delusions, somnolence, and dystonia. Dyskinesia serves as the primary reason for considering other medical and surgical interventions for the treatment of PD, as it is found to frequently limit the dose which can be used [3–5]. Dopamine agonists such as ropinirole, pramipexole, and rotigotine activate dopaminergic receptors in the central nervous system, alleviating Parkinson's symptoms. Though they alleviate symptoms, they are comparatively less potent than levodopa. They are commonly used because they have a longer half-life and are less likely to cause dyskinesias. The lower risk of dyskinesias may be because they are less potent D2 receptor stimulators. Anticholinergic medications such as trihexyphenidyl and benztropine are ineffective in treating bradykinesia but may help with rigidity, dystonia, and tremor. Side effects associated with anticholinergic medications include dry eyes, urinary retention, dry mouth, memory problems, and hallucinations. In the

Targeting Angiogenesis, Inflammation and Oxidative Stress in Chronic Diseases. https://doi.org/10.1016/B978-0-443-13587-3.00016-3

elderly, it is advised to use these agents with caution. Antipsychotic medications are sometimes required to treat the symptoms of hallucination and paranoid delusions that occur in PD patients. Apart from this, cognitive behavioral therapy has also been shown to be beneficial in PD patients [3,5].

2 Epidemiology and prevalence of PD

The incidence and prevalence of PD increase with age; only 1% of people from the whole population get affected over the age of 65 [6]. The onset of PD symptoms that appears before the age of 40 is known as early-onset Parkinson's disease (EOPD). This is further divided into two types: juvenile-onset (occurring before the age of 21) and young-onset (YOPD, occurring in the age range of 21–40 years) [7]. In most populations, men are twice as likely as women to have PDS [8,9]. Female sex hormones are found to have a protective effect against the progression of PD. The prevalence of PD varies by country; it is higher in Europe and North America than in West Africa and Asia. The prevalence rate in India is the lowest in the world (70 per 100,000 normal populations). However, the Parsi community in Mumbai (India) has the world's highest prevalence of PD (328 per 100,000 population) [10].

3 Apoptosis, oxidative stress, inflammation, and angiogenesis in the pathophysiology of Parkinson's disease

The pathogenesis of PD involves mainly two hypotheses: dopaminergic and nondopaminergic. As PD is associated with the neurodegeneration of dopaminergic neurons including mainly D1 and D2 subtypes, the D1 and D2 receptors are found highly abundant in the striatum and are mostly associated with PD pathophysiology since they are activated by the dopaminergic pathway which begins in the substantia nigra (SNc) and ends in the caudate and putamen [3]. The pathological hallmark of PD is depigmentation of the SNc and locus coeruleus with neuronal loss in the pars compacta of the SNc [1,2]. The disturbance in the movement of PD is thought to be caused by the progressive depletion of nigrostriatal neurons, situated in the SNc of the brainstem. Dopamine is the primary neurotransmitter for communication with other regions of the basal ganglia. This would support the fact that the site of the neurodegenerative process in PD is not just restricted till the SNc [4]. On the other hand, the nondopaminergic hypothesis states the role of other neurons apart from dopaminergic which comprises noradrenergic neurons present in locus coeruleus, cholinergic neurons in the nucleus basalis of Meynert and serotonergic neurons in the midline raphe [5]. Several theories could explain neuronal degradation in Parkinson's disease. Apart from the above two hypothesis, recent advances focusing on novel factors involved in the pathogenesis of PD comprise apoptosis, immunological mechanisms, proteolysis defects, oxidative stress, mitochondrial dysfunction, iron metabolism disorders, and protein misfolding [4,5]. Till date, the etiology and pathogenesis of PD are not fully elucidated, as such no model has been developed that can recapitulate all the pathological features of PD. Among these, four main mechanisms, apoptosis, oxidative stress, inflammation, and angiogenesis, have been primarily focused regarding advancements in the herbal drugs for the betterment of PD conditions.

3.1 Role of apoptosis in Parkinson's disease

Cells possess multiple mechanisms to determine their fate when faced with unfavorable internal or external conditions. These functions are genetically predetermined and referred to as programmed cell death (PCD) or apoptosis. Research has found that improper regulation of PCD is a primary cause of neurodegenerative illnesses like PD. Recent histological studies of PD patients' brains show that the dopaminergic neurons in the nigra do die. Additionally, studies using the MPTP model of PD suggest that dopaminergic cell death occurs via apoptosis [6,7].

3.2 Role of oxidative stress in Parkinson's disease

Another primary cause of neurodegeneration in PD is oxidative stress. Since the nigrostriatal area is associated with higher energy consumption and low levels of the antioxidants such as glutathione, dopaminergic neurons are more vulnerable to oxidative damage [10]. The demand for high energy consumption in dopaminergic neurons, mitochondria's main producer of adenosine triphosphate (ATP), has been proposed to play a role in the development of PD. Mitochondria, via oxidative phosphorylation (OXPHOS), produces mainly ATP. On the other hand, OXPHOS also leads to the accumulation of reactive oxygen species (ROS), as electrons constantly leak through the electron transport chain [9]. Typically, mitochondrial antioxidative systems detoxify ROS, balancing damaging radical production and antioxidative protection. An increase in oxidative stress increases put macromolecules in the mitochondria vulnerable to oxidative damage. As a consequence, the function of the mitochondria is disrupted as macromolecules start getting accumulated due to oxidative stress. This eventually results in the leakage of cytochrome c from the mitochondria and the activation of cell apoptosis, as seen in PD dopaminergic neuronal death [11,12].

Mitochondria are the chief energy-producing centers in almost all eukaryotic cells, generating ATP [13]. Over the last few decades, it has been observed that mitochondrial dysfunction particularly oxidative stress has played a major role in the pathogenesis of PD. A synthetic opioid MPTP produced during the production of 1-methyl-4-phenyl-4propionoxypiperidine (MPPP), works by interfering with mitochondrial electron transport chain components, exacerbating them to be transformed into a toxic cation known as 1-methyl-4-phenyl pyridinium (MPP$^+$) via monoamine oxidase B enzymatic action [14,15]. This MPP$^+$ induces oxidative stress in neurons with limiting ATP production thereby leading to an increase in intracellular calcium levels. Very frequently, it has been encountered that the metabolites of MPTP have been seen in the SNc region of Parkinson's patients, which causes the inactivation of the ETC complex [16]. Total oxygen consumption of brain is approximately 20%; thus, a significant amount of this oxygen is indoctrinated into ROS in neurons and glia from various sources. Main source for the production of ROS in mitochondria is ETC, but other sources include NADPH oxidase, monoamine oxidase, other nitric oxide (NO), and flavoenzymes [17].

Oxidative damage to mitochondrial DNA impairs the cell DNA of the respiratory chain, resulting in complex I inhibition [10]. ROS-mediated protein oxidation, DNA damage, and rise in levels of malondialdehyde, thiobarbituric acid, and 4-hydroxynonenal (HNE) have also been reported in PD patients in SNc and striatum. Aging increases alteration in the transcriptional pathways, including nuclear factor erythroid 2-related factor 2, nuclear factor kappa B (NF-κB), mitogen-activated protein kinase, glycogen synthase kinase 3, and decreased activity of glutathione, superoxide dismutase, and catalase [18]. It is well known that oxidative damage impairs protein ubiquitination and degradation by the proteasome [19]. Three mechanisms in proteolysis defects remove nonfunctional and abnormal proteins: the autophagy-lysosomal pathway, the ubiquitin-proteasome system, and molecular chaperones. This pathway is primarily in charge of removing α-synuclein protein. Cell death is due to the build-up of abnormal proteins, which can misfold, aggregate, and block normal molecular pathways when these mechanisms are suppressed as depicted in Fig. 1. Genes such as PINK1, parkin, and DJ-1, UCH-L1 normally regulate mitochondrial and ubiquitin-proteasome system (UPS) function. DJ-1 mutations increase oxidative and nitrosative stress, which leads to dopamine oxidation. Changes in mitochondrial function also activate several major pathways, accelerating oxidative stress and α-synuclein aggregation and eventually leading to dopaminergic neurodegeneration in PD [20]. Protein aggregation and misfolding are the most common molecular phenomenon and causative factors in PD pathogenesis, even though the underlying mechanism is unknown. Examples include PARK2, SNCA, PINK1, DJ-1, and LRRK2. Mutations in their gene cause the SNpc in the midbrain to frequently misfold [19,21]. Lewy bodies are aggregated proteins present in nerve cells of specific brain regions; these are considered to be the primary hallmark features of Parkinson's disease (PD) [22]. Despite being the most abundant protein in LBs, α-synuclein has also been linked to other Lewy neurites. In normal physiological conditions, α-synuclein is a protein that has unfolded structure and is unlikely to transform into fibrils however in extremely high conditions—such as acidic P^H and high temperature, it gets transformed into partially folded conformation or intermediate, which intensively promotes the formation of α-synuclein fibrils; thus, a model based on α-synuclein fibrillation has been proposed by Uversky et al., and observation was made that aggregated α-synuclein forms were seen in a spherical shape, amyloid fibrils, metal ions and have been shown to accelerate α-synuclein aggregation, lending support to the environment-induced pathogenesis of PD [23].

Age factor and environmental factors both are contributing toward mitochondrial dysfunctionality, due to which there is an increase in oxidative stress and ROS. Due to oxidative stress, dopaminergic neurons start degrading and leads to neurodegeneration. Elevated ROS is responsible for hypoxia condition which activates the hypoxia-inducible factor-1 alpha (HIF-1α), it stimulates RBC production and VEGF which promotes angiogenesis. Age factors and genetic mutation both are

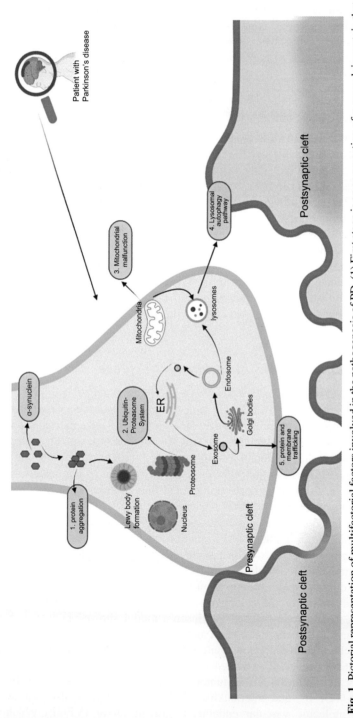

Fig. 1 Pictorial representation of multifactorial factors involved in the pathogenesis of PD. (1) First stage is an aggregation of α-synuclein protein that is the initiation of Lewy body formation; (2) Malfunction of ubiquitin system; (3) mitochondrial dysfunction leading to neuronal cell apoptosis; (4) lysosomal autophagy pathway activation degrading the macromolecule present in the cytoplasm; and (5) protein and membrane trafficking causes other debris and aggregated protein to spread all over, which leads to increase in pathological alteration in this disease.

contributing toward the dysfunctionality of α-synuclein; this is the reason for the formation of Lewy bodies that are contributing to the pathological symptoms of PD. Due to the dysfunctionality of α-synuclein, oxidative stress increases which will lead to neurodegeneration as well as regulating the neuroinflammation pathway as shown in Fig. 2.

3.3   Role of inflammation in Parkinson's disease

Inflammation has been known to be involved in the development and progression of Parkinson's disease (PD), although the exact mechanism is not yet clear. Genetic studies have revealed that changes in the expression of certain genes coding for the human leukocyte antigen (HLA) complex, which is involved in immune function [24]. Animal models of PD have been developed using compounds such as MPTP, 6-OHDA, paraquat, and rotenone, which induce damage to the nigrostriatal dopamine pathway and activate microglia and proinflammatory cytokines. These animal models have provided insights into the inflammatory response in the central nervous system (CNS) in PD. Furthermore, studies have shown increased levels of peripheral cytokines that act on the endothelial cells of the blood-brain barrier, leading to increased vascular permeability. While inflammation is implicated in PD, more research is needed to understand its precise role and whether it accelerates the progression of the disease [25].

3.4   Role of angiogenesis in Parkinson's disease

The phenomenon of the formation of new blood vessels from a preexisting vasculature is termed angiogenesis or neovascularization. Angiogenesis plays a crucial role in growth and development, ensuring an adequate supply of oxygen, nutrients, and hormones associated with the maintenance of whole-body homeostasis. Physiologically seen, it is stimulated during wound healing and development of the fetus; however, alteration of the normal physiological process due to irregular and/or anarchic vascular pattern leads to different pathophysiology and thus diseases [18]. The immature vessels in the brain are probably dearth of exhaustive characteristics of the blood-brain barrier (BBB) which comprises tight junction development, pericytes recruitment, and glia limitans (glial limiting membrane formation) and thus suggesting that the phenomenon of angiogenesis can disrupt the blood-brain barrier integrity and thus allowing the peripheral molecules, immune cells access to brain parenchyma contributing to ongoing neuroinflammation [18,26]. Concerning this, an association between BBB dysfunction and angiogenesis was established in a study by Carvey and colleagues where toxin-induced Parkinson's model showed punctate areas of FITC-LA leakage in the substantia nigra and striatum were shown colocalized with β3 integrin which is a marker for angiogenesis [26]. Furthermore, Parkinson's patients show an elevated number of activated microglia and activated microglia release proinflammatory cytokines such as tumor necrosis factor (TNF-α), transforming growth factor (TGF-β), and interleukins (IL-1β). Similarly, it also releases the proangiogenic molecule vascular endothelial growth factor (VEGF) which is a

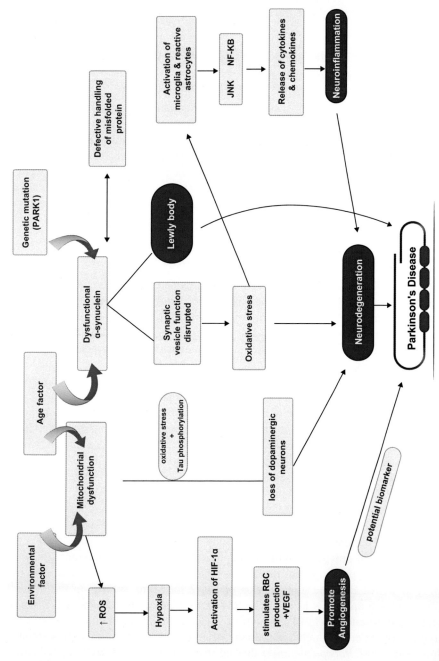

Fig. 2 Flowchart depicting pathways involved in the pathogenesis of PD. *PD*, Parkinson's disease; *ROS*, reactive oxygen species.

well-known regulator of vasculogenesis and angiogenesis assuring blood supply when either the tissue is in hypoxic condition and/or the blood supply is inadequate, and this VEGF is raised in SNpc of Parkinson's subjects [18,26].

4 Advancement in herbal drugs for the treatment of Parkinson's disease

Several studies have shown the beneficial effects of herbal drugs with limited adverse effects and have been proven to show anti-Parkinson activity. Below are discussed advancements in the treatment of PD with mechanism of action peculiarly mediated through oxidative stress, angiogenesis, and apoptosis. The herbal drugs have been explained which include *Sutellaria baicalensis* (Baicalein), *Curcuma longa* (Turmeric), *Withania somnifera* (Ashwagandha), *Bacopa monnieri* (Brahmi), *Mucuna pruriens* (Velvet bean), *Panax ginseng* (Ginseng), *Centellaas asiatica* Indian pennywort) as summarized in Table 1.

4.1 Role of Bacopa monnieri in Parkinson's disease

B. monnieri (BM) with the common name Brahmi named after Brahma (Hindu God) belongs to the family *Scrophulariaceae*. This herbal drug is familiar for its antioxidant, antiinflammatory, antidepressant, and antimicrobial activity. It is a creeping perennial herb that grows in warm wetlands and is native to India and Australia having small and oblong leaves and purple flowers. Dammarane type of triterpenoid saponins called bacosides, novel saponins called bacosides I–XII, alkaloids which comprise of brahmine, nicotine, herpestine in addition to D-mannitol, apigenin, her saponin, cucurbitacins, monnierasides I–III, plantainoside B are the main constituents of BM [13,55]. It is used as a potential neuronal tonic in the management of many neurological disorders. In PD, the mechanism considered responsible for providing neuroprotection includes increasing dopamine levels, decreasing oxidative stress, favoring an antiapoptotic state and improving behavioral deficits as shown in Fig. 3 [55]. The extract of BM also bears the potential to attenuate α-synuclein aggregates by activation of HSP-70 expression, a stress buffer protein [56,57].

It has been shown that the treatment with BM stabilizes the disrupted apoptotic homeostasis in PD. In this study, it was shown that the MPTP-induced model of PD shifts toward a more proapoptotic nature due to the decrease in sensitivity of antiapoptotic Bcl-2 and upregulation of proapoptotic Bax leading to dopaminergic neuronal loss. This proapoptotic state was stabilized with BM treatment by downregulation of Bax and upregulation of Bcl-2 therefore, setting up the role of BM in regulating apoptosis in PD [57]. BM extract can correct motor defects in a rotenone-induced rat model of Parkinson's by decreasing the oxidative stress in various regions of the brain which comprise the striatum, substantia nigra, hippocampus, and brain stem [58]. *Bacopa monnieri* showed strong antioxidant activity by reducing the levels of malondialdehyde (MDA) and nitrites, and increasing antioxidant levels such as SOD, CAT, GR, and GPx in MPTP-induced Parkinsonian mice in the

Table 1 Summary of herbal drugs with their mechanism of action in the treatment of Parkinson's disease.

S. no.	Botanical name	Family	Plant part	Mechanism of action	References
1	*Scutellaria baicalensis* (Baicalein)	*Lamiaceae*	Roots	Suppresses ROS pileup and ATP depletion, improves impaired spontaneous motor activity and prevents DA depletion, reduces cell apoptosis with a rise in the number of TH-positive neurons, increases DA and 5-hydroxytryptamine levels, and inhibits synuclein aggregation in cells	[27–36]
2	*Curcuma longa* (Turmeric)	*Zingiberaceae*	Rhizomes	Reduces ROS production, inhibits caspase-3 activation and inflammatory markers, and prevents expression of TH-positive neurons, inhibiting JNK hyperphosphorylation and mitochondrial dysfunction	[19,37–48]
3	*Withania somnifera* (Ashwagandha)	*Solanaceae*	Roots	Antiapoptotic action, decreases iNOS and GFAP expression, enhances cell viability, alters the oxidative stress proteins expression, regulates redox balance, restores striatal dopamine content, improves locomotor activity and muscle coordination	[26,49–54]
4	*Bacopa monnieri* (Brahmi)	*Scrophulariaceae*	Leaves	Regulates apoptosis (Bax downregulation and Bcl-2 upregulation), decreases oxidative stress, correct motor deficits, and increases the number of dopaminergic neurons and fibers	[13,27,55–59]
5	*Mucuna pruriens* (Velvet bean)	*Fabaceae*	Seeds	Mitochondria levels restored to normal, and TH expression increased	[50,60–63]

Continued

Table 1 Continued

S. no.	Botanical name	Family	Plant part	Mechanism of action	References
6	*Panax ginseng* (Ginseng)	*Araliaceae*	Roots	Inhibit Cdk5 overexpression and cleavage of p35 into p25 in SN and STN region of brain, increases TH+ and decreases TUNEL+ ratio, antiapoptotic action, JNK signaling cascade activation, restores nitric oxide signaling, cytoprotection against oxidative stress	[2–5,15,28,64–66]
7	*Centella asiatica* (Indian pennywort)	*Apiaceae*	Leaves	Protection of mitochondrial activity; reduces oxidative stress; reduces cell apoptosis; and increases TH-positive neurons	[67–70]

ROS, reactive oxygen species; *ATP*, adenosine triphosphate; *JNK*, c-Jun N-terminal kinase, *TH*, tyrosine hydroxylase; *iNOS*, inducible nitric oxide synthase; *GFAP*, glial fibrillary expression protein; *Cdk*, cyclin-dependent kinase; *SN*, substantia nigra, *STN*, subthalamic nucleus; *TUNEL*, terminal deoxynucleotidyl transferase dUTP nick end labeling.

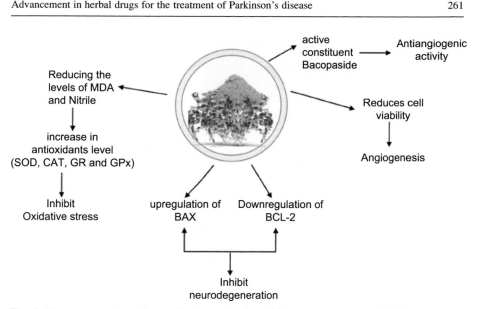

Fig. 3 Figure summarizes the mechanism of action of *Bacopa monnieri*. It inhibits oxidative stress by increasing antioxidants levels, reduces malondialdehyde and nitrite levels, upregulating Bax and downregulating Bcl-2 and leads to angiogenesis.

nigrostriatal region. This increased oxidative stress in the MPTP model is responsible for the loss of dopaminergic neurons determined by TH-immunoreactivity which is the characteristic feature of dopaminergic neurons and fibers. The ethanolic extract of BM was found to increase the number of dopaminergic neurons and fibers in comparison with the untreated Parkinson's group [58,59]. In an in vitro study it was found that bacopaside II constituent of BM acts as an antiangiogenic agent, and thus, it considerably decreased the cell viability of endothelial cell lines 2H11, 3B11, and HUVEC [27], and angiogenesis is found to play a role in the pathophysiology of PD [27,55–59].

4.2 Role of Withania somnifera in Parkinson's disease

Withania somnifera (Ws), a woody shrub, is a part of the *Solanaceae* family also known by "Winter Cherry," "Indian Ginseng," "Ashwagandha" as roots of it share similar properties with the Chinese Ginseng. It is one of the official drug which is mentioned in the Indian Pharmacopoeia (1985) and has also been used in Ayurveda for more than 5000 years. It is used for the management of many neurological disorders such as depression, epilepsy, poor memory, and neurodegenerative diseases. Furthermore, it also invigorates the regeneration of neurons and the reconstruction of synapses [26,49,50]. The various actions shown by Ws in preclinical studies include antioxidant, antiinflammatory, antimicrobial, antistress, antitumor, antidiabetic, cardioprotective, and neuroprotective activities as shown in Fig. 4. It has the potential to reduce ROS, improves mitochondrial function, enhances endothelial function,

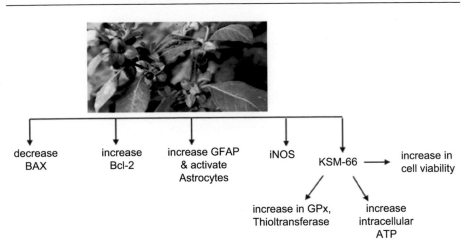

Fig. 4 Figure summarizes the mechanism of action of *Withania somnifera* as antiapoptotic, decreases iNOS expression and GAFP overexpression, enhances cell viability, alters the oxidative stress proteins expression, regulates redox balance, restores striatal dopamine content, and improves locomotor activity and muscle coordination.

regulates apoptosis, and inhibits NF-κB transcription, MAPK signaling pathways, and angiogenic, and reduces ER stress [50].

MB-PQ Parkinsonian mice showed significant antiapoptotic effects when treated with Ws further contributing to improved behavioral response. It regulated the hampered apoptotic homeostasis in Parkinsonian mice by decreasing Bax protein levels (proapoptotic) and elevating the levels of Bcl-2 protein (antiapoptotic) [51,52]. The study performed by Prakash and colleagues showed that Ws in addition to anti-apoptotic activity can also reduce the expression of inducible nitric oxide synthase (iNOS) at both the transcriptional and translational levels. Also, it reduced glial fibrillary acidic protein (GFAP) levels in activated astrocytes which is a neuro-inflammatory marker and also linked to iNOS activity implicating antioxidant property of Ws [26,50,51]. Another study assessed the effect of KSM-66, a Ws extract with withanolides as the main active constituent with antioxidant protection in SH-SY5Y in 6-hydroxydopamine (6-OHDA) PD model. The parameters assessed included percent cell viability, oxidative stress protein expression, ATP levels, antioxidant enzyme activities, and detection of protein glutathionylation. KSM-66 was evidenced to enhance cell viability, glutathione peroxidase, thiol transferase enzyme activities, and intracellular ATP levels and also altered the oxidative stress proteins expression namely peroxiredoxin I, VGF and Vimentin and regulated redox balance by reducing glutathionylated protein levels [26]. Capability to restore the striatal content of dopamine and its metabolites through antioxidant effects in the 6-OHDA rat model of PD was further elucidated. Improvement in locomotor activity and muscle coordination was illustrated in the study [53]. Furthermore, Ws has been shown to normalize the catecholamine content in the midbrain of MPTP intoxicated mice by reducing oxidative stress associated with improved functional activity [52]. Withaferin A showed potential interaction with VEGF and was found have potent anti-VEGF activity. This was proposed to be a probable mechanism of Withaferin A for controlling angiogenesis [50]. Also, Ws was demonstrated to have a potential antiangiogenic activity in a study by Rajani Mathur and colleagues [54].

4.3 Role of Ginseng in Parkinson's disease

The term panax came from the Greek word "Panakos" which means "cure all." The widely studied varieties include *Panax quinquefolius* L. (American Ginseng), *Panax ginseng* C.A. Meyer (Chinese, Korean, Asian, and Oriental ginseng), *Panax quinquefolius* L. (American Ginseng), *Panax japonicas* C.A. Meyer (Japanese Ginseng), *Panax notoginseng* Burk (Sanqi ginseng), and *Panax vietnamensis* Ha et Grushv (Vietnamese ginseng). Out of these *P. ginseng*, *P. quinquefolius*, and *P. notoginseng* are the chief commercial species. The growth of Ginseng usually takes place in cold, moist, and shaded areas of woodland and forest and is placed in 35 countries in eastern Asia and eastern North America [5]. The various constituents of Ginseng include ginsenosides, minerals like calcium, magnesium and manganese, amino acids, vitamins B and C and peptides, etc. [2,4]. It possesses a plethora of therapeutic benefits such as cardioprotective, hepatoprotective, immunomodulation, hypoglycemic, anticancer, antioxidant, neurotransmitter modulation, and cognitive performance enhancer as shown in Fig. 5 [5]. Various pieces of evidence reports *P. ginseng* and

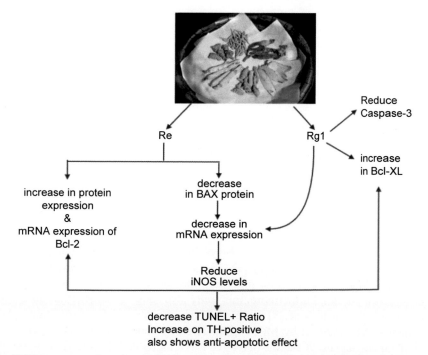

Fig. 5 Figure summarizes the mechanism of action of Ginseng: inhibits Cdk5 overexpression and cleft of p35 into p25 in SN and ST region of the brain, increase TH[+] and decreases TUNEL[+] ratio, antiapoptotic, JNK signaling cascade activation, restore nitric oxide signaling, cytoprotection against oxidative stress.

its pharmacologically active constituents; ginsenosides perform a wide variety of actions in the brain including benefits in several cultures and animal models of PD. It was shown to increase in cell survival, rescuing neurons from the neurotoxic damage, etc., by its antiinflammatory and antioxidant properties [5]. Furthermore, it has been shown to modulate several biochemical markers recognized as prime for Parkinson's initiation and progression [3].

Korean red ginseng possesses the ability to alleviate behavioral impairments and protect dopaminergic neurons, inhibit Cdk5 overexpression, and cleft of p35 into p25 in SNc and ST regions of the brain. It is thought that Cdk5 under normal conditions supports neuronal survival by its phosphorylation via p35. However, MPTP enhances p35 degradation to p25 causing Cdk5 hyperactivation, and this hyperactivity of Cdk5 is known to be involved in PD pathogenesis and is shown in the brains of PD patients causing abnormal hyperphosphorylation of cytoskeletal proteins and hampered antioxidant defense ultimately leading to neuron death. Thus, this study suggested that Korean red ginseng can mitigate neuronal death in the nigrostriatal system by the above-mentioned mechanisms [15]. A study conducted by Xu and colleagues in 2005 showed that the apoptosis of SNc neurons in mice can be prevented by ginseng compounds in the MPTP (1-methyl-4- phenyl-1,2,3,6-tetrahydropyridine) model of PD. The outcome of the study revealed that pretreatment with ginsenoside Re possesses neuroprotection, i.e., a rise in TH^+ and a reduce in $TUNEL^+$ ratio and this neuroprotection offered by ginsenoside Re was achieved by the increase in protein and mRNA expression of Bcl-2 and reduction in expression of Bax protein and mRNA along with iNOS and caspase-3 activation [28]. Similarly, another ginsenoside Rg1 possessed protective activity against MPTP-induced apoptosis in vivo, wherein pretreatment with 5 and 10 mg/kg dose resulted in enhanced Nissl staining and TH^+ neurons and reduced $TUNEL^+$ neurons in the zona compacta region of SNc probably by decreasing caspase-3, Bax, and iNOS and increasing Bcl-xl and Bcl-2 (49). Similarly, oral administration of ginseng extract, G115 in β-sitosterol β-D-glucoside (BSSG)-induced PD model showed protection against BSSG-induced apoptosis and dopaminergic cell death in SNc which is a main characteristic of PD as evidenced by attenuated caspase-3 activation and TUNEL labeling, and this nigral neuronal protection was further indicated by alleviation of locomotor abnormalities. Also, this study was the first to show beneficial action on synucleinopathy by diminishing the α-synuclein aggregates [3].

One of the neuroprotective mechanisms by which ginsenosides Rg1 can alleviate the dopaminergic neuron loss in the SNc of Parkinsonian mice is by its antioxidant activity through alteration of Bcl-2, iNOS, GSH, and T-SOD comprising both Cu/ZnSOD and Mn-SOD along with inhibiting the activation of JNK signaling (c-Jun NH2-terminal kinase). Furthermore, this study also showed that Rg1 has an antioxidant activity similar to N-acetyl cysteine which is an established protectant against SNc neuron loss induced by MPTP [65]. A study conducted by Kyung-Hee Kim and colleagues suggested that the defective PINK1-LRPPRC/Hsp90-Hsp60-complex IV signaling in PINK1 null neuronal cells can be corrected by ginsenosides Re by restoring nitric oxide signaling supporting the use of ginsenosides Re in enhancing dopaminergic neuron survival in different types of familial PD as it was found that mitochondrial complex IV activity was diminished due to downregulation of chaperones Hsp60 and its upstream regulators namely Hsp90 and Leucine-rich pentatricopeptide repeat (LRPPRC) containing in PINK1 null dopaminergic neurons which were rectified by ginsenoside Re through NO signaling. Also, PINK 1 (PARK6) mutations in humans are associated with autosomal recessive types of familial PD [66]. Similarly, another ginsenoside Rb1 is demonstrated to provide cytoprotection in SH-SY5Y human dopaminergic cell line against oxidative stress induced by 6-OHDA (hydroxydopamine) possibly by enhancing HO-1 (heme oxygenase-1) via ER-related P13K/Akt/Nrf-2-dependent pathway thereby considerably decreasing caspase-3 activation and associated cell death [65,66].

4.4   Role of Centella asiatica in Parkinson's disease

Active constituents of *Centella asiatica* include madecassoside, asiaticoside, madeccasic acid, asiatic acid, essential oils, amino acids, and other compounds like vellarin, etc. [67,68]. ECa233 is a standardized extract of *C. asiatica* consisting madecassoside and asiaticoside in the ratio $1.5 \pm 0.5{:}1$ in $\geq 80\%$ concentration [69]. ECa233 by modulating oxidative stress more specifically by protecting against mitochondrial complex I activity, decreasing lipid peroxidation and increasing antioxidant enzyme expression protected against dopaminergic neuronal cell death and impaired locomotor activity in a rotenone-induced Parkinson's model as depicted in Fig. 6. Mitochondrial activity protection probably reduces oxidative stress. Furthermore, this study considered axonal regeneration also as a protective mechanism other than mitochondrial protection and antioxidant action [69]. In other studies, asiaticoside was found to exert neuroprotection by the mechanism of recovering redox balance in

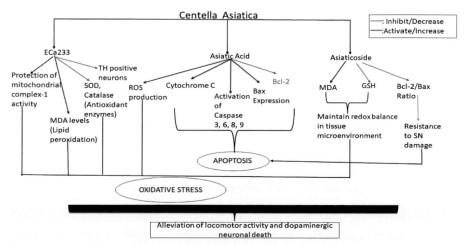

Fig. 6 Figure summarizes the mechanism of action of *Centella asiatica*. It comprises three major active constituents: ECa233, asiatic acid, and asiaticoside. Asiatic acid is responsible for carrying down apoptosis, while the rests are involved in oxidative stress. PD pathogenesis involves both oxidative stress and apoptosis.

the tissue microenvironment more specifically by reducing MDA and enhancing GSH [70]. A study with asiatic acid showed that its pretreatment can suppress apoptosis by upregulating antiapoptotic and downregulating proapoptotic events, i.e., by decreasing cytochrome c, Bax, activation of caspase 9, 8, 6, 3, and increasing bcl-2 [69]. One of the studies showed that asiaticoside can enhance the Bcl-2/Bax ratio which serves as a deciding factor in whether a cell will initiate apoptosis or not, and an increased ratio means apoptosis prevention and the resistance to damage of SN was also enhanced [70].

4.5   Role of Baicalein in Parkinson's disease

It is a flowering species plant and flavonoid compound that belongs to the *Lamiaceae* family [28]. It is isolated from decoction of dried root of *Scutellaria baicalensis* [28], and that particular decoction in the common language is known as Huang qin decoction. The chemical nature of baicalin is hydrosoluble. The blood-brain barrier (BBB), which is very robust, can also be penetrated by it. Due to this particular property, it exhibits remarkable neuroprotective effects in various in vitro models [29]. Baicalin is hydrolyzed into baicalein by β-glucosidases after entering the human body [30]. To lower the levels of nitric oxide (NO) and COX-2 [31], a Schiff base formation is required with β-amyloid or α-synuclein, an oxidized variant of baicalein (baicalein quinone); this could aid in the inhibition of accumulation of disease-specific amyloid proteins [32,33], as the CNS shows evidence of antioxidant activity from the ethanolic extract of *Scutellaria baicalensis*, which is commonly known to scavenge free radicals as an antioxidant. Baicalein acts by inhibiting lipoxygenase and the NF-κB signaling pathways in glial cells, which suppresses the expression of iNOS and the production of nitric oxide (NO), ultimately leading to a reduction in chronic neuroinflammation. Additionally, baicalein promotes neurogenesis and differentiation of neural cells by enhancing the signaling pathway of brain-derived neurotrophic factor (BDNF) [34,35]. As mentioned above that baicalein exhibits a neuroprotective effect in various in vitro models, one of the well-established models is the PD model 6-OHDA-induced cellular model; it was proven to promote neuronal growth, reduction in cell apoptosis. Baicalein reduces TH-positive neurons and improves impaired motor function; rats treated with baicalein have increased levels of DA and 5-hydroxytryptamine, by inhibiting the basal ganglia's ability to lose dopamine [27,36]. Recently, a study very well demonstrated that baicalein also has a protective effect against oxidative stress-related injury. As we are very well aware of the fact that oxidative stress plays an important role in the pathogenesis of neurodegenerative diseases and Parkinson's is one of them [27]. On further investigation, it has been explored that baicalein protects mitochondria from autophagy by inhibiting miR-30b and initiating the NIX/BNIP3 pathway. It also helps in reviving neuronal health and protects against Parkinson's disease [71]. Based on the above therapeutic action and pharmacological effect, baicalein can be considered to be an effective herb in the treatment of PD as shown in Fig. 7.

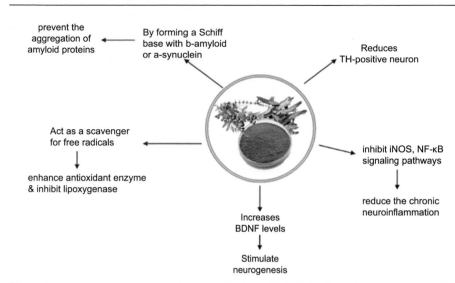

Fig. 7 Figure summarizes the mechanism of action of baicalein. It works by scavenging free radicals, increasing antioxidants enzymes, suppressing lipoxygenase, and inhibiting the NF-κB signaling pathways in glial cells and inducible nitric oxide synthase (iNOS) expression, increases the BDNF levels which stimulates neurogenesis and prevents β-amyloid protein aggregation.

4.6   Role of Mucuna pruriens in Parkinson's disease

It is commonly known as the "velvet bean" and has been studied for its anti-Parkinson activity since the 19th century. It belongs to the *Fabaceae* family; it is an annual and perennial legume that has a variety of therapeutic properties that help in the treatment of Parkinson's like its antioxidant nature and antiinflammatory. Apart from this, it can be useful as an antiepileptic antimicrobial agent [60]. In Ayurvedic principles, it is mentioned that *M. pruriens* (Mp) is a "Rasyana." A "Rasyana" is a drug or nondrug intervention having properties like replenishing or restoring the system/body organ/ specific tissues [60]. The main phenolic compound in Mp is L-DOPA (5%); it is discovered by using the HPTLC method [61]. In 1937, for the very first time, L-DOPA was isolated from the seeds of Mp, and since there it comes to highlight that it can be a very potential therapeutic agent for Parkinson's patients. In recent studies of paraquat-induced mouse model of Parkinson, it has been observed that it provides neuroprotection [50,62,63] and also enhances catecholamine levels that help in improving the mice's motor activity. It stimulates antioxidant potential in the nigrostriatal region. On treatment with Mp, the tyrosine hydroxylase (TH) expression in the SN and striatal regions, as well as iNOS and GFAP expression levels were recovered in MPTP-treated animals [63]; through this study, it was encountered that it downregulates the expression of nitric oxide. Another study revealed that ursolic acid found in Mp has potent antiinflammatory properties; it lowers the expression of NF-kB, COX, IL-6, and TNF-α. Both Mp and UA have antiinflammatory properties; it works by inhibiting the expression of iNOS; thus, the levels of proinflammatory

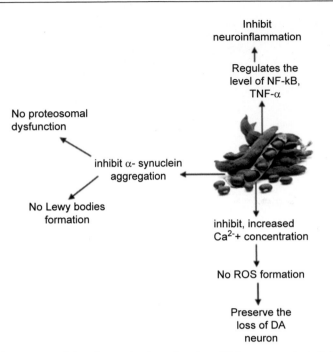

Fig. 8 Flowchart summarizing the mechanism of action of *Mucuna pruriens*. It inhibits the formation of synuclein aggregation, Ca^{2+} concentration, and thus inhibits the formation of Lewy bodies, proteasomal dysfunction, and ROS. It also regulates NF-κB and TNF-levels, which inhibit neuroinflammation.

cytokines such as NF-kB, COX, TNF-α, and IL-6 regulate [62,63]. As a result, Mp has been shown to have potent immunomodulatory activity in PD as depicted in Fig. 8.

4.7   Role of Curcuma longa in Parkinson's disease

Curcuma longa is a flowering plant of the ginger family, *Zingiberaceae*, whose rhizomes are used to treat a variety of disorders. It is well known for its antioxidant, antiseptic, antiinflammatory, antibacterial, and antitumor agent properties as shown in Fig. 9 [37]. The gathering of the protein is a distinctive characteristic and the predominant reason for PD, and in the case of Parkinson, the presence of the protein α-synuclein has increased, it is insoluble, and it gets clumped together and forms a Lewy-body in the neuron. A molecular dynamic simulation technique was utilized in a recent investigation to explore how curcumin influences the α-synuclein oligomer. The study revealed that curcumin interfered with the general properties of the s-oligomer, causing a reduction in its structural stability. In addition, curcumin aids in preventing the accumulation of α-synuclein oligomers and reduces intracellular ROS accumulation and caspase-3 activation induced by α-synuclein in SH-SY5Y cells. Curcumin also prevents the loss of TH-positive neurons and DA depletion

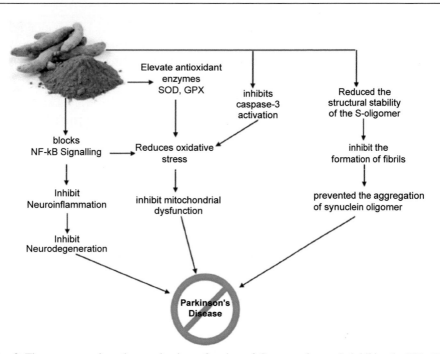

Fig. 9 Figure summarizes the mechanism of action of *Curcuma longa*. It inhibits the NF-κB signaling pathway, reduces oxidative stress, reduces the stability of the S-oligomer structure, which is responsible for the formation of fibrils and prevents α-synuclein aggregation.

caused by MPTP [38,39]. It downregulates cytokines, nitrite, inflammatory markers, and iNOS present in the striatum [40]. Curcumin protects DA neurons from the toxicity that is induced by MPTP [41]. It inhibits the expression of the GFAP and iNOS; thus, there is an increase in dopamine and tyrosine hydroxylase activity [42]. The impact of curcumin nanoparticles on three significant signaling pathways (Nrf2, NF-κB, and Akt/pTau) aligns with its antiinflammatory and antioxidant characteristics, as these pathways play a crucial role in regulating oxidative stress, tau protein phosphorylation, and inflammation pathways [43]. On performing postmortem, nanocurcumin has shown visible results that it will reduce oxidative stress and apoptosis [44]. Curcumin protects neurons from oxidation by restoring the mitochondrial membrane potential, enhancing the activity of Cu–Zn superoxide dismutase, and inhibiting intracellular ROS production [45]. In a PD siRNA-mediated PINK1 knockdown model, it has been demonstrated that curcumin can safeguard against cell death and mitochondrial dysfunction [46]. The role of oxidative stress in PD pathology is suggested by degenerative processes such as nitric oxide and mitochondrial toxicity [19,47]. A study examined the effects of curcumin on mitochondrial dysfunction in a paraquat-induced toxicity model. The results indicated that administering curcumin to the cell model before paraquat treatment improved both maximal and ATP-associated respiration without altering respiratory capacity. However, administering curcumin to fibroblasts after paraquat treatment did not enhance mitochondrial respiration across the three parameters of maximal respiration, ATP-associated respiration, and spare respiratory capacity. These observations suggest that curcumin may have a preventive impact when administered before the onset of PD [48].

5 Nanotechnological aspects for the management of Parkinson's disease

In recent years, there has been a growing interest in utilizing nanotechnologies for treating PD. One promising approach involves the use of nanoencapsulated drugs to enhance the effectiveness of macromolecules that target the brain. Nanomaterials have unique physical and chemical properties, and because of their smaller size, they can be administered through various routes, not just locally but also systemically and intranasally [47]. However, developing formulations that allow nanoparticles (NPs) to pass through the BBB after systemic administration without causing harmful effects in the brain or elsewhere in the body remains a challenge. Creating biocompatible NPs is also necessary to develop optimal particle-based treatments [19]. To this end, natural and synthetic polymers, lipids, and inorganic metal-based materials are used to formulate NPs for CNS delivery [48].

Nanoparticles have been tested for various purposes, including providing sustained release of traditional PD treatments, avoiding the immune system, and facilitating entry into the CNS. They have also been modified with functional elements to enhance entry across mucosal surfaces, target specific cell types, or release the active drug only in specific circumstances. Nanotherapies offer potential benefits beyond improving the pharmacokinetic properties of traditional treatments to deliver new therapies that could target specific intracellular pathways, such as oxidative stress or inflammation or even specific genes [72].

Initially, early research on nanoparticles for PD aimed to improve the release profile of dopamine and other dopaminergic medications, such as levodopa, dopamine agonists, and monoamine oxidase B inhibitors [73]. For instance, implanting dopamine-loaded liposomes directly into the striatum of a rat model of PD maintained elevated levels for 25 days after implantation [74,75]. Recent research has focused on intranasal formulations of levodopa-loaded PLGA nanoparticles [76], which provide motor benefits that can persist for at least a week. Polymeric nanoparticles offering extended release have been successfully demonstrated in preclinical models with various dopaminergic medications. However, further evaluation of the efficacy, safety, biodistribution, and pharmacokinetics of each nanoparticle is necessary before human studies can be conducted [74,75].

6 Conclusion and future prospectives

The Global Burden of Diseases, Injuries, and Risk Factors Study (GBD) documented that in 2017, there were 1.02 million cases of PD. Also, from 1990 to 2016, the age standardized rate (ASR) prevalence of PD has enhanced by 21.7%. Furthermore, evidence suggests that the burden of PD will significantly increase in future intensifying the need to efficiently manage this disease. Various treatment approaches both pharmacological and nonpharmacological are currently available for Parkinson's management yet there is no cure available for the same. In recent years, researchers are actively investigating the potential of herbal products to effectively manage PD. Indeed, many herbs have proven themselves beneficial and more effective than synthetic drugs. Furthermore, India concerning floristic diversity is one of the chief nations in the world where approximately 54% of its land is used for farming. Medicinal herbs render important for meeting the world's various healthcare needs, and they are preferred for a long time due to numerous benefits like safety, acceptance, relatively fewer side effects than synthetic drugs and synergistic action as herbal extracts usually contain more than one constituent. As suggested by various pieces of evidence mentioned in this chapter, it is evident that medicinal herbs play an important role in PD through various actions like antiapoptotic, antioxidant, regulating mitochondrial and neuronal dysfunction, etc., and now, the role of herbs in controlling dysregulated angiogenesis can also be seen.

PD is considered to be a degenerative progressive disease, and so far no model as such has been developed that recapitulates all the hallmark features of PD. There are multiple targets on which the drugs are working; however, as of now the drugs that are available in the market have limited effect on the pathology of PD, the benefit of herbal medicine is that they are made up of combining multiples herbs and that are targeting multiple targets, the herbs that are listed above show their significant effect, and further investigation has been going on to cure this disease fully. Thus, it can be concluded that medicinal herbs bear a plethora of benefits and scope in future for the effective management of Parkinson's disease, further studies about their precise mechanisms, formulation, scalability, etc., if done might prove as a breakthrough in Parkinson's research.

Acknowledgments

The authors would like to thank Amity Institute of Pharmacy (AIP), Amity University Uttar Pradesh, Noida Campus 201313, India, for providing facilities for the completion of this review.

References

[1] Amro MS, Teoh SL, Norzana AG, Srijit DJ. The potential role of herbal products in the treatment of Parkinson's disease. La Clinicaterapeutica 2018;169(1):e23–33. https://doi.org/10.7417/T.2018.2050.
[2] Hayes MT. Parkinson's disease and parkinsonism. Am J Med 2019;132(7):802–7. https://doi.org/10.1016/j.amjmed.2019.03.001.
[3] Gerfen CR. Molecular effects of dopamine on striatal-projection pathways. Trends Neurosci 2000;23:S64–70. https://doi.org/10.1016/s1471-1931(00)00019-7.
[4] Del Tredici K, Rüb U, De Vos RA, Bohl JR, Braak H. Where does Parkinson disease pathology begin in the brain? J Neuropathol Exp Neurol 2002;61(5):413–26. https://doi.org/10.1093/jnen/61.5.413.

[5] González-Burgos E, Fernandez-Moriano C, Gómez-Serranillos MP. Potential neuro-protective activity of Ginseng in Parkinson's disease: a review. J Neuroimmune Pharmacol 2015;10:14–29. https://doi.org/10.1007/s11481-014-9569-6.

[6] Goldman SM, Tanner C. Etiology of Parkinson's disease. In: Jankovic J, Tolosa E, editors. Parkinson's disease and movement disorders. London, UK: Williams and Wilkins; 1998. p. 133–58.

[7] Schrag A, Schott JM. Epidemiological, clinical, and genetic characteristics of early-onset parkinsonism. Lancet Neurol 2006;5(4):355–63. https://doi.org/10.1016/S1474-4422(06)70411-2.

[8] Baldereschi M, Di Carlo A, Rocca WA, Vanni P, Maggi S, Perissinotto E, Grigoletto F, Amaducci L, Inzitari D. Parkinson's disease and parkinsonism in a longitudinal study: two-fold higher incidence in men. Neurology 2000;55(9):1358–63. https://doi.org/10.1212/wnl.55.9.1358.

[9] Van Den Eeden SK, Tanner CM, Bernstein AL, Fross RD, Leimpeter A, Bloch DA, Nelson LM. Incidence of Parkinson's disease: variation by age, gender, and race/ethnicity. Am J Epidemiol 2003;157(11):1015–22. https://doi.org/10.1093/aje/kwg068.

[10] Aryal S, Skinner T, Bridges B, Weber JT. The pathology of Parkinson's disease and potential benefit of dietary polyphenols. Molecules 2020;25(19):4382. https://doi.org/10.3390/molecules25194382.

[11] Kung HC, Lin KJ, Kung CT, Lin TK. Oxidative stress, mitochondrial dysfunction, and neuroprotection of polyphenols with respect to resveratrol in Parkinson's disease. Biomedicine 2021;9(8):918. https://doi.org/10.3390/biomedicines9080918.

[12] Khan TA, Hassan I, Ahmad A, Perveen A, Aman S, Quddusi S, Alhazza IM, Ashraf GM, Aliev G. Recent updates on the dynamic association between oxidative stress and neuro-degenerative disorders. CNS Neurol Disord Drug Targets 2016;15(3):310–20. https://doi.org/10.2174/1871527315666160202124518.

[13] Yin R, Xue J, Tan Y, Fang C, Hu C, Yang Q, Mei X, Qi D. The positive role and mechanism of herbal medicine in Parkinson's disease. Oxid Med Cell Longev 2021;2021. https://doi.org/10.1155/2021/9923331.

[14] Verma AK, Raj J, Sharma V, Singh TB, Srivastava S, Srivastava R. Epidemiology and associated risk factors of Parkinson's disease among the north Indian population. Clin Epidemiol Glob Health 2017;5(1):8–13. https://doi.org/10.1016/j.cegh.2016.07.003.

[15] Van Kampen JM, Baranowski DB, Shaw CA, Kay DG. Panax ginseng is neuroprotective in a novel progressive model of Parkinson's disease. Exp Gerontol 2014;50:95–105. https://doi.org/10.1016/j.exger.2013.11.012.

[16] Dias V, Junn E, Mouradian MM. The role of oxidative stress in Parkinson's disease. J Parkinsons Dis 2013;3(4):461–91. https://doi.org/10.3233/JPD-130230.

[17] Jenner P. Oxidative stress in Parkinson's disease. Ann Neurol 2003;53(S3):S26–38. https://doi.org/10.1002/ana.10483.

[18] Desai Bradaric B, Patel A, Schneider JA, Carvey PM, Hendey B. Evidence for angiogenesis in Parkinson's disease, incidental Lewy body disease, and progressive supranuclear palsy. J Neural Transm 2012;119:59–71. https://doi.org/10.1007/s00702-011-0684-8.

[19] Henchcliffe C, Beal MF. Mitochondrial biology and oxidative stress in Parkinson disease pathogenesis. Nat Clin Pract Neurol 2008;4(11):600–9. https://doi.org/10.1038/ncpneuro0924.

[20] Levy OA, Malagelada C, Greene LA. Cell death pathways in Parkinson's disease: proximal triggers, distal effectors, and final steps. Apoptosis 2009;14:478–500. https://doi.org/10.1007/s10495-008-0309-3.

[21] Doke RR, Pansare PA, Sainani SR, Bhalchim VM, Rode KR, Desai SR. Natural products: an emerging tool in Parkinson's disease therapeutics. IP Indian J Neurosci 2019;5(3):95–105. https://doi.org/10.18231/j.ijn.2019.014.

[22] Ehrlich G, Dhru D. Ayurvedic medicine in neurology. Integr Neurol 2020;327.
[23] Brown GC. Control of respiration and ATP synthesis in mammalian mitochondria and cells. Biochem J 1992;284(1):1–3. https://doi.org/10.1042/bj2840001.
[24] Rabhi C, Arcile G, Cariel L, Lenoir C, Bignon J, Wdzieczak-Bakala J, Ouazzani J. Antiangiogenic-like properties of fermented extracts of ayurvedic medicinal plants. J Med Food 2015;18(9):1065–72. https://doi.org/10.1089/jmf.2014.0128.
[25] Anglade P, Vyas S, Javoy-Agid F, Herrero MT, Michel PP, Marquez J, Mouatt-Prigent A, Ruberg M, Hirsch EC, Agid Y. Apoptosis and autophagy in nigral neurons of patients with Parkinson's disease. Histol Histopathol 1997;12(1):25–31.
[26] Saha S, Islam MK, Shilpi JA, Hasan S. Inhibition of VEGF: a novel mechanism to control angiogenesis by *Withania somnifera*'s key metabolite Withaferin A. In Silico Pharmacol 2013;1:1–9. https://doi.org/10.1186/2193-9616-1-11.
[27] Cheng Y, He G, Mu X, Zhang T, Li X, Hu J, Xu B, Du G. Neuroprotective effect of baicalein against MPTP neurotoxicity: behavioral, biochemical and immunohistochemical profile. Neurosci Lett 2008;441(1):16–20. https://doi.org/10.1016/j.neulet.2008.05.116.
[28] Wang ZL, Wang S, Kuang Y, Hu ZM, Qiao X, Ye M. A comprehensive review on phytochemistry, pharmacology, and flavonoid biosynthesis of Scutellaria baicalensis. Pharm Biol 2018;56(1):465–84. https://doi.org/10.1080/13880209.2018.1492620.
[29] Chen CJ, Raung SL, Liao SL, Chen SY. Inhibition of inducible nitric oxide synthase expression by baicalein in endotoxin/cytokine-stimulated microglia. Biochem Pharmacol 2004;67(5):957–65. https://doi.org/10.1016/j.bcp.2003.10.010.
[30] Liu C, Wu J, Xu K, Cai F, Gu J, Ma L, Chen J. Neuroprotection by baicalein in ischemic brain injury involves PTEN/AKT pathway. J Neurochem 2010;112(6):1500–12. https://doi.org/10.1111/j.1471-4159.2009.06561.x.
[31] Jeong K, Shin YC, Park S, Park JS, Kim N, Um JY, Go H, Sun S, Lee S, Park W, Choi Y. Ethanol extract of Scutellaria baicalensis Georgi prevents oxidative damage and neuroinflammation and memorial impairments in artificial senescense mice. J Biomed Sci 2011;18(1):1–2. https://doi.org/10.1186/1423-0127-18-14.
[32] Zhu M, Rajamani S, Kaylor J, Han S, Zhou F, Fink AL. The flavonoid baicalein inhibits fibrillation of α-synuclein and disaggregates existing fibrils. J Biol Chem 2004;279(26):26846–57. https://doi.org/10.1074/jbc.M403129200.
[33] Hong DP, Fink AL, Uversky VN. Structural characteristics of α-synuclein oligomers stabilized by the flavonoid baicalein. J Mol Biol 2008;383(1):214–23. https://doi.org/10.1016/j.jmb.2008.08.039.
[34] Chen YC, Shen SC, Chen LG, Lee TJ, Yang LL. Wogonin, baicalin, and baicalein inhibition of inducible nitric oxide synthase and cyclooxygenase-2 gene expressions induced by nitric oxide synthase inhibitors and lipopolysaccharide. Biochem Pharmacol 2001;61(11):1417–27. https://doi.org/10.1016/s0006-2952(01)00594-9.
[35] Allen SJ, Watson JJ, Shoemark DK, Barua NU, Patel NK. GDNF, NGF and BDNF as therapeutic options for neurodegeneration. Pharmacol Ther 2013;138(2):155–75. https://doi.org/10.1016/j.pharmthera.2013.01.004.
[36] Mu X, He GR, Yuan X, Li XX, Du GH. Baicalein protects the brain against neuron impairments induced by MPTP in C57BL/6 mice. Pharmacol Biochem Behav 2011;98(2):286–91. https://doi.org/10.1016/j.pbb.2011.01.011.
[37] Mythri B, R, M Srinivas Bharath M. Curcumin: a potential neuroprotective agent in Parkinson's disease. Curr Pharm Des 2012;18(1):91–9. https://doi.org/10.2174/138161212798918995.
[38] Harish G, Venkateshappa C, Mythri RB, Dubey SK, Mishra K, Singh N, Vali S, Bharath MS. Bioconjugates of curcumin display improved protection against glutathione depletion mediated oxidative stress in a dopaminergic neuronal cell line: implications for

Parkinson's disease. Bioorg Med Chem 2010;18(7):2631–8. https://doi.org/10.1016/j.bmc.2010.02.029.

[39] Jagatha B, Mythri RB, Vali S, Bharath MS. Curcumin treatment alleviates the effects of glutathione depletion in vitro and in vivo: therapeutic implications for Parkinson's disease explained via in silico studies. Free Radic Biol Med 2008;44(5):907–17. https://doi.org/10.1016/j.freeradbiomed.2007.11.011.

[40] Kamelabad MR, Sardroodi JJ, Ebrahimzadeh AR, Ajamgard M. Influence of curcumin and rosmarinic acid on disrupting the general properties of alpha-synuclein oligomer: molecular dynamics simulation. J Mol Graph Model 2021;107, 107963. https://doi.org/10.1016/j.jmgm.2021.107963.

[41] Wang MS, Boddapati S, Emadi S, Sierks MR. Curcumin reduces α-synuclein induced cytotoxicity in Parkinson's disease cell model. BMC Neurosci 2010;11(1):1. https://doi.org/10.1186/1471-2202-11-57.

[42] Ojha RP, Rastogi M, Devi BP, Agrawal A, Dubey GP. Neuroprotective effect of curcuminoids against inflammation-mediated dopaminergic neurodegeneration in the MPTP model of Parkinson's disease. J Neuroimmune Pharmacol 2012;7:609–18. https://doi.org/10.1007/s11481-012-9363-2.

[43] Sharma N, Nehru B. Curcumin affords neuroprotection and inhibits α-synuclein aggregation in lipopolysaccharide-induced Parkinson's disease model. Inflammopharmacology 2018;26:349–60. https://doi.org/10.1007/s10787-017-0402-8.

[44] Yavarpour-Bali H, Ghasemi-Kasman M, Pirzadeh M. Curcumin-loaded nanoparticles: a novel therapeutic strategy in treatment of central nervous system disorders. Int J Nanomedicine 2019;4449–60. https://doi.org/10.2147/IJN.S208332.

[45] Siddique YH, Naz F, Jyoti S. Effect of curcumin on lifespan, activity pattern, oxidative stress, and apoptosis in the brains of transgenic Drosophila model of Parkinson's disease. Biomed Res Int 2014;2014. https://doi.org/10.1155/2014/606928.

[46] van der Merwe C, van Dyk HC, Engelbrecht L, van der Westhuizen FH, Kinnear C, Loos B, Bardien S. Curcumin rescues a PINK1 knock down SH-SY5Y cellular model of Parkinson's disease from mitochondrial dysfunction and cell death. Mol Neurobiol 2017;54:2752–62. https://doi.org/10.1007/s12035-016-9843-0.

[47] Yana MH, Wang X, Zhu X. Mitochondrial defects and oxidative stress in Alzheimer disease and Parkinson disease. Free Radic Biol Med 2013;62:90–101. https://doi.org/10.1016/j.freeradbiomed.2012.11.014.

[48] Abrahams S, Miller HC, Lombard C, van der Westhuizen FH, Bardien S. Curcumin pretreatment may protect against mitochondrial damage in LRRK2-mutant Parkinson's disease and healthy control fibroblasts. Biochem Biophys Rep 2021;27, 101035. https://doi.org/10.1016/j.bbrep.2021.101035.

[49] Wongtrakul J, Thongtan T, Kumrapich B, Saisawang C, Ketterman AJ. Neuroprotective effects of Withania somnifera in the SH-SY5Y Parkinson cell model. Heliyon 2021;7(10), e08172. https://doi.org/10.1016/j.heliyon.2021.e08172.

[50] Prakash J, Yadav SK, Chouhan S, Singh SP. Neuroprotective role of Withania somnifera root extract in Maneb–Paraquat induced mouse model of parkinsonism. Neurochem Res 2013;38:972–80. https://doi.org/10.1007/s11064-013-1005-4.

[51] Prakash J, Chouhan S, Yadav SK, Westfall S, Rai SN, Singh SP. Withania somnifera alleviates parkinsonian phenotypes by inhibiting apoptotic pathways in dopaminergic neurons. Neurochem Res 2014;39:2527–36. https://doi.org/10.1007/s11064-014-1443-7.

[52] Dar NJ, Hamid A, Ahmad M. Pharmacologic overview of Withania somnifera, the Indian ginseng. Cell Mol Life Sci 2015;72:4445–60. https://doi.org/10.1007/s00018-015-2012-1.

[53] Mathur R, Gupta SK, Singh N, Mathur S, Kochupillai V, Velpandian T. Evaluation of the effect of Withania somnifera root extracts on cell cycle and angiogenesis. J Ethnopharmacol 2006;105(3):336–41. https://doi.org/10.1016/j.jep.2005.11.020.

[54] Giri MA, Bhalke RD, Prakash KV, Kasture SB. Evaluation of Camellia sinensis, Withania somnifera and their combination for antioxidant and antiparkinsonian effect. J Pharm Sci Res 2020;12(8):1093–9.

[55] Russo A, Borrelli F. Bacopa monniera, a reputed nootropic plant: an overview. Phytomedicine 2005;12(4):305–17. https://doi.org/10.1016/j.phymed.2003.12.008.

[56] Aguiar S, Borowski T. Neuropharmacological review of the nootropic herb Bacopa monnieri. Rejuvenation Res 2013;16(4):313–26. https://doi.org/10.1089/rej.2013.1431.

[57] Singh B, Pandey S, Yadav SK, Verma R, Singh SP, Mahdi AA. Role of ethanolic extract of Bacopa monnieri against 1-methyl-4-phenyl-1, 2, 3, 6-tetrahydropyridine (MPTP) induced mice model via inhibition of apoptotic pathways of dopaminergicneurons. Brain Res Bull 2017;135:120–8. https://doi.org/10.1016/j.brainresbull.2017.10.007.

[58] Palethorpe HM, Tomita Y, Smith E, Pei JV, Townsend AR, Price TJ, Young JP, Yool AJ, Hardingham JE. The aquaporin 1 inhibitor bacopaside II reduces endothelial cell migration and tubulogenesis and induces apoptosis. Int J Mol Sci 2018;19(3):653. https://doi.org/10.3390/ijms19030653.

[59] Fatima U, Roy S, Ahmad S, Al-Keridis LA, Alshammari N, Adnan M, Islam A, Hassan MI. Investigating neuroprotective roles of Bacopa monnieri extracts: mechanistic insights and therapeutic implications. Biomed Pharmacother 2022;153, 113469. https://doi.org/10.1016/j.biopha.2022.113469.

[60] Rai SN, Chaturvedi VK, Singh P, Singh BK, Singh MP. Mucuna pruriens in Parkinson's and in some other diseases: recent advancement and future prospective. 3Biotech 2020;10:1. https://doi.org/10.1007/s13205-020-02532-7.

[61] Mohapatra S, Ganguly P, Singh R, Katiyar CK. Estimation of levodopa in the unani drug Mucuna pruriens Bak. and its marketed formulation by high-performance thin-layer chromatographic technique. J AOAC Int 2020;103(3):678–83. https://doi.org/10.5740/jaoacint.19-0288.

[62] Yadav SK, Prakash J, Chouhan S, Westfall S, Verma M, Singh TD, Singh SP. Comparison of the neuroprotective potential of Mucuna pruriens seed extract with estrogen in 1-methyl-4-phenyl-1, 2, 3, 6-tetrahydropyridine (MPTP)-induced PD mice model. Neurochem Int 2014;65:1–3. https://doi.org/10.1016/j.neuint.2013.12.001.

[63] Yadav SK, Rai SN, Singh SP. Mucuna pruriens reduces inducible nitric oxide synthase expression in parkinsonian mice model. J Chem Neuroanat 2017;80:1. https://doi.org/10.1016/j.jchemneu.2016.11.009.

[64] Hou JP. The chemical constituents of ginseng plants. Am J Chin Med 1977;5(02):123–45. https://doi.org/10.1142/s0147291777000209.

[65] Chen XC, Zhou YC, Chen Y, Zhu YG, Fang F, Chen LM. Ginsenoside Rg1 reduces MPTP-induced substantia nigra neuron loss by suppressing oxidative stress 1. Acta Pharmacol Sin 2005;26(1):56–62. https://doi.org/10.1111/j.1745-7254.2005.00019.x.

[66] Nataraj J, Manivasagam T, Justin Thenmozhi A, Essa MM. Neuroprotective effect of asiatic acid on rotenone-induced mitochondrial dysfunction and oxidative stress-mediated apoptosis in differentiated SH-SYS5Y cells. Nutr Neurosci 2017;20(6):351–9. https://doi.org/10.1080/1028415X.2015.1135559.

[67] Hwang YP, Jeong HG. Ginsenoside Rb1 protects against 6-hydroxydopamine-induced oxidative stress by increasing heme oxygenase-1 expression through an estrogen receptor-related PI3K/Akt/Nrf2-dependent pathway in human dopaminergic cells. Toxicol Appl Pharmacol 2010;242(1):18–28. https://doi.org/10.1016/j.taap.2009.09.009.

[68] Günther B, Wagner H. Quantitative determination of triterpenes in extracts and phyto-preparations of Centella asiatica (L.) urban. Phytomedicine 1996;3(1):59–65. https://doi.org/10.1016/S0944-7113(96)80011-0.

[69] Teerapattarakan N, Benya-aphikul H, Tansawat R, Wanakhachornkrai O, Tantisira MH, Rodsiri R. Neuroprotective effect of a standardized extract of Centella asiatica ECa233 in rotenone-induced parkinsonism rats. Phytomedicine 2018;44:65–73. https://doi.org/10.1016/j.phymed.2018.04.028.

[70] Xu CL, Wang QZ, Sun LM, Li XM, Deng JM, Li LF, Zhang J, Xu R, Ma SP. Asiaticoside: attenuation of neurotoxicity induced by MPTP in a rat model of parkinsonism via maintaining redox balance and up-regulating the ratio of Bcl-2/Bax. Pharmacol Biochem Behav 2012;100(3):413–8. https://doi.org/10.1016/j.pbb.2011.09.014.

[71] Chen M, Peng L, Gong P, Zheng X, Sun T, Zhang X, Huo J. Baicalein mediates mitochondrial autophagy via miR-30b and the NIX/BNIP3 signaling pathway in Parkinson's disease. Biochem Res Int 2021;2021. https://doi.org/10.1155/2021/2319412.

[72] Torres-Ortega PV, Saludas L, Hanafy AS, Garbayo E, Blanco-Prieto MJ. Micro-and nano-technology approaches to improve Parkinson's disease therapy. J Control Release 2019;295:201–13. https://doi.org/10.1016/j.jconrel.2018.12.036.

[73] Baskin J, Jeon JE, Lewis SJ. Nanoparticles for drug delivery in Parkinson's disease. J Neurol 2021;268:1981–94. https://doi.org/10.1007/s00415-020-10291-x.

[74] Di Stefano A, Sozio P, Iannitelli A, Marianecci C, Santucci E, Carafa M. Maleic-and fumaric-diamides of (O, O-diacetyl)-L-Dopa-methylester as anti-Parkinson prodrugs in liposomal formulation. J Drug Target 2006;14(9):652–61. https://doi.org/10.1080/10611860600916636.

[75] Yang X, Zheng R, Cai Y, Liao M, Yuan W, Liu Z. Controlled-release levodopa methyl ester/benserazide-loaded nanoparticles ameliorate levodopa-induced dyskinesia in rats. Int J Nanomedicine 2012;19:2077–86. https://doi.org/10.2147/IJN.S30463.

[76] Gambaryan PY, Kondrasheva IG, Severin ES, Guseva AA, Kamensky AA. Increasing the efficiency of Parkinson's disease treatment using a poly(lactic-co-glycolic acid) (PLGA) based L-DOPA delivery system. Exp Neurobiol 2014;23(3):246. https://doi.org/10.5607/en.2014.23.3.246.

Targeting angiogenesis, inflammation, and oxidative stress in depression

Ansab Akhtar and Shubham Dwivedi
School of Health Sciences & Technology, UPES, Dehradun, Uttarakhand, India

1   Introduction

Depression or major depressive disorder (MDD) is a psychiatric disorder characterized by lower mood, loss of interest, psychomotor retardation, suicidal tendency, disrupt sleep and appetite, sexual dysfunction, and melancholia. Depression is a usual, troubling associate of medical conditions and is counted as one of the major socioeconomic burdens worldwide. Depression is estimated to affect 280 million people, i.e., 3.8% population globally with 5% and 5.7% among adults and elderlies, respectively [1]. Fig. 1 depicts the region-wise prevalence of depression. Depression is suggested to be the significant cause of disability as per the years lived with disability (YLDs) measures. Moreover, MDD is ranked second in disability adjusted life years (DALY) involving all age groups [2]. The one-half of the patients suffering from depression do not receive appropriate treatment due to poor diagnosis, or unavailability of trained professional and efficient treatment strategies which further surges disease burden. Majority of clinical antidepressants used are chemically derived through synthesis and may take around 6–8 weeks to deliver evident therapeutic responses [3].

Depression can be divided as reactive and endogenous depression. Reactive depression is the outcome of a stressor; however, endogenous depression takes place in the absence of stress. Although the reactive and endogenous depression are known to have varying etiology, the mechanistic role of presence or absence of stress in MDD is not very clear [4]. The literature suggests that patients with endogenous depression do respond positively to tricyclic antidepressants (TCAs) than selective serotonin reuptake inhibitors (SSRIs) [5]. The risk factors associated to depression are numerous, including genetic, biological, cultural, economic, and social considerations. However, the environmental impacts elicit the symptoms of depression. The associated risk factors, comorbidities and symptoms of depression are exhibited in Fig. 2.

Certain evidence suggests the involvement of reactive oxygen species (ROS) or nitrogen species (RNS) and damage by oxidative and nitrosative stress, including lipid peroxidation, damage to DNA, protein and lowered levels of antioxidants and antioxidant enzymes in the pathophysiology of depression [6].

At molecular level, depression is characterized by aberrations in six entwined pathways: [2] elevated proinflammatory cytokines levels representing inflammatory

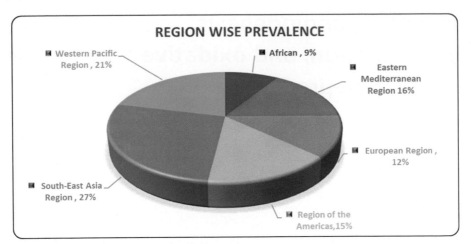

Fig. 1 Region-wise prevalence of depression. The graph clearly signifies that developed and developing countries are more prone to depression.

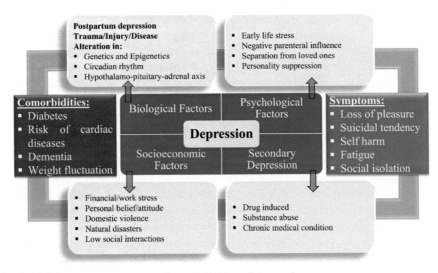

Fig. 2 Risk factors, symptoms, and comorbidities of depression.

pathways [3]. Activated cell-mediated immune system evident by high neopterin and interferon-γ production [4]. Rise in oxidative and nitrosative stress leading to impairment in mitochondrial function, and gene or protein level [5]. Decreased levels of key antioxidants, such as coenzyme Q10, zinc, vitamin E, glutathione, and glutathione peroxidase [6]. Damage to mitochondria and mitochondrial DNA and reduced activity of respiratory chain enzymes and ATP production [7]. Neuroprogression, which is the progressive process of neurodegeneration, apoptosis, and reduced neurogenesis and neuronal plasticity [7].

2 Pathogenesis with emphasis on molecular pathways involved

Depression is a psychiatric disorder, which majorly occurs due to neurochemical. The neurochemicals involved are serotonin, dopamine, norepinephrine, etc. In addition, several other factors such as oxidative stress, neuroinflammation, and angiogenesis have found to have a place in the disease pathology in last few decades. These features signify multifactorial etiology of depression. Furthermore, these factors are interrelated, and one factor is influenced and affected by another factor. This further makes the depression pathophysiology a bit more complicated, and this needs attention in terms of therapeutic target, cellular and molecular signaling pathway [8–10].

2.1 Oxidative stress and depression

Since depression causation is inclined toward neurochemical alterations rather than neurodegeneration, it was thought less likely to be influenced by oxidative stress markers. In spite of that, lipid peroxidation and protein carbonylation signifying oxidatively damaged lipid products and protein adducts in brain regions of substantia nigra, lateral habenula, cerebral cortex, and hippocampus have been reported [11].

2.1.1 Antioxidant enzymes

Glutathione peroxidase, superoxide dismutase, reduced glutathione, and catalase are the either antioxidant enzymes or antioxidant markers which are accountable for scavenging reactive oxygen species like superoxide ion and hydroxyl ion can have a role in diminishing the deteriorating features of depression. The reason behind the positive role of these antioxidant enzymes is evidence from several previous studies [12]. Targeting these antioxidant enzymes by elevating their levels or enhancing the expression could be the promising options in the treatment of depression.

2.1.2 Nrf2 and depression

Nuclear factor erythroid 2-related factor 2 (Nrf2) is one of the well-known factors in regulating oxidative stress by inhibiting or resisting oxidant agents. Nrf2 also regulated cytoprotective antioxidant genes which are dysregulated in depression [13]. In a preclinical study, Nrf2-null mice were found to be to show increased immobility time in forced swim test indicating depressive-like phenotype. Nrf2/HO-1 complex plays an essential role in reversing depressive symptoms [14]. In addition, Nrf2 and BDNF have been found to interplay corroborating its role in depression [15]. In this regard, improving the oxidative balance through Nrf2/HO-1 pathway is additional opportunity to cure depressive pathology.

2.1.3 BDNF and oxido-nitrosative stress system in depression

Oxidative stress can lead to the destruction of neuronal integrity possibly due to DNA damage or lipid peroxidation of the cell membrane [16]. Oxidative stress is well evident in depression and downregulation of glutathione, reduction of GSH-Px, vitamin C and rise in lipid peroxidation, nitric oxide (NO) are associated with stress-induced behavioral depression [17]. Master redox-sensitive factor, i.e., Nrf2, possesses a significant part in the activation of endogenous antioxidants, and a preclinical study demonstrated that low levels of BDNF significantly decrease the translocation of Nrf2, thereby inhibiting the activation of antioxidants, leading to neuronal damage [18]. In developing hypothalamic neuronal cells, BDNF was also found to downregulate molecular oxidative damage as well as apoptotic process on alcohol administration [19]. These phenomena inflicted by BDNF suggest its antioxidant-like features [20]. Beside the above-mentioned findings, a study conducted on cisplatin-mediated ROS generation infers that BDNF has the potential to decline the ROS production [21]. In vitro study of rat cortical neurons also revealed that ROS transactivates the TrkB via Zn^{2+} and Src family kinase (SFK)-dependent pathway but independent of neurotrophin pathway [22]. Still, a particular pathway has not yet been revealed which can completely create an association between TrkB and ROS.

In the brain, nitric oxide acts as a messenger paracrine molecule. Pathobiology of depression is associated with L-arginine which is responsible for the synthesis of nitric oxide by NOS enzyme, whereas BDNF and NOS inhibitors are linked with antidepressant like behavior [23]. Administration of NOS inhibitors (aminoguanidine) revealed the panicolytic effect in dorsal periaqueductal gray (dPAG) region of brain [24].

However, action of nitric oxide (NO), cGMP, is synthesized, and further, PKG mediates the downregulation of BDNF [25]. Surprisingly, hippocampal cells demonstrated the decreased release of BDNF by NO donor [26]. A clinical study conducted to investigate the link between BDNF and oxido-nitrosative stress in peritoneal dialysis (PD) patients with elevated depressive-like symptoms which suggested the lowering of BDNF levels in the presence of oxido-nitrosative stress [27]. But, in vitro study on the mammalian brain was conducted which reflected that BDNF positively enhances the stimulation of nNOS in the differentiation of neural cells which suggested a positive feedback loop in neurogenesis [28]. Therefore, keeping the above studies into an account, it is concluded that the NO-BDNF pathway has a peculiar mechanism.

2.2 Neuroinflammation and depression

Proinflammatory cytokines like interleukins (IL-6, IL-1β), tumor necrosis factor-α (TNF-α) have widely been determined to cause neuroinflammation. Furthermore, its role in depression has also been documented. The inflammatory pathways like nuclear factor kappa B (NF-κB), mitogen-activated protein kinase (MAPK) have also been found to play a role in depression pathology [29].

2.2.1   NF-κB and depression

NF-κB and its p53 subunit have the cytokine-releasing characteristics. Additionally, this nuclear factor is provoked upon low levels of mood elevating neurochemicals. This is another indication of inflammation-mediated depressive pathology. So, the partial or complete inhibition of NF-κB in central and peripheral regions can resist the stressful symptoms [30]. Furthermore, NF-κB and BDNF also have an inversely proportional relation, indicating decreased activation of NF-κB leads to increased activation of BDNF [31].

2.2.2   Microglia-BDNF interplay in depression

In CNS, microglia belong to the glia system contributing to both supportive and protective neuronal functions [32]. Neuronal injury in CNS results in the elevation of microglia numbers which can be characterized by elevated M1/M2 expression of CD 11, CD 68, and CD 86 [33]. Induction of inflammation is the prominent factor for the microglial promulgation, and thus, preclinical evidence reflects that CUMS model of depression significantly increases the microglia expression (Iba) and simultaneously reduces the BDNF levels [34]. This suggests that inflammation results in increased microglial expression which inversely regulates the levels of BDNF impacting neurogenesis and cell survival and are depicted in Fig. 3.

2.3   Angiogenesis

Angiogenesis is the process of formation of new blood vessels from the already exist. Both pro- and antiangiogenic factors play a significant role in the pathogenesis of several diseases including cancer and brain disorders like depression and Alzheimer's disease. Proangiogenic factors such as vascular endothelial growth factor, a cytokine, angiopoietins, fibroblast growth factor, BDNF, transforming growth factor, platelet-derived endothelial growth factor decline the apoptotic process and enhances the neuron formation. Depression, however, is more associated with neurochemical fluctuations rather than neurodegeneration; still, the angiogenesis process plays a crucial role in depression. The process of angiogenesis also helps in wound healing, injury recovery, and capillary formation [10].

3   Drugs targeting angiogenesis, inflammation, and oxidative stress in the management/treatment depression

3.1   Drugs targeting oxidative stress

3.1.1   Melatonin

Melatonin is a natural hormone released I response to dark from pineal gland of brain. It is responsible for inducing sleep, thereby could be the reason for the attenuating the depressive-like symptoms as sleeplessness is one of the signs of depression. Lack of

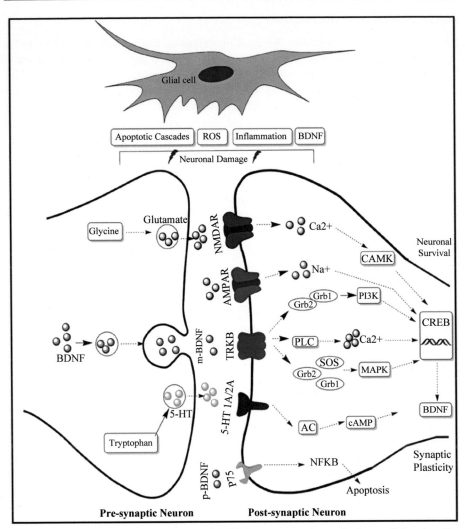

Fig. 3 Interaction of m-BDNF isoforms with TrkB receptor activates the intracellular signaling cascades. Homodimerization and phosphorylation of TrkB occur after binding of m-BDNF with TrkB and subsequently allowing the translocation into the cellular membrane lipid rafts which is rich in cholesterol and sphingolipids. This homodimerization activates the following pathways: PI3K/Akt, MAPK, and PLC-ɣ. Further, these activated pathways lead to the phosphorylation of CREB at serine 133 and activate CREB-mediated gene transcription. This phosphorylated CREB modulates the expression of the BDNF accordingly. Serotonin, glutamate, and microglia have a significant role in the modulation of CREB.

this hormone might originate from some comorbidity or any other underlying mechanism. However, synthetic melatonin can also be directly administered to reverse depression. The effect of melatonin against depression could be because its antioxidant effects [35–37].

3.1.2 Resveratrol

Sources of resveratrol include grapes, blueberries, peanuts, etc. It has already been reported in cancer and cardiovascular diseases. Due to its phenolic nature and robust antioxidant property, it could also have action against major depressive disorder [38].

3.1.3 Salidroside

Salidroside propagates anxiolytic and antidepressant activity, thereby mitigating the stressful life condition. The antiinflammatory and antioxidant actions of salidroside have widely been reported [39].

3.1.4 Pterostilbene

Pterostilbene is chemically related to resveratrol and also derived from plants. It is anticipated to have similar tendency as of resveratrol in terms of its antioxidant nature. Red grapes and red wines are the main sources of pterostilbene [40].

3.1.5 Punicalagin

It is ellagitannin in nature and found in pomegranate. This plant product possesses potent antioxidant potential. It is available as alpha and beta isomers. It is most often used in cardiovascular disorders. It has also been studied in male rats to target AMPK-PGC-1α/Nrf2 cascade pathway thereby making a way to dual nature as antiinflammatory and antioxidant actions. It also improves mitochondrial functioning suggesting a proportional relationship between oxidative stress and mitochondrial dysfunction [41].

3.2 Drugs targeting inflammation

3.2.1 Minocycline

It is a tetracycline category of antibiotic. It has recently been investigated in several inflammatory conditions. Even neuroinflammatory brain diseases like Alzheimer's disease and Parkinson's disease also has been reported to get benefitted from minocycline owing to its antiinflammatory action [42]. It also possesses additional antioxidant property [43].

3.2.2 Statins

Statins like simvastatin and rosuvastatin are usually applied as cholesterol lowering agents. However, these has also been reported to have their influence in brain diseases like depression. Statins exerts angiogenesis property thereby wresting depressive havoc. It triggers antiinflammatory action as well [44]. It also gives rise to beneficial effects in stroke by promoting synaptogenesis and neurogenesis [45].

3.2.3 Nonsteroidal antiinflammatory drugs (NSAIDs)

Nonsteroidal antiinflammatory drugs like aspirin and ibuprofen are well-known antiinflammatory agents. These are also the potential drugs to be repurposed against depression as depressive pathology can have the neuroinflammatory factor as its etiology [46,47].

3.2.4 Curcumin

Curcumin is a well-established naturally occurring antiinflammatory moiety. It is obtained from *Curcuma longa*. Peripheral inflammatory diseases like rheumatoid arthritis and osteoarthritis complications have already been reported to be alleviated. However, central inflammation happening in depression is not widely studied in regard to curcumin [48,49].

3.2.5 Berberine

Berberine has earlier been reported in type 2 diabetic patients functioning as the regulator of glucose and lipid metabolism and reversing insulin resistance at the same time. Depression, however, is not associated with insulin resistance, still association of insulin resistance with oxidative stress can make berberine a potential candidate [50]. In a preclinical study, it protects against apoptosis and oxidative damage [51]. This is also a potent NF-κB inhibitor, thus, possessing antiinflammatory property [52].

3.2.6 4-O-methylhonokiol

It is a neolignane and phenolic in nature. It decreases lipid peroxidation and in turn oxidative stress. It also inhibits NF-κB signaling its antiinflammatory action [53].

3.2.7 Sodium orthovanadate (SOV)

In some of the recently reported findings, SOV has been depicted to be a strong TrkB agonist and BDNF activator. This signifies its angiogenesis property as BDNF is a proangiogenic marker. SOV has also been revealed to show insulin-sensitizing action. Additionally, it reduces oxidative stress, mitochondrial dysfunction, and neuroinflammation. These effects collectively be beneficial in depressive patients, and depression is associated with distorted angiogenesis and markedly high oxidative stress added with neuroinflammation [54,55].

3.3 Drugs targeting angiogenesis

3.3.1 Tetrandrine

It is benzylisoquinoline alkaloid and calcium channel blocker. Its mechanism of action makes it appropriate for blood vessel disorder by improving angiogenesis. Since depression is also characterized by lower level of blood and blood vessel formation, hence angiogenesis agent like tetrandrine could be proved to be helpful [56]. It also suppresses NLRP3-mediated inflammation [57] and oxidative stress.

3.3.2   Nicotine

Earlier, nicotine was found to accelerate tumor growth and enhance cancer chances. However, in moderate dose, dosage regimen and with proper delivery system, it can promote controlled angiogenesis in brain. This might make it acceptable for depression combat. It can promote neovascularization in the damaged part of the brain, thereby reversing the damaged-induced depressive pathology. Besides, nicotine acts through nicotinic acetylcholine receptors, strengthening synaptic plasticity and memory formation.

3.3.3   Electrical stimulation

Deep brain stimulation (DBS) through electric current has broadly been investigated to induce angiogenesis and blood vessel remodeling. Since blood vessel formation and organization make the blood flow more rapid and continuous, this technique has been clinically utilized in depressed patients when other treatment options fail.

4   Strategic drug delivery approaches

Although several new compounds based on molecular pathway are identified for depression every year, very few are effective in preclinical studies with less than 1% of clinical translational. Most of these drugs are denied entry into CNS guarded by blood-brain barrier (BBB). The BBB plays an important role in homeostasis and maintains the suitable microenvironment for optimum functioning of CNS. BBB comprises the endothelial cells coating the microvessels reaching the brain, which possess the low permeability and thus forming the key interface among the blood and brain. Moreover, the presence of specialized transporters, enzymes, and receptors further controls the entrance of undesirable cellular and molecular components across this layer [58]. Although the BBB is significant for maintaining the CNS microenvironment, but is also accountable for failure of multiple approaches in drug discovery. The existing strategies in treatment of depression are limited due to incapability of crossing BBB. Moreover, available drugs either have low bioavailability or the adverse effects associated are high. BBB prohibits CNS entry of all large molecule therapies such as viral vectors for gene delivery, antibodies, peptides, or recombinant proteins and 98% of small molecule [59].

The drug delivery systems are meant to deliver and maintain therapeutic drug concentrations to defined biological site. The traditional forms of delivery systems like pills or injections partially fail to achieve the expected effect due to poor bioavailability, toxicity associated with medicine and dose-dependent adverse effects [60]. Modern strategies in drug delivery may offer chemical stability to active ingredients, enhance pharmacological activities, and restrict the adverse effects [61].

In case of neurological disorder such as depression, it becomes important that the drug maintains the optimal concentration in CNS for sufficient time to achieve desirable therapeutic effect. Nevertheless, the drug concentration in other organs/tissue should remain insignificant to avoid adverse effects. To accomplish the optimum response, nanoformulations can be utilized as an opportunity. Nanocarriers or drug delivery strategies have abilities to deliver the drug in targeted and specific site, hence increasing the efficacy. Nanoparticles may enhance the bioavailability and sustained delivery of antidepressants, subsequently reducing the frequency and concentration of drug leading to less side effects in depression patients. The smaller size ($\leq$100 nm) of nanomedicine may penetrate the BBB with ease, thus increasing the bioavailability in brain. Furthermore, targeting specific pharmacological receptors with nanoparticles complex boosts therapeutic efficacy [62]. The modern-day drug delivery strategies for crossing BBB have been discussed with pros and cons.

4.1 Oral drug delivery system

The oral route being the most preferable delivery system for repeated dosing with advantages such as easy administration, good patient compliances, and cost-effective. However, the environment of gastrointestinal tract (GIT) may hamper the bioavailability and stability of orally administered drugs. The efficacy of neuroeffective drug in GIT is majorly influenced by enzymatic activities, pH, hepatic first pass metabolism, and restricted ability to cross BBB [63]. The factors associated with GIT can be improved with slight changes in formulation; however, to enhance CNS availability of the drug, inclusion of biopolymers can be helpful. The biopolymers like polysaccharides are reliable tool for antidepressant drug delivery due to its advantages such as small size, biocompatibility biodegradability, sustained release profiles, and limited toxicity [64]. Apart from biopolymers, use of chitosan nanoparticles and PEGylated oral nanoparticles is also suggested. The properties of chitosan such as no/little toxicity, biodegradability, biocompatibility, and mucoadhesive nature makes them a vehicle choice for drug delivery in neurological treatments [65]. PEGylated polymeric nanoparticles permeate the brain due to covalent bonding between PEG and polymer which blocks the preleaking of the active ingredient from the PEGylated polymeric nanoparticles [66].

4.2 Parenteral drug delivery system

The drug delivery through parenteral route is beneficial in improving bioavailability, avoid first pass metabolism, and unfailing doses; thus, boosting the efficiency of antidepressants. Few studies demonstrate formulation and characterization of chitosan-based nanoparticles loaded with sertraline for the depression remedy. The reports suggest improved half-life and the entrapment rate of drug with four times more plasma bioavailability of nanoparticles loaded. Another example is parenterally administered L-tyrosine-loaded nanoparticles which has been indicated to improves the efficacy of L-tyrosine in rat model of depression [67].

4.3 Nasal drug delivery systems

The intranasal drug delivery route is a noninvasive technique for direct transport of drugs to brain which is not possible with oral route. The drug delivery to brain occurs through trigeminal nerves and olfactory pathway which are safe and efficient. In nasal cavity, trigeminal nerve control respiratory zone and sensation with the help of three branches which includes the ophthalmic, maxillary, and mandibular nerve. However, ophthalmic and maxillary nerves help in transmission of signals from nose to brain. Thus, these two branches can serve as the suitable target for brain drug delivery [68,69]. The composition of olfactory pathway includes the olfactory epithelium, lamina propria, and olfactory bulb. The distribution of olfactory bulb to various brain regions makes it a suitable target for quick and efficient brain delivery [70]. The intranasal administration of drug not only delivers the drug to the brain but also to cerebrospinal fluid, interstitial spaces of brain, and perivascular spaces. The excellent bioavailability, patient acceptance, quick maintenance of therapeutic levels with avoiding gastrointestinal degradation of drug advocates the utility of intranasal drug delivery in neurological disorders such as depression [71]. However, the intranasal drug delivery should be investigated for physiochemical properties, nasal enzymatic degradation, nasal membrane permeability of drug, viscosity, and pharmacokinetic parameters of formulation along with environmental pH, mucociliary clearance in nasal cavity [71].

5 Conclusion and future perspectives

Depression is one of the leading causes of disability globally and is a major contributes to the socioeconomic burdens. Although the lots of antidepressant drugs have been approved by the Food and Drug Administration (FDA) however, 10%–30% of the patients may not respond to the standard antidepressant therapies [72]. The poor or no response is often linked with loss of function, compromised quality of life, suicidal tendency, self-harm behavior, and loss of pleasure. The reasons associated with failure of therapy may vary however, can be overcome by strategies such as optimization, switching, combination, augmentation, and somatic therapies [72]. The other cause of antidepressant failure includes multifactorial pathogenesis of depression and limitations of drug delivery approaches. The current antidepressant therapies mostly work through monoaminergic pathways of depression. The specific mechanism of existing therapeutics includes monoamine oxidase inhibition and reuptakes inhibition of serotonin and noradrenaline.

Despite decades of research and well-established concepts on inflammatory and oxidative/nitrosative stress in depression, the discovery of drugs targeting these pathways is not available. On the other hand, crossing BBB has been challenging for large numbers of drugs which have been proven efficacious in preclinical studies. The strategies to target delivery in brain has always been a scorching topic of research till date.

The advancement in diagnostic approaches and neuroimaging has helped in understanding the complex pathophysiology of depression. The scientific and clinical studies have revealed that depression is not just distortion of monoaminergic systems. The other key factors governing the etiology of depression includes genetic vulnerability, hypothalamo-pituitary-adrenal (HPA) axis dysregulation, abnormalities in immune and biochemical systems [73]. Recently, immunological mechanisms such as inflammatory and oxidative stress have gained lot of attention in drug discovery for depression [3,74,75]. Ample of studies report the elevated inflammatory and proinflammatory mediators along with oxidative stress associated with MDD. To directly target inflammation, use of NSAIDs such as aspirin and celecoxib is suggested to be common approach. Moreover, the natural products/drugs, such as vitamin D, omega-3 fatty acid which are reported to possess antiinflammatory properties, can be thought of being the possible therapies for depression. Nevertheless, the reports confirming the mechanistic antiinflammatory or antioxidant effect of natural/synthetic compounds are limited and inconclusive with mixed success and failure anecdotes. Thus, these therapies are mostly used as an add-on therapy to available antidepressants [76]. However, the monotherapies with antiinflammatory drugs were unsuccessful to elevate the symptoms of depression [77].

Another factor influencing the success rate of therapies for neurological disorders includes poor drug delivery approaches to brain. BBB acts a potential barrier for the drug to reach brain. Interesting approaches using nanobased drug delivery platforms have been explored for targeted delivery of antidepressant drugs. This chapter summarizes various delivery systems, routes, and nanomaterials which can be helpful in improving the efficacy and lowering the adverse effect or toxicity of the drugs in consideration. The mentioned necessities can be achieved by focusing on properties such as development of sustained release formulations, improved BBB permeability, target-based delivery, improved bioavailability, and high specificity to target organ.

Altogether, the strategies toward discovery of novel drug molecules against depression as well as modulation of formulation/drug delivery approaches among antidepressant therapy are the much-needed areas to overcome the shortcomings for therapeutic approaches against depression. However, ample scientific literature suggests the significance of proinflammatory mediators and oxidative stress in depression. Moreover, the studies also advocate the increased efficacy of antidepressant therapy with novel drug delivery approaches. Unfortunately, neither antiinflammatory/antioxidant drug nor nanobased formulations has been approved to date by USFDA. Thus, it is essential to understand the reasons for the low translational rate of both drug discovery and delivery approaches against MDD, and clinical translation of aforementioned strategies should be taken into future consideration.

Acknowledgement

Authors would like to thank and acknowledge University of Petroleum and Energy Studies for providing facilities and opportunities to write this book chapter.

References

[1] World Health organization. Depressive disorder (depression) [Fact sheet]; 2021. https://www.atsdr.cdc.gov/docs/limitingenvironmentalexposures_factsheet-508.pdf.

[2] Planchez B, Surget A, Belzung C. Animal models of major depression: drawbacks and challenges. J Neural Transm 2019;126(11):1383–408. Springer-Verlag Wien.

[3] Sulakhiya K, et al. Honokiol abrogates lipopolysaccharide-induced depressive like behavior by impeding neuroinflammation and oxido-nitrosative stress in mice. Eur J Pharmacol 2015;744:124–31.

[4] Keers R, Uher R. Gene-environment interaction in major depression and antidepressant treatment response. Curr Psychiatry Rep 2012;14(2):129–37.

[5] Malki K, et al. The endogenous and reactive depression subtypes revisited: integrative animal and human studies implicate multiple distinct molecular mechanisms underlying major depressive disorder. BMC Med 2014;12:1.

[6] Maes M, Galecki P, Chang YS, Berk M. A review on the oxidative and nitrosative stress (O&NS) pathways in major depression and their possible contribution to the (neuro) degenerative processes in that illness. Prog Neuropsychopharmacol Biol Psychiat 2011;35(3):676–92.

[7] Maes M, Fišar Z, Medina M, Scapagnini G, Nowak G, Berk M. New drug targets in depression: inflammatory, cell-mediate immune, oxidative and nitrosative stress, mitochondrial, antioxidant, and neuroprogressive pathways and new drug candidates-Nrf2 activators and GSK-3 inhibitors. Inflammopharmacology 2012;20(3):127–50.

[8] Mohammadi AB, Torbati M, Farajdokht F, Sadigh-Eteghad S, Fazljou SMB, Vatandoust SM, et al. Sericin alleviates restraint stress induced depressive-and anxiety-like behaviors via modulation of oxidative stress, neuroinflammation and apoptosis in the prefrontal cortex and hippocampus. Brain Res 2019;1715:47–56.

[9] Salehpour F, Farajdokht F, Cassano P, Sadigh-Eteghad S, Erfani M, Hamblin MR, et al. Near-infrared photobiomodulation combined with coenzyme Q10 for depression in a mouse model of restraint stress: reduction in oxidative stress, neuroinflammation, and apoptosis. Brain Res Bull 2019;144:213–22.

[10] Boldrini M, Hen R, Underwood MD, Rosoklija GB, Dwork AJ, Mann JJ, et al. Hippocampal angiogenesis and progenitor cell proliferation are increased with antidepressant use in major depression. Biol Psychiatry 2012;72:562–71.

[11] Karunasinghe RN, Lipski J. Oxygen and glucose deprivation (OGD)-induced spreading depression in the substantia nigra. Brain Res 2013;1527:209–21.

[12] Katrenčíková B, Vaváková M, Paduchová Z, Nagyová Z, Garaiova I, Muchová J, et al. Oxidative stress markers and antioxidant enzymes in children and adolescents with depressive disorder and impact of omega-3 fatty acids in randomised clinical trial. Antioxidants 2021;10:1256.

[13] Subba R, Ahmad MH, Ghosh B, Mondal AC. Targeting NRF2 in type 2 diabetes mellitus and depression: efficacy of natural and synthetic compounds. Eur J Pharmacol 2022; 174993.

[14] Robledinos-Antón N, Fernández-Ginés R, Manda G, Cuadrado A. Activators and inhibitors of NRF2: a review of their potential for clinical development. Oxid Med Cell Longev 2019;2019, 9372182.

[15] Mendez-David I, Tritschler L, El Ali Z, Damiens M-H, Pallardy M, David DJ, et al. Nrf2-signaling and BDNF: a new target for the antidepressant-like activity of chronic fluoxetine treatment in a mouse model of anxiety/depression. Neurosci Lett 2015;597:121–6.

[16] Maria Michel T, Pulschen D, Thome J. The role of oxidative stress in depressive disorders. Curr Pharm Des 2012;18:5890–9.

[17] Eren I, Nazıroğlu M, Demirdaş A, Çelik Ö, Uğuz AC, Altunbaşak A, et al. Venlafaxine modulates depression-induced oxidative stress in brain and medulla of rat. Neurochem Res 2007;32:497–505.

[18] Bouvier E, Brouillard F, Molet J, Claverie D, Cabungcal JH, Cresto N, et al. Nrf2-dependent persistent oxidative stress results in stress-induced vulnerability to depression. Mol Psychiatry 2017;22:1701–13.

[19] Frühauf-Perez PK, Temp FR, Pillat MM, Signor C, Wendel AL, Ulrich H, et al. Spermine protects from LPS-induced memory deficit via BDNF and TrkB activation. Neurobiol Learn Mem 2018;149:135–43.

[20] Mehrpouya S, Nahavandi A, Khojasteh F, Soleimani M, Ahmadi M, Barati M. Iron administration prevents BDNF decrease and depressive-like behavior following chronic stress. Brain Res 2015;1596:79–87.

[21] Carroll BJ. Monoamine precursors in depression: clinical trials and theoretical implications. Comm Contemp Psychiat 1971;1:87–94.

[22] Huang YZ, McNamara JO. Neuroprotective effects of reactive oxygen species mediated by BDNF-independent activation of TrkB. J Neurosci 2012;32:15521–32.

[23] Donato F, de Gomes MG, Goes ATR, Filho CB, Del Fabbro L, Antunes MS, et al. Hesperidin exerts antidepressant-like effects in acute and chronic treatments in mice: possible role of l-arginine-NO-cGMP pathway and BDNF levels. Brain Res Bull 2014;104:19–26.

[24] Ribeiro DE, Casarotto PC, Spiacci Jr A, Fernandes GG, Pinheiro LC, Tanus-Santos JE, et al. Activation of the TRKB receptor mediates the panicolytic-like effect of the NOS inhibitor aminoguanidine. Prog Neuropsychopharmacol Biol Psychiat 2019;93:232–9.

[25] Canossa M, Giordano E, Cappello S, Guarnieri C, Ferri S. Nitric oxide down-regulates brain-derived neurotrophic factor secretion in cultured hippocampal neurons. Proc Natl Acad Sci U S A 2002;99:3282–7.

[26] Canossa M, Giordano E, Cappello S, Guarnieri C, Ferri S. Nitric oxide down-regulates brain-derived neurotrophic factor secretion in cultured hippocampal neurons. Proc Natl Acad Sci 2002;99:3282–7.

[27] Eraldemir FC, Ozsoy D, Bek S, Kir H, Dervisoglu E. The relationship between brain-derived neurotrophic factor levels, oxidative and nitrosative stress and depressive symptoms: a study on peritoneal dialysis. Ren Fail 2015;37:722–6.

[28] Cheng A, Wang S, Cai J, Rao MS, Mattson MP. Nitric oxide acts in a positive feedback loop with BDNF to regulate neural progenitor cell proliferation and differentiation in the mammalian brain. Dev Biol 2003;258:319–33.

[29] Kim Y-K, Na K-S, Myint A-M, Leonard BE. The role of pro-inflammatory cytokines in neuroinflammation, neurogenesis and the neuroendocrine system in major depression. Prog Neuropsychopharmacol Biol Psychiat 2016;64:277–84.

[30] Xu X, Piao HN, Aosai F, Zeng XY, Cheng JH, Cui YX, et al. Arctigenin protects against depression by inhibiting microglial activation and neuroinflammation via HMGB1/TLR4/NF-κB and TNF-α/TNFR1/NF-κB pathways. Br J Pharmacol 2020;177:5224–45.

[31] Caviedes A, Lafourcade C, Soto C, Wyneken U. BDNF/NF-κB signaling in the neurobiology of depression. Curr Pharm Des 2017;23:3154–63.

[32] Ginhoux F, Lim S, Hoeffel G, Low D, Huber T. Origin and differentiation of microglia. Front Cell Neurosci 2013;7:45.

[33] Zhu C, Xu J, Lin Y, Ju P, Duan D, Luo Y, et al. Loss of microglia and impaired brain-neurotrophic factor signaling pathway in a comorbid model of chronic pain and depression. Front Psych 2018;9.

[34] Liu L-L, Li J-M, Su W-J, Wang B, Jiang C-L. Sex differences in depressive-like behaviour may relate to imbalance of microglia activation in the hippocampus. Brain Behav Immun 2019;81:188–97.

[35] Adamczyk-Sowa M, Pierzchala K, Sowa P, Mucha S, Sadowska-Bartosz I, Adamczyk J, et al. Melatonin acts as antioxidant and improves sleep in MS patients. Neurochem Res 2014;39:1585–93.

[36] Hickie IB, Rogers NL. Novel melatonin-based therapies: potential advances in the treatment of major depression. Lancet 2011;378:621–31.

[37] Cardinali DP, Srinivasan V, Brzezinski A, Brown GM. Melatonin and its analogs in insomnia and depression. J Pineal Res 2012;52:365–75.

[38] Moore A, Beidler J, Hong MY. Resveratrol and depression in animal models: a systematic review of the biological mechanisms. Molecules 2018;23:2197.

[39] Xiong Y, Wang Y, Xiong Y, Teng L. Protective effect of Salidroside on hypoxia-related liver oxidative stress and inflammation via Nrf2 and JAK2/STAT3 signaling pathways. Food Sci Nutr 2021;9:5060–9.

[40] McCormack D, McFadden D. A review of pterostilbene antioxidant activity and disease modification. Oxid Med Cell Longev 2013;2013.

[41] Cao K, Xu J, Pu W, Dong Z, Sun L, Zang W, et al. Punicalagin, an active component in pomegranate, ameliorates cardiac mitochondrial impairment in obese rats via AMPK activation. Sci Rep 2015;5:1–12.

[42] Leite LM, Carvalho AGG, Tavares Ferreira PL, Pessoa IX, Gonçalves DO, de Araújo LA, et al. Anti-inflammatory properties of doxycycline and minocycline in experimental models: an in vivo and in vitro comparative study. Inflammopharmacology 2011;19:99–110.

[43] Pabreja K, Dua K, Sharma S, Padi SS, Kulkarni SK. Minocycline attenuates the development of diabetic neuropathic pain: possible anti-inflammatory and anti-oxidant mechanisms. Eur J Pharmacol 2011;661:15–21.

[44] Dulak J, Józkowicz A. Anti-angiogenic and anti-inflammatory effects of statins: relevance to anti-cancer therapy. Curr Cancer Drug Targets 2005;5:579–94.

[45] Chen J, Zhang ZG, Li Y, Wang Y, Wang L, Jiang H, et al. Statins induce angiogenesis, neurogenesis, and synaptogenesis after stroke. Ann Neurol 2003;53:743–51.

[46] Day RO, Graham GG. Republished research: non-steroidal anti-inflammatory drugs (NSAIDs). Br J Sports Med 2013;47:1127.

[47] Eyre HA, Air T, Proctor S, Rositano S, Baune BT. A critical review of the efficacy of non-steroidal anti-inflammatory drugs in depression. Prog Neuropsychopharmacol Biol Psychiat 2015;57:11–6.

[48] Ng QX, Koh SSH, Chan HW, Ho CYX. Clinical use of curcumin in depression: a meta-analysis. J Am Med Dir Assoc 2017;18:503–8.

[49] Fusar-Poli L, Vozza L, Gabbiadini A, Vanella A, Concas I, Tinacci S, et al. Curcumin for depression: a meta-analysis. Crit Rev Food Sci Nutr 2020;60:2643–53.

[50] Yin J, Xing H, Ye J. Efficacy of berberine in patients with type 2 diabetes mellitus. Metabolism 2008;57:712–7.

[51] Seth E, Ahsan AU, Kaushal S, Mehra S, Chopra M. Berberine affords protection against oxidative stress and apoptotic damage in F1 generation of wistar rats following lactational exposure to chlorpyrifos. Pesticide Biochem Physiol 2021;179, 104977.

[52] Fernandes MA, Custodio JB, Santos MS, Moreno AJ, Vicente JA. Tetrandrine concentrations not affecting oxidative phosphorylation protect rat liver mitochondria from oxidative stress. Mitochondrion 2006;6:176–85.

[53] Uddin MS, Hasana S, Ahmad J, Hossain M, Rahman M, Behl T, et al. Anti-neuroinflammatory potential of polyphenols by inhibiting NF-κB to halt Alzheimer's disease. Curr Pharm Des 2021;27:402–14.

[54] Joshi A, Akhtar A, Saroj P, Kuhad A, Sah SP. Antidepressant-like effect of sodium orthovanadate in a mouse model of chronic unpredictable mild stress. Eur J Pharmacol 2022;919, 174798.

[55] Akhtar A, Bishnoi M, Sah SP. Sodium orthovanadate improves learning and memory in intracerebroventricular-streptozotocin rat model of Alzheimer's disease through modulation of brain insulin resistance induced tau pathology. Brain Res Bull 2020;164:83–97.

[56] Tang SW, Tang WH, Leonard BE. Multitarget botanical pharmacotherapy in major depression: a toxic brain hypothesis. Int Clin Psychopharmacol 2017;32:299–308.

[57] Wang J, Guo M, Ma R, Wu M, Zhang Y. Tetrandrine alleviates cerebral ischemia/reperfusion injury by suppressing NLRP3 inflammasome activation via Sirt-1. PeerJ 2020;8, e9042.

[58] Pandit R, Chen L, Götz J. The blood-brain barrier: physiology and strategies for drug delivery. Adv Drug Deliv Rev 2020;165–6. Elsevier B.V., pp. 1–14.

[59] Pardridge WM. The blood-brain barrier: bottleneck in brain drug development. NeuroRX 2005;2:3–14.

[60] Zorkina Y, et al. Nano carrier drug delivery systems for the treatment of neuropsychiatric disorders: advantages and limitations. Molecules 2020;25(22).

[61] Li C, et al. Recent progress in drug delivery. Acta Pharmaceut Sin B 2019;9(6):1145–62. Chinese Academy of Medical Sciences.

[62] Mulvihill JJE, Cunnane EM, Ross AM, Duskey JT, Tosi G, Grabrucker A. Drug delivery across the blood-brain barrier: recent advances in the use of nanocarriers. Nanomedicine (Lond) 2020;15(2):205–14. [Online]. Available: https://hdl.handle.net/10344/8570.

[63] Lin CH, Chen CH, Lin ZC, Fang JY. Recent advances in oral delivery of drugs and bioactive natural products using solid lipid nanoparticles as the carriers. J Food Drug Anal 2017;25(2):219–34. Elsevier Taiwan LLC.

[64] Mutingwende FP, Kondiah PPD, Ubanako P, Marimuthu T, Choonara YE. Advances in nano-enabled platforms for the treatment of depression. Polymers 2021;13(9).

[65] Rajput R, Kumar S, Nag P, Singh M. Fabrication and characterization of chitosan based polymeric escitalopram nanoparticles. J Appl Pharm Sci 2016;6(7):171–7.

[66] Lahkar S, Das MK. Surface modified polymeric nanoparticles for brain targeted drug delivery. Curr Trends Biotechnol Pharm 2013;7(4):914–31. https://www.researchgate.net/publication/258154290.

[67] Mutingwende FP, Kondiah PPD, Ubanako P, Marimuthu T, Choonara YE. Advances in nano-enabled platforms for the treatment of depression. Polymers 2021;13(9).

[68] Ruigrok MJR, de Lange ECM. Emerging insights for translational pharmacokinetic and pharmacokinetic-pharmacodynamic studies: towards prediction of nose-to-brain transport in humans. AAPS J 2015;17(3):493–505.

[69] Johnson NJ, Hanson LR, Frey WH. Trigeminal pathways deliver a low molecular weight drug from the nose to the brain and orofacial structures. Mol Pharmacol 2010;7(3):884–93.

[70] Khan AR, Liu M, Khan MW, Zhai G. Progress in brain targeting drug delivery system by nasal route. J Control Release 2017;268:364–89. Elsevier B.V.

[71] Xu J, Tao J, Wang J. Design and application in delivery system of intranasal antidepressants. Front Bioeng Biotechnol 2020;8. Frontiers Media S.A.

[72] Al-Harbi KS. Treatment-resistant depression: therapeutic trends, challenges, and future directions. Patient Prefer Adherence 2012;6:369–88.

[73] Otte C, et al. Major depressive disorder. Nat Rev Dis Primers 2016;2:16065.

[74] Jangra A, et al. Honokiol abrogates chronic restraint stress-induced cognitive impairment and depressive-like behaviour by blocking endoplasmic reticulum stress in the hippocampus of mice. Eur J Pharmacol 2016;770.

[75] Jangra A, et al. Sodium phenylbutyrate and edaravone abrogate chronic restraint stress-induced behavioral deficits: implication of oxido-nitrosative, endoplasmic reticulum stress cascade, and neuroinflammation. Cell Mol Neurobiol 2017;37:1.

[76] Rabiei Z, Rabie S. A review on antidepressant effect of medicinal plants. Bangladesh J Pharmacol 2017;12(1):1–11. Bangladesh Pharmacological Society.

[77] Eyre HA, Baune BT. Anti-inflammatory intervention in depression. JAMA Psychiatry 2015;72(5):511. American Medical Association.

Role of inflammation, angiogenesis and oxidative stress in developing epilepsy

Pranay Wal[a], Himangi Vig[a], Sulaiman Mohammed Alnaseer[b], Mohd Masih Uzzaman Khan[c], Arun Kumar Mishra[d], and Tapan Behl[e]
[a]PSIT-Pranveer Singh Institute of Technology (Pharmacy), Kanpur, Uttar Pradesh, India, [b]Department of Pharmacology and Toxicology, Unaizah College of Pharmacy, Qassim University, Qassim, Saudi Arabia, [c]Department of Pharmaceutical Chemistry and Pharmacognosy, Unaizah College of Pharmacy, Unaizah, Saudi Arabia, [d]Pharmacy Academy, IFTM University, Moradabad, Uttar Pradesh, India, [e]Amity School of Pharmaceutical Sciences, Amity University, Mohali, Punjab, India

1 Introduction

Individuals of different socioeconomic levels and ages, and locations are prone to epilepsy, which is among the most prevalent neurological disorders. The neurological condition of epilepsy is distinguished by a continuous tendency to cause seizures and also the neurobiological, intellectual, psychological, and social implications of seizure recurrence [1]. Among the most widespread neurological conditions, it can afflict people at any stage of life. In developed nations, 34% of the population will experience epilepsy at some point in their lives. In nations with few resources, the probability is greater [2]. Recurrent epilepsy was among the four most severely disabled of 220 medical illnesses in the Global Burden of Disease 2010 survey [3]. It is characterized by several indicators, including abrupt paroxysmal bouts of neuronal electrical discharge which often cause multiple kinds of clinically evident convulsions [4]. About 65 million individuals globally are thought to have epilepsy, making it the fourth most common neurological illness [5]. Epilepsy is a universal condition that affects people of any age and gender. Due to the greater incidence of strokes, neurological disorders, and tumors, males are more inclined to develop epilepsy than women, while the disease generally peaks in elderly people [6]. Repeated uncontrolled seizures of mega-factored etiology are induced by the persistent disease's abnormal dynamism of neuronal systems, which produces abnormally synchronically discharged neurons [7]. Several of the typical symptoms include substantial neuropathological abnormalities in the hippocampus fluctuations in mental state, poor motor coordination, multiple medical conditions, prejudice, and additional neural processes [8]. The majority of seizures (whether focal or generalized) begin suddenly, and their repercussions (such as automatisms, behavioral arrest, hyperactivity, autonomic disturbance, decreased mental activity, and lack of awareness) are frequently linked to

Targeting Angiogenesis, Inflammation and Oxidative Stress in Chronic Diseases. https://doi.org/10.1016/B978-0-443-13587-3.00014-X

CNS injuries and psychological impairments. Alongside age and geography, there are several warning signs for the onset of epilepsy. The onset of epilepsy is most frequently linked to inherited, developmental, and familial disorders in childhood, teenagers, and the beginning of adulthood. Epilepsy can be brought on by head injuries, CNS diseases, and tumors, which can happen to anyone at any stage. Medical investigations have shown that one of the main causes of epileptic attacks, whether brought on by a collision or neural disorders, is the collapse of the blood-brain barrier (BBB) [9]. Additionally, it has been suggested that increased micro-pinocytosis, expansion of the basal membrane, a reduction in the number of mitochondria in endothelial cells as a whole, and the existence of unusual tightly bound junctions are additional associated factors that can be identified by ultrastructural analyses of epileptic cells in humans [10]. The idea that events that trigger cause epilepsy and a wide range of cognitive deficits are supported by several wounds, lesions, or procedures which unintentionally damage the BBB [11]. The fact that hereditary mitochondrial abnormalities like myoclonic epilepsy frequently manifest provides indicative of the significance of mitochondrial in epilepsy. Reactive oxygen species (ROS) are produced mostly by mitochondria [12]. The oxidative stress caused by the generation of ROS impairs the performance of mitochondria by damaging nuclear DNA, mitochondrial barriers, and respiratory chain proteins, including mitochondrial DNA (mtDNA) [13]. It is becoming clear that oxidative damage brought on by high levels of ROS in the mitochondria plays a crucial role in the development of epileptogenesis and seizure onset. There is growing research that suggests inflammation may both contribute to and result in epilepsy. In surgically eliminated neural tissue from individuals who had refractory epilepsies, especially temporal lobe epilepsy or cortical dysplasia-related epilepsy, some inflammation-related mediators have been identified found [14]. The discovery that neurological inflammation is possible in epilepsies not typically associated with immunological disorders raised the likelihood that a few epilepsies may be predisposed to chronic inflammation regardless of the initial injury or trigger, as opposed to only having been a complication of a particular underlying inflammation etiology [15]. Despite the abundance of antiseizure drugs, many people still have seizures that recur and fail to improve with medication. In this situation, nanomedicine represents an intriguing strategy to improve the bioavailability of antiepileptic drugs (AEDs) in the CNS. The selection and accessibility of the deepest brain area and safeguarding of the medication's molecular makeup result in improved drug effectiveness when many chemical components are encapsulated concurrently in some regulated systems for drug delivery [16]. Other methods have been used for successfully managing tough-to-manage epileptic syndromes, including immunotherapy, which primarily involves the administration of immunoglobulin, steroids, and adrenocorticotropic hormone [17]. Antioxidant therapy may be an important tactic in the fight against the neurological disorder brought on by epilepsy. The maintenance of oxidative imbalance and regulation in the brain is thus essential, and antioxidants play a crucial role in this process. The significance of antioxidants in clinical as well as preclinical forms of epilepsy has attracted tremendous attention over the past two decades [18,19]. Here, we will discuss in depth how epilepsy is caused by microvascular damage, which includes inflammation and oxidative stress, as well as current developments in drug delivery systems for treating it.

2 Epilepsy

The word used in healthcare for repeated, spontaneous seizures is "epilepsy". Every one of the various manifestations of epilepsy indicates underlying neurological conditions [20].

Approximately 75% of incidents of epilepsy begin in the toddler stage because the growing nervous system is more susceptible to seizures. A range of clinical traits collectively referred to as "epilepsy syndrome" include identical seizure kinds, age of start, triggering events, inheritance, and prognosis [21]. A medical diagnosis of epilepsy ought to be created if each of the conditions that follow is met: a minimum of two unprovoked (or reflex) epileptic seizures that happen more than 24 h apart; a single unprovoked (or reflex) seizure as well as a likelihood of subsequent seizures that is a minimum of 60% higher than the average risk of recurring after a pair unprovoked seizure for another 10 years [22]. A minimum of one epileptic seizure must occur for someone to be diagnosed with epilepsy, according to the definition approved by the International League Against Epilepsy (ILAE) in 2014 [23]. A paroxysmal change in neurological functioning known as a "seizure" is brought on by an exaggerated, aggressive-synchronous fire of neurons in the brain [24]. Epileptic seizures can include the momentary absence of consciousness, localized posturing or jerking of only one limb, ocular or additional sensory indicators, bodily trembling disorientation, and difficulties reacting, according to what regions of the brain are implicated [25]. According to where there is any motor interaction and if consciousness persists or is diminished (focal aware or focal impaired awareness seizures), seizures are subsequently categorized. Myoclonic jerks, automatism, or tonic-clonic action are examples of particular indications that can be utilized to additionally categorize the seizure [26]. Epilepsies have been categorized in a variety of ways based on the form of seizures; the most prevalent varieties are listed in Table 1 in the following section [27,28].

2.1 Prevalence and epidemiology

According to the regional variation of potential risk and pathophysiological variables, the percentage of seizures at assessment, and whether active epilepsy (present prevalence) or cases in relapse (lifetime prevalence) are taken into account, the incidence of epilepsy varies greatly among nations [29]. Epilepsy rates and prevalence differ significantly across rich- and poor-income countries, with poor-income nations being significantly more impacted. This disparity is caused by many variables, including a greater likelihood of head injury and a shortage of appropriate care. Progressive discovery concerning the genealogy of epilepsy, it will be possible to assess whether variations in genes have any impact on this variation [30]. According to data from the World Health Organization (WHO) for 2019, epilepsy is to blame for over 13 million disability-adjusted life years (DALYs), additionally above 0.5% of the global burden of disease (GBD). Individuals of every generation, gender, race, socioeconomic class, and geographic place are impacted. The lifetime prevalence of epilepsy is 7.6 per 1000 people. Depending on age, it exhibits a bimodal distribution, featuring maxima in the

Table 1 Epilepsies categorization based on the sort of seizures.

Types	Categories	Characteristics
Generalized seizures (origin involving both hemispheres of the brain)	Grand mal epilepsy/tonic-clonic seizures	• The cortical discharge starts in the local brain area but spreads rapidly to both the hemispheres • Lasts 1–2 min • Tongue biting • Respiratory depression • Involuntary defecation/urination
	Petit mal epilepsy/absence seizures	• Most prevalent in kids • Lasts 1–2 min • Loss of consciousness • No convulsions
	Atonic seizures	• Loss of consciousness • Relaxation of all muscles • Patient may fall
Partial seizures (unilateral localized brain origin that could spread to a limited or large area)	Simple partial	• lasts 30–90s • localized sensory disturbances such as pin pricks • Visual/auditory hallucinations • Patient is conscious and attentive of the attack • Affects mood and behavior
	Complex partial	• Attacks of bizarre and confused behavior • state of sleep and aimless motion • Emotional changes • Lasts 1–2 min • Impairment of consciousness • Patient has no recollection of the attack
Status epilepticus	—	• Repetitive seizures with no recovery in between • Life-threatening • Medical emergency
Myoclonic seizures	—	• Repetitive muscle contractions involving one body part or whole body • May coexist with other seizures

youngest people and people over 60. When compared to high-income countries (HIC), which have a lower prevalence of epilepsy (48.9), low income has a higher rate (139 per 100,000 person-years) [31]. Epilepsy affects between 50.4 and 81.7 per 100,000 people each year [32]. Nevertheless, 25% of individuals who build epilepsy during childhood are still included in the ongoing fatalities at the age of 45 [33]. According to several studies, the usual prevalence of epilepsy across Europe and the US is between 24 and 82/100,000 people per year and 44–162/100,000 people per year, correspondingly [34]. 2.2 and 10.4 cases of epilepsy have been recorded stated per 1000 people in India's various parts [35]. Numerous epidemiological investigations conducted in India starting in 1964 have found rates of prevalence that range from 1.3 to 11.9 per 1000 people [36–38]. According to the WHO, if epilepsy is correctly identified and treated, about 70% of individuals who currently have seizures might become free of seizures [39]. Epilepsy shows a bimodal spectrum, with a single peak occurring between the aged of five and nine and a second peak occurring at the average age of 80 [40]. Epilepsy does not affect people of all ages similarly. Epilepsy incidence does not alter based on gender [41].

3   Microvascular pathogenesis involved in epilepsy

The development of the majority of diseases is a complicated and poorly understood aspect of individual physiology. Our current clinical paradigm looks for particular reasons for individual diseases, yet there is a rising understanding that the majority of diseases may be caused by underlying processes that are pathological. Several neurological disorders have a secondary impact from epilepsy in addition to being their primary pathological conditions. Vascular abnormalities are frequently linked to epileptic seizures, and in certain instances, vascular defects may result in epilepsy. Up to 30% of epileptic individuals exhibit medication resistance to several currently accessible AEDs, which indicates that epilepsy ought to be assigned to additional brain cells, like glial cells and vascular cells, in addition to neurons.

The BBB, which strictly controls the interchange of chemicals across the brain parenchyma and the flowing blood, is made up mostly of astrocytes, pericytes, and endothelial cells. According to some theories, BBB malfunction, particularly barrier loss, worsens the course of epileptic convulsions, and vice versa, epileptic episodes cause barrier leakage [42]. In this sense, vascular modifications, including seizure-induced angiogenesis, and microcirculatory abnormalities, are likely prevalent in epilepsy. Vascular endothelial growth factor (VEGF), seems to be essential. As a result, epilepsy may be a disorder marked by abnormal cortical microvasculature, wherein VEGF is essential. Brain seizures can be caused by abnormal control of glial activities. Glial abnormalities, such as glial scarring, different gliomas, and persistently agitated astrocytes and microglia, are prone to create epileptic hotspots in the cerebral cortex [43]. Enhanced stimulation of neurons and inflammatory reactions can be the processes by which glial cells induce epileptogenesis. During the past 10 years, the fundamental contributions of inflammatory mechanisms to epilepsy are being elucidated

[44]. Astrocytes are a biological component of neuroinflammation that have a particular connection to the BBB framework and can respond to impulses generated by damaged neurons or active microglia. They can make an important impact on tissue recovery, as seen in the instance of glial growth of scars that are kept to encourage axon regeneration. Yet, long-lasting chronic insults may encourage the reactivation of cellular processes that support the inflammatory characteristics of brain-resident cells, leading to an untoward reaction that could harm the central nervous system (CNS) [45].

The etiology of epilepsy includes a significant function for inflammation and oxidative damage. There is currently much proof that mitochondrial oxidative stress and epilepsy are related. While mitochondrial damage is linked to some inherited epilepsies, minimal is understood regarding how it operates in acquired epilepsies. Emerging evidence points to mitochondrial oxidative stress and malfunction as important causes of seizures as well as potential contributors to epileptogenesis.

Age is a risk factor for epilepsy, and mitochondrial oxidative stress is a major cause of aging and age-related degenerative illness, raising the possibility that mitochondrial malfunction plays a further role in the onset of seizures [46]. Fig. 1 illustrates the connections between disruption of the BBB, oxidative stress, and inflammation in the etiology of epilepsy.

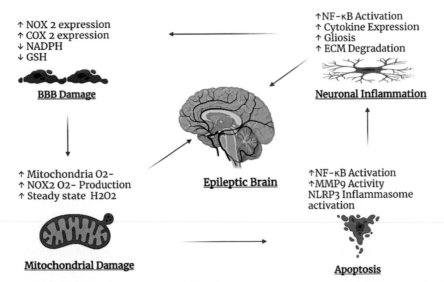

Fig. 1 Shows the associations among BBB disruption, oxidative stress, and inflammation in the etiology of epilepsy. *NOX2*: NADPH oxidase 2; *COX2*: cyclo-oxygenase 2; *NADPH*: nicotinamide adenine dinucleotide phosphate; *GSH*: glutathione; H_2O_2: hydrogen peroxide; *NF-κ B*: nuclear factor kappa B; *MMP9*: matrix metalloproteinase; *ECM*: extracellular matrix; *NLRP3*: nucleotide-binding domain, leucine-rich-containing family, pyrin domain-containing-3.

The main microvascular dysfunctions implicated in the pathophysiology of epilepsy, including angiogenesis, oxidative stress, and inflammation, are further discussed in detail.

3.1 BBB dysfunction

Pericytes, and astrocytes, basement membrane, endothelial cells make up the BBB. Endothelial cells that exist in the BBB create rigid junctions that control paracellular flux, and pericytes, which also control blood circulation and infiltration of immune cells, make up the basement membrane, which is made up of the extracellular matrix discharged by endothelial cells as well as is connected to vascular signaling. There has been extensive research on the connection involving epilepsy and BBB malfunction, particularly BBB permeability. Within the neural networks of epilepsy patients as well as in animal models, BBB disruption has been identified [47]. It had been proposed that some substances, including matrix metalloproteinase (MMP-9), cause BBB failure in epilepsy. MMP-9 has been reported to cut apart extracellular matrix and break down tight junction proteins.

Another chemical linked to BBB failure in brain tissue is VEGF [48]. In rodents, VEGF treatment caused BBB permeability. The permeability was positively correlated with elevated MMP-9 expression in ischemic brain tissue [49]. In addition, platelet-derived growth factor (PDGF) as well as its beta receptor (PDGFRb) have been linked to BBB failure. PDGFRb is produced in endothelial cells, while PDGF is released by platelets and macrophages. In epilepsy, PDGFRb amplification in endothelial cells led to the formation of a pericyte-microglia scar [50]. BBB breakdown is hypothesized to be brought on by inflammatory cytokines [51]. A BBB problem could be the cause of aberrant neuronal activity. Leucocyte infiltration into the cerebral cortex and serum protein outflow are both brought on by BBB damage. These exogenous mediators of inflammation can reduce seizure thresholds [52], which may change channel sensitivity, neurotransmitter uptake or discharge, and glia-associated modulation of extracellular surroundings [53]. A further indication that inflammation could boost neuronal excitability is by causing the BBB to collapse since cytokines are being demonstrated to do so by rupturing tight connections, stimulating endothelial inducible nitric oxide synthase (iNOS), and triggering matrix metalloproteinases [54]. The viability of endothelial cells might be aided by ROS, which also boosts P-gap expression. Additionally, ROS may cause a spike in lipid peroxidation that causes the BBB to separate, resulting in a decrease in P-gap expression or action [55].

3.2 Angiogenesis

Either biological or developmental, as well as pathologic settings can be used to investigate angiogenesis, the creation of new vessels developing from existing vessels. Multiple investigations indicate a connection between VEGF-linked BBB dysfunction and angiogenesis in the epileptic brain. In KA-treated slice cultures, the tight junction-related proteins ZO-1 and VEGF were increased, while ZO-1 was downregulated. AntiVEGF therapy tempered the rise in vascular density and branch count as well as the decline in ZO-1 expression. Additionally, the researchers discovered that the proto-oncogene tyrosine-protein kinase Src pathway, which is activated by VEGF/VEGFR2 signaling, is likely what causes angiogenesis and malfunction of the BBB [56]. The primary cause of VEGF overexpression in general is hypoxia, although it is unknown if this pathway also pertains to epilepsy. On the contrary, it was recently demonstrated that VEGF and hypoxia-inducible factor 1a (HIF-1a) are expressed together in the human brain's temporal cortex as well as hippocampus cells [57], in addition to in the coriaria lactone rat model [58]. Additionally, VEGF influences vessel permeation by momentarily making mature, functional vessels leakier after VEGF discharge [59]. In both the hippocampus of rodents and the brains of patients with epilepsy, seizures cause VEGF expression and BBB disruption [56]. In mice, the level of disrupted BBB is correlated with the frequency of seizures, and BBB damage can cause seizures in rodents.

Furthermore, astrocytes absorb plasma albumin leaks, resulting in K^+ channel reduction and impaired extracellular K^+ removal. Increased K^+ levels depolarize neurons, causing them to become more excitable [60]. Additionally, disruption of the BBB is made worse, and epileptogenic inflammatory processes are sustained by astrocytes and microglia that have been triggered in reaction to BBB leaking [61], Fig. 2. These cells discharge cytokines and VEGF. AntiVEGF antibodies can reduce the effects of seizures on VEGFR2 signaling and angiogenesis while improving BBB rigidity, according to a study of hippocampal slices [62].

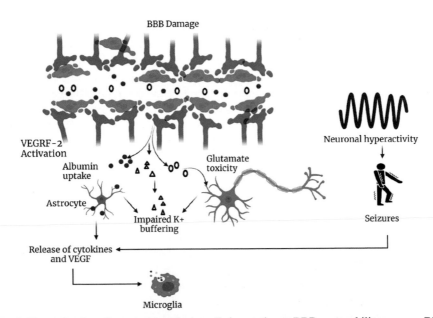

Fig. 2 The activation of astrocytes and microglia in reaction to BBB permeability worsens BBB disruption and maintains epileptogenic inflammation through the production of cytokines and VEGF.

3.3 Inflammation

Neuroinflammation is a state of inflammation that affects neural systems; it may be brought on by a number of external or endogenous causes [63]. An infection, trauma to the brain, toxic substances, autoimmune disorders, aging, polluted air, smoking habits, and damage to the spinal cord are just a few of the triggers that can cause neuroinflammation. These variables additionally encourage the creation of cytokines and chemokines, that promote cell growth and survival. More than 40 distinct interleukins (IL) kinds are among them, Fig. 3. These were formerly believed to be generated by just leukocytes [64] but were later discovered to be generated by various types of cells. Recently, prevalent processes linking neuroinflammation and epilepsy were discovered [65]. Remarkable elevation of genes expressed in inflammation chains, observed in patients, is seen in numerous models of chemically and electronically generated seizures [66]. In surgically eliminated brain tissue from individuals with persistent epilepsies, many mediators of inflammation, notably TLE and cortical dysplasia-related epilepsy, have been identified found [67]. The discovery that brain inflammation is possible in epilepsies not previously associated with immune dysfunction raised the likelihood that persistent inflammation may be a primary feature of certain epilepsies, independent of the first insult or trigger, as opposed to being merely a side effect of a particular that underlie inflammatory or auto immune etiology [68]. Chemical mediators and receptors for neuroinflammation that are important for understanding how they contribute to epileptogenesis are listed in Table 2.

Amplification of Toll-like receptors (TLRs) is linked to epileptogenesis and also to various other diseases that result in secondary epileptic phenotype [78]. TLRs, which are elements upstream of IL-1, do, in reality, control the innate immune response.

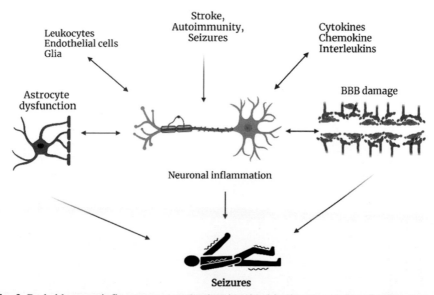

Fig. 3 Probable neuroinflammatory mechanism involved in the progression of epilepsy.

Table 2 List of neuro-inflammation-related chemical mediators and receptors with neurological molecular implications to epileptogenesis.

S. no.	Mediator/receptor	Origin	Mechanism	Reference
1.	Cytokines	generated by glial cells and neurons following brain inflammation	Regulate inflammatory procedures, develop deleterious synaptic alterations and neuronal hyper-excitability	[9]
2.	Interleukin-1β	Found in stimulated microglia and astrocytes	Increases the emission of glutamate from astrocytes while decreasing glutamate re-uptake, boosting glutamate accessibility in neuronal synapses and contributing to neuronal hyper-excitability	[69]
3.	Tumor necrosis factor-α	Generated by stimulating microglia and astrocytes	Controls N-cadherin, a binding molecule that governs the development and organization of excitation and inhibition synapses. TNF-α has been shown to enhance the microglial discharge of glutamate by upregulating glutaminase and gap junctions in microglia. It also activates AMPA receptors, enhancing glutamatergic transmission. Elevated AMPA receptors facilitate calcium overabsorption, which causes neurotoxicity	[70,71]
4.	Interleukin-6	Present in small amounts in the CNS, however activation of astrocytes and microglia can result in greater production of IL-6	Upregulation of IL-6 reduces hippocampus neurogenesis and enhances gliosis, resulting in parameters that might lead to epileptogenesis	[72]
5.	Prostaglandins	Mainly released by astrocytes and microglia and are produced from arachidonic acid	PGE2 activates EP3 on astrocytes, boosting the discharge of glutamate and triggering hyper-excitability and neural death	[73]
6.	Platelet-activating factor (PAF)	Phospholipid mediator generated by endothelial, lamina propria, and inflammatory cells	Increases the discharge of glutamate and acts as a retrograde mediator for the long-term growth of synapses, resulting in a boost in the exchange of signals across neurons	[74]
7.	Matrix metalloproteinase-9	Predominantly released by astrocytes and microglia in the cerebellum, hippocampus, and cerebral cortex	Rises in reaction to the depolarization of neurons and elevation of additional inflammatory mediators like IL-Iβ and chemokines. Prolonged MMP-9 levels promote dendritic spine weakening and extension, resulting in alterations in morphology in synapses and decreased plasticity of synapses	[75,76]
8.	Toll-like receptors (TLR 1, 2, 3)	Predominantly mediated by microglia and astrocytes	TLRs cause the release of cytokines like IL-1 along with other inflammatory mediators that contribute to epileptogenesis	[77]

Transmembrane receptors, which are particularly abundant on the outermost layer of macrophages and dendritic cells, respectively, recognize pathogens once they have entered their host and start a localized inflammatory response [79]. Additionally, a number of hyperacetylated molecules, including the chromatin component known as "high-mobility box 1 group protein" (HMGB), which emerges at necrosis, can enhance the activation of the TLR [80]. Additional factors include cyclo-oxygenase 2, tumor necrosis factor-alpha (TNF-α), transforming growth factor beta (TGF-β), and thrombospondin (TSP-1) [53]. The pentraxin family (PTXs) has recently been linked to the immunological response that promotes epilepsy. White blood cells release PTX3 in the nervous system in reaction to signals related to inflammation, wherein it is synthesized. It engages in extracellular matrix interaction, contributes to AMPA receptor remodeling, and controls circuit excitation. It has been demonstrated that elevated levels of inflammation have consequences for the extracellular matrix, boosting the expression of the redox-sensitive MMP-9 in epilepsy [81]. Variations in the extracellular framework may ultimately have an impact on the equilibrium among stimulation and inhibition as well as synaptic plasticity as a result of MMP-9 stimulation of the receptor for advanced glycation end-products (RAGE), which ultimately results in the production of numerous cytokines [82]. In individuals with epilepsy, proinflammatory triggers like fever cause and worsen seizures [83]. Additionally, among individuals with treatment-resistant epileptic seizures that received surgical resection to eradicate the seizure focus, signs of cerebral inflammation have been discovered to be connected to a variety of pathological aetiologies [84]. Proinflammatory particles, active astrocytosis, triggered microglia, and additional inflammatory markers have been identified in the dissected hippocampi of TLE individuals [85,86].

3.3.1 Inflammation and immunity

The triggering of both innate and adaptive immunity is necessary for the dynamic process of inflammation. It is characterized by either the generation of a range of inflammatory biomarkers by tissue-resident immune cells or immune cells that are moving in the circulation. Both adaptive and innate immunity are reported to be connected to epilepsy, and it is assumed that microglia, astrocytes, and neurons take part in immune-type processes that lead to brain inflammation [87]. Many different inflammatory mediators play an aspect in the stimulation of innate immunity and the switch from it to adaptive immunity, but cytokines—polypeptides that function as soluble mediators of inflammation—play a crucial part. These compounds comprise ILs, TNF-α, TGF-β, interferons, and ILs. To facilitate interaction among effector and target cells after an immunological response or tissue injury, cytokines are generated by immunological-competent, endothelial, glial, and neuronal cells in the CNS [44]. When cytokines are released, they engage any number of appropriate receptors in an interaction. Gene transcription, cytokine precursor breakdown by certain proteolytic enzymes, cellular discharge, receptor signaling, and other processes can all affect cytokine production [88,89].

Multiple mechanisms have been discovered found to reduce the inflammatory reaction, highlighting the value of strict regulation for maintaining homeostasis and preventing injury. Generation of proteins that contend with cytokines for binding their receptors, such as IL-1 receptor antagonist protein (IL-1RA), [90] as well as decoy receptors which adhere cytokines and chemokines but are unable to message, act as atomic retains to prevent like ligands from binding to biologically functioning receptors, are examples of regulatory processes [91].

3.3.2 Inflammation and stress

Distress and immunity seem to be inextricably linked, and it has been shown that seizures activate the stress axis [92,93]. Corticotropin-releasing hormone (CRH), which is released by the hypothalamus in response to stress, is finally followed by a systemic production of glucocorticoids (GCs), adrenaline, and norepinephrine, all having a significant impact on immunological and inflammatory responses. In contrast, following inflammatory reactions, immunological molecules, particularly IL-1, IL-6, and TNF-α, boost CRH production, stimulating the hypothalamic-pituitary-adrenal (HPA) axis including the sympathetic nervous system. By exerting its effects on its individual receptors on immune cells as well as via the receptor-mediated activities of GCs on target immunological tissues, CRH modifies immune/inflammatory responses [94]. By blocking leukocytes from migrating from the bloodstream into extravascular regions, lowering the build-up of monocytes and granulocytes at areas of inflammation, and inhibiting the synthesis and activity of numerous cytokines and inflammatory agents, GCs have immunosuppressive effects. Despite the fact that stress has traditionally been thought of as being immunosuppressive, research suggests that stress hormones may differently influence the differentiation of CD4+ T cells onto the T helper (Th)1 and Th2 trends, as well as the type 1 and type 2 cytokines that are secreted as a result. Interferon, IL-2, and TNF are the main type 1 cytokine secreted by Th1 cells, regulating cellular immunity. Th2 cells, however, produce type 2 cytokines, such as IL-4, IL-10, and IL-13, that boost humoral immunity (mast cells, eosinophils, and B cells) [95]. Th cells can connect with microglia as they approach the brain, and these microglia can effectively excite Th1 and Th2 cells again and discharge inflammatory mediators [96].

3.4 Oxidative stress

Neurodegenerative illnesses frequently include chronic inflammation, that aggravates the pathophysiology by stimulating microglia, producing proinflammatory cytokines like IL-1 and IL-6, and creating ROS. Collectively, these incidents influence the propagation of oxidative stress (OS), which results in the BBB's breakdown and functioning as well as changes to the brain's microenvironment [97]. With hypoxia, ROS, particularly superoxide anion, might lead to vascular endothelial cell malfunction in the BBB, and in OS, large amounts of superoxide anion generated exceed the metabolizing power of superoxide dismutase (SOD) [98]. The fact that hereditary mitochondrial illnesses such as myoclonic epilepsy frequently develop provides proof for the importance of mitochondrial contribution to epilepsy. Mitochondria are significant producers of ROS [99]. ROS production has a negative impact on mitochondrial elements including mitochondrial DNA (mtDNA), and its membranes, respiratory chain protein molecules, and nuclear DNA, resulting in decreased activity of the mitochondria [100]. The oxidative stress caused by ROS excess production in mitochondria has emerged as a significant element in epileptogenesis and seizure development. Any chemical entity possessing any number of unpaired electrons in the exterior orbit is considered a free radical. Reactive species (RSs), also known as ROS and reactive nitrogen species (RNS), comprise hydrogen peroxide (H_2O_2), singlet oxygen, superoxide (O_2), nitric oxide (NO), hydroxyl radical (HO) [101]. Throughout aerobic metabolism, a negligible fraction of the oxygen absorbed by mitochondria leakage through electron transport chain (ETC) complexes I and III to form O_2, which can then be converted to other RSs like H_2O_2 and OH [102]. While O_2 doesn't qualify as a strong oxidant in and of its own, it can oxidize particular targets like the labile iron-sulfur center of aconitase(s) [103].

Antagonizing oxidative stress in mitochondria by antioxidants has been shown to reduce or postpone the onset of illness in a range of experimental epileptic models [104]. In addition, we explore the pathogenesis of mitochondrial oxidative damage in epilepsy.

3.4.1   Mitochondrial homeostasis

Massive quantities of superoxide ($O_2{}^{\cdot-}$) are produced by mitochondrial respiration through numerous biochemical reactions, such as electron venting from the ETC throughout the tricarboxylic acid (TCA) cycle and mitochondrial oxidative phosphorylation (OXPHOS), making mitochondria one of the biggest locations of the generation of ROS inside the cell. In fact, it was previously proven that two of the ETC complexes, notably complexes I (CI) and CIII, play a part in the generation of oxygen [105]. A number of matrix proteins and complexes, comprising TCA cycle enzymes like pyruvate dehydrogenase, aconitase, and α-ketoglutarate dehydrogenase, can produce mitochondrial ROS (mtROS) [106]. Homeostatic redox state in mitochondria is regulated by mitochondria-associated antioxidant defenses in addition to mtROS production [107]. Free radical production and scavenger processes tightly preserve the mtROS equilibrium physiologically. Nevertheless, the overproduction of ROS, particularly in the brain mitochondria, seems to be after neurological disorders like epilepsy and causes oxidative damage [108].

3.4.2   Mitochondrial DNA oxidative damage

Damage to DNA is a possibility after ROS stimulation. It is widely acknowledged that mtDNA codes for crucial components of OXPHOS, the process through which mitochondria produce energy while preserving the equilibrium of ROS in the cell [109]. Therefore, oxidative mtDNA damage leads to OXPHOS imperfections which result in inadequate required energy for a cell and an excess of ROS, developing a vicious process. Numerous human mitochondrial ailments are thought to be triggered by oxidative damage to mtDNA, especially when it affects ETC elements [110]. According to epilepsy studies, the creation of ROS during seizures damages mtDNA and lowers the activity of mitochondrial ETC subunits encoded by mtDNA, particularly CI, CIII, CIV, and CV. It suggests that mtDNA is especially susceptible to oxidative damage from ROS [111].

3.4.3   Seizure-induced oxidative damage

Recent research has revealed that oxidative stress, which affects several metabolic functions, may increase excitability in neurons and could potentially be a direct cause of epilepsy. A decline in mitochondrial redox state that results in irreversible oxidation, mtDNA destruction, an inability of mitochondrial restoration, and reduced oxidative phosphorylation and generation of ATP by mitochondria are some of the underlying processes that could become active during this "latent period" as chronic epilepsy evolves. Although dropped ATP synthesis and subsequently ATP-dependent Na^+/K^+ ATPase are adequate to boost neuronal excitability, additional causes that boost neuronal excitability could include reduced levels and/or operation of astrocytic glutamate transporters and glutamate synthetase because of oxidative damage and redox control of ion channels. Essential proteins associated with regulating neuronal excitability may be altered during seizures, rendering them more vulnerable to breakdown by intracellular proteolytic processes and raising the risk of seizures [112].

3.4.4   Apoptosis

According to reports, a number of variables can affect cell apoptosis and oxidative stress of the mitochondria in epilepsy. Mitochondrial Ca^{2+} build-up has the capacity to increase the synthesis of NADH and ROS amid low-Mg^{2+}-induced epileptiform action [113]. Additionally, oxidative damage and a number of neurological disorders are linked to mitochondrial fission. The mitochondrial fission protein Drp1 is inhibited by the mitochondrial division inhibitor 1 (mdivi-1) in the pilocarpine-induced epilepsy rat model, resulting in higher activity of SOD, lowering the levels of cytochrome c and caspase-3, and improving neuron survival. According to those findings, there is a direct correlation between the beginning of epileptic episodes and mitochondrial fission, and inhibiting mitochondrial fission prevents neuronal death by inhibiting the mtROS/cytochrome c system [114].

4   Diagnosis

The individual having seizures and any witnesses can accurately and methodically describe the seizure to make the diagnosis of epilepsy, which may not require any more testing. There are a lot of misdiagnoses due to clinicians' unfortunate ignorance of the semiology that distinguishes epileptic seizures from other illnesses including convulsive hypotension and psychotic nonepileptic episodes [115]. The inclusion of multiple factors and several investigations, based on the apparent problem, is required for accurate identification of the basic epilepsy disorder. The following criteria must be met: interictal EEG, 12-lead ECG to exclude out cardiac anomalies age at development, seizure category, and neuropsychiatric condition.

Especially for children who show usual abnormalities such as juvenile absence epilepsy, juvenile myoclonic epilepsy, or selflimiting epilepsy in childhood, a brain MRI is typically required. When particular reasons have been identified, blood testing, lumbar punctures, and other studies may be beneficial [116]. The discovery of fresh forms of autoimmune encephalitis, which have been related to antiGABAB receptors, antiNMDA receptors, glioma inactivated 1 protein (LRIP) [117], and antibodies to Kv1 potassium channel-complex proteins leucine-rich among additional significant detection innovations across the past 10 years, include enhanced visualization equipment and the implementation of epilepsy aimed protocols for image collection and evaluation, permitting for the identification of formerly undetected subtle epileptogenic lesions [118].

5   Pharmacological treatment

Antiepileptic medications are commonly used as the initial choice therapy for epileptic seizures. When repeated seizures arise, familial relationships are discovered, or irregularities on EEG lead to an epilepsy condition, such therapies are undertaken [119]. With adequate medical therapy, approximately 70% of individuals attain seizure independence, with outcomes ranging according to epileptic syndrome, the root cause, and additional variables. Regardless of prognostic considerations, the majority of individuals who achieve seizure-free status react to the first medicine prescribed [120].

In every patient, no particular antiepileptic medicine is optimal for initial therapy. According to the UK National Institute for Health and Care Excellence regulations, [121] alternatives to therapy should include seizure category, condition, and other factors such as gender, age, and comorbidities. Table 3 lists some traditional AEDs as well as their mechanisms of action.

6   Recent advances in micro dysfunction management

6.1   Nanoparticles-based drug delivery

The BBB, a microvascular barrier that protects the CNS from systemic blood flow, limits medicinal medication's access to the CNS [122]. As a result, novel ways for medicinal medication targeted to brain tumors are required. Chemical disturbance of the BBB employing vasoactive substances that trigger an inflammatory response or hyperosmolar materials that harm endothelial cells [123], centered ultrasound-mediated reversible impairment employing thermal destruction or acoustic cavitation, intranasal administration, repression of efflux pump inhibiting agents [124], virus vectors including adeno-associated viruses, and cellular "Trojan horse" proteins that adhere to the BBB and enables movement across transporters [125]. Nanoparticles (NPs) are a revolutionary technique for traversing the BBB, containing a wide range of chemicals with distinct physiochemical attributes that enable tailored treatment to be administered to brain tumors. Because of their small dimensions, low toxicology,

Table 3 A list of both conventional and innovative AEDs, together with their respective seizure types and method of action.

S. no.	Conventional drugs	Newer drugs	Types of seizures	Action
1.	Carbamazepine Phenytoin Primidone	Lamotrigine Gabapentin	Complex partial	Ion channel (Ca^{2++}, K^+, Na^+) modulator
2.	Phenobarbital Phenytoin Primidone Valproate	Pregabalin Retigabine Lamotrigine Gabapentin	Partial and generalized seizures	Enhance GABAergic transmission and Ion channel (Ca^{2++}) modulator
3.	Clonazepam Valproate Ethosuximide		Absence seizures	Ion channel (Ca^{2++}) modulator
4.	Valproate		Myoclonic seizures	Acts on GABA (γ-aminobutyric acid) levels, blocking voltage-gated ion channels, and inhibiting histone deacetylase
5.	Carbamazepine Phenobarbital Phenytoin Primidone Valproate		Tonic-clonic seizures	Ion channel (Ca^{2++}, K^+, Na^+) modulator and (γ-aminobutyric acid) levels
6.		Perampanel	Generalized seizures	Selective postsynaptic excitatory neurotransmission inhibitors
7.		Felbamate	Focal seizures	Multiple

and regulated release of drug characteristics, NPs are especially beneficial. Furthermore, the surface of their cells can be manipulated easily with proteins targeting specific receptors to localize the delivery of drugs [126,127].

NPs are microscopic molecules with sizes extending from 1 to 1000 nm. Their small dimensions make it easier for them to pass through the BBB [128], and investigations indicate that as NP size drops, permeation increases via BBB gaps, with basically no permeation above 200 nm [129]. To speed up the rate of circulation or attach to endothelium receptors, some ligands, such as proteins, peptides, antibodies, and surfactants, can be coupled to the edges of NPs. An increase in ligand density improves polyvalency and avidity, which increases the likelihood that endothelial cells will internalize the compound [130]. To enhance transcytosis, additional particles, such as proteins and peptides, might be applied on NP surfaces. These compounds can target certain cellular receptors on the BBB in addition to a particular tissue [131].

NPs are often given intravenously. Nevertheless, this technique remains more pervasive than intravascular administration and is constrained by the quick turnover of cerebrospinal fluid [132]. Intraventricular administration may be used to enhance the delivery of drugs and bypass the BBB. Drugs have been delivered directly intraparenchymal, however, this method necessitates surgical procedures on individuals. While ciliary activity reduces clearance, intranasal administration of NPs via the olfactory neural route has also been investigated [133]. When NPs reach the BBB, they may migrate by using a variety of techniques, like passive diffusion, carrier transportation, and adsorptive- and receptor-mediated transcytosis, among others. NP transfer is aided by methods that compromise BBB integrity [134]. Biodegradable NPs have gained popularity as treatment options for epilepsy in the past few years. Biodegradable nanomaterials can be tailored to organically decompose into nontoxic bioproducts after they reach the target site while staying durable at off-target locations [135] some of which are highlighted in Table 4.

The delivery of AEDs via the intranasal (IN) approach is an emerging field of research, as seen by the benzodiazepine nasal sprays currently being tested in clinical studies for the medical management of seizures [145]. Several studies have been

Table 4 Examples of nanoparticle loaded drug delivery in the management of epilepsy [136].

S. no.	Nanoparticle	Loaded drug	Benefits	Reference
1.	Solid lipid nanoparticle	Alprazolam	Increase in bioavailability	[137]
		Carbamazepine	Increased drug entrapment Higher anticonvulsant efficacy	[138]
		Clonazepam	Increased in vitro permeability	[139]
2.	Nanostructured lipid carriers	Carbamazepine	Increased drug loading and increased efficacy	[140]
		Clonazepam	100% of the entrapped drug release	[141]
		Lamotrigine	Increased brain concentration and efficacy	[142]
3.	Polymeric nanoparticle	Carbamazepine	30 times more effective than the free drug	[143]
		Catechin hydrate	Significant brain biodistribution	[143]
		Oxcarbamazepine	Reduced dose frequency	[144]

conducted to compare the impact of AEDs administered via traditional and IN routes [146]. The initial results are intriguing and encouraging. To get directly into the brain, novel devices (Optinose, Bi-Directional technology) are being created and distributed to channel the medicine to the nasal region in the uppermost region of the nose [147].

6.2 Antiinflammatory and immunotherapy

Immune-modulating and antiinflammatory medications may be helpful therapies for all or some types of epilepsy. Immune processes and inflammation do play a part in the development of seizures. With variable degrees of efficacy, medications such as ACTH, corticosteroids, and intravenous immunoglobulin (IVIg) have been successfully used for the management of seizures and/or epilepsy [148]. Table 5 highlights the alleged immunological therapeutic mechanisms of numerous medications used for managing epilepsy.

6.3 Antioxidant-based therapy

The association underlying oxidative strain and neuronal death in epilepsy has been recognized, which has spurred an intense curiosity in creating antioxidation strategies to shield neurons against oxidative harm during epileptic seizures [153]. Many antioxidants that specifically target the mitochondria have been created and developed, including Mito-CP, MitoPBN, Mito-TEMPO, Mito-peroxidase, MitoSOD, MitoQ

Table 5 Examples of immunosuppressive and antiinflammatory drugs that may be effective in the management of epilepsy [149].

S. no.	Therapeutics	Mechanism	Reference
1.	Corticosteroids	Diapedesis is prevented, which prevents lymphocytes from entering damaged areas. Synthesis of IL-2 is also reduced, and leukocyte activity—primarily that of helper T lymphocytes—as in addition to endothelial cell activity is also inhibited	[150]
2.	Plasmapheresis	change elements in cells and remove humoral immune system elements from the body, increasing the production of antiinflammatory substances while inhibiting the stimulation of innate cells and the generation of proinflammatory cytokines	[151]
3.	Human immunoglobulin G	limits the generation of proinflammatory cytokines, the stimulation of innate cells, as well as soluble substances like IL-6 and TNF-α that are brought about by this factor	[152]

(ubiquinone), Mito-apocynin, and Mito-VitE (vitamin E) [154]. There are numerous papers about MitoQ's neuroprotective properties against epilepsy among these substances [155]. The stimulation of cAMP response element-binding protein by MitoQ therapy has been shown to greatly improve cognitive and memory impairment in pilocarpine-induced mouse models of temporal lobe epilepsy [156]. It has been established that MitoQ's therapeutic impact is most likely accomplished by lowering superoxide production and decreasing neuronal death. MitoQ's favorable effect in the amelioration of memory loss is connected with epileptic pathophysiology. These findings suggest MitoQ is an effective neuroprotective treatment for epilepsy and related comorbidities. Additionally, AEOL 11207, a lipophilic metalloporphyrin, exhibits outstanding antiseizure properties and substantially enhances this phenotype, pointing to the drug's potential as a treatment for epilepsy [112]. Table 6 lists the many types of antioxidants that can effectively manage ROS-induced damage in epilepsy.

Table 6 Antioxidants listed along with information about how they help treat epilepsy caused harm from ROS.

S. no.	Antioxidant	Mechanism	Reference
1.	Vitamin A	Increases the level of total glutathione and reduces the gap junction within the BBB	[157]
2.	Vitamins B12	Plays an important part in myelin sheath generation and maintenance, and downregulates TNF-α levels	[158]
3.	Vitamin C	Effectively remove free radicals and boost SOD Also activates inhibitory receptors GABA and decreases the inhibition of caspase-3	[159]
4.	Vitamin E	Increase catalase activity and reduces lipid peroxidation and also limits blood-brain barrier damage	[160]
5.	Vitamin K	Reduce the degradation of glutathione	[161]
6.	Glutathione	Inactivates ROS directly	[162]
7.	N-acetylcysteine	Reduces the excitotoxicity caused by N-methyl-D-aspartate (NMDA)	[163]
8.	Buspirone	Increased SOD and catalase activity, considerable decrease in lipid peroxidation amount and nitrite content, and inhibition of apoptosis	[164]
9.	Ubiquinone	Lowers lipid peroxidation markers, limits oxidative stress-induced apoptosis, and slows the mitochondrial permeability transition	[165]
10.	α-Lipoic acid	elevates intracellular glutathione, suppresses TNF-α induced ROS production, GSH decrease, and apoptosis, and eliminates singlet oxygen	[166]

7 Conclusion and future prospective

Among the most prevalent neurological conditions in the world, epilepsy affects a million individuals globally. Epilepsy is a symptomatic complex with numerous risk elements and a strong genetic propensity. The foundations of a diagnosis are an exhaustive medical record. Additional investigations can be used to figure out its origin and prognosis. Comorbidities are becoming more widely accepted as significant prognosis and aetiological indicators. BBB, that strictly controls the movement of chemicals across the brain parenchyma and the blood flow. Certain hypotheses contend that BBB dysfunction, in particular barrier loss, exacerbates the course of epileptic convulsions, and that barrier leakage results from epileptic episodes. In this way, microcirculatory irregularities and vascular changes, such as seizure-induced angiogenesis, are probably common in epilepsy. By encouraging adult neurogenesis, increased VEGF/VEGFR signaling may help to lessen cognitive deficits following epileptic convulsions. The importance of the role played by inflammatory mediators in epilepsy vulnerability and epileptogenesis has been highlighted by preclinical and clinical findings. Strong and widespread inflammatory reactions reduce the seizure threshold, improve the excitability of neurons, raise BBB permeation, and encourage epileptogenesis in the brain. Both drug-resistant epileptic patients and numerous preclinical epilepsy models exhibit persistent inflammation. The cascade processes of frequent seizures can be combated most effectively by preventing neuroinflammation. Future research should focus on whether therapy with antiinflammatory medications can prevent the onset of seizures in individuals who are predisposed to epilepsy.

Epilepsy, characterized by recurring unexplained seizures, is also fundamentally influenced by mitochondrial malfunction. The fact that hereditary mitochondrial abnormalities like myoclonic epilepsy frequently manifest provides proof for the significance of mitochondria in epilepsy. ROS are frequently produced by mitochondria. The generation of ROS impairs the activity of mitochondria by negatively affecting nuclear DNA in addition to mtDNA, mitochondrial membranes, and respiratory chain proteins. It is becoming clear that oxidative damage brought on by high levels of ROS in the mitochondria contributes a major part to the development of epileptogenesis and seizure onset. To reduce side effects, improve compliance among patients, in particular during chronic medical care, and penetrate the BBB, that's an important obstacle for CNS therapy, it is required not just to find new medications but additionally to redefine consolidated treatment. Nose-to-brain administration is currently an emerging method of overcoming BBB. Antiseizure drugs' effectiveness as therapies can be improved by biodegradable NPs, which also provide a sustained drug release. Biodegradable NPs surface functionalization may help to increase the permeation of the BBB. Promising approaches to enhance the pharmacokinetics of AEDs include polymeric and lipid nanoparticles. Recent understanding of epilepsy's immunological response enables us to improve early therapy measures, preventing neurological squeals in individuals. Usually, it is advised to begin the treatment. This action will stop the immune system from responding continuously, resolving the neurological signs that come along with neuropsychological autoimmune epilepsy. In simple terms, although new treatment approaches have been made as a result of increased knowledge of how inflammation contributes to the development of resistant epilepsy, numerous compounds still have not been investigated as potential anticonvulsant medications. Therefore, to enhance the current therapy for this neurological condition, research centered on the possibility of clinical use of antiinflammatory therapies is required. Antioxidant therapies that target mitochondrial biological energies and oxidative damage may be effective in both the management of epilepsy and the prevention of its onset.

Further study in this field will help us comprehend how oxidative damage to the mitochondria and malfunction bring towards the disease progression of TLE as well as how treatment strategies focusing on mitochondrial biological energies might be utilized to treat patients who are resistant to antiepileptic medications currently available.

Acknowledgments

I'm highly obliged to the PSIT-Pranveer Singh Institute of Technology (Pharmacy) department for giving me the opportunity in writing this chapter also thankful for their constant guidance and support.

Conflict of interest

None.

Funding

None.

References

[1] Falco-Walter JJ, Scheffer IE, Fisher RS. The new definition and classification of seizures and epilepsy. Epilepsy Res 2018;139:73–9.

[2] Beghi E, Hesdorffer D. Prevalence of epilepsy—an unknown quantity. Epilepsia 2014;55 (7):963–7.

[3] Perucca P, Scheffer IE, Kiley M. The management of epilepsy in children and adults. Med J Aust 2018;208(5):226–33.

[4] Scheffer IE, Berkovic S, Capovilla G, Connolly MB, French J, Guilhoto L, Hirsch E, Jain S, Mathern GW, Moshé SL, Nordli DR. ILAE classification of the epilepsies: position paper of the ILAE Commission for Classification and Terminology. Epilepsia 2017;58 (4):512–21.

[5] Scorza FA, Cavalheiro EA, Costa JC. Sudden cardiac death in epilepsy disappoints, but epileptologists keep faith. Arq Neuropsiquiatr 2016;74:570–3.

[6] Beghi E. The epidemiology of epilepsy. Neuroepidemiology 2020;54(2):185–91.

[7] Najm IM. Mapping brain networks in patients with focal epilepsy. Lancet Neurol 2018;17(4):295–7.

[8] Fisher RS, Bonner AM. The revised definition and classification of epilepsy for neurodiagnostic technologists. Neurodiagn J 2018;58(1):1.

[9] Vezzani A. Epilepsy and inflammation in the brain: overview and pathophysiology: epilepsy and inflammation in the brain. Epilepsy Curr 2014;14(2 Suppl):3–7.

[10] van Vliet EA, Ndode-Ekane XE, Lehto LJ, Gorter JA, Andrade P, Aronica E, et al. Long-lasting blood-brain barrier dysfunction and neuroinflammation after traumatic brain injury. Neurobiol Dis 2020;145, 105080.

[11] Marchi N, Banjara M, Janigro D. Blood-brain barrier, bulk flow, and interstitial clearance in epilepsy. J Neurosci Methods 2016;260:118–24.

[12] Li X, Ruan C, Zibrila AI, Musa M, Wu Y, Zhang Z, et al. Children with autism spectrum disorder present glymphatic system dysfunction evidenced by diffusion tensor imaging along the perivascular space. Medicine 2022;101(48), e32061.

[13] Yogeswaran S, Muthumalage T, Rahman I. Comparative reactive oxygen species (ROS) content among various flavored disposable vape bars, including cool (iced) flavored bars. Toxics 2021;9(10):235.

[14] Bradley L. Rehabilitation following anti-NMDA encephalitis. Brain Inj 2015;29(6): 785–8.

[15] Du Y, Brennan FH, Popovich PG, Zhou M. Microglia maintain the normal structure and function of the hippocampal astrocyte network. Glia 2022;70(7):1359–79.

[16] Bonilla L, Esteruelas G, Ettcheto M, Espina M, García ML, Camins A, et al. Biodegradable nanoparticles for the treatment of epilepsy: from current advances to future challenges. Epilepsia Open 2022;7:S121–32.

[17] Ranjan S, Quezado M, Garren N, Boris L, Siegel C, Lopes Abath Neto O, et al. Clinical decision making in the era of immunotherapy for high grade-glioma: report of four cases. BMC Cancer 2018;18:1–9.

[18] Geronzi U, Lotti F, Grosso S. Oxidative stress in epilepsy. Expert Rev Neurother 2018;18 (5):427–34.

[19] Azam F, Prasad VV, Thangavel NM. Targeting oxidative stress component in the therapeutics of epilepsy. Curr Top Med Chem 2012;12(9):994–1007.

[20] Hirabayashi H, Hirabayashi K, Wakabayashi M, Murata T. A case of diagnosis of occipital lobe epilepsy complicated by right hemianopsia associated with left occipital lobe cerebral infarction. Case Rep Ophthalmol 2022;13:141–6.

[21] Stafstrom CE, Carmant L. Seizures and epilepsy: an overview for neuroscientists. Cold Spring Harb Perspect Med 2015;5(6), a022426.

[22] Fisher RS, Acevedo C, Arzimanoglou A, Bogacz A, Cross JH, Elger CE, et al. ILAE official report: a practical clinical definition of epilepsy. Epilepsia 2014;55(4):475–82.

[23] Abramovici S, Bagić A. Epidemiology of epilepsy. Handb Clin Neurol 2016;138:159–71.

[24] Falco-Walter J. Epilepsy—definition, classification, pathophysiology, and epidemiology. Semin Neurol 2020;40(6):617–23.

[25] Johnson EL. Seizures and epilepsy. Med Clin 2019;103(2):309–24.

[26] Fisher RS, Cross JH, French JA, Higurashi N, Hirsch E, Jansen FE, Lagae L, Moshé SL, Peltola J, Roulet Perez E, Scheffer IE. Operational classification of seizure types by the international league against epilepsy: position paper of the ILAE Commission for Classification and Terminology. Z Epileptol 2018;31:272–81.

[27] Tripathi KD. Essentials of medical pharmacology. 8th ed. JP Medical Ltd.; 2019. p. 438–52.

[28] Thijs RD, Surges R, O'Brien TJ, Sander JW. Epilepsy in adults. Lancet 2019;393 (10172):689–701.

[29] Kanner AM. Management of psychiatric and neurological comorbidities in epilepsy. Nat Rev Neurol 2016;12(2):106–16.

[30] Téllez-Zenteno JF, Hernández-Ronquillo L. A review of the epidemiology of temporal lobe epilepsy. Epilepsy Res Treat 2012;2012:1–5.

[31] World Health Organization. Epilepsy: a public health imperative. World Health Organization; 2019. ISBN: 978-92-4-151593-1:147.

[32] Beghi E, Giussani G, Nichols E, Abd-Allah F, Abdela J, Abdelalim A, et al. Global, regional, and national burden of epilepsy, 1990–2016: a systematic analysis for the global burden of disease study 2016. Lancet Neurol 2019;18(4):357–75.

[33] Joutsa J, Rinne JO, Hermann B, Karrasch M, Anttinen A, Shinnar S, et al. Association between childhood-onset epilepsy and amyloid burden 5 decades later. JAMA Neurol 2017;74(5):583–90.

[34] Szaflarski M, Wolfe JD, Tobias JG, Mohamed I, Szaflarski JP. Poverty, insurance, and region as predictors of epilepsy treatment among US adults. Epilepsy Behav 2020;107, 107050.

[35] Dhiman V, Menon GR, Kaur S, Mishra A, John D, Vishnu MV, et al. A systematic review and meta-analysis of prevalence of epilepsy, dementia, headache, and Parkinson disease in India. Neurol India 2021;69(2):294.

[36] Singh G, Sander JW. The global burden of epilepsy report: implications for low-and middle-income countries. Epilepsy Behav 2020;105, 106949.

[37] Levira F, Thurman DJ, Sander JW, Hauser WA, Hesdorffer DC, Masanja H, et al. Epidemiology Commission of the international league against epilepsy. Premature mortality of epilepsy in low-and middle-income countries: a systematic review from the mortality task force of the international league against epilepsy. Epilepsia 2017;58(1):6–16.

[38] Watila MM, Balarabe SA, Ojo O, Keezer MR, Sander JW. Overall and cause-specific premature mortality in epilepsy: a systematic review. Epilepsy Behav 2018;87:213–25.

[39] Ngugi AK, Bottomley C, Fegan G, Chengo E, Odhiambo R, Bauni E, et al. Premature mortality in active convulsive epilepsy in rural Kenya: causes and associated factors. Neurology 2014;82(7):582–9.

[40] Johnson EL, Krauss GL, Lee AK, Schneider AL, Dearborn JL, Kucharska-Newton AM, et al. Association between midlife risk factors and late-onset epilepsy: results from the atherosclerosis risk in communities study. JAMA Neurol 2018;75(11):1375–82.

[41] Sen A, Jette N, Husain M, Sander JW. Epilepsy in older people. Lancet 2020;395 (10225):735–48.

[42] Ogaki A, Ikegaya Y, Koyama R. Vascular abnormalities and the role of vascular endothelial growth factor in the epileptic brain. Front Pharmacol 2020;11:20.

[43] Liu Z, Yang C, Meng X, Li Z, Lv C, Cao P. Neuroprotection of edaravone on the hippocampus of kainate-induced epilepsy rats through Nrf2/HO-1 pathway. Neurochem Int 2018;112:159–65.

[44] Nazarinia D, Karimpour S, Hashemi P, Dolatshahi M. Neuroprotective effects of Royal Jelly (RJ) against pentylenetetrazole (PTZ)-induced seizures in rats by targeting inflammation and oxidative stress. J Chem Neuroanat 2023;129, 102255.

[45] Sharma AA, Szaflarski JP. In vivo imaging of neuroinflammatory targets in treatment-resistant epilepsy. Curr Neurol Neurosci Rep 2020;20:1–3.

[46] Liang LP, Waldbaum S, Rowley S, Huang TT, Day BJ, Patel M. Mitochondrial oxidative stress and epilepsy in SOD2 deficient mice: attenuation by a lipophilic metalloporphyrin. Neurobiol Dis 2012;45(3):1068–76.

[47] Vanstone JR, Smith AM, McBride S, Naas T, Holcik M, Antoun G, et al. DNM1L-related mitochondrial fission defect presenting as refractory epilepsy. Eur J Hum Genet 2016;24 (7):1084–8.

[48] Peng Y, Chu S, Yang Y, Zhang Z, Pang Z, Chen N. Neuroinflammatory in vitro cell culture models and the potential applications for neurological disorders. Front Pharmacol 2021;12, 671734.

[49] Jiang S, Xia R, Jiang Y, Wang L, Gao F. Vascular endothelial growth factors enhance the permeability of the mouse blood-brain barrier. PLoS One 2014;9(2), e86407.

[50] van Vliet EA, Marchi N. Neurovascular unit dysfunction as a mechanism of seizures and epilepsy during aging. Epilepsia 2022;63(6):1297–313.

[51] Giannoni P, Claeysen S, Noe F, Marchi N. Peripheral routes to neurodegeneration: passing through the blood-brain barrier. Front Aging Neurosci 2020;12:3.

[52] Yang J, He F, Meng Q, Sun Y, Wang W, Wang C. Inhibiting HIF-1α decreases expression of TNF-α and caspase-3 in specific brain regions exposed kainic acid-induced status epilepticus. Cell Physiol Biochem 2016;38(1):75–82.

[53] MacAllister WS, Murphy H, Coulehan K. Serial neuropsychological evaluation of children with severe epilepsy. J Pediatr Epilepsy 2017;6(1):037–43.

[54] Vezzani A, Friedman A, Dingledine RJ. The role of inflammation in epileptogenesis. Neuropharmacology 2013;69:16–24.

[55] Merelli A, Repetto M, Lazarowski A, Auzmendi J. Hypoxia, oxidative stress, and inflammation: three faces of neurodegenerative diseases. J Alzheimers Dis 2021;82(s1):S109–26.

[56] Morin-Brureau M, Rigau V, Lerner-Natoli M. Why and how to target angiogenesis in focal epilepsies. Epilepsia 2012;53:64–8.

[57] van Lanen RH, Melchers S, Hoogland G, Schijns OE, Zandvoort MA, Haeren RH, et al. Microvascular changes associated with epilepsy: a narrative review. J Cereb Blood Flow Metab 2021;41(10):2492–509.

[58] Long Q, Fan C, Kai W, Luo Q, Xin W, Wang P, et al. Hypoxia inducible factor-1α expression is associated with hippocampal apoptosis during epileptogenesis. Brain Res 2014;1590:20–30.

[59] Stone NL, England TJ, O'Sullivan SE. A novel transwell blood brain barrier model using primary human cells. Front Cell Neurosci 2019;13:230.

[60] Gorter JA, Aronica E, Van Vliet EA. The roof is leaking and a storm is raging: repairing the blood-brain barrier in the fight against epilepsy. Epilepsy Curr 2019;19(3):177–81.

[61] Castañeda-Cabral JL, Colunga-Durán A, Ureña-Guerrero ME, Beas-Zárate C, de los Angeles Nuñez-Lumbreras M, Orozco-Suárez S, et al. Expression of VEGF-and tight junction-related proteins in the neocortical microvasculature of patients with drug-resistant temporal lobe epilepsy. Microvasc Res 2020;132, 104059.

[62] Yu X, Ji C, Shao A. Neurovascular unit dysfunction and neurodegenerative disorders. Front Neurosci 2020;14:334.

[63] DiSabato DJ, Quan N, Godbout JP. Neuroinflammation: the devil is in the details. J Neurochem 2016;139:136–53.

[64] Rodríguez-Gómez JA, Kavanagh E, Engskog-Vlachos P, Engskog MK, Herrera AJ, Espinosa-Oliva AM, et al. Microglia: agents of the CNS pro-inflammatory response. Cell 2020;9(7):1717.

[65] Löscher W, Potschka H, Sisodiya SM, Vezzani A. Drug resistance in epilepsy: clinical impact, potential mechanisms, and new innovative treatment options. Pharmacol Rev 2020;72(3):606–38.

[66] Kim I, Mlsna LM, Yoon S, Le B, Yu S, Xu D, et al. A postnatal peak in microglial development in the mouse hippocampus is correlated with heightened sensitivity to seizure triggers. Brain Behav 2015;5(12), e00403.
[67] Borham LE, Mahfoz AM, Ibrahim IA, Shahzad N, Alrefai AA, Labib AA, et al. The effect of some immunomodulatory and anti-inflammatory drugs on Li-pilocarpine-induced epileptic disorders in Wistar rats. Brain Res 2016;1648:418–24.
[68] Riazi K, Galic MA, Kentner AC, Reid AY, Sharkey KA, Pittman QJ. Microglia-dependent alteration of glutamatergic synaptic transmission and plasticity in the hippocampus during peripheral inflammation. J Neurosci 2015;35(12):4942–52.
[69] Mishra A, Bandopadhyay R, Singh PK, Mishra PS, Sharma N, Khurana N. Neuroinflammation in neurological disorders: pharmacotherapeutic targets from bench to bedside. Metab Brain Dis 2021;36(7):1591–626.
[70] Horiuchi H, Parajuli B, Wang Y, Azuma YT, Mizuno T, Takeuchi H, et al. Interleukin-19 acts as a negative autocrine regulator of activated microglia. PLoS One 2015;10(3), e0118640.
[71] Mazarati AM, Lewis ML, Pittman QJ. Neurobehavioral comorbidities of epilepsy: role of inflammation. Epilepsia 2017;58:48–56.
[72] Gruol DL. IL-6 regulation of synaptic function in the CNS. Neuropharmacology 2015;96:42–54.
[73] Bachtell R, Hutchinson MR, Wang X, Rice KC, Maier SF, Watkins LR. Targeting the toll of drug abuse: the translational potential of Toll-like receptor 4. CNS Neurol Disord Drug Targets 2015;14(6):692–9.
[74] Rana A, Musto AE. The role of inflammation in the development of epilepsy. J Neuroinflammation 2018;15:1–2.
[75] Tao H, Gong Y, Yu Q, Zhou H, Liu Y. Elevated serum matrix metalloproteinase-9, interleukin-6, hypersensitive C-reactive protein, and homocysteine levels in patients with epilepsy. J Interf Cytokine Res 2020;40(3):152–8.
[76] Breviario S, Senserrich J, Florensa-Zanuy E, Garro-Martínez E, Díaz Á, Castro E, et al. Brain matrix metalloproteinase-9 activity is altered in the corticosterone mouse model of depression. Prog Neuro-Psychopharmacol Biol Psychiatry 2023;120, 110624.
[77] Gross A, Benninger F, Madar R, Illouz T, Griffioen K, Steiner I, et al. Toll-like receptor 3 deficiency decreases epileptogenesis in a pilocarpine model of SE-induced epilepsy in mice. Epilepsia 2017;58(4):586–96.
[78] Li D, Wu M. Pattern recognition receptors in health and diseases. Signal Transduct Target Ther 2021;6(1):291.
[79] Balaji R, Subbanna M, Shivakumar V, Abdul F, Venkatasubramanian G, Debnath M. Pattern of expression of toll like receptor (TLR)-3 and-4 genes in drug-naïve and antipsychotic treated patients diagnosed with schizophrenia. Psychiatry Res 2020;285, 112727.
[80] Ravizza T, Terrone G, Salamone A, Frigerio F, Balosso S, Antoine DJ, et al. High mobility group box 1 is a novel pathogenic factor and a mechanistic biomarker for epilepsy. Brain Behav Immun 2018;72:14–21.
[81] Bronisz E, Kurkowska-Jastrzębska I. Matrix metalloproteinase 9 in epilepsy: the role of neuroinflammation in seizure development. Mediat Inflamm 2016;2016:1–14.
[82] Mazuir E, Fricker D, Sol-Foulon N. Neuron-oligodendrocyte communication in myelination of cortical GABAergic cells. Life 2021;11(3):216.
[83] Faini G, Aguirre A, Landi S, Lamers D, Pizzorusso T, Ratto GM, et al. Perineuronal nets control visual input via thalamic recruitment of cortical PV interneurons. elife 2018;7, e41520.

[84] Garcia-Curran MM, Hall AM, Patterson KP, Shao M, Eltom N, Chen K, et al. Dexamethasone attenuates hyperexcitability provoked by experimental febrile status epilepticus. eNeuro 2019;6(6), ENEURO.0430-19.2019.

[85] Lorigados Pedre L, Morales Chacón LM, Pavón Fuentes N, Robinson Agramonte MD, Serrano Sánchez T, Cruz-Xenes RM, et al. Follow-up of peripheral IL-1β and IL-6 and relation with apoptotic death in drug-resistant temporal lobe epilepsy patients submitted to surgery. Behav Sci 2018;8(2):21.

[86] Shmakova AA, Rubina KA, Rysenkova KD, Gruzdeva AM, Ivashkina OI, Anokhin KV, et al. Urokinase receptor and tissue plasminogen activator as immediate-early genes in pentylenetetrazole-induced seizures in the mouse brain. Eur J Neurosci 2020;51(7):1559–72.

[87] Vezzani A, Balosso S, Ravizza T. Neuroinflammatory pathways as treatment targets and biomarkers in epilepsy. Nat Rev Neurol 2019;15(8):459–72.

[88] Dai D, Yuan J, Wang Y, Xu J, Mao C, Xiao Y. Peli1 controls the survival of dopaminergic neurons through modulating microglia-mediated neuroinflammation. Sci Rep 2019;9(1):8034.

[89] Woodcock TM, Frugier T, Nguyen TT, Semple BD, Bye N, Massara M, et al. The scavenging chemokine receptor ACKR2 has a significant impact on acute mortality rate and early lesion development after traumatic brain Inj. PLoS One 2017;12(11), e0188305.

[90] Mantovani A, Dinarello CA, Molgora M, Garlanda C. Interleukin-1 and related cytokines in the regulation of inflammation and immunity. Immunity 2019;50(4):778–95.

[91] Tiberio L, Del Prete A, Schioppa T, Sozio F, Bosisio D, Sozzani S. Chemokine and chemotactic signals in dendritic cell migration. Cell Mol Immunol 2018;15(4):346–52.

[92] Chakrabarti S, Jana M, Roy A, Pahan K. Upregulation of suppressor of cytokine signaling 3 in microglia by cinnamic acid. Curr Alzheimer Res 2018;15(10):894–904.

[93] Takahashi S, Sakamaki M, Ferdousi F, Yoshida M, Demura M, Watanabe MMI, et al. Ethanol extract of Aurantiochytrium mangrovei 18w-13a strain possesses anti-inflammatory effects on murine macrophage RAW264 cells. Front Physiol 2018;9:1205.

[94] van Campen JS, Hompe EL, Velis DN, Otte WM, van der Berg F, Jansen FE, et al. Stress sensitivity of seizures influences the relation between cortisol fluctuations and interictal epileptiform discharges in people with epilepsy. Stress Childhood Epilepsy 2015;137.

[95] Jones K, Snead III OC, Boyd J, Go C. Adrenocorticotropic hormone versus prednisolone in the treatment of infantile spasms post vigabatrin failure. J Child Neurol 2015;30(5):595–600.

[96] Amabebe E, Anumba DO. Psychosocial stress, cortisol levels, and maintenance of vaginal health. Front Endocrinol 2018;9:568.

[97] Heir R, Stellwagen D. TNF-mediated homeostatic synaptic plasticity: from in vitro to in vivo models. Front Cell Neurosci 2020;14, 565841.

[98] Sun MS, Jin H, Sun X, Huang S, Zhang FL, Guo ZN, et al. Free radical damage in ischemia-reperfusion injury: an obstacle in acute ischemic stroke after revascularization therapy. Oxidative Med Cell Longev 2018;2018:1–17.

[99] Nasser M, Bejjani F, Raad M, Abou-El-Hassan H, Mantash S, Nokkari A, et al. Traumatic brain injury and blood-brain barrier cross-talk. CNS Neurol Disord Drug Targets 2016;15(9):1030–44.

[100] Zhao Q, Liu J, Deng H, Ma R, Liao JY, Liang H, et al. Targeting mitochondria-located circRNA SCAR alleviates NASH via reducing mROS output. Cell 2020;183(1):76–93.

[101] Shi T, Dansen TB. Reactive oxygen species induced p53 activation: DNA damage, redox signaling, or both? Antioxid Redox Signal 2020;33(12):839–59.

[102] McElroy PB, Liang LP, Day BJ, Patel M. Scavenging reactive oxygen species inhibits status epilepticus-induced neuroinflammation. Exp Neurol 2017;298:13–22.

[103] Fulton RE, Pearson-Smith JN, Huynh CQ, Fabisiak T, Liang LP, Aivazidis S, et al. Neuron-specific mitochondrial oxidative stress results in epilepsy, glucose dysregulation and a striking astrocyte response. Neurobiol Dis 2021;158, 105470.

[104] Patel M. A metabolic paradigm for epilepsy. Epilepsy Curr 2018;18(5):318–22.

[105] Ralta A, Prakash A, Kumar MP, Kumar R, Sarma P, Bhatia A, et al. Neuroprotective effect of *Celastrus paniculatus* seed extract on epilepsy and epilepsy-associated cognitive deficits. Basic Clin Neurosci 2023;14(1):155–66.

[106] Parsons AL, Bucknor E, Castroflorio E, Soares TR, Oliver PL, Rial D. The interconnected mechanisms of oxidative stress and neuroinflammation in epilepsy. Antioxidants 2022;11(1):157.

[107] Semwal P, Painuli S, Anand J, Martins NC, Machado M, Sharma R, et al. The neuroprotective potential of endophytic fungi and proposed molecular mechanism: a current update. Evid Based Complement Alternat Med 2022;2022:1–12.

[108] He LY, Hu MB, Li RL, Zhao R, Fan LH, Wang L, et al. The effect of protein-rich extract from *Bombyx batryticatus* against glutamate-damaged PC12 cells via regulating γ-aminobutyric acid signaling pathway. Molecules 2020;25(3):553.

[109] Kim MH, Lee HJ, Lee SR, Lee HS, Huh JW, et al. Peroxiredoxin 5 inhibits glutamate-induced neuronal cell death through the regulation of calcineurin-dependent mitochondrial dynamics in HT22 cells. Cell Mol Biol 2019;39(20), e00148-19.

[110] Cameron AM, Castoldi A, Sanin DE, Flachsmann LJ, Field CS, Puleston DJ, et al. Inflammatory macrophage dependence on NAD+ salvage is a consequence of reactive oxygen species-mediated DNA damage. Nat Immunol 2019;20(4):420–32.

[111] Shemiakova T, Ivanova E, Grechko AV, Gerasimova EV, Sobenin IA, Orekhov AN. Mitochondrial dysfunction and DNA damage in the context of pathogenesis of atherosclerosis. Biomedicine 2020;8(6):166.

[112] Yang N, Guan QW, Chen FH, Xia QX, Yin XX, et al. Antioxidants targeting mitochondrial oxidative stress: promising neuroprotectants for epilepsy. Oxidative Med Cell Longev 2020;2020:1–14.

[113] Pearson-Smith JN, Patel M. Metabolic dysfunction and oxidative stress in epilepsy. Int J Mol Sci 2017;18(11):2365.

[114] Andreasen M, Nedergaard S. Differential role of oxidative stress in synaptic and non-synaptic in vitro ictogenesis. J Neurophysiol 2023;129:999–1009.

[115] Nordgaard J, Melchior T. Long-term arrhythmia detection using an implantable loop recorder in patients receiving psychotropic medication. JAMA Psychiatry 2022;79 (1):77–8.

[116] Dalmau J, Armangué T, Planagumà J, Radosevic M, Mannara F, Leypoldt F, et al. An update on anti-NMDA receptor encephalitis for neurologists and psychiatrists: mechanisms and models. Lancet Neurol 2019;18(11):1045–57.

[117] Petit-Pedrol M, Armangue T, Peng X, Bataller L, Cellucci T, Davis R, et al. Encephalitis with refractory seizures, status epilepticus, and antibodies to the GABAA receptor: a case series, characterisation of the antigen, and analysis of the effects of antibodies. Lancet Neurol 2014;13(3):276–86.

[118] Gozzelino L, Kochlamazashvili G, Baldassari S, Mackintosh AI, Licchetta L, Iovino E, et al. Defective lipid signalling caused by mutations in PIK3C2B underlies focal epilepsy. Brain 2022;145(7):2313–31.

[119] Kaeberle J. Epilepsy disorders and treatment modalities. NASN Sch Nurse 2018;33 (6):342–4.

[120] Tomson T, Battino D, Perucca E. Valproic acid after five decades of use in epilepsy: time to reconsider the indications of a time-honoured drug. Lancet Neurol 2016;15(2):210–8.

[121] Chen Z, Brodie MJ, Liew D, Kwan P. Treatment outcomes in patients with newly diag-
      nosed epilepsy treated with established and new antiepileptic drugs: a 30-year longitu-
      dinal cohort study. JAMA Neurol 2018;75(3):279–86.
[122] Hendricks BK, Cohen-Gadol AA, Miller JC. Novel delivery methods bypassing the
      blood-brain and blood-tumor barriers. Neurosurg Focus 2015;38(3):E10.
[123] Mathew EN, Berry BC, Yang HW, Carroll RS, Johnson MD. Delivering therapeutics to
      glioblastoma: overcoming biological constraints. Int J Mol Sci 2022;23(3):1711.
[124] Fine JM, Stroebel BM, Faltesek KA, Terai K, Haase L, Knutzen KE, et al. Intranasal
      delivery of low-dose insulin ameliorates motor dysfunction and dopaminergic cell death
      in a 6-OHDA rat model of Parkinson's disease. Neurosci Lett 2020;714, 134567.
[125] Eng ME, Imperio GE, Bloise E, Matthews SG. ATP-binding cassette (ABC) drug trans-
      porters in the developing blood-brain barrier: role in fetal brain protection. Cell Mol Life
      Sci 2022;79(8):415.
[126] Wang H, Chao Y, Zhao H, Zhou X, Zhang F, Zhang Z, et al. Smart nanomedicine to
      enable crossing blood-brain barrier delivery of checkpoint blockade antibody for immu-
      notherapy of glioma. ACS Nano 2022;16(1):664–74.
[127] Parodi A, Rudzińska M, Deviatkin AA, Soond SM, Baldin AV, Zamyatnin Jr AA.
      Established and emerging strategies for drug delivery across the blood-brain barrier in
      brain cancer. Pharmaceutics 2019;11(5):245.
[128] Zhang W, Mehta A, Tong Z, Esser L, Voelcker NH. Development of polymeric
      nanoparticles for blood-brain barrier transfer—strategies and challenges. Adv Sci
      2021;8(10):2003937.
[129] Johnsen KB, Bak M, Melander F, Thomsen MS, Burkhart A, Kempen PJ, Andresen TL,
      Moos T. Modulating the antibody density changes the uptake and transport at the
      blood-brain barrier of both transferrin receptor-targeted gold nanoparticles and liposomal
      cargo. J Control Release 2019;295:237–49.
[130] Alkilany AM, Zhu L, Weller H, Mews A, Parak WJ, Barz M, et al. Ligand density on
      nanoparticles: a parameter with critical impact on nanomedicine. Adv Drug Deliv Rev
      2019;143:22–36.
[131] Georgieva JV, Hoekstra D, Zuhorn IS. Smuggling drugs into the brain: an overview of
      ligands targeting transcytosis for drug delivery across the blood-brain barrier. Pharma-
      ceutics 2014;6(4):557–83.
[132] Ceña V, Játiva P. Nanoparticle crossing of blood-brain barrier: a road to new therapeutic
      approaches to central nervous system diseases. Nanomedicine 2018;13(13):1513–6.
[133] Chaichana KL, Pinheiro L, Brem H. Delivery of local therapeutics to the bRn: working
      toward advancing treatment for malignant gliomas. Ther Deliv 2015;6(3):353–69.
[134] Zhu X, Jin K, Huang Y, Pang Z. Brain drug delivery by adsorption-mediated transcytosis.
      In: Brain targeted drug delivery system. Academic Press; 2019. p. 159–83.
[135] Su S, Kang PM. Systemic review of biodegradable nanomaterials in nanomedicine.
      Nanomaterials 2020;10(4):656.
[136] Jabir NR, Tabrez S, Firoz CK, Kashif Zaidi S, Baeesa SS, Hua Gan S, et al. A synopsis of
      nano-technological approaches toward anti-epilepsy therapy: present and future research
      implications. Curr Drug Metab 2015;16(5):336–45.
[137] Yasir M, Chauhan I, Zafar A, Verma M, Alruwaili NK, Noorulla KM, et al. Glyceryl
      behenate-based solid lipid nanoparticles as a carrier of haloperidol for nose to brain deliv-
      ery: formulation development, in-vitro, and in-vivo evaluation. Braz J Pharm Sci
      2023;58, e20254.
[138] Qushawy M, Prabahar K, Abd-Alhaseeb M, Swidan S, Nasr A. Preparation and evalu-
      ation of carbamazepine solid lipid nanoparticle for alleviating seizure activity in
      pentylenetetrazole-kindled mice. Molecules 2019;24(21):3971.

[139] Leyva-Gómez G, González-Trujano ME, López-Ruiz E, Couraud PO, Wekslerg B, Romero I, et al. Nanoparticle formulation improves the anticonvulsant effect of clonazepam on the pentylenetetrazole-induced seizures: behavior and electroencephalogram. J Pharm Sci 2014;103(8):2509–19.

[140] Deshkar SS, Jadhav MS, Shirolkar SV. Development of carbamazepine nanostructured lipid carrier loaded thermosensitive gel for intranasal delivery. Adv Pharm Bull 2021;11 (1):150.

[141] Montoto SS, Sbaraglini ML, Talevi A, Couyoupetrou M, Di Ianni M, Pesce GO, et al. Carbamazepine-loaded solid lipid nanoparticles and nanostructured lipid carriers: physicochemical characterization and in vitro/in vivo evaluation. Colloids Surf B: Biointerfaces 2018;167:73–81.

[142] Hersom M, Helms HC, Pretzer N, Goldeman C, Jensen AI, Severin G, et al. Transferrin receptor expression and role in transendothelial transport of transferrin in cultured brain endothelial monolayers. Mol Cell Neurosci 2016;76:59–67.

[143] Hersh AM, Alomari S, Tyler BM. Crossing the blood-brain barrier: advances in nanoparticle technology for drug delivery in neuro-oncology. Int J Mol Sci 2022;23(8):4153.

[144] Lalatsa A, Schatzlein AG, Uchegbu IF. Strategies to deliver peptide drugs to the brain. Mol Pharm 2014;11(4):1081–93.

[145] Anraku Y, Kuwahara H, Fukusato Y, Mizoguchi A, Ishii T, Nitta K, et al. Glycaemic control boosts glucosylated nanocarrier crossing the BBB into the brain. Nat Commun 2017;8(1):1001.

[146] Singh SK, Mishra DN. Nose to brain delivery of galantamine loaded nanoparticles: in-vivo pharmacodynamic and biochemical study in mice. Curr Drug Deliv 2019;16 (1):51–8.

[147] Rabinowicz AL, Carrazana E, Maggio ET. Improvement of intranasal drug delivery with Intravail alkylsaccharide excipient as a mucosal absorption enhancer aiding in the treatment of conditions of the central nervous system. Drugs R D 2021;21(4):361–9.

[148] Greco A, Rizzo MI, De Virgilio A, Conte M, Gallo A, Attanasio G, et al. Autoimmune epilepsy. Autoimmun Rev 2016;15(3):221–5.

[149] Jang Y, Kim DW, Yang KI, Byun JI, Seo JG, No YJ, et al. Clinical approach to autoimmune epilepsy. J Clin Neurol 2020;16(4):519.

[150] Kattepur AK, Patil D, Shankarappa A, Swamy S, Chandrashekar NS, Chandrashekar P, Prabhu S, Gopinath KS. Anti-NMDAR limbic encephalitis-a clinical curiosity. World J Surg Oncol 2014;12:1–5.

[151] McKeon A, Dubey D, Flanagan E, Pittock S, Zekeridou A. Autoimmune psychosis. Lancet Psychiatry 2020;7(2):122.

[152] Das M, Karnam A, Stephen-Victor E, Gilardin L, Bhatt B, Kumar Sharma V, Rambabu N, et al. Intravenous immunoglobulin mediates anti-inflammatory effects in peripheral blood mononuclear cells by inducing autophagy. Cell Death Dis 2020;11(1):50.

[153] Munguía-Martínez MF, Nava-Ruíz C, Ruíz-Díaz A, Díaz-Ruíz A, Yescas-Gómez P, Méndez-Armenta M. Immunohistochemical study of antioxidant enzymes regulated by Nrf2 in the models of epileptic seizures (KA and PTZ). Oxidative Med Cell Longev 2019;2019:1–8.

[154] Ramis MR, Esteban S, Miralles A, Tan DX, Reiter RJ. Protective effects of melatonin and mitochondria-targeted antioxidants against oxidative stress: a review. Curr Med Chem 2015;22(22):2690–711.

[155] Berzabá-Evoli E, Tejas-Juárez JG, Gómez-Crisóstomo NP, De la Cruz-Hernández EN, Martínez-Abundis E. Chemicals with mitochondrial targets for the treatment of neurodegenerative disorders. Annu Res Rev Biol 2017;21:1–9.

[156] Xing J, Han D, Xu D, Li X, Sun L. CREB protects against temporal lobe epilepsy associated with cognitive impairment by controlling oxidative neuronal damage. Neurodegener Dis 2019;19(5–6):225–37.

[157] Zalkhani R, Moazedi A. Basic and clinical role of vitamins in epilepsy. JBRMS 2020;6 (2):104–14.

[158] Dubaj C, Czyż K, Furmaga-Jabłońska W. Vitamin B12 deficiency as a cause of severe neurological symptoms in breast fed infant—a case report. Ital J Pediatr 2020;46(1):1–6.

[159] Ullah I, Badshah H, Naseer MI, Lee HY, Kim MO. Thymoquinone and vitamin C attenuates pentylenetetrazole-induced seizures via activation of GABA B1 receptor in adult rats cortex and hippocampus. NeuroMol Med 2015;17:35–46.

[160] Elfakhri KH, Duong QV, Langley C, Depaula A, Mousa YM, Lebeouf T, Cain C, Kaddoumi A. Characterization of hit compounds identified from high-throughput screening for their effect on blood-brain barrier integrity and amyloid-β clearance: in vitro and in vivo studies. Neuroscience 2018;379:269–80.

[161] Farhadi Moghaddam B, Fereidoni M. The effect of Menaquinone-4 administration on the expression of proinflammatory factors following transient global cerebral ischemia/reperfusion in the hippocampus of male Wistar rat. JSUMS 2021;28(1):116–22.

[162] Ekezie JC, Okoromah CA, Lesi FE. Serum vitamin E levels in children and adolescents with epilepsy at a tertiary hospital in Nigeria. SN Compr Clin Med 2020;2:2278–87.

[163] Motaghinejad M, Motevalian M, Shabab B, Fatima S. Effects of acute doses of methylphenidate on inflammation and oxidative stress in isolated hippocampus and cerebral cortex of adult rats. J Neural Transm 2017;124:121–31.

[164] Ahmad M, Wadaan MA. Ameliorating effects of proglumide on neurobehavioral and biochemical deficits in animal model of status epilepticus. Pak J Pharm Sci 2014;27 (6):1945–51.

[165] Hass DT, Barnstable CJ. Mitochondrial uncoupling protein 2 knock-out promotes mitophagy to decrease retinal ganglion cell death in a mouse model of glaucoma. J Neurosci 2019;39(18):3582–96.

[166] Sztolsztener K, Hodun K, Chabowski A. α-Lipoic acid ameliorates inflammation state and oxidative stress by reducing the content of bioactive lipid derivatives in the left ventricle of rats fed a high-fat diet. Biochim Biophys Acta Mol basis Dis 2022;1868(9), 166440.

Diabetic retinopathy: Stressing the function of angiogenesis, inflammation and oxidative stress

14

Pranay Wal[a], Ankita Wal[a], Divyanshi Gupta[a], Shubhrajit Mantry[b], Kiran Chandrakant Mahajan[b], Shruti Rathore[c], and Tapan Behl[d]
[a]PSIT-Pranveer Singh Institute of Technology (Pharmacy), Kanpur, Uttar Pradesh, India, [b]Sgmspm's Sharadchandra Pawar College of Pharmacy, Pune, Maharashtra, India, [c]LCIT School of Pharmacy, Bilaspur, Chhattisgarh, India, [d]Amity School of Pharmaceutical Sciences, Amity University, Mohali, Punjab, India

Abbreviations

AGEs	angiogenesis
CRP C	reactive protein
CWs	cotton wool spots
DM	diabetes mellitus
DME	diabetic macular edema
DR	diabetic retinopathy
EXs	exudates
HEMs	hemorrhages
ICAM	intracellular adhesion molecule
IL	interleukin
MAs	microaneurysms
MCP-1	monocyte chemoattractant protein
MIP	macrophage inflammatory protein
NF-κB	nuclear factor kappa-light-chain-enhancer of activated B cells
NO	nitric oxide
NPAs	nonperfusion regions
NPD	nonproliferative diabetes retinopathy
NPDR	nonproliferative diabetic retinopathy
NSAID	nonsteroidal antiinflammatory drug
PASCAL	pattern scanning laser
PDR	proliferative diabetes retinopathy
PDR	proliferative diabetic retinopathy
PKC	protein kinase c
PUFAs	poly unsaturated fatty acids
RAS	renin-angiotensin system
RBRB	blood-retinal barrier
ROS	reactive oxygen species
TNF	tumor necrosis factor

Targeting Angiogenesis, Inflammation and Oxidative Stress in Chronic Diseases. https://doi.org/10.1016/B978-0-443-13587-3.00002-3

UV   ultraviolet
VCAM   vascular cell adhesion molecule
VEGF   vascular endothelial growth factor
VE-PTP   vascular endothelial protein tyrosine phosphatase
WHO   World Health Organization

1   Introduction

The prevalence of diabetes mellitus is rising globally. By 2035, the World Health Organization projects that 592 million people will have diabetes worldwide [1,2]. The body's inability to regulate the secreted insulin causes diabetes. Type I diabetes, for which the cause is unclear, and type II diabetes, which can be identified by the symptoms shown in imaging technology, are the two distinct subtypes of diabetes [3]. Numerous indications, including microaneurysms, cotton wool spots, exudates, HEMs, etc., can be used to identify type II diabetes. In ophthalmology, it may result in Diabetic Retinopathy if neglected for longer than 5 years [4]. Chronic diabetes and insufficient blood glucose control cause observable clinical defects in the retinal fundus, which are the hallmarks of diabetic retinopathy. Although virtually all diabetics experience diabetic retinopathy (DR) at some time, even those with well-controlled hyperglycemia can delay the onset and progression of DR with appropriate systemic glucose management [5–7]. One of the main pathologies of DR is retinal nonperfusion areas, which are caused by retinal capillary closure [8,9]. The difference between normal retinal function and diabetic retinopathy is seen in Fig. 1 in the following section.

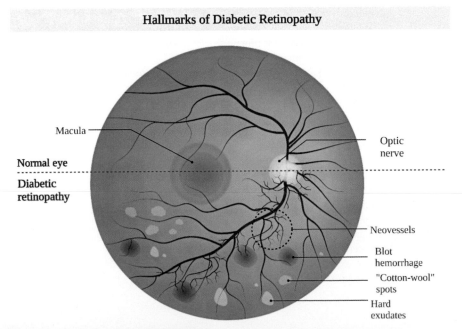

Fig. 1 Normal retina and diabetic retina.

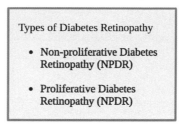

Fig. 2 Diabetic retinopathy stages.

Although virtually all diabetics experience DR at some time, even those with well-controlled hyperglycemia can delay its onset and progression with adequate systemic glucose control.

Diabetic retinopathy manifests as vascular abnormalities in the retina and is divided into two types as NonProliferative Diabetes Retinopathy and Proliferative Diabetes Retinopathy as shown in Fig. 2 [5]. Two important discoveries may be seen in the retinal vasculature of NPDR, the initial stage of DR: increased vascular permeability and capillary obstruction. Even though the patients are asymptomatic, fundus photography can still identify retinal abnormalities such as microaneurysms, hemorrhages, and hard exudates. Neovascularization is a defining feature of the PDR stage of DR. At this point, tractional retinal detachment, or bleeding into the vitreous from newly formed aberrant arteries might cause the patients to lose considerable amounts of their vision [10].

Several pieces of literature claim that there are several methods for identifying DR and judging the severity of the condition. The severity of the ailment can be assessed using the grading system shown in Fig. 3 [11]. Along with numbers, a table displaying the various DR stages is also present. During the early stage, there is less or no chance of damage in the retinal images, microaneurysms can be seen in one of the four quadrants in mild NPDR. On fundus imaging with moderate NPDR, microaneurysms, hemorrhages, and exudates can be detected. In PDR, neovascularization—the growth of new blood vessels—occurs after microvascular abnormalities in the retina and venous beading in severe NPDR [12]. The metabolic changes brought on by diabetes that affect the retina and result in a range of molecules and pathways that are briefly described in the following section in the pathophysiology of DR are the focus of this study. The schematic graphic in Fig. 4 illustrates these important sequences of events.

2 Pathophysiology

The pathophysiology results in the endpoints of diabetic retinopathy that pose a danger to eyesight. Numerous studies have demonstrated that glial cells initially identify metabolic changes, which leads to glial dysfunction, which then results in inflammation and metabolic abnormalities which lead to neuronal death [13]. The primary characteristics of early microvascular anomalies and the ultimate stage of neovascularization, such as blood-retinal barrier collapse, vasoregression, and subsequent hypoxia, have an impact

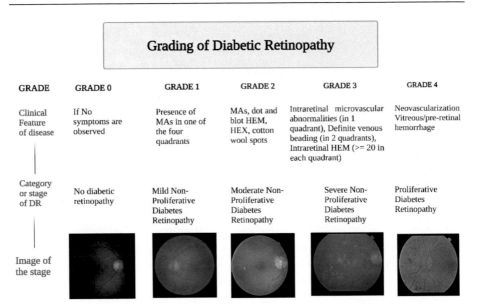

Grading of Diabetic Retinopathy					
GRADE	GRADE 0	GRADE 1	GRADE 2	GRADE 3	GRADE 4
Clinical Feature of disease	If No symptoms are observed	Presence of MAs in one of the four quadrants	MAs, dot and blot HEM, HEX, cotton wool spots	Intraretinal microvascular abnormalities (in 1 quadrant), Definite venous beading (in 2 quadrants), Intraretinal HEM (>= 20 in each quadrant)	Neovascularization Vitreous/pre-retinal hemorrhage
Category or stage of DR	No diabetic retinopathy	Mild Non-Proliferative Diabetes Retinopathy	Moderate Non-Proliferative Diabetes Retinopathy	Severe Non-Proliferative Diabetes Retinopathy	Proliferative Diabetes Retinopathy
Image of the stage					

Fig. 3 Stages of diabetic retinopathyas well as various diabetic retinopathy grades.

on neurodegeneration as well. As seen in mentioned previously Fig. 3, ophthalmoscopy is utilized to stage the severity of DR in line with the vascular lesions. It is often believed that hyperglycemia is the main cause of the retinal damage associated with DR since it activates and dysregulates a variety of metabolic pathways [14,15]. High glucose concentrations inside cells may increase the flux through glycolytic pathways, causing protein kinase C to be activated as well as the polyol, polymerase, and hexosamine pathways to be activated. These pathways then increase the production of reactive oxygen species and nonenzymatic glycosylation, which leads to a high level of advanced glycation end products. Numerous studies suggest that an increase in the fluxes through these hyperglycemia-induced pathways may set off a chain of events that may damage the diabetic retina and lead to DR, including an increase in oxidative stress, an increase in apoptosis, an increase in the inflammatory response, and an increase in angiogenesis as shown in Fig. 4 [16–18].

2.1 Inflammation pathway

The pathophysiology of DR is heavily influenced by inflammation [19,20]. Inflammation, a widespread inflammatory response to injury that encompasses a wide variety of molecular and functional mediators, involves the gathering and activation of leukocytes. Many of the molecular mediators and functional changes that are characteristic of inflammation have been discovered in the retinas of diabetic animals or people, as well as retinal cells under diabetes conditions. As early as 3 days following the introduction of diabetes in rats, increased leukocyte adhesion was found in the retinal vasculature [21–25]. Additionally, the researchers discovered a geographical correlation between enhanced leukostasis and endothelium degradation as well as BRB

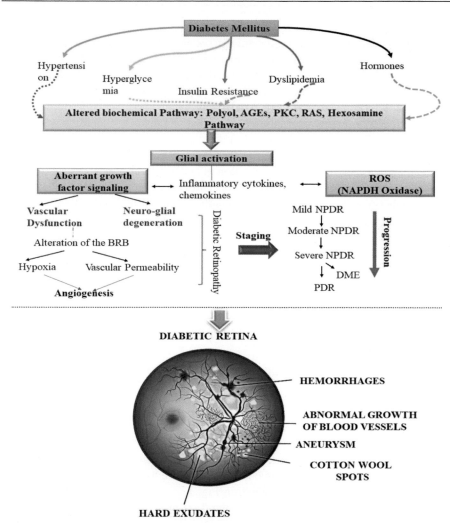

Fig. 4 General characteristics of diabetes-related retinopathy: Among the mediators that diabetes creates include growth factors, hormones, oxidative stress, angiogenesis, and inflammatory biomarkers.

dysfunction. The role of inflammation factors in the development of diabetic retinopathy is critical. Leukostasis is first to blame for the persistent low-grade inflammation, according to a review of the literature on screening models in both humans and animals. Through the Fas (CD95)/Fas ligand pathway, increased levels of leukostasis are to blame for endothelium damage, BRB deficits, endothelial cell death, and BRB breakdown [26,27]. The expression levels of inflammatory cytokines such as tumor necrosis factor-alpha, interleukin 6, IL-8, and IL-1 were significantly greater in diabetic patients and were associated with the severity of DR [28]. Also thought to have a role in the emergence and escalation of retinal inflammation in DR is the malfunctioning of retinal glial cells. The retina's glial cells, which consist of astrocytes, Müller cells, and microglia, oversee sustaining the retina's structure and maintaining homeostasis. Hyperglycemic stress causes microglia to activate, which is followed by an increase in the production of TNF-, IL-6, MCP-1, and VEGF [29]. Müller cells and astrocytes are associated with the amplification of inflammatory responses because they produce proinflammatory cytokines afterward.

Diabetes has been associated with leukostasis, which is mediated by adhesion molecules and involves leukocyte-endothelium adhesion. Leukocyte adhesion was shown to be higher in diabetic rats and humans, and the expression of b2-integrins was raised [30,31]. The severity of DR is correlated with plasma levels of VCAM-1 and E-selectin expression in patients. Production of adherent leukocytes was significantly reduced as a result of genetic ICAM-1 abnormality [32]. Additionally, selectins, intercellular adhesion molecule-1, vascular cell adhesion molecule, and other endothelial cell adhesion molecules have been found to be more abundant in diabetes patients and animals [33–35]. By blocking CD18 or ICAM-1, antiCD18 F(ab9)2 fragments or antibodies decreased retinal leukostasis and vascular lesions in diabetic rats [36]. Additionally, it has been shown that chemokines, which regulate the activation and recruitment of leukocytes, play a part in the pathogenesis of DR. Several chemokines, including monocyte chemotactic protein-1, macrophage inflammatory protein-1alpha, and MIP-1, have been linked to increased levels in diabetic patients [37]. MCP-1 mutation reduces retinal vascular leakage in diabetic mice [38].

2.2 Role of renin-angiotensin system (RAS)

There is strong evidence that vascular endothelial growth factor-A controls pathologic ocular angiogenesis in diabetic retinopathy (VEGF-A). The primary VEGF-A producers in the retina are ganglion cells, Müller cells, and cells of the retinal pigment epithelium. Highly affinity VEGF receptors have been discovered to be present in retinal endothelial cells and pericytes [39,40]. In a morphogenic process, the angiogenic switch mechanism can activate the ocular vasculature to create new capillaries. The structure of the VEGF protein is shown in Fig. 5.

Along with the synthesis of VEGF brought on by hypoxia, angiogenesis may be controlled by angiogenic inducers and inhibitors. Research has shown how important it is to balance angiogenesis inducers and inhibitors to prevent neovascularization in diabetic retinopathy. Studies have frequently just looked at the effects of angiogenesis inducers. In this study, the effects of endostatin and PF-4—two potential angiogenesis inhibitors—on capillary development in diabetic retinopathy were examined. The amounts of VEGF, one of the most important angiogenic stimulators, endostatin, and PF-4 in the vitreous plasma of diabetic patients (angiogenic inhibitors). The results of this research demonstrated a significant association between ocular levels of VEGF and endostatin and the severity of diabetic retinopathy [41]. The levels of VEGF and endostatin in the vitreous and plasma did not appear to be correlated. Although endostatin levels were equivalent to those in the plasma, VEGF levels in the vitreous were much higher than those in the latter. Although the exact relationship between VEGF and angiogenic inhibitors (endostatin and PF-4) could not be determined by this investigation, it does suggest a correlation between VEGF expression and the presence of these substances. These results [42–44] suggest that the interaction

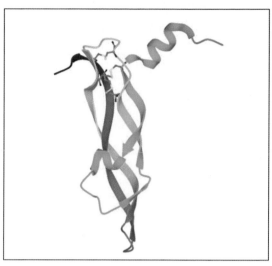

Fig. 5 VEGF molecule: *"Vascular Endothelial Growth Factor protein."*

between VEGF and angiogenic inhibitors may regulate the formation of angiogenesis in diabetic retinopathy.

2.3 Role of oxidative stress

The term "oxidative stress" refers to an imbalance in the production and elimination of free radicals. Oxidative stress is caused by an excess of free radicals that are not eliminated by scavenging enzymes [45,46]. Another way to describe the sickness is a disruption in dynamic redox balance, which causes oxidative stress and eventually affects the cells of target organs such as the heart, kidneys, and retina. The retina is vulnerable to oxidative stress due to the large concentration of rapidly oxidized PUFAs in the membranes of its outer photo receptor segment and the likelihood of persistent exposure to ROS-producing visible light or UV. Docosahexaenoic acid, arachidonic acid, and oleic acid are the primary PUFAs in the human retina's outer photoreceptor segments, accounting for roughly 50%, 8%, and 10% of total fatty acids, respectively [47]. These PUFAs are essential for the structure and function of the retina, but because they are susceptible to oxidative damage, they are also excellent targets for the ROS that cause lipid peroxidation. In addition to harming visual cells, lipid peroxidation of PUFAs in the retina impacts the retina's physiological health. By interacting with biological macromolecules (DNA and proteins) in the retina, lipid peroxidation byproducts such as hydroxy hexenal and hydroxynonenal cause deficits in photoreceptor cells and abnormalities in the retinal pigment epithelium [48–51]. The presence of high oxygen tension in the retina stimulates ROS generation and lipid peroxidation. Both the activity of the visual imaging system and an active metabolism necessitate a significant consumption of oxygen [52]. The impact of oxidative stress in the development of DR is critical. A substantial buildup of ROS may induce tissue damage inside and around retinal arteries, eventually resulting to DR. AGEs (hyperglycemia-induced oxidative damage in the retina) are generated by four major

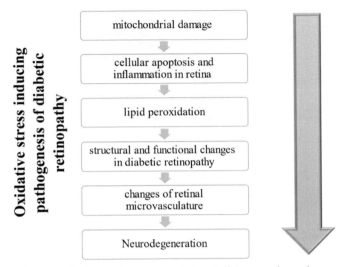

Fig. 6 Process involved in the oxidative stress induced diabetes retinopathy.

metabolic abnormalities, according to research: the protein kinase C (PKC) pathway, polyol pathway flow, hexosamine route, and intracellular AGE synthesis [53–56].

Excess ROS in DR has been linked to metabolic abnormalities, irregular epigenetic changes, and hyperglycemia-mediated mitochondrial dysfunction [57]. Furthermore, abnormal nuclear factor activity, such as highly activated nuclear factor-B (NF-B) and decreased activity of Nrf2 (also known as NFE2L2, nuclear factor erythroid 2 related factor, and hyperinsulinemia-mediated the mitochondrial dysfunction), has been linked to these symptoms. Fig. 6 summarizes the pathogenetic implications of the effects of oxidative stress in the onset of diabetic retinopathy.

3   Current and novel treatment approaches for diabetes retinopathy

Diabetes Retinopathy causes pathway upregulation and enzymatic stimulation, and the formation of complex glycosylated end products. Each of these pathways can cause oxidative stress, inflammation, and angiogenesis, thereby initiating the pathogenesis of diabetic retinopathy, which is characterized by the degeneration of endothelial cells in the vascular system, tight junctions as well as cellular basement membrane thickening, eventually leading to hypoxia and neovascularization. Using antiVEGF medicine, steroids, advanced laser therapy, and vitrectomy is the most successful treatment for diabetic retinopathy. Pan-retinal photocoagulation has been found to be effective in preventing the progressive stage of diabetic retinopathy.

The possible therapies for diabetic retinopathy are summarized in Table 1.

Table 1 List of all possible treatments for diabetes retinopathy.

Category	Class of drugs	Name of drugs	Status of treatment	Clinical uses	Complications	Reference
Antiangiogenic agents						
	AntiVascular endothelial growth factor (VEGF)	*Bevacizumab*	Off-label use	• Better median visual acuity over laser • More central retinal thickness reduction	1. Inflammation 2. Increased IOP 3. Vitreous hemorrhage	[58]
		Ranibizumab	Approved (FDA)	• Greater reduction in central retinal thickness over laser in treating DME • Greater BCVA improvements	1. Inflammation 2. Increased IOP 3. Vitreoushemorrhage	[59,60]
		Aflibercept	Approved (FDA)	• Greater BCVA improvements over laser in treating DME	1. Inflammation 2. Increased IOP 3. Vitreoushemorrhage	[61]
		Pegaptanib	Approved (FDA)	• Greater BCVA improvement over laser in treating DME	1. Inflammation 2. Increased IOP 3. Conjunctival hemorrhage	[62]
	Nonspecific antiangiogenic	Tie-2-activator	In Phase 2 trial (In progress)	—	—	[63]
		"Squalamine"	In Phase 2 trial	Shows reduction in central retinal thickness	1. Diabetes worsens retinal edoema 2. Lower visual acuity	[64]

Continued

Table 1 Continued

Category	Class of drugs	Name of drugs	Status of treatment	Clinical uses	Complications	Reference
		"Nesvacumab"	In clinical trial	No distinction between aflibercept monotherapy and combinations of Nesvacumab and aflibercept to support phase 3 development	Comparing it to other antiangiogenic drugs, there were no additional safety flags found	[65]
		"antiang-2 + antiVEGF" (Bispecific Antibody)	In clinical trial	higher improvement in BCVA and a larger decrease in CRT compared to ranibizumab in DME patients	Good toleration and no new safety signs were noticed	[65]
Antiinflammatory drugs						
	IL-6 inhibitor	"EBI-031" (clinicaltrials. gov ID: NCT02842541)	In clinical trial	–	–	[66]
	IL-6 receptor inhibitor	"Tocilizumab"	In clinical trial	–	–	[67]
	Integrin inhibitor	"Luminate"	Phase 2b trial	The mean change was noninferior.	No intraocular inflammation and toxicity reported	[68]
	Intravitreal steroids	"Triamcinolone"	Approved (FDA)	Triamcinolone + prompt → more improvements than laser	Increase IOP Vitreous hemorrhage	[69]

	"DEX implant"	Approved (FDA)	Patients with DME who had CRT showed higher decrease and BCVA improvement compared to the sham group	• Cataract surgery • Increase IOP • Vitreous hemorrhage	[70,71]
	"Iluvien" (0.2 mg)	Approved (FDA)	Over a two-year period, individuals with DME showed greater BCVA improvement than the sample treated	• Cataract surgery • Increase IOP	[72]

Laser photocoagulation

Traditional	"Focal or grid laser"	Alternative for DME	Reduced chronic macular edoema, increased likelihood of visual recovery, and reduced risk of mild vision loss	Loss of visual activity	[73]
	"Pan-retinal photocoagulation"	PDR adjuvant therapy for high-risk consequences	Reduces the progression and risk of retinopathy	Loss of peripheral visual activity	[74]
Laser approaches	"PASCAL"	A way to approved	Reduced treatment time and precise laser control	—	[75]
	"D-MPL"	Under testing	Put a limit on collateral injury	—	[76]

Continued

Table 1 Continued

Category	Class of drugs	Name of drugs	Status of treatment	Clinical uses	Complications	Reference
		"NAVILAS"	Under testing	High laser pinpoint precision	–	[77]
Alternative treatments						
	Cardiolipin inhibitor	"MTP-131"	In clinical trial	–	–	[14]
	Mitochondria specific antioxidant	"ALA"	Under testing	Improved contrast sensitivity.	–	[15]
	Antioxidant	"Lutein"	Under testing	Improvement in persons with DR's vision	–	[16]

3.1   Anti-VEGF therapy

Ranibizumab and Nesvacumab, according to reports, considerably improve the condition of individuals with progressive diabetic retinopathy. There was also less of a need for vitrectomy and less peripheral vision field loss. These drugs have also improved the ratings on the diabetic retinopathy severity scale. Medications such as pegaptanib, ranibizumab, aflibercept, and intravitreal bevacizumab are now being studied in clinical trials for diabetic retinopathy. The CLARITY (Clinical Efficacy and Mechanistic Evaluation of Aflibercept for Proliferative Diabetic Retinopathy) research compared pan-retinal Photocoagulation with aflibercept for patients with PDR without DME [78,79]. According to a study group, ranibizumab reduced neovascularization and maintained better visual acuity over PRP treatment throughout the first 12 months of the PRIDE research, but it is not sustained for more than 24 months in real-world scenarios. These findings show that PRP may not be as effective as antiVEGF therapy in delaying the progression of DR, but antiVEGF therapy requires constant monitoring, whereas general laser photocoagulation has a long-term effect on the treated eyes. Patients who stop visiting the clinic risk having their retinopathy worsen since antiVEGF treatment requires regular follow-up appointments. If a patient is on antiVEGF medication, the prognosis is said to be worse than photocoagulation [80]. Because antiVEGF medications are expensive and necessitate frequent clinic visits, both financial and patient-specific considerations like as visit compliance should be considered while regulating PDR. New drugs that selectively target members of the Ang-Tie2 signaling pathway have also been created. AKB-9778, a small molecule that activates Tie2 and lowers vascular permeability, inhibits vascular endothelial protein tyrosine phosphatase (VE-PTP), a negative Tie2 regulator [81]. Nesvacumab suppresses Ang-2, which lowers vascular permeability by activating Tie2 as shown in Fig. 7. In phase 2 research with DME patients, Nesvacumab and the VEGF inhibitor aflibercept are being investigated [82]. However, the drawbacks and side effects of antiVEGF treatment are a major source of concern. Because antiVEGF medications have a limited half-life, monthly or biweekly injections are required to maintain efficacy. With frequent intravitreal injections, the risk of endophthalmitis, an uncommon adverse effect of the procedure, may increase. Three cases (0.08%) of injection-related endophthalmitis were described in the five-year DRCR.net Protocol I study after 3973 injections [83]. The use of antiVEGF medications in clinical practice was limited by financial constraints and patients' poor compliance. Furthermore, the use of high-dose antiVEGF medications should be carefully scrutinized [84], even though VEGF may have a neuroprotective impact.

3.2   Antiinflammatory therapy

Corticosteroids are powerful antiinflammatory drugs that specifically target the tight-junction "*protein phosphorylation*," "*VEGF*," "*TNF*," and "*chemokines*," which are mediators in the pathogenesis of DME. Intravitreal corticosteroids are now being employed in clinical studies for DME therapy utilizing off-label triamcinolone

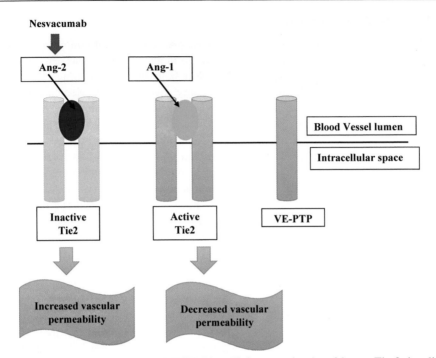

Fig. 7 New approaches to suppress VEGF: "As a Tie2 antagonist, Ang-2 boosts Tie-2 signaling and reduces vascular permeability, whereas Nesvacumab blocks its effect. When Ang-1 suppresses a negative regulator of tie2, tie-2 signaling is triggered."

acetonide, intravitreal fluocinolone acetonide implants, and intravitreal dexamethasone implants [85]. A study of several studies has revealed that IL-6, one of the most significant proinflammatory cytokines found in the retinal of Diabetic Retinopathy patients has been identified as a possible target for treatment. An IL-6 receptor antibody called tocilizumab has been created. Since the inflammatory process has been recognized as a crucial factor in the pathophysiology of DR, much effort has been undertaken to target it. By significantly reducing cellular swelling and accelerating fluid reabsorption, fluocinolone acetonide, dexamethasone, and other corticosteroids, such as triamcinolone acetonide, reduce vascular permeability and restore BRB integrity in DR. Effective antiinflammatory medications such as corticosteroids also target several mediators, including as "adhesion molecules," chemokines," and "inflammatory-molecules." A possible integrin antagonist is also being researched [86,87]. Numerous integrin receptors are inhibited by the integrin inhibitor Luminate. It produced encouraging results in lowering macula edema and promoting visual gain in a phase 2b trial for DME. Strong glucocorticoids like TA have been shown to lessen central macular thickness and neovascularization by inhibiting ICAM-1, TNF-, and IL-6 in hypoxic conditions [88,89]. Canakinumab, a selective IL-1 antibody, can stabilize retinal neovascularization and reduce macular edema in PDR patients.

3.3 Antioxidant therapy

The primary enzymatic source of ROS, NOX, is intimately linked to the promotion of pathogenic neovascularization in the retina by hyperglycemia. As of now, NOX1, NOX2, and NOX4 are the isoforms of the NOX family that has been most thoroughly researched in connection to DR, and several inhibitors have been found to stop the development of DR. Apocynin and diphenyleneiodonium are NOX inhibitors. Both have ROS-decreasing actions that are independent of NOX, despite their therapeutic effects also reducing VEGF and ROS levels. As a result, it is challenging to determine their therapeutic impact on DR based just on NOX inhibition. The creation of NOX inhibitors may enhance the management of DR.

3.4 Polyphenols

The most prevalent antioxidants in our diet are polyphenols as well as Natural polyphenols are secondary plant metabolites that may be found in large quantities in fruits, vegetables, and whole grains, as well as in meals and drinks that are made from these ingredients, such as chocolate, wine, olive oil, and tea. The polyphenol most often used and researched is resveratrol. Resveratrol was given orally to prevent streptozotocin-nicotinamide-induced diabetic mice by reducing oxidative stress and inflammatory cytokines caused by hyperglycemia via NRF2-Keap1 signaling. Localized hypoxia is brought on by diabetes and ROS and is accompanied by increased mRNA expression levels of proapoptotic proteins, HIF-1, and AMPK phosphorylation, which trigger cell death and VEGF synthesis. Resveratrol, epigallocatechin gallate, curcumin, and quercetin are a few more compounds that have been shown to have a protective impact against diabetic retinopathy. They all share the ability to function as antioxidants against oxidative stress following hypoxia as well as the ability to effectively halt pathological angiogenesis by reducing VEGF levels [90,91]. Therefore, it has been proposed that the administration of polyphenols can prevent the development of neoangiogenesis in DR.

3.4.1 Cardiolipin-targeting peptide

Recent investigations have revealed that the cardiolipin-targeting peptide may be crucial in triggering retinal neuron death because reduced cardiolipin, a phospholipid found in the inner mitochondrial membrane, the cardiolipin-targeting peptide may have an impact on cell death and a protective impact on visual function [92,93].

3.5 Laser treatment

PASCAL is a pattern scanning new laser technique which use to minimize retinal damage and reduce the treatment time. This technique involves micropulse diode which treat diabetes retinopathy with reduced collateral damage. Now-a-days Photocoagulation laser is the standard laser treatment for progressive diabetes retinopathy and has the most diverse future developments. These laser treatment

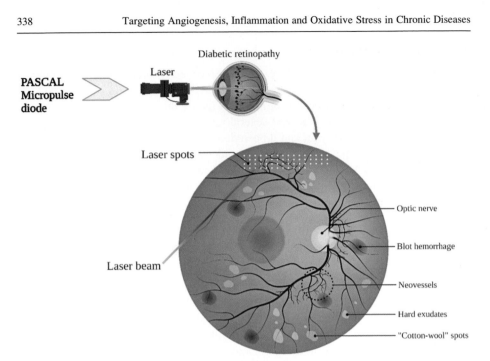

Fig. 8 Diabetic retinopathy laser photocoagulation treatment.

decreases the DME and reduced macular edema. Direct closure of leaky micro-aneurysms, a decrease in retinal blood flow connected to fewer retinal tissues and improved oxygenation, and stimulation of the retinal pigment epithelium (RPE) are all potential contributing factors. To reduce adverse effects, new laser treatments are being developed. Fig. 8 illustrates the laser treatment through different techniques.

3.6   Alternative and conventional treatments

Supplemental treatment for DR has recently drawn increased attention. The pathogenic abnormalities in DR have been regulated by drugs such as curcumin, resveratrol, palmitoylethanolamide, and melatonin due to their potential antioxidant as well as antiinflammatory and antiproliferative effects. In a study by Fan and Lei [94] shows that the active constituent of turmeric results in vascular alterations and neurodegeneration which results in the restoration of microvascular and blood vessels. According to this study reduced production of TNF-factor and VEGF protein are responsible for this restoration. Resveratrol, an antioxidant and neuroprotective molecule, does wonders to improve DR. Retinal ganglion cells are prevented from degenerating in mouse retinas when diabetes has been induced by resveratrol, which blocks the formation of caspase-3 and caspase-8. Palmitoylethanolamide, an endogenous cell-protective lipid, reduces the levels of cytokines and other substances that cause inflammation by preventing reactive gliosis and oxidative stress in glial cells.

4 Recent advances in application of nanotechnology in the treatment of diabetes retinopathy

Nowadays, a lot of nanotechnology is applied to treat diabetes complications. According to reports, nanoparticles can affect neovascularization [95]. Yang [96] reported using nanoparticles to treat diabetes complications by reducing the production of inflammatory cytokines and other expressions. In this study, the crocetin-loaded PLGA nanoparticle formulation exhibits antiinflammatory and antifibrosis properties. Numerous applications of nanotechnology have been reported in the treatment of diabetes complications, including the improvement of drug effect [96], reduction of blood pressure [97], inhibition of retinal neovascularization [98], inhibition of VEGF protein [99], improvement of heart function [100], induction of collagen secretion [101], increased ALP activity [101], extracellular matrix mineralization, improvement of wood contraction in diabetes foot [102], and increase in serum testosterone level. Silver nanoparticles, gold nanoparticles, silicate nanoparticles, and titanium dioxide nanoparticles are used to treat diabetes retinopathy. These nanoparticles work in a variety of ways, including by promoting phosphorylation, VEGFR-2 autophosphorylation, and blocking numerous signaling pathways. Nanoparticles are discovered to be an invasive way in many ocular disorders, and deal with diverse ocular barriers, according to a recent study by Mukherjee S et al. 2023 [103]. The study looked at the nanoparticle for the diabetic retinopathy as a medication delivery agent. Additionally, by lengthening drug molecules' retention times and slowing down their release, nanoparticles boost bioavailability. As nanotechnology advances, a growing number of novel nanomaterials will be developed and utilized in medical procedures, advancing contemporary medicine, generating fresh insights, and making advancements to preventing and curing illnesses.

5 Conclusion

Diabetic retinopathy, often known as a microvascular disease condition which has two major causes one of which is inflammation and another is retinal neurodegeneration. Angiogenesis, inflammation, and oxidative stress were highlighted specifically in this study's summary of the molecular causes of diabetic retinopathy and associated treatment options. Also concluded the alternate treatment as well as laser, surgical techniques, and recent advances of nanoparticles in diabetic retinopathy management. The creation of drugs that specifically target the molecules in these pathways in identification of new potential therapies for DR. It was found that intravitreal "antiVEGF" drugs have taken the place of alternative therapies for DME and PDR. However, due to frequent injection requirements, high cost, and poor patient compliance, the use of this medicine is restricted. It was also concluded from the mentioned previously review article that the Laser technique (photocoagulation) plays a big role in the adjuvant treatment of DR. Corticosteroids have demonstrated therapeutic benefit that is resistant to medication or patients that don't respond to antiVEGF therapy. For individuals with significant vision loss, it is still difficult to obtain driving or reading vision with the available therapy choices. If patients with DR are to live better lives, the number of intravitreal injections, dose, and length of existing treatment regimens, as well as techniques for combination therapy, must all be adjusted.

6   Future perspectives

Even though a significant quantity of research has led to ineffective DR diagnostic techniques. However, the national health services haven't yet acquired the majority of the models. Several gaps in the study need to be filled to automate the system as shown in Fig. 9.

6.1   Medical or retinal imaging

A computer-aided diagnosis tool is used for medical imaging of diabetic complications patients. Screening of fundus-images with various medical imaging tools has attracted worldwide attention in diabetes retinopathy. Images are captured using a smartphone and a handheld 20-diopter condensing lens in these techniques, and various commercial developers have produced hardware and software add-ons to make fundus photography more feasible. These methods rely on the installation of extra lens components that are aligned with the smartphone camera and typically include

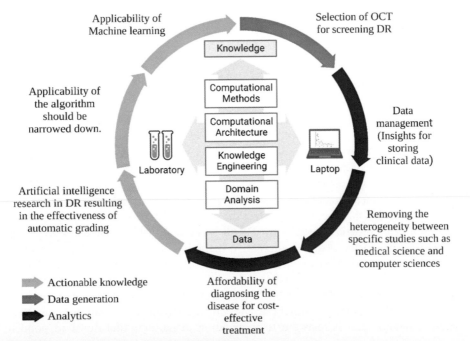

Fig. 9 Future direction for diabetes retinopathy.

specialized software that allows for the acquisition, tagging, and secure sharing of fundus pictures [104]. According to recent medical imaging research, binocular slit-lamp ophthalmoscopy is regarded as a standard approach for DR diagnosis or screening.

6.2 Machine learning or artificial intelligence

Deep learning and other modern machine learning (ML) approaches, which have shown positive clinical outcomes in the detection of images and machine translation, have been widely adopted in a range of industries, including social media, telecommunications, cybersecurity, and medicine. Deep learning carries on the long heritage of independently and collaboratively analyzing retinal images. Machine learning diagnosis of diabetic retinopathy is a significant technological advancement that has the potential to significantly increase the efficacy and accessibility of DR screening programs, especially in poorer nations where the availability of specialized physicians is limited and patient identification is critical to receiving prompt treatment and preventing a serious and disabling vision loss [45]. These artificial intelligence methods have been shown to reduce costs, boost diagnostic precision, and broaden patient access to DR screening.

References

[1] Guariguata L, Whiting DR, Hambleton I, Beagley J, Linnenkamp U, Shaw JE. Global estimates of diabetes prevalence for 2013 and projections for 2035. Diabetes Res Clin Pract 2014;103(2):137–49. https://doi.org/10.1016/j.diabres.2013.11.002.

[2] Nanditha A, Ma RCW, Ramachandran A, Snehalatha C, Chan JCN, Chia KS, et al. Diabetes in Asia and the Pacific: implications for the global epidemic. Diabetes Care 2016;39 (3):472–85. Available from: https://care.diabetesjournals.org/content/39/3/472.

[3] Singh A, Dutta MK. A robust zero-watermarking scheme for tele-ophthalmological applications. J King Saud Univ Comput Inf Sci 2017;32(8):895–908. https://doi.org/10.1016/j.jksuci.2017.12.008.

[4] Williams R, Airey M, Baxter H, Forrester J, Kennedy-Martin T, Girach A. Epidemiology of diabetic retinopathy and macular oedema: a systematic review. Eye 2004;18(10):963–83. Available from: https://www.nature.com/articles/6701476.pdf?origin=ppub.

[5] Shah CA. Diabetic retinopathy: a comprehensive review. Indian J Med Sci 2008;62 (12):500–19. Available from: https://pubmed.ncbi.nlm.nih.gov/19265246.

[6] Cheung CY, Sabanayagam C, Law AK, Kumari N, Ting DS, Tan G, et al. Retinal vascular geometry and 6 year incidence and progression of diabetic retinopathy. Diabetologia 2017;60(9):1770–81. https://doi.org/10.1007/s00125-017-4333-0.

[7] Zhao S, Li T, Li J, Lu Q, Han C, Wang N, et al. miR-23b-3p induces the cellular metabolic memory of high glucose in diabetic retinopathy through a SIRT1-dependent signalling pathway. Diabetologia 2015;59(3):644–54. https://doi.org/10.1007/s00125-015-3832-0.

[8] Kowluru RA. Diabetic retinopathy, metabolic memory and epigenetic modifications. Vis Res 2017;139:30–8. https://doi.org/10.1016/j.visres.2017.02.011.

[9] Jin K, Pan X, You K, Wu J, Liu Z, Cao J, et al. Automatic detection of non-perfusion areas in diabetic macular edema from fundus fluorescein angiography for decision making using deep learning. Sci Rep 2020;10(1):15138. Available from: https://www.nature.com/articles/s41598-020-71622-6.

[10] Nagpal D, Panda SN, Malarvel M, Pattanaik PA, Zubair Khan M. A review of diabetic retinopathy: datasets, approaches, evaluation metrics and future trends. J King Saud Univ Comput Inf Sci 2021. https://doi.org/10.1016/j.jksuci.2021.06.006.

[11] Kaur J, Mittal D. A generalized method for the segmentation of exudates from pathological retinal fundus images. Biocybern Biomed Eng 2018;38(1):27–53. https://doi.org/10.1016/j.bbe.2017.10.003.

[12] Ting DSW, Cheung CY, Nguyen Q, Sabanayagam C, Lim G, Lim ZW, et al. Deep learning in estimating prevalence and systemic risk factors for diabetic retinopathy: a multiethnic study. npj Digit Med 2019;2(1). https://doi.org/10.1038/s41746-019-0097-x.

[13] Tong N, Wang LY, Gong H, Pan L, Yuan F, Zhou Z. Clinical manifestations of supralarge range nonperfusion area in diabetic retinopathy. Int J Clin Pract 2022;2022:1–7. https://doi.org/10.1155/2022/8775641.

[14] Kowluru RA, Chan PS. Oxidative stress and diabetic retinopathy. Exp Diabetes Res 2007;2007:1–12. https://doi.org/10.1155/2007/43603.

[15] Obrosova IG, Kador PF. Aldose reductase/polyol inhibitors for diabetic retinopathy. Curr Pharm Biotechnol 2011;12(3):373–85. https://doi.org/10.2174/138920111794480642.

[16] Yamagishi S, Maeda S, Matsui T, Ueda S, Fukami K, Okuda S. Role of advanced glycation end products (AGEs) and oxidative stress in vascular complications in diabetes. Biochim Biophys Acta Gen Subj 2012;1820(5):663–71. Available from: https://www.sciencedirect.com/science/article/pii/S0304416511000638.

[17] Ola MS, Nawaz MI, Siddiquei MM, Al-Amro S, Abu El-Asrar AM. Recent advances in understanding the biochemical and molecular mechanism of diabetic retinopathy. J Diabetes Complicat 2012;26(1):56–64. https://doi.org/10.1016/j.jdiacomp.2011.11.004.

[18] Miyamoto K, Khosrof S, Bursell SE, Rohan R, Murata T, Clermont AC, et al. Prevention of leukostasis and vascular leakage in streptozotocin-induced diabetic retinopathy via intercellular adhesion molecule-1 inhibition. Proc Natl Acad Sci 1999;96(19):10836–41. https://doi.org/10.1073/pnas.96.19.10836.

[19] Yuuki T, Kanda T, Kimura Y, Kotajima N, Tamura J, Kobayashi I, et al. Inflammatory cytokines in vitreous fluid and serum of patients with diabetic vitreoretinopathy. J Diabetes Complicat 2001;15(5):257–9. https://doi.org/10.1016/s1056-8727(01)00155-6.

[20] Schröder S, Palinski W, Schmid-Schönbein GW. Activated monocytes and granulocytes, capillary nonperfusion, and neovascularization in diabetic retinopathy. Am J Pathol 1991;139(1):81–100. Available from: https://www.ncbi.nlm.nih.gov/pmc/articles/PMC1886150/.

[21] Sharp PJ, Manivannan A, Xu H, Forrester JV. The scanning laser ophthalmoscope—a review of its role in bioscience and medicine. Phys Med Biol 2004;49(7):1085–96. https://doi.org/10.1088/0031-9155/49/7/001.

[22] Joussen AM, Poulaki V, Mitsiades N, Cai W, Suzuma I, Pak J, et al. Suppression of Fas-FasL-induced endothelial cell apoptosis prevents diabetic blood-retinal barrier breakdown in a model of streptozotocin-induced diabetes. FASEB J 2002;17(1):76–8. https://doi.org/10.1096/fj.02-0157fje.

[23] Barouch FC, Miyamoto K, Allport JR, Fujita K, Bursell SE, Aiello LP, et al. Integrin-mediated neutrophil adhesion and retinal leukostasis in diabetes. Invest Ophthalmol Vis Sci 2000;41(5):1153–8. Available from: https://pubmed.ncbi.nlm.nih.gov/10752954/.

[24] Chibber R, Ben-Mahmud BM, Coppini DV, Christ ER, Kohner EM. Activity of the glycosylating enzyme, core 2 GlcNAc (beta1,6) transferase, is higher in polymorphonuclear leukocytes from diabetic patients compared with age-matched control subjects: relevance to capillary occlusion in diabetic retinopathy. Diabetes 2000;49(10):1724–30. https://doi.org/10.2337/diabetes.49.10.1724.

[25] Kasza M, Meleg J, Várdai J, Nagy B, Szalai E, Damjanovich J, et al. Plasma E-selectin levels can play a role in the development of diabetic retinopathy. Graefes Arch Clin Exp Ophthalmol 2017;255(1):25–30. https://doi.org/10.1007/s00417-016-3411-1.

[26] Forrester JV, Kuffova L, Delibegovic M. The role of inflammation in diabetic retinopathy. Front Immunol 2020;11. https://doi.org/10.3389/fimmu.2020.583687.

[27] Limb GA, Hickman-Casey J, Hollifield RD, Chignell AH. Vascular adhesion molecules in vitreous from eyes with proliferative diabetic retinopathy. Invest Ophthalmol Vis Sci 1999;40(10):2453–7. Available from: https://pubmed.ncbi.nlm.nih.gov/10476819/.

[28] Abcouwer SF. Müller cell–microglia cross talk drives neuroinflammation in diabetic retinopathy. Diabetes 2017;66(2):261–3. https://doi.org/10.2337/dbi16-0047.

[29] Sorrentino FS, Allkabes M, Salsini G, Bonifazzi C, Perri P. The importance of glial cells in the homeostasis of the retinal microenvironment and their pivotal role in the course of diabetic retinopathy. Life Sci 2016;162:54–9. https://doi.org/10.1016/j.lfs.2016.08.001.

[30] Joussen AM, Poulaki V, Le ML, Koizumi K, Esser C, Janicki H, et al. A central role for inflammation in the pathogenesis of diabetic retinopathy. FASEB J 2004;18(12):1450–2. https://doi.org/10.1096/fj.03-1476fje.

[31] Suzuki Y, Nakazawa M, Suzuki K, Yamazaki H, Miyagawa Y. Expression profiles of cytokines and chemokines in vitreous fluid in diabetic retinopathy and central retinal vein occlusion. Jpn J Ophthalmol 2011;55(3):256–63. https://doi.org/10.1007/s10384-011-0004-8.

[32] Prieur X, Mok CYL, Velagapudi VR, Núñez V, Fuentes L, Montaner D, et al. Differential lipid partitioning between adipocytes and tissue macrophages modulates macrophage lipotoxicity and M2/M1 polarization in obese mice. Diabetes 2011;60(3):797–809. https://doi.org/10.2337/db10-0705.

[33] Rangasamy S, McGuire PG, Franco Nitta C, Monickaraj F, Oruganti SR, Das A. Chemokine mediated monocyte trafficking into the retina: role of inflammation in alteration of the blood-retinal barrier in diabetic retinopathy. PLoS One 2014;9(10), e108508. https://doi.org/10.1371/journal.pone.0108508.

[34] Uysal KT, Wiesbrock SM, Marino MW, Hotamisligil GS. Protection from obesity-induced insulin resistance in mice lacking TNF-α function. Nature 1997;389 (6651):610–4. https://doi.org/10.1038/39335.

[35] Lumeng CN, Bodzin JL, Saltiel AR. Obesity induces a phenotypic switch in adipose tissue macrophage polarization. J Clin Investig 2007;117(1):175–84. https://doi.org/10.1172/JCI29881.

[36] Madhumitha H, Mohan V, Deepa M, Babu S, Aravindhan V. Increased Th1 and suppressed Th2 serum cytokine levels in subjects with diabetic coronary artery disease. Cardiovasc Diabetol 2014;13(1). https://doi.org/10.1186/1475-2840-13-1.

[37] Koleva-Georgieva DN, Sivkova NP, Terzieva D. Serum inflammatory cytokines IL-1β, IL-6, TNF-α and VEGF have influence on the development of diabetic retinopathy. Folia Med 2011;53(2). https://doi.org/10.2478/v10153-010-0036-8.

[38] Boss JD, Singh PK, Pandya HK, Tosi J, Kim C, Tewari A, et al. Assessment of neurotrophins and inflammatory mediators in vitreous of patients with diabetic retinopathy. Invest Ophthalmol Vis Sci 2017;58(12):5594–603. Available from: https://pubmed.ncbi.nlm.nih.gov/29084332/.

[39] Miller JW, Adamis AP, Shima DT, D'Amore PA, Moulton RS, O'Reilly MS, et al. Vascular endothelial growth factor/vascular permeability factor is temporally and spatially correlated with ocular angiogenesis in a primate model. Am J Pathol 1994;145(3):574–84. Available from: https://pubmed.ncbi.nlm.nih.gov/7521577/.

[40] Aiello LP, Avery RL, Arrigg PG, Keyt BA, Jampel HD, Shah ST, et al. Vascular endothelial growth factor in ocular fluid of patients with diabetic retinopathy and other retinal disorders. N Engl J Med 1994;331(22):1480–7. https://doi.org/10.1056/NEJM199412013312203.

[41] Hanahan D, Folkman J. Patterns and emerging mechanisms of the Angiogenic switch during tumorigenesis. Cell 1996;86(3):353–64. Available from: https://www.cell.com/cell/fulltext/S0092-8674(00)80108-7.

[42] Tolentino MJ. Vascular endothelial growth factor is sufficient to produce Iris neovascularization and neovascular glaucoma in a nonhuman primate. Arch Ophthalmol 1996;114(8):964. https://doi.org/10.1001/archopht.1996.01100140172010.

[43] Ivan M, Kondo K, Yang H, Kim W, Valiando J, Ohh M, et al. HIFalpha targeted for VHL-mediated destruction by proline hydroxylation: implications for O2 sensing. Science 2001;292(5516):464–8. Available from: https://pubmed.ncbi.nlm.nih.gov/11292862/.

[44] Jaakkola P, Mole DR, Tian YM, Wilson MI, Gielbert J, Gaskell SJ, et al. Targeting of HIF-α to the von Hippel-Lindau ubiquitylation complex by O2-regulated prolyl hydroxylation. Science 2001;292(5516):468–72. Available from: https://science.sciencemag.org/content/292/5516/468?hwshib2=authn%3A1606035370%3A20201121%253A591f1acd-3e6e-4683-9b23-33075a26c33e%3A0%3A0%3A0%3ARpcgPWqrx8eAHm%2Be0hubvw%3D%3D.

[45] Poprac P, Jomova K, Simunkova M, Kollar V, Rhodes CJ, Valko M. Targeting free radicals in oxidative stress-related human diseases. Trends Pharmacol Sci 2017;38(7):592–607. https://doi.org/10.1016/j.biopha.2017.11.009.

[46] Prasad S, Gupta SC, Tyagi AK. Reactive oxygen species (ROS) and cancer: role of antioxidative nutraceuticals. Cancer Lett 2017;387:95–105. https://doi.org/10.1016/j.canlet.2016.03.042.

[47] Halliwell B. Free radicals and antioxidants – quo vadis? Trends Pharmacol Sci 2011;32(3):125–30. https://doi.org/10.1016/j.tips.2010.12.002.

[48] Kang Q, Yang C. Oxidative stress and diabetic retinopathy: molecular mechanisms, pathogenetic role and therapeutic implications. Redox Biol 2020;37, 101799. https://doi.org/10.1155/2007/43603.

[49] Sui A, Chen X, Demetriades AM, Shen J, Cai Y, Yao Y, et al. Inhibiting NF-κB signaling activation reduces retinal neovascularization by promoting a polarization shift in macrophages. Invest Ophthalmol Vis Sci 2020;61(6):4. https://doi.org/10.1167/iovs.61.6.4.

[50] Sunilkumar S, Toro AL, McCurry CM, VanCleave AM, Stevens SA, Miller WP, et al. Stress response protein REDD1 promotes diabetes-induced retinal inflammation by sustaining canonical NF-κB signaling. J Biol Chem 2022;298(12), 102638. https://doi.org/10.1016/j.jbc.2022.102638.

[51] Zhong Q, Mishra M, Kowluru RA. Transcription factor Nrf2-mediated antioxidant defense system in the development of diabetic retinopathy. Invest Opthalmol Vis Sci 2013;54(6):3941. https://doi.org/10.1167/iovs.13-11598.

[52] Miller RG, Orchard TJ. Understanding metabolic memory: a tale of two studies. Diabetes 2020;69(3):291–9. https://doi.org/10.2337/db19-0514.

[53] Kato M, Natarajan R. Epigenetics and epigenomics in diabetic kidney disease and metabolic memory. Nat Rev Nephrol 2019;15(6):327–45. Available from: https://www.ncbi.nlm.nih.gov/pmc/articles/PMC6889804/.

[54] Chen Z, Miao F, Paterson AD, Lachin JM, Zhang L, Schones DE, et al. Epigenomic profiling reveals an association between persistence of DNA methylation and metabolic memory in the DCCT/EDIC type 1 diabetes cohort. Proc Natl Acad Sci 2016;113(21), E3002-11. https://doi.org/10.1073/pnas.1603712113.

[55] Jemt E, Persson Ö, Shi Y, Mehmedovic M, Uhler JP, Dávila López M, et al. Regulation of DNA replication at the end of the mitochondrial D-loop involves the helicase TWINKLE and a conserved sequence element. Nucleic Acids Res 2015;43(19):9262–75. Available from: https://academic.oup.com/nar/article/43/19/9262/2528179.

[56] Zhang YL, Zhang J, Jiang N, Lu YH, Wang L, Xu SH, et al. Immunosuppressive polyketides from Mantis-associated Daldiniaeschscholzii. J Am Chem Soc 2011;133(15):5931–40. https://doi.org/10.1021/ja110932p.

[57] Kowluru RA, Abbas SN. Diabetes-induced mitochondrial dysfunction in the retina. Invest Opthalmol Vis Sci 2003;44(12):5327. https://doi.org/10.1167/iovs.03-0353.

[58] Sun Jennifer K, Jampol Lee M. The diabetic retinopathy clinical research network (DRCR.net) and its contributions to the treatment of diabetic retinopathy. Ophthalmic Res 2019;62(4):225–30. https://doi.org/10.1159/000502779.

[59] Mitchell P, Bandello F, Schmidt-Erfurth U, Lang GE, Massin P, Schlingemann RO, et al. The RESTORE study. Ophthalmology 2011;118(4):615–25. https://doi.org/10.1016/j.ophtha.2011.01.031.

[60] Massin P, Bandello F, Garweg JG, Hansen LL, Harding SP, Larsen M, et al. Safety and efficacy of Ranibizumab in diabetic macular edema (RESOLVE study): a 12-month, randomized, controlled, double-masked, multicenter phase II study. Diabetes Care 2010;33(11):2399–405. https://doi.org/10.2337/dc10-0493.

[61] Sultan MB, Zhou D, Loftus J, Dombi T, Ice KS. A phase 2/3, multicenter, randomized, double-masked, 2-year trial of Pegaptanib sodium for the treatment of diabetic macular edema. Ophthalmology 2011;118(6):1107–18. https://doi.org/10.1016/j.ophtha.2011.02.045.

[62] Heier JS, Korobelnik JF, Brown DM, Schmidt-Erfurth U, Do DV, Midena E, et al. Intravitreal Aflibercept for diabetic macular edema: 148-week results from the VISTA and VIVID studies. Ophthalmology 2016;123(11):2376–85. Available from: https://www.sciencedirect.com/science/article/abs/pii/S0161642016307382.

[63] Campochiaro PA, Brown DM, Pearson A, Chen S, Boyer D, Ruiz-Moreno J, et al. Sustained delivery Fluocinolone Acetonide vitreous inserts provide benefit for at least 3 years in patients with diabetic macular edema. Ophthalmology 2012;119(10):2125–32. https://doi.org/10.1016/j.ophtha.2012.04.030.

[64] Wroblewski J, Hu AY. Topical Squalamine 0.2% and intravitreal Ranibizumab 0.5 mg as combination therapy for macular edema due to branch and central retinal vein occlusion: an open-label, randomized study. Ophthalmic Surg Lasers Imaging Retina 2016;47(10):914–23. https://doi.org/10.3928/23258160-20161004-04.

[65] Search of: diabetes retinopathy – List Results – ClinicalTrials.gov [Internet]. clinicaltrials.gov. Available from https://clinicaltrials.gov/ct2/results?cond=diabetes+retinopathy&term=&cntry=&state=&city=&dist=https://clinicaltrials.gov/ct2/results?cond=diabetes+retinopathy&term=&cntry=&state=&city=&dist=.

[66] Eleven Biotherapeutics. An open label, multi-center, safety study of intravitreal EBI-031, an interleukin-6 (IL-6) inhibitor, administered as single and repeat injections in subjects with diabetic macular edema (DME). clinicaltrials.gov; 2016. Available from: https://clinicaltrials.gov/ct2/show/NCT02842541.

[67] University of Nebraska, Genentech, Inc. Evaluation of the safety, tolerability and efficacy of Ranibizumab and tocilizumab in eyes with diabetic macular edema. clinicaltrials.gov; 2018. Available from: https://clinicaltrials.gov/ct2/show/NCT02511067.

[68] LLC AO. Allegro Ophthalmics announces last patient enrolled in PACIFIC phase 2b clinical trial of Luminate® for non-proliferative diabetic retinopathy. Allegro Ophthalmics - from theory to therapy; 2016. Available from: https://www.allegroeye.com/allegro-

ophthalmics-announces-last-patient-enrolled-in-pacific-phase-2b-clinical-trial-of-luminate-for-non-proliferative-diabetic-retinopathy/.

[69] Sun Jennifer K, Jampol Lee M. The diabetic retinopathy clinical research network (DRCR.net) and its contributions to the treatment of diabetic retinopathy. Ophthalmic Res 2019;62(4):225–30. https://doi.org/10.1159/000502779.

[70] Boyer DS, Yoon YH, Belfort R, Bandello F, Maturi RK, Augustin AJ, et al. Three-year, randomized, sham-controlled trial of dexamethasone intravitreal implant in patients with diabetic macular edema. Ophthalmology 2014;121(10):1904–14. Available from: https://www.ncbi.nlm.nih.gov/pubmed/24907062.

[71] Pacella F, Romano MR, Turchetti P, Tarquini G, Carnovale A, Mollicone A, Mastromatteo A, Pacella E. An eighteen-month follow-up study on the effects of intravitreal dexamethasone implant in diabetic macular edema refractory to anti-VEGF therapy. Int J Ophthalmol 2016;9:1427–32. https://doi.org/10.18240/ijo.2016.10.10.

[72] Campochiaro PA, Brown DM, Pearson A, Ciulla T, Boyer D, Holz FG, et al. Long-term benefit of sustained-delivery fluocinolone acetonide vitreous inserts for diabetic macular edema. Ophthalmology 2011;118(4):626–635.e2. https://doi.org/10.1016/j.ophtha.2010.12.028.

[73] Treatment techniques and clinical guidelines for photocoagulation of diabetic macular edema: early treatment diabetic retinopathy study report number 2. Ophthalmology 1987;94(7):761–74. Available from: https://www.sciencedirect.com/science/article/abs/pii/S0161642087335274.

[74] Photocoagulation treatment of proliferative diabetic retinopathy: the second report of diabetic retinopathy study findings. Ophthalmology 1978;85(1):82–106. Available from: https://pubmed.ncbi.nlm.nih.gov/345173/.

[75] Blumenkranz MS, Yellachich D, Andersen DE, Wiltberger MW, Mordaunt D, Marcellino GR, et al. Semiautomated patterned scanning laser for retinal photocoagulation. Retina 2006;26(3):370–6. https://doi.org/10.1097/00006982-200603000-00024.

[76] Vujosevic S, Martini F, Convento E, Longhin E, Kotsafti O, Parrozzani R, et al. Subthreshold laser therapy for diabetic macular edema: metabolic and safety issues. Curr Med Chem 2013;20(26):3267–71. https://doi.org/10.2174/09298673113209990030.

[77] Kernt M, Kampik A, Neubauer, Langer, Wolf, Kozak, et al. Navigated macular laser decreases retreatment rate for diabetic macular edema: a comparison with conventional macular laser. Clin Ophthalmol 2013;121. https://doi.org/10.2147/OPTH.S38559.

[78] Nguyen QD, Brown DM, Marcus DM, Boyer DS, Patel S, Feiner L, et al. Ranibizumab for diabetic macular edema. Ophthalmology 2012;119(4):789–801.

[79] Gross JG, Glassman AR, Jampol LM, Inusah S, Aiello LP, Antoszyk AN, et al. Panretinal photocoagulation vs intravitreous ranibizumab for proliferative diabetic retinopathy. JAMA 2015;314(20):2137. https://doi.org/10.1001/jama.2015.15217.

[80] Noma H, Mimura T, Yasuda K, Shimura M. Role of inflammation in diabetic macular edema. Ophthalmologica 2014;232(3):127–35.

[81] Sulaiman RS, Merrigan S, Quigley J, Qi X, Lee B, Boulton ME, et al. A novel small molecule ameliorates ocular neovascularisation and synergises with anti-VEGF therapy. Sci Rep 2016;6(1), 25509.

[82] Campochiaro PA, Khanani AM, Singer MB, Patel SJ, Boyer DS, Dugel PU, et al. Enhanced benefit in diabetic macular edema from AKB-9778 Tie2 activation combined with vascular endothelial growth factor suppression. Ophthalmology 2016;123(8):1722–30. https://doi.org/10.1016/j.ophtha.2016.04.025.

[83] Elman MJ, Bressler NM, Qin H, Beck RW, Ferris FL, Friedman SM, et al. Expanded 2-year follow-up of Ranibizumab plus prompt or deferred laser or triamcinolone plus prompt laser for diabetic macular edema. Ophthalmology 2011;118(4):609–14.

[84] Kurihara T, Westenskow PD, Bravo S, Aguilar E, Friedlander M. Targeted deletion of Vegfa in adult mice induces vision loss. J Clin Investig 2012;122(11):4213–7. https://doi.org/10.1172/JCI65157.

[85] Pacella F, Agostinelli E, Carlesimo SC, Nebbioso M, Secondi R, Forastiere M, et al. Management of anterior chamber dislocation of a dexamethasone intravitreal implant: a case report. J Med Case Rep 2016;10(1). https://doi.org/10.1186/s13256-016-1077-2.

[86] Callanan DG, Gupta S, Boyer DS, Ciulla TA, Singer MA, Kuppermann BD, et al. Dexamethasone intravitreal implant in combination with laser photocoagulation for the treatment of diffuse diabetic macular edema. Ophthalmology 2013;120(9):1843–51.

[87] Jain N, Stinnett SS, Jaffe GJ. Prospective study of a fluocinolone acetonide implant for chronic macular edema from central retinal vein occlusion. Ophthalmology 2012;119(1):132–7.

[88] Jeong S, Ku SK, Bae JS. Anti-inflammatory effects of pelargonidin on TGFBIp-induced responses. Can J Physiol Pharmacol 2017;95(4):372–81. https://doi.org/10.1139/cjpp-2016-0322.

[89] Behl Y, Krothapalli P, Desta T, Roy S, Graves DT. FOXO1 plays an important role in enhanced microvascular cell apoptosis and microvascular cell loss in type 1 and type 2 diabetic rats. Diabetes 2009;58(4):917–25. https://doi.org/10.2337/db08-0537.

[90] Kashyap MP, Singh AK, Siddiqui MA, Kumar V, Tripathi VK, Khanna VK, et al. Caspase cascade regulated mitochondria mediated apoptosis in monocrotophos exposed PC12 cells. Chem Res Toxicol 2010;23(11):1663–72. https://doi.org/10.1021/tx100234m.

[91] Rodríguez ML, Pérez S, Mena-Mollá S, Desco MC, Ortega ÁL. Oxidative stress and microvascular alterations in diabetic retinopathy: future therapies. Oxidative Med Cell Longev 2019;2019:1–18. https://doi.org/10.1155/2019/4940825.

[92] Alam NM, Mills WC, Wong AA, Douglas RM, Szeto HH, Prusky GT. A mitochondrial therapeutic reverses visual decline in mouse models of diabetes. Dis Model Mech 2015;8(7):701–10. Available from: https://pubmed.ncbi.nlm.nih.gov/26035391/.

[93] Staurenghi G, Ye L, Magee MH, Danis RP, Wurzelmann J, Adamson P, et al. Darapladib, a lipoprotein-associated phospholipase A2 inhibitor, in diabetic macular edema. Ophthalmology 2015;122(5):990–6. https://doi.org/10.1016/j.ophtha.2014.12.014.

[94] Fan F, Lei M. Mechanisms underlying curcumin-induced neuroprotection in cerebral ischemia. Front Pharmacol 2022;13. https://doi.org/10.3389/fphar.2022.893118.

[95] He Y, Al-Mureish A, Wu N. Nanotechnology in the treatment of diabetic complications: a comprehensive narrative review. J Diabetes Res 2021;2021:1–11. Available from: https://www.ncbi.nlm.nih.gov/pmc/articles/PMC8110427/.

[96] Yang X. Design and optimization of crocetin loaded PLGA nanoparticles against diabetic nephropathy via suppression of inflammatory biomarkers: a formulation approach to preclinical study. Drug Deliv 2019;26(1):849–59. https://doi.org/10.1080/10717544.2019.1642417.

[97] Ahad A, Raish M, Al-Jenoobi FI, Al-Mohizea AM. Sorbitane monostearate and cholesterol based niosomes for oral delivery of telmisartan. Curr Drug Deliv 2018;15(2):260–6. https://doi.org/10.2174/1567201814666170518131934.

[98] Fangueiro JF, Silva AM, Garcia ML, Souto EB. Current nanotechnology approaches for the treatment and management of diabetic retinopathy. Eur J Pharm Biopharm 2015;95:307–22. https://doi.org/10.1016/j.ejpb.2014.12.023.

[99] Kim JH, Kim MH, Jo DH, Yu YS, Lee TG, Kim JH. The inhibition of retinal neovascularization by gold nanoparticles via suppression of VEGFR-2 activation. Biomaterials 2011;32(7):1865–71. https://doi.org/10.1016/j.biomaterials.2010.11.030.

[100] Gurunathan S, Lee KJ, Kalishwaralal K, Sheikpranbabu S, Vaidyanathan R, Eom SH. Antiangiogenic properties of silver nanoparticles. Biomaterials 2009;30(31):6341–50. https://doi.org/10.1016/j.biomaterials.2009.08.008.

[101] Enomoto M, Ishizu T, Seo Y, Kameda Y, Suzuki H, Shimano H, et al. Myocardial dysfunction identified by three-dimensional speckle tracking echocardiography in type 2 diabetes patients relates to complications of microangiopathy. J Cardiol 2016;68 (4):282–7. https://doi.org/10.1016/j.jjcc.2016.03.007.

[102] Zhu YS, Sun ZJ, Han Q, Liao LZ, Wang J, Bian C, et al. Human mesenchymal stem cells inhibit cancer cell proliferation by secreting DKK-1. Leukemia 2009;23(5):925–33. https://doi.org/10.1038/leu.2008.384.

[103] Mukherjee S, Panda P, Mishra M. Nanoparticles as drug delivery agents for managing diabetic retinopathy. Elsevier eBooks; 2023. p. 329–64. https://doi.org/10.1016/B978-0-12-820557-0.00014-X.

[104] Ghosh K, Capell BC. The senescence-associated secretory phenotype: critical effector in skin cancer and aging. J Investig Dermatol 2016;136(11):2133–9. https://doi.org/10.1016/j.jid.2016.06.621.

Targeting the role of angiogenesis, inflammation and oxidative stress in pathogenesis of glaucoma: Strategic nanotechnology based drug delivery approaches

15

Neelam Sharma[a], Neha Tiwary[a], Sukhbir Singh[a], Sumeet Gupta[b], Tapan Behl[c], and Gaurav Malik[a]

[a]Department of Pharmaceutics, MM College of Pharmacy, Maharishi Markandeshwar (Deemed to be University), Mullana-Ambala, Haryana, India, [b]Department of Pharmacology, MM College of Pharmacy, Maharishi Markandeshwar (Deemed to be University), Mullana-Ambala, Haryana, India, [c]Amity School of Pharmaceutical Sciences, Amity University, Mohali, Punjab, India

1 Introduction

Glaucoma is a common eye illness which can cause permanent blindness if left untreated. It is driven by elevated intraocular pressure (IOP), which destroyed the optic nerve and impairs vision [1]. The term "glaucoma" refers to a group of ocular disorders with multisystem etiologies that are linked by a clinically distinctive optic neuropathy with potentially progressive, clinically noticeable differences at the optic nerve head (ONH), including focal or generalized neuro-retinal rim thinned with excavation and expansion of the optic cup, signifying neurodegeneration of retinal ganglion cell (RGC) axons and deformation of the lamina cribrosa (LC) corresponding widespread and localized nerve-fiber-bundle pattern peripheral vision loss may not be detected in early stages [2]. A homoeostatic equilibrium between the production and expulsion of aqueous humor (AH), the intraocular fluid, regulates the pressure inside the eye. The AH generation rate in a healthy human eye is 2.5–2.8 min, and the complete volume is replaced every 100 min [3]. There is only acute angle-closure glaucoma has well-defined symptoms; all other types of chronic glaucoma are unknown mostly symptomless [4]. Glaucoma is classified as primary, secondary, or coupled with observable comorbidities, depending on how the drainage system looks and whether it is blocked [5]. Glaucoma can affect anyone regardless of age, with early-onset cases demonstrating Mendelian inheritance before the age of 40 and adult-onset cases evolving after that, inheriting complex traits. In general, gene

Targeting Angiogenesis, Inflammation and Oxidative Stress in Chronic Diseases. https://doi.org/10.1016/B978-0-443-13587-3.00011-4

variants that contribute to adult-onset glaucoma are widespread and have smaller biological consequences than mutations in genes that cause early-onset glaucoma [6]. The optic nerve in humans is composed of the 1.2 million RGCs that comprise its axons [7]. The optic nerve axons in glaucoma are destroyed, which causes the death of RGCs [8]. The ONH is considered to be the major area of destruction, with secondary effects on the lateral geniculate nucleus and optic radiations to the visual cortex [9]. RGCs in a normal eye receive consistent supplies of neurotrophic factors from the brain via retrograde axonal transport and from the retinal Muller glia. Among them is the brain-derived growth factor, a member of the neurotrophin family of growth factors, Nerve growth factor, Neurotrophin-3, Neurotrophin-4/5, and receptors bind to the p75-neurotrophin receptor, and the tropomyosin-receptor kinases [7]. The two basic subtypes are angle-closure and open-angle. The IOP rises during these, helping in distinguishing them [10]. Angle-closure is a condition of ocular anatomy marked by the closing of the drainage angle caused by the iris's appositional or synechial proximity to the trabecular meshwork, obstructing the iris's ability to access aqueous fluid [11]. It is caused by the iris's occlusion of the trabecular meshwork, which impairs aqueous outflow and raises IOP. Iris (pupil block), ciliary body (plateau iris), lens, and forces posterior to the lens are the four anatomical levels on which the angle-closure process is based [12,13]. These mechanisms are each characterized by unique clinical symptoms. Once the relevant mechanism is identified, the underlying pathology can be specifically treated. When there is the pupillary block, the area of iridolenticular contact experiences resistance to aqueous flow through the pupil. As a result, aqueous flow from the anterior chamber to the nonpigmented ciliary epithelium in the posterior chamber is constrained. Raising the pressure gradient between the anterior and posterior chambers, causes the anterior folding of the iris, narrowing of the angle, and acute or chronic iridotrabecular contact [14]. Primary angle-closure glaucoma (PACG) and secondary angle-closure glaucoma are further classifications for angle-closure glaucoma. Appositional angle-closure is not caused by primary angle-closure. Secondary angle-closures feature recognizable mechanisms for iris apposition to the angle or closure of the angle by direct obstruction of the trabecular meshwork [15]. Acute or chronic PACG can be distinguished based on the time or abruptness of the onset. The whole trabecular meshwork is rapidly and fully covered by the iris during the acute onset stage of PACG, which causes an abrupt, symptomatic increase in IOP. The condition will progress to the chronic stage if it does not resolve on its own or with medical treatment. The symptoms of a subacute attack are milder, and the IOP may return on its own [16]. A range of topical and systemic medications, such as adrenergic, both anticholinergic and cholinergic, antidepressant and antianxiety, sulfa-based, and anticoagulant medicines, can have the potentially blinding adverse effect of acute angle-closure glaucoma [17]. In secondary angle-closure glaucoma, the underlying cause may work to move the crystalline lens forward, resulting in the pupillary block, or it may act locally, affecting the iris and angle factors [18]. Open-angle glaucoma (OAG) can be classified into subtypes primary open-angle glaucoma (POAG), normal-tension glaucoma (NTG), and secondary OAG [19]. NTG is the term that is used to describe glaucoma with disc destruction at statistically "normal" IOP. It is quite simple to suspect or identify glaucoma and begin treatment when the IOP is high. When the IOP is normal yet there are changes to the optic disc or visual field that resemble glaucoma, a dilemma occurs [20]. When IOP is within the normal range (21 mmHg, or below the statistical upper limit of normal range), a condition known as NTG is present [21]. The pathophysiology of POAG and NTG is similar, with many of the risk factors falling into the mechanical and vascular categories. Anatomical variances of the LC, higher LC displacement, thinner LC, higher IOP, and translaminar pressure dynamics are all mechanical risk factors [22]. Elevated IOP and/ or ocular neuropathy are symptoms of secondary OAG [23]. In the beginning, new epidemiologic data have shown the prevalence and distribution of OAG in numerous communities. Both the scientific and public health ramifications of these data are significant. Epidemiologic data are also offering etiology-related hints by highlighting risk variables. Additionally, they enable the identification of high-risk populations that can be targeted for early detection, such as those who have glaucoma sufferers in their families [24]. With 8% of blindness worldwide due to glaucoma in 2010, it is the most common cause of permanent blindness. 64.3 million people were affected by glaucoma in 2013, and that number will rise to 76.0 million in 2040 and then to 111.8 million in 2040 [25]. In the world, the prevalence of POAG is 2.2%, affecting 57.5 million people in total. 7.8 million individuals across Europe have POAG, and the prevalence is 2.51% overall. The most common kind of glaucoma in the UK is POAG, which mostly affects people of African and Caribbean heritage and affects 2% of people over 40 and 10% of those over 75. Just 0.17% of adults under 40, particularly those of East Asian heritage, have PACG, a less prevalent condition [23]. 12 million instances of glaucoma, or around one-fifth of all cases worldwide, are thought to exist in India. Although POAG accounts for almost two-thirds of occurrences in the Caucasian population, open-angle and closed-angle glaucoma are equally prevalent in the Indian population. In rural south India, 1.7% of people aged 40 and beyond were assessed to have POAG, according to the Adverse childhood experiences study (ACES). The prevalence (3.5%) in the urban south India-Chennai Glaucoma Study was comparatively higher. More critically, it was discovered that more than 90% of glaucoma cases went untreated and were only discovered during the survey (93% in ACES and 98.6% in the Chennai Glaucoma Study) [26]. Based on the information available, we anticipate that there are about 11.2 million people who are 40 years or older. There may be 27.6 million people who have primary angle-closure illness [27]. Angiogenesis is a process, which is described as the development of new blood vessels from preexisting vasculature, is responsible for a wide range of physiological functions, including growth and differentiation, wound healing, and pathological conditions like neoplasia and complex ocular diseases, which cause profound vision loss [28,29]. Angiogenesis is a multistep, complicated process. The basement membrane is broken down by extracellular matrix-degrading enzymes released by active endothelial cells, allowing for cell migration and proliferation and the development of solid endothelial cell capillary tubes. The eye's pathologic angiogenesis can impair vision [30]. In both healthy and unhealthy tissue formation and function, angiogenesis is a significant process. Vascular endothelial growth factors and receptors (VEGF-VEGFRs), the Ephrin-Eph receptor pathway, the Delta-Notch pathway, the angiopoietin-Tie receptor axis, insulin-like growth factor-1 (IGF-1), interleukin 8

(IL-8), platelet-derived growth factor (PDGF), and the Netrins-UNC5R pathway are just a few of the numerous factors and systems that are VEGF-VEGFR system is one of the most effective [31]. VEGF-A is a major regulator of neovascularization. Through alternative splicing, a single gene can produce several VEGF-A isoforms. In the retina, VEGF-A165 is the proangiogenic isoform that is expressed the most frequently. Recently, antiangiogenic VEGF-A sister isoforms have also been discovered. VEGF-A165b, a human antiangiogenic VEGF-A isoform, has been demonstrated to inhibit VEGF-A induced [32]. An acute and persistent inflammatory process is defined by the production of proinflammatory mediators and the infiltration of various types of inflammatory cells into the ischemic tissue through the intercellular space between vascular endothelial cells. Inflammation typically occurs in response to ischemic injury [33]. An imbalance between the production and elimination of ROS leads to oxidative stress. By causing DNA damage, excessive ROS levels cause apoptosis in a variety of cell types. H_2O_2 has shown that ROS are messengers for activating NF-κB. The transcription factor NF-κB is activated by ROS, and its activation leads to the emergence of a variety of substances, including inflammatory cytokines like IL-1, IL-6 and TNF-α [34]. Microglial exosomes from the retina subjected to high hydrostatic pressure (EHP) boosted the responsiveness of naive microglial cells, including higher levels of proinflammatory cytokines (TNF-α and IL-1), expression of MHC-II molecules, increased microglia motility, phagocytic effectiveness, and proliferation [35].

2 Role play of angiogenesis, inflammation, and oxidative stress in the pathophysiology of glaucoma

For many years, it was assumed that the only cause of glaucoma was a persistently elevated IOP greater than 21mmHg. The majority of persons affected by POAG, the most prevalent kind of glaucoma, have elevated IOP. As a result, this glaucoma subtype is a member of the high-tension glaucoma group (HTG) [4]. When the "aqueous humor" fluid accumulates in the front of the eye, it can cause glaucoma. Increased IOP causes irreparable damage to the optic nerve and RGCs due to either excessive production or inadequate drainage of the aqueous fluid [36]. High IOP can cause ischemia, as can vascular dysfunction from a lack of autoregulatory function or vasospasm. It is possible for ischemia to cause pathologic abnormalities in glaucoma due to mechanical (IOP) and vascular (autoregulatory deficiencies or vasospasm) causes [37]. Some people have excessive IOP, also known as ocular hypertension (OHT) but do not have glaucoma. When there is no visual field defect and there is a pattern of consistently increased IOP [38]. The ONH's structural alterations are one of glaucoma's defining characteristics. RGC axons aggregate and exit the eye at the optic disc (or simply "disc") in the ONH [36]. The main oxidant in metabolic processes intended to produce energy from the oxidation of various organic compounds is oxygen. Potentially harmful reactive oxygen species are produced as a byproduct of cells' metabolism of oxygen (ROS) [38]. IOP should be lowered to decrease glaucoma progression. The traditional first-line medications are β-blockers and prostaglandin analogs because of their efficacy and tolerability. These medications lower IOP by reducing aqueous production and boosting uveoscleral aqueous outflow, respectively. Carbonic anhydrase inhibitors, cholinergic agonists, and 2-adrenoceptor agonists are additional antihypertensive glaucoma drugs [29]. Sympathomimetics which activate α-receptors enhance AH outflow and perhaps its development, whilst sympatholytics that inhibit β-receptors decrease AH formation and result in lower IOP [3]. The placental growth factor, VEGF-A, VEGF-B, VEGF-C, and VEGF-D are all members of the VEGF class of molecules. VEGF-A is the primary mediator of proangiogenic signaling. Endothelial and nonendothelial cells that express isoform-specific VEGF receptors (neuropilins) are where the VEGF binds. Several treatments have been developed to reduce VEGF and improve the treatment of various ocular disorders [39]. A gradual loss of RGCs, weakening of the retinal nerve fiber layer, and cupping of the ONH are all symptoms of glaucomatous degenerative disease. Particularly in the aged retina, a variety of stressors, including prolonged mechanical stress brought on by a high IOP, hypoxia/ischemia, and oxidative stress, along with a lack of neurotrophic factors, may result in RGC malfunction and death. The regulation of the local protective immune response, which is controlled by glial cells and the complement system, may fail under conditions of increasing stress over a long period. This failure could result in a neuroinflammatory degenerative process that speeds up the progression of the disease [35]. Multiple locations can experience neuroinflammation associated with glaucoma, such as the (i) posterior compartments of the eye (such as the retina and ONH), (ii) optic tract and brain (such as the superior colliculus and lateral geniculate), and (iii) peripherally in the blood, bone marrow, or other tissues [40]. Tumor necrosis factor (TNF) levels in glaucoma patients were higher than in healthy individuals. Along with proinflammatory interleukins, TNF-α is one of the main proinflammatory growth factors (ILs). Patients with glaucoma, for instance, have significantly higher levels of IL-6 and IL-8. Some IL family members, including IL-7, IL-10, and IL9-22, have antifibrotic actions on cells. When IL-7 binds to its receptor, it activates signaling that blocks the TGF/Smad signal and inhibits fibroblasts' expression of the extracellular matrix [41]. Since mitochondria are a cell's principal energy source and the main site of ROS production, they are a major feature for minimizing oxidative stress. RGC apoptosis in glaucoma has been linked to an increase in membrane permeability and a decrease in mitochondrial membrane potential. The mitochondrial-mediated RGC death process can be triggered by glaucoma-related stimuli such as hypoxia, TNF-α, and oxidative stress [42]. The effect of oxidative stress, angiogenesis and inflammation on Glaucoma is described below in Fig. 1.

3 Classification of therapeutic treatments available for management of glaucoma

Nowadays, various drugs are available for therapeutics of glaucoma and other ocular disorders. Fig. 2 represents the classification of drugs available for the treatment of glaucoma which includes β-adrenergic blockers, α-adrenergic agonists, prostaglandin analogs, carbonic anhydrase inhibitors, and miotic. Topical beta-blockers, adrenergic

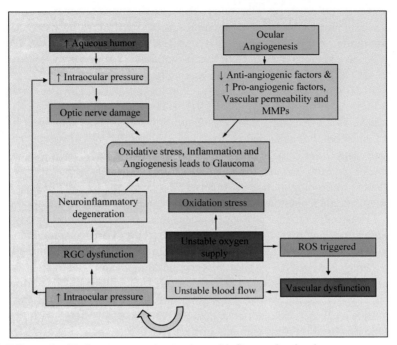

Fig. 1 Effect of oxidative stress, angiogenesis, and inflammation in glaucoma.

drugs, miotics, and oral carbonic anhydrase inhibitors have often been used in the treatment of glaucoma (CAIs). Nevertheless, new types of ocular hypotensive drugs, such as prostaglandins, local CAIs, and α2-adrenergic agents, have recently been added to the therapeutic arsenal available for the medical treatment of glaucoma. The mechanism of action of various classes of drugs used for the management of hypertension is given in Fig. 3. Table 1 illustrates the pharmacokinetics of drugs utilized for the management of glaucoma.

3.1 Prostaglandin

The first prostaglandin agonists used to treat glaucoma were prostaglandin F2 analogs. Due to their effectiveness and good tolerability, they established a high standard for the competition, which led to a long period with few new hypotensive drugs being developed [54]. Many people are affected by glaucoma, and treatments for it rely on the use of prostaglandin (PG) analogs, carbonic anhydrase (CA) inhibitors, and adrenergic agonists/antagonists. Derivatives of prostaglandin F2a (PGF2a), such as latanoprost 15 and unoprostone 16 have been used in medicine for over 15 years. The majority of the novel PG analogs have been marketed in recent years as PGF2a serves as the primary molecule in antiglaucoma medications since it is the therapeutically relevant derivatives 15 and 16 (mimics of these naturally occurring eicosanoids) are powerful glaucoma medications [55]. The use of prostaglandin analogs as the first-line of treatment for glaucoma arises from their potent IOP-lowering properties, once-daily administration, and

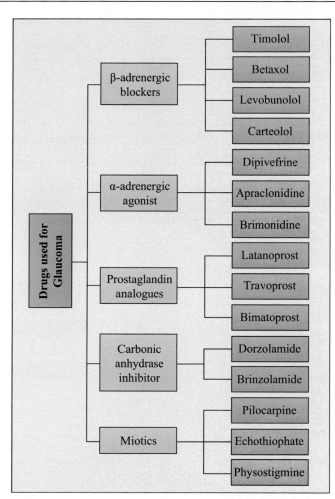

Fig. 2 Classification of drugs used for the management of glaucoma.

negligible side effects. There are nine prostaglandin receptors: PGIP receptor, PGFP receptor, thromboxane A2 receptor, PGE receptor 1–4 (EP1–4), and PGD receptor 1–2 (DP1–2). In addition to relaxing the ciliary muscle and increasing AH outflow, PGF2a, and prostaglandin analogs also disturb the turnover of the extracellular matrix by binding to EP and FP receptors in the ciliary muscle [56].

3.2 β-Adrenergic blockers

A popular class of medications for the treatment of glaucoma is called β-adrenoceptor antagonists. They have been a cornerstone of glaucoma treatment and a popular first-line therapy since they were originally introduced for ocular use in 1979. AH production is decreased by β-adrenoceptor antagonists, most likely by inhibiting

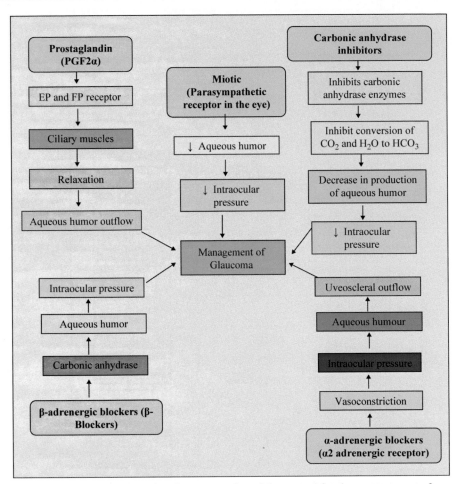

Fig. 3 Mechanism of action of various categories of drugs used for the management of glaucoma.

catecholamine-stimulated cyclic adenosine monophosphate synthesis in the ciliary epithelium [57]. Among the β-blockers with a topically acting approach there are cardioselective medicines that solely block β 1-receptors and nonselective treatments that target both β1- and β2-adrenoceptors in the treatment of glaucoma. Blockers were once the most widely used first-line topical glaucoma drug, but during the 1990s, the usage of more effective prostaglandin analogs replaced them as the mainstay of treatment [58].

3.3 α-Adrenergic blockers

Adrenergic receptors are cell membrane receptors that are a part of the receptor super-family that is G-protein related. Adrenergic receptors mediate sympathetic nervous system actions and trigger the "fight or flight" response. There are two subtypes of

Table 1 Recapitulation of the pharmacokinetics of drugs used in the management of glaucoma.

Drug	Absorption	Distribution	Metabolism	Excretion	Marketed drug	Reference
Latanoprost	Topical route	1–2 h	Liver	2–3 h	Xalatan	[43]
Travoprost	Topical route	30 min	Kidney, liver, and lung	1 h	Travatan Z, Alcon	[44]
Bimatoprost	Topical route	20 min	Blood	45 min	Lumigan	[45]
Timolol	Topical route	2 h	Liver	4–5 h	Timoptic	[46]
Betaxol	Topical route	2 h	Liver	14–22 h	Kerlone, Betoptic	[47]
Carteolol	Topical route		CYP2D6		Mikelan, Teoptic, Ocupress	[48]
Apraclonidine	Topical route	2 h		8 h	Iopidine, Alcon	[49]
Bromonidine	Topical route	2 h	Liver	120 h	Alphagan, Allergan, Inc., Irvine	[50]
Dorzolamide	Topical route	2 h	Systemic circulation	>4 months	Trusopt	[51]
Brinzolamide	Topical route	30 min–2 h		3–5 h	Azopt	[52]
Pilocarpine	Topical route	1 h	Plasma	24 h	Vuity	[53]

adrenergic receptors: excitatory α receptors and inhibitory receptors. Alpha-agonists in the eye activate alpha-1 receptors, which stimulate the iris dilator and Muller's muscles to contract, causing mydriasis and lid retraction. Alpha-1 stimulation also induces vasoconstriction, which reduces aqueous output and restricts blood flow to the ciliary muscle [59]. Vasoconstriction, inhibition of norepinephrine release, and inhibition of insulin secretion are all effects of alpha-2 receptor activation. Alpha-2 agonists lower IOP by increasing uveoscleral outflow and constricting the afferent ciliary process vasculature, which reduces the generation of AH [60].

3.4 Carbonic anhydrase inhibitors (CAIs)

Enzymes called CAs play an important role in many physiological processes, including bone resorption, calcification, tumorigenicity, CO_2 and bicarbonate transport, pH and CO_2 homeostasis, electrolyte secretion in a variety of tissues and organs, and biosynthetic reactions (such as gluconeogenesis, lipogenesis, and ureagenesis) [61]. Due to their capacity to lower AH secretion and consequently lower IOP, CAIs are traditionally employed in the management of glaucoma in the eyes. Additionally, before surgery, CAIs are utilized for acute angle-closure glaucoma and chronic OAG. Similarly, CAIs lower the amount of CSF fluid produced, which makes them effective for treating idiopathic intracranial hypertension [62].

3.5 Miotic

The preferred third line of treatment is miotics, cholinergic agonists, or alternatively parasympathomimetics. Miotics like pilocarpine and carbacol are frequently prescribed. The medication is applied locally using drops. By tightening the trabecular meshwork fibers, they accelerate the rate at which AH is evacuated from the eye. When pilocarpine is used with latanoprost in humans, it has been demonstrated to have a synergistic effect since it helps to reduce uveosclera outflow by constricting the ciliary muscle. High dosages of Physostigmine have been reported to be efficient for preventing ocular hypotension brought on by latanoprost [63].

4 Nanotechnological aspects as a paradigm in therapeutics of glaucoma

Nanotechnology approaches like niosomes, liposomes, polymeric nanoparticles (PNPs), nanogel, nano emulsion, and ocular insert have been investigated for the management of glaucoma in recent years. Table 2 provides a detailed description of their outcomes and clinical significance.

4.1 Niosomes

Niosomes are vesicles made primarily of hydrated nonionic surfactants, along with cholesterol (CHOL) or its derivatives in many circumstances. Niosomes can encapsulate both hydrophilic and lipophilic molecules due to their distinctive structural

Table 2 Recent advancement in nanocarrier based delivery of medications used in glaucoma.

Dosage form	Technique	Excipients	Outcome	Ref.
Latanoprost				
Thermosensitive hydrogel containing latanoprost and curcumin-loaded nanoparticles (CUR-NPs)	Emulsion evaporation method	PLGA, curcumin, chitosan	The study showed CUR-NPs, reduce the oxidative stress-mediated damage in TM cells by reducing the genes related to inflammation, the production of ROS, and apoptosis. By using the (transferase dUTP nick end labeling) TUNEL test, the DNA fragmentation amount of TM cells was assessed. When compared to the H or HG groups, the number of TUNEL-positive cells was considerably lower in the HGC group 9.38%. On days 1, 4, and 7, the total latanoprost release percentage from the hydrogel was 2.47 ± 1.35, 13.48 ± 2.97, and $23.63 \pm 2.41\%$, respectively	[64]
Timolol and latanoprost loaded mPEG-PLA micelles-laden contact lenses (CLs-M)	Thin film hydration	mPEG2000-PLA2400	The absorbance of CLs-M was maintained by micelles with ultra-small particle size and narrow particle size distribution. Timolol and latanoprost were slowly released from CLs-M over the course of 120–144h. The BA of timolol and latanoprost was 2.2- and 7.3-fold greater for the CLs-M than for the eye drops 0.5% timolol and, respectively, and a significant reduction in IOP with 9.8-fold PA was achieved for CLs-M compared with 0.5% timolol and 0.005% latanoprost eye drops in an in vivo PK study	[65]

Continued

Table 2 Continued

Dosage form	Technique	Excipients	Outcome	Ref.
Latanoprost-loaded phytantriol cubosomes (CubLnp)	Top-down method	Phytantriol, acetonitrile and formic acid	CubLnp is used in the pharmacotherapy of chronic therapies for glaucoma as it requires less frequent administration and a lower dose of the medicine, lowering the risk of side effects and eventually increasing patient compliance	[66]
Poly(lactide)/monomethoxy-poly(ethyleneglycol) (PLA-PEG) nanoparticles	Emulsification-solvent evaporation technique	D,L-Lactide	The study showed Group A was considerably affected by the hypotensive effects of LA-loaded NPs throughout the research, and its IOP values remained significantly lower than those of the other groups. While LA levels in group A gradually grew, LA AH concentrations in group B progressively dropped over time. On day 6, LA levels in group A were greater than in group B is 344 ± 73.5 and 228 ± 41.01 ng/mL, respectively	[67]
Niosomes	Reverse-phase evaporation technique	Cholesterol, Pluronic F-127 (PL)	The study demonstrates that latanoprost was gradually released from the niosome suspension, with a 50% release rate in 24h and a 100% release rate in 96h. After 24h, just 24% of the latanoprost in the PL niosomal gel had been released. This demonstrates that the niosomal gel preparation's latanoprost ocular bioavailability was more than four times higher	[68]

Biodegradable nanosheet	Layer-by-layer method	Chitosan	The IOP reduction rates of 0.25 µg/cm^2 LBNS and 0.005% latanoprost ophthalmic solution were significant only for the first day after application, whereas those of 2.5 lg/cm^2 LBNS were $-27.0\% \pm 14.8\%$, $-22.0\% \pm 16.7\%$, $-25.8\% \pm 18.0\%$, $-22.7\% \pm 20.9\%$, and $-6.6\% \pm 17.0\%$, at 1, 2, 4, 7, and 9 days, respectively. Similarly, the 25 µg/cm^2 LBNS decreased IOP	[69]
Travoprost				
Travoprost-loaded NPs (Trav-NPs)	Hybridization with fluorophore-DNA conjugates		Successfully demonstrates the use of a DNA-based drug delivery system in the treatment of glaucoma, In the Trav -NP group, 10 out of 20 (50%) eyes showed a drop in IOP, but only 6 out of 18 (33%) eyes in the Travoprost group did. The average delta IOP for all animals receiving Trav -NP was -0.9 mmHg, whereas it was 5.3 mmHg for those receiving the pure medication	[70]
Bimatoprost				
Bimatoprost nanoparticles loaded pH-sensitive in situ gel	High-shear homogenization and probe sonication method	Bimatoprost, Precirol ATO 5, Soya Lecithin, Acrypol 941	Gel's nanoparticles increased precorneal residence duration and prolonged medication release. Extended drug release was demonstrated with the improved formulation BIM-SLN4 for up to 19h which is 99.43%	[71]

Continued

Table 2 Continued

Dosage form	Technique	Excipients	Outcome	Ref.
Bimatoprost loaded ocular inserts	Solvent casting method	Chitosan	After a single application, BIM-loaded inserts reduced IOP for 4 weeks, but IOP levels in the placebo and untreated groups remained noticeably high. Only throughout the daily therapy period were eye drops effective RGC counting and damage to the optic nerve head cupping were both affected by IOP. BIM-loaded inserts offered sustained BIM release and appear to be a potential glaucoma management solution	[72]
Bimatoprost loaded nanovesicular (BMT-NV)	Ethanol-injection technique	Acetonitrile	With a single subconjunctival injection of BMT-NV-GEL-IM in rats, BMT was released in vitro for 10 days is 80.23% while the IOP was lowered for over 2 months. IOP was reduced over 5 days by a single topical drop of BMT-NV-GEL. Direct evidence of good sustained delivery can be seen in the presence of considerable diffuse fluorescence in internal eye tissues observed by confocal microscopy following in vivo administration as a subconjunctival implant even after 2 months and eye drops up to 1 week	[73]

Bimatoprost loaded microemulsion laden contact lens		Hydroxyl ethylmethacrylate, ethyleneglycol dimethacrylate Irgacure	The cumulative release of bimatoprost from the bimatoprost-microemulsion-soaked contact lenses (ME) batches was 29.7, 42.5, and 59.25 µg from ME-25, ME-50, and ME-75, respectively, according to the data on bimatoprost release from the batches. With bimatoprost microemulsion (ME) batches, the release profile improved as evidenced by the sustained high release rate of 33 ng/h up to 96 h	[74]
Timolol				
Self-assembling elastin-like hydrogels for timolol delivery	Recombinant technology	Glucose, NaCl	Comparing the two systems, the (EIS)x2 system exhibits a more sustained release, with the percentage release after 8 h being 80.39% for the (EI)x2 system and 40.04% TM for the (EIS)x2 system, respectively	[75]
Niosomal gels loaded with Timolol Maleate (TM)	Thin film hydration method	Sodium carboxymethylcellulose, Span20, 40	The best entrapment efficiency and percentage of medication released within 8 h were demonstrated by formulation F6 (containing span 60:cholesterol in molar ratio 160:60) and F7 (containing span 60:tween 40: cholesterol in molar ratio 80:80:60)	[76]
Stimuli-sensitive hydrogel	Timolol maleate and brimonidine tartrate	Benzalkonium chloride, brimonidine tartrate, poly acrylic acid	The study showed that developed hydrogel viscosity is within the ideal range of 25–55 cps. At the end of 8 h, the drug was released up to 90%. The hydrogel membranes were discovered to be clean and nonirritating. Marketed	[77]

Continued

Table 2 Continued

Dosage form	Technique	Excipients	Outcome	Ref.
Timolol loaded composite ocular films	Solvent casting method	Glucose, gelatin, albumin, HPMC	formulation demonstrated a drop in IOP of up to 14 mmHg after 5 h, and after drug elimination, F2 and F6 continued to have a sustained impact for up to 12 h F1 had the largest cumulative release, at 71.59%, while F4 displayed the lowest release, at 41.48% in 8 h	[78]
Timolol maleate gelatinized core liposomes	Thin film hydration method	Gelatin A bloom 300, cholesterol, potassium dihydrogen phosphate	This novel formulation demonstrated decreased drug leakage and increased compatibility with the corneal layers, evoking a stronger and longer-lasting pharmacological response linked to a safe histological profile and greater stability. In F7 formulation IOP measurements yielded the greatest result	[79]
Mucoadhesive and responsive nanogels	Precipitation/ dispersion free radical polymerization	Acrylic acid, N-isopropylacrylamide, ammonium persulfate, N,N'-methylenebisacrylamide, mucin	The final outcome of timolol integration into poly(NIPA-co-AAc) (80:20) NGs is significant since a greater reduction in IOP may be observed in the first 10 h of therapy compared to a commercial eye drop. These tests found that after 48 h of exposure, the poly(NIPA-co-AAc) (80:20) NGs-timolol formulation did not irritate the eye's anterior portion	[80]

Betaxol				
Niosomes	Thin film hydration technique followed by sonication	Carbopol, cholesterol	The marketed eye drops showed a maximum reduction in IOP at 2–3h, and the niosomal gel (N6F6) showed an IOP decline that lasted from 0 to 6h. In normal and glaucomatous rabbits, the N6F6 formulation group showed the greatest IOP decrease (9.47 ± 1.0 and 10.8 ± 1.09 mmHg), compared to the control group receiving commercial eye drops (5.89 ± 0.74 and 6.57 ± 0.47 mmHg), respectively	[81]
Montmorillonite/chitosan nanoparticles	Ionic gelation	1-(4, 5-Dimethylthiazol-2-yl)-3, 5-diphenylformazan	The cell vitality of BH-Mt/CS NPs was 50% at a dosage of 30μL, indicating more cytotoxicity than that of Mt/CS NPs, whereas the cell viability of Mt/CS NPs was 80% at 50μL for 120min	[82]
Levobunolol				
Eudragit nanoparticle-laden contact lenses	Nanoprecipitation methodology	Eudragit S100	The lenses appeared transparent during the whole test, indicating the presence of medicine at the nanoscale after Eudragit S100 was dissolved, as opposed to the contact lenses with high burst release, which released $91.12\% \pm 0.900$ of the LB in 3 days. After 1, 2, and 3 months, the drug release from NP-loaded contact lenses was measured as $1.210\% \pm 0.109$, $1.894\% \pm 0.161$, and $2.381\% \pm 0.091$ of leaching, respectively	[83]

Continued

Table 2 Continued

Dosage form	Technique	Excipients	Outcome	Ref.
Poly(ε-caprolactone) (PCL) microparticles of Levobunolol HC1	Solvent evaporation technique	PCL, PVA	The microparticles' ability to release drugs appears to have two stages, with the first release happening quickly and the second happening more slowly. In comparison to the free drug introduced into the gel and the free microparticle, drug release was slower when the microparticle was incorporated into the thermally reversible gel (Pluronic F127)	[84]
Carteolol				
Gel formulation		Gelrite, theophylline anhydrous	The created in situ gel formulation demonstrated promise for usage as delivery method for Carteolol HC, which has a higher ocular bioavailability	[85]
Magnesium hydroxide (MH) nanoparticles	Bead mill method	MH, mannitol, MC	The instillation of nCMFC allows for the delivery of large concentrations of dissolved carteolol into the aqueous humor. For water-soluble medicines, a combination with MH nanoparticles may increase corneal penetration	[86]
Dipivefrine				
Thermoresponsive gels	Cold method	Poloxamer-407, and Poloxamer-188	Study showed nearly 99% of the DV was released from the DV-AqS in less than 4 h, whereas it took 8 h for about 89% of the DV to be released from F8. When compared to DV-AqS in rabbits, F8 had a successful, continuous, and superior ability to lower IOP. It was also nonirritating to the eyes	[87]

Brimonidine

Nano vesicular	Thin film hydration technique	Cholesterol, Carbopol 940	Preliminary IOP reduction was 16.77 ± 1.25 mmHg, and drug activity persisted after another 7.5 h, when the optimized nanovesicular formulation was recorded to have reduced IOP to 6.3 ± 1.48 mmHg. The in vitro release investigation made it clear that this formulation released its 68% concentration in 3.5 h. IOP decreased rapidly (16.77 ± 1.25 to 7.36 ± 1.40 mmHg after 3.5 h) and then continued to decrease over time	[88]
Microspheres	Solvent evaporation methods	PLA	Brimonidine-loaded microspheres delivered with extreme precision using a microneedle into the supraciliary region were able to do so for 1 month	[89]
Nanoparticle	Double emulsion-solvent evaporation technique	Eudragit RL 100 and Eudragit RS 100	They were discovered to extend drug release in vitro and extend the effectiveness of IOP reduction in vivo, making them a viable carrier in the development of enhanced drug delivery systems for the treatment of glaucoma	[90]
Ocular inserts		Magnesium stearate, Eudragits RL 100 or Eudragit RS 100	The effects of eye implants When an ocular insert carrying 1 mg of drug was compared to 2–3 drops of eye drop preparation (0.33 mg of BRT), it can be concluded that one administration of an ocular insert providing enough medication to keep IOP under control is significantly favored than four times per day	[91]

Continued

Table 2 Continued

Dosage form	Technique	Excipients	Outcome	Ref.
Silicone contact lenses	High-speed homogenization technique	Hydroxyl ethyl methacrylate, Irgacure, dimethyl acrylamide, siloxane	The NLC-30%-OA-CL batch of rabbit tear fluid demonstrated significant brimonidine content at all time periods for up to 144 h. The pharmacodynamic investigation revealed that the NLC-30%-OA-CL batch produced a persistent IOP decrease for 144 h as opposed to 6 h with the eye drop solution	[92]
Dorzolamide				
In situ gel of chitosan nanoparticles	Ionotropic gelation method	Chitosan, tripolyphosphate (TPP)	Compared to traditional eye drops, an in situ gel of dorzolamide hydrochloride loaded nanoparticles provides a more intensive glaucoma treatment with improved patient compliance	[93]
Chitosan/hydroxyethyl cellulose inserts	Solvent/casting technique	Hydroxyethyl cellulose, chitosan	According to in vitro findings, DI released 75% of the dorzolamide that had been entrapped in just 3 h. However, ex vivo biodistribution studies and scintigraphic images showed that more than 50% of 99mTc-dorzolamide remained in the eye after 18 h of DI administration, whereas only about 30% remained in the eye following drop instillation. After a single dosage, DI significantly reduced blood pressure for 2 weeks, although IOP levels in the placebo and untreated groups remained high	[94]

PLGA nanoparticles	Double emulsification-solvent evaporation technique	vitamin *E*-TPGS, Poly(lactide-*co*-glycolide)	After a single topical instillation into the eye, DZ-P-NPs, and DZ-T-NPs dramatically decreased intraocular pressure by 22.81% and 29.12%, respectively	[95]
Dorzolamide loaded 6-*O*-carboxymethyl chitosan nanoparticles	Ionic gelation method	Sodium tripolyphosphate, agarose, and monochloroacetic acid	When compared to CSNPs, DRZ-loaded OCM-CSNPs would be a better alternative to the glaucoma eye drops currently on the market because they had improved absorption and reduced pulse entry	[96]
Nano emulsions	Water titration method	Isopropyl myristate (IPM), triacetin	It provides more intensive glaucoma treatment, fewer daily treatments, and more patient compliance	[97]
Brinzolamide				
Chitosan-pectin mucoadhesive nanocapsules	Polyelectrolyte complex coacervation method	Sodium chloride, calcium chloride dihydrate	The early burst release ($23.51 \pm 0.45\%$ within 1h) and sustained drug release of $68.42 \pm 1.22\%$ at 6h and $76.89 \pm 0.25\%$ at 8h, respectively, were accomplished by CPNCs (BCP3), whereas the marketed formulation exhibited $51.4\% \pm 1.26$ in 1h and $98.39\% \pm 1.28$ at 6h, respectively. When compared to the commercially available formulation of BNZ ($63.25 \pm 1.8\,\mu g/cm^2$) on an 8-h study, the transcorneal permeation of optimized CPNCs (BCP3) was able to cross the cornea with a better permeation rate ($82.51 \pm 2.4\,\mu g/cm^2$)	[98]

Continued

Table 2 Continued

Dosage form	Technique	Excipients	Outcome	Ref.
Nanoliposomes	Thin film dispersion method	Soybean phosphatidylcholine, cholesterol	When compared to the commercially available formulation (BRZ-Sus) (10mg/mL BRZ), BCL (1mg/mL BRZ) had a sustained and enhanced intraocular pressure reduction efficacy and had a 9.36-fold increase in the apparent permeability coefficient	[99]
Pilocarpine				
Niosomal gels	Ether injection technique	Span (20, 40, 80), cholesterol, cust bean gum and carbopol 934	The pilocarpine HCl-containing niosomal gels are promising ocular carriers for the treatment of glaucoma. The G2 formulation's relative bioavailability was 2.64 times greater than that of commercial Pilopine HS gel in terms of its ability to reduce intraocular pressure (IOP). After 8h, the total amount of drug that migrated from the niosomal gel formulation ranged from $50.13 \pm 0.81\%$ to $62.89 \pm 2.21\%$	[100]
Chitosan-g-poly(N-isopropylacrylamide) copolymers intracameral	Grafting of carboxylic end-capped PNIPAA	N-isopropylacrylamide and N-hydroxy succinimide, chitosan	Injectable biodegradable thermogels based on chitosan have the potential to be used as intracameral biomaterials for better delivery carrier performance and prolonged release of antiglaucoma drugs	[101]
Liquid crystal nanoparticles	Top-down method	Poloxamer 407, glyceryl monoolein	CNP measures 202.28 ± 19.32 nm in size, and its encapsulation effectiveness is 61.03%. According to the in vitro release profiles, PN-loaded LCNPs could maintain sustained release for 8h. The apparent permeability coefficient of PN-loaded LCNPs was 2.05 times greater than that of commercial eye drops, according to an ex vivo corneal permeation analysis	[102]

makeup. This is accomplished by entrapping. Although the lipophilic chemicals are enclosed by their partitioning into the lipophilic domain of the bilayers, the hydrophilic substances are either in the vesicular aqueous core or adsorbed on the bilayer surfaces [103]. The components of niosomes, which are biodegradable, nonimmunogenic, and biocompatible, are small lamellar structures with diameters between 10 and 1000 nm. Niosomes have proven to be a promising drug delivery system with the potential to lessen medication adverse effects and improve therapeutic efficacy for a variety of disorders [104]. Niosomes may contain a wide range of pharmacological moieties, which include hydrophilic, lipophilic, and amphiphilic substances. By adjusting the vesicle's composition, size lamellarity, surface charge, tapping volume, and concentration, vesicle properties may be changed. The medication can release in a steady, regulated manner [105].

4.2   Liposomes

Liposomes are spherical vesicles made of one or perhaps more phospholipid bilayers or lamellae around an aqueous core. Most typically, the number of bilayers and the size of the vesicles (small, big, and enormous) are used to describe liposomes. There are two different ways to make liposomes: (i) in bulk, where phospholipids are transferred from an organic phase into an aqueous phase and (ii) in films, where lipid films are first coated on a substrate and then hydrated to produce liposomes [106]. Dr. Alec D. Bangham of the Babraham Institute in Cambridge, England, initially described liposomes in 1961 (published in 1964). This was found by Bangham and R.W. Horne when they were experimenting with the institute's new electron microscope by applying negative stains to dried phospholipids [107]. The biodegradability, reduced systemic toxicity, targeted administration, protection of sensitive molecules, and better pharmacokinetic effects of systemic liposomes as therapeutic formulations are their key benefits. However, their benefits for topical applications come from their demonstrated ability to minimize serious incompatibilities and side effects that may arise from the unfavorably high systemic drug absorption, the significantly increased drug accumulation at the necessary site, and their capacity to produce the desired results [108].

4.3   Polymeric nanoparticles

The medication is dissolved, entrapped, encapsulated, or linked to a nanoparticle matrix in the PNPs, which range in size from 10 to 1000 nm. These PNPs are made up of biocompatible and biodegradable polymers. Depending on the technique of preparation, one can produce nanoparticles, nanospheres, or nano capsules [109]. Therapeutics like small molecule medicines and biologics are often contained in the core of polymeric NPs during the selfassembly process that produces them. By allowing for the modification of physicochemical characteristics (size, surface charge, hydrophobicity), selfassembly conditions, and drug release qualities, polymers, and selfassembly provide significant design freedom [110]. Many substances, including carbon nanotubes, ceramics, lipids, and polymers, can be used to make them. The best combination of properties can be found in polymer materials, as opposed to other compositions: they are stable, allow high loading of a variety of agents, control drug release kinetics, are simple to modify to display a variety of surface-attached ligands, and many polymers have a long history of safe use in humans [111].

4.4 Nanogel

Nanogels are characterized as nanoscale particles made of polymer networks that have undergone physical or chemical crosslinking and have swelled in a suitable solvent. Initially, the term "nanogel" was used to describe cross-linked bifunctional networks containing a polyion and a nonionic polymer for polynucleotide delivery [112]. Nanogels can be created through heterogeneous polymerization of monomers or from polymer precursors. Cross-link, comprising physical cross-link and chemical cross-link, is the main component of nanogel fabrication. Since the polymer contains hydrophilic groups like —OH, —CONH—, —CONH$_2$—, and —SO$_3$H, nanogels can absorb large volumes of water or biological fluids while retaining their structural integrity [113]. Chemical crosslinking and physical selfassembly are two types of nanogel production that may be categorized according to the many architectures and building blocks found in nanogels. Via the covalent crosslinking between functional groups on polymer chains, the nanogel created by chemical crosslinking shows better stability compared to the nanogel created by physical crosslinking. Meanwhile, noncovalent interactions, such as hydrogen bonds, the Van der Waals force, hydrophobic contacts, host-guest interactions, and electrostatic interactions, are frequently responsible for the reversible connections of physically cross-linked nanogels [114].

4.5 Nanoemulsion

The delivery strategy for pharmaceutical compounds is a nanoemulsion, a colloidal particle system of submicron size range. Their sizes range from 10 to 1000 nm. These carriers have an amorphous, lipophilic, negatively charged surface and are solid spheres [115]. Nanoemulsions are generally spherical in shape and are a collection of dispersed particles. It can be utilized for medicines, biomedical assistance, and transportation. They hold significant potential for the future of biotechnologies, cosmetics, diagnostics and pharmacological treatments. Mini-emulsion, extremely fine emulsion and submicron emulsion (SME) are all used interchangeably. By adding an appropriate substance known as emulsifying agent, stability is produced in nanoemulsions, which are heterogeneous mixtures of lipid and aqueous phases [116]. The main distinction between emulsions and nanoemulsions is that the earlier, having good kinetic stability, are inherently thermodynamically unstable and will ultimately phase split. Emulsions seem foggy, but Nanoemulsions are transparent or translucent, which is another significant distinction [117]. There are two techniques for producing nanoemulsions: a mechanical device-based high-energy technique and a chemical potential-based low-energy technique. The high-energy technique employs a rotor/stator system for high-shear stirring, as well as ultrasonication, a high-pressure homogenizer, and, in particular, microfluidization and membrane emulsification. The low-energy method includes solvent diffusion, phase inversion temperature, phase inversion composition, selfemulsification, and even spontaneous emulsification in nonequilibrium [118].

4.6   Ocular insert

Ocular inserts are described as sterile preparations with thin, multilayered, drug-impregnated, solid, or semisolid consistency devices that are inserted into conjunctival sacs or cul-de-sacs and whose size and design are specifically intended for ocular use. They consist of a polymeric support that may or may not contain drug(s), with the latter being integrated as a dispersion or a solution in the polymeric support. Topical or therapeutic uses are possible for the inserts [119]. They have several benefits, involving extended ocular residency and sustained medicine delivery into the eye. The implant has a body part that fits perfectly within the lachrymal canaliculus of the eyelid. The inserts are characterized as insoluble, soluble, or bio erodible depending too how soluble they are. The drug's diffusion, osmosis, and bioerosion all play a role in how easily the insert releases the medication [120].

5   Conclusion and future perspectives

The current study highlights glaucoma therapy's existing treatment plans and discusses prospective future targets as well as measures to safeguard and enhance the survival and regeneration of RGCs. In animal glaucoma models, it has been demonstrated that targeting several pathways increases the survival of RGCs. The current version gives a summary of the IOP-lowering drugs that are currently on the market. In addition, prospective new therapeutic targets for neuroprotective and IOP-lowering therapy are suggested. The development of personalized medication and continuous drug delivery systems are two prospective trends in glaucoma treatment that are discussed. Future personalized medicine shows potential for predicting the best therapy and preventative options for each individual glaucoma patient by taking into account the individual genetic variability, environmental factors, and lifestyle factors of each person and can also enhance the possibility of a sustained drug delivery system.

Acknowledgments

The authors would like to thank the Department of Pharmaceutics, MM College of Pharmacy, Maharishi Markandeshwar (Deemed to be University), Mullana-Ambala, Haryana, India 133207 for providing facilities for the completion of this chapter.

References

[1] Lee DA, Higginbotham EJ. Glaucoma and its treatment: a review. Am J Health Syst Pharm 2005;62(7):691–9. https://doi.org/10.1093/ajhp/62.7.691.

[2] Casson RJ, Chidlow G, Wood JP, Crowston JG, Goldberg I. Definition of glaucoma: clinical and experimental concepts. Clin Exp Ophthalmol 2012;40(4):341–9. https://doi.org/10.1111/j.1442-9071.2012.02773.x.

[3] Vaajanen A, Vapaatalo H. Local ocular renin–angiotensin system—a target for glaucoma therapy? Basic Clin Pharmacol Toxicol 2011;109(4):217–24. https://doi.org/10.1111/j.1742-7843.2011.00729.x.

[4] Rieck J. The pathogenesis of glaucoma in the interplay with the immune system. Invest Ophthalmol Vis Sci 2013;54(3):2393–409. https://doi.org/10.1167/iovs.12-9781.

[5] Cvenkel B, Kolko M. Current medical therapy and future trends in the management of glaucoma treatment. J Ophthalmol 2020;2020. https://doi.org/10.1155/2020/6138132.

[6] Wiggs JL, Pasquale LR. Genetics of glaucoma. Hum Mol Genet 2017;26(R1): R21–7. https://doi.org/10.1093/hmg/ddx184.

[7] Dahlmann-Noor AH, Vijay S, Limb GA, Khaw PT. Strategies for optic nerve rescue and regeneration in glaucoma and other optic neuropathies. Drug Discov Today 2010;15 (7–8):287–99. https://doi.org/10.1016/j.drudis.2010.02.007.

[8] Rodríguez-Muela N, Germain F, Mariño G, Fitze PS, Boya P. Autophagy promotes survival of retinal ganglion cells after optic nerve axotomy in mice. Cell Death Differ 2012;19(1):162–9. https://doi.org/10.1038/cdd.2011.88.

[9] Vrabec JP, Levin LA. The neurobiology of cell death in glaucoma. Eye 2007;21(1): S11–4. https://doi.org/10.1038/sj.eye.6702880.

[10] Greco A, Rizzo MI, De Virgilio A, Gallo A, Fusconi M, De Vincentiis M. Emerging concepts in glaucoma and review of the literature. Am J Med 2016;129(9):1000.e7. https://doi.org/10.1016/j.amjmed.2016.03.038.

[11] Tarongoy P, Ho CL, Walton DS. Angle-closure glaucoma: the role of the lens in the pathogenesis, prevention, and treatment. Surv Ophthalmol 2009;54(2):211–25. https://doi.org/10.1016/j.survophthal.2008.12.002.

[12] Nongpiur ME, Ku JY, Aung T. Angle closure glaucoma: a mechanistic review. Curr Opin Ophthalmol 2011;22(2):96–101. https://doi.org/10.1097/ICU.0b013e32834372b9.

[13] He M, Foster PJ, Johnson GJ, Khaw PT. Angle-closure glaucoma in East Asian and European people. Different diseases? Eye 2006;20(1):3–12. https://doi.org/10.1038/sj.eye.6701797.

[14] Amerasinghe N, Aung T. Angle-closure: risk factors, diagnosis and treatment. Prog Brain Res 2008;173:31–45. https://doi.org/10.1016/S0079-6123(08)01104-7.

[15] Patel K, Patel S. Angle-closure glaucoma. Dis Mon 2014;60(6):254–62. https://doi.org/10.1016/j.disamonth.2014.03.005.

[16] Sun X, Dai Y, Chen Y, Yu DY, Cringle SJ, Chen J, Kong X, Wang X, Jiang C. Primary angle closure glaucoma: what we know and what we don't know. Prog Retin Eye Res 2017;57:26–45. https://doi.org/10.1016/j.preteyeres.2016.12.003.

[17] Lachkar Y, Bouassida W. Drug-induced acute angle closure glaucoma. Curr Opin Ophthalmol 2007;18(2):129–33. https://doi.org/10.1097/ICU.0b013e32808738d5.

[18] Parivadhini A, Lingam V. Management of secondary angle closure glaucoma. J Curr Glaucoma Pract 2014;8(1):25. https://www.ncbi.nlm.nih.gov/pmc/articles/PMC4741163/#.

[19] Allison K, Patel D, Alabi O. Epidemiology of glaucoma: the past, present, and predictions for the future. Cureus 2020;12(11). https://doi.org/10.7759/cureus.11686.

[20] Senthil S, Nakka M, Sachdeva V, Goyal S, Sahoo N, Choudhari N. Glaucoma mimickers: a major review of causes, diagnostic evaluation, and recommendations. Semin Ophthalmol 2021;36(8):692–712. Taylor & Francis https://doi.org/10.1080/08820538.2021.1897855.

[21] Chen MJ. Normal tension glaucoma in Asia: epidemiology, pathogenesis, diagnosis, and management. Taiwan J Ophthalmol 2020;10(4):250. https://www.ncbi.nlm.nih.gov/pmc/articles/PMC7787092/#.

[22] Choi J, Kook MS. Systemic and ocular hemodynamic risk factors in glaucoma. Biomed Res Int 2015;2015. https://doi.org/10.1155/2015/141905.

[23] Muacevic A, Adler J, Allison K, Patel D, Alabi O. Epidemiology of glaucoma: the past, present, and predictions for the future. Cureus 2021;12(11). https://doi.org/10.7759/cureus.11686.

[24] Leske MC. Open-angle glaucoma—an epidemiologic overview. Ophthalmic Epidemiol 2007;14(4):166–72. https://doi.org/10.1080/09286580701501931.

[25] Wang W, He M, Li Z, Huang W. Epidemiological variations and trends in health burden of glaucoma worldwide. Acta Ophthalmol 2019;97(3):e349–55. https://doi.org/10.1111/aos.14044.

[26] Saxena R, Singh D, Vashist P. Glaucoma: an emerging peril. Indian J Community Med 2013;38(3):135. https://www.ncbi.nlm.nih.gov/pmc/articles/PMC3760320/.

[27] George R, Ramesh SV, Vijaya L. Glaucoma in India: estimated burden of disease. J Glaucoma 2010;19(6):391–7. https://doi.org/10.1097/IJG.0b013e3181c4ac5b.

[28] Taurone S, Ripandelli G, Pacella E, Bianchi E, Plateroti AM, De Vito S, Plateroti P, Grippaudo FR, Cavallotti C, Artico M. Potential regulatory molecules in the human trabecular meshwork of patients with glaucoma: immunohistochemical profile of a number of inflammatory cytokines. Mol Med Rep 2015;11(2):1384–90. https://doi.org/10.3892/mmr.2014.2772.

[29] Kim M, Lee C, Payne R, Yue BY, Chang JH, Ying H. Angiogenesis in glaucoma filtration surgery and neovascular glaucoma: a review. Surv Ophthalmol 2015;60(6):524–35. https://doi.org/10.1016/j.survophthal.2015.04.003.

[30] Dreyfuss JL, Giordano RJ, Regatieri CV. Ocular angiogenesis. J Ophthalmol 2015;2015. https://doi.org/10.1155/2015/892043.

[31] Pożarowska D, Pożarowski P. The era of anti-vascular endothelial growth factor (VEGF) drugs in ophthalmology, VEGF and anti-VEGF therapy. Cent Eur J Immunol 2016;41(3):311–6. https://doi.org/10.5114/ceji.2016.63132.

[32] Ergorul C, Ray A, Huang W, Darland D, Luo ZK, Grosskreutz CL. Levels of vascular endothelial growth factor-A165b (VEGF-A165b) are elevated in experimental glaucoma. Mol Vis 2008;14:1517. https://www.ncbi.nlm.nih.gov/pmc/articles/PMC2518529/#.

[33] Vohra R, Tsai JC, Kolko M. The role of inflammation in the pathogenesis of glaucoma. Surv Ophthalmol 2013;58(4):311–20. https://doi.org/10.1016/j.survophthal.2012.08.010.

[34] Saccà SC, Izzotti A. Focus on molecular events in the anterior chamber leading to glaucoma. Cell Mol Life Sci 2014;71(12):2197–218. https://doi.org/10.1007/s00018-013-1493-z.

[35] Baudouin C, Kolko M, Melik-Parsadaniantz S, Messmer EM. Inflammation in glaucoma: from the back to the front of the eye, and beyond. Prog Retin Eye Res 2021;83:100916. https://doi.org/10.1016/j.preteyeres.2020.100916.

[36] Shinozaki Y, Koizumi S. Potential roles of astrocytes and Müller cells in the pathogenesis of glaucoma. J Pharmacol Sci 2021;145(3):262–7. https://doi.org/10.1016/j.jphs.2020.12.009.

[37] Harris A, Rechtman E, Siesky B, Jonescu-Cuypers C, McCranor L, Garzozi HJ. The role of optic nerve blood flow in the pathogenesis of glaucoma. Ophthalmol Clin N Am 2005;18(3):345–53. https://doi.org/10.1016/j.ohc.2005.04.001.

[38] Mozaffarieh M, Grieshaber MC, Flammer J. Oxygen and blood flow: players in the pathogenesis of glaucoma. Mol Vis 2008;14:224. https://www.ncbi.nlm.nih.gov/pmc/articles/PMC2267728/#.

[39] Andrés-Guerrero V, Perucho-González L, García-Feijoo J, Morales-Fernández L, Saenz-Francés F, Herrero-Vanrell R, Júlvez LP, Llorens VP, Martínez-de-la-Casa JM, Konstas AG. Current perspectives on the use of anti-VEGF drugs as adjuvant therapy in glaucoma. Adv Ther 2017;34(2):378–95. https://doi.org/10.1007/s12325-016-0461-z.

[40] Williams PA, Marsh-Armstrong N, Howell GR, Bosco A, Danias J, Simon J, Di Polo A, Kuehn MH, Przedborski S, Raff M, Trounce I. Neuroinflammation in glaucoma: a new opportunity. Exp Eye Res 2017;157:20–7. https://doi.org/10.1016/j.exer.2017.02.014.

[41] Yamanaka O, Kitano-Izutani A, Tomoyose K, Reinach PS. Pathobiology of wound healing after glaucoma filtration surgery. BMC Ophthalmol 2015;15(1):19–27. https://doi.org/10.1186/s12886-015-0134-8.

[42] Baltmr A, Duggan J, Nizari S, Salt TE, Cordeiro MF. Neuroprotection in glaucoma—is there a future role? Exp Eye Res 2010;91(5):554–66. https://doi.org/10.1016/j.exer.2010.08.009.

[43] Alm A. Latanoprost in the treatment of glaucoma. Clin Ophthalmol 2014;8:1967. https://www.ncbi.nlm.nih.gov/pmc/articles/PMC4196887/#.

[44] Ulc M.P., Etobicoke O.N.. Pr Mylan-Travoprost Z.

[45] Curran MP, Orman JS. Bimatoprost/timolol. Drugs Aging 2009;26(2):169–84. https://doi.org/10.2165/0002512-200926020-00008.

[46] Volotinen M, Hakkola J, Pelkonen O, Vapaatalo H, Mäenpää J. Metabolism of ophthalmic timolol: new aspects of an old drug. Basic Clin Pharmacol Toxicol 2011;108(5):297–303. https://doi.org/10.1111/j.1742-7843.2011.00694.x.

[47] Sharma R, Shastri N, Sadhotra P. β-Blockers as glaucoma therapy. JK Sci 2007;9:42–5.

[48] Henness S, Harrison TS, Keating GM. Ocular carteolol. Drugs Aging 2007;24(6):509–28. https://doi.org/10.2165/00002512-200724060-00007.

[49] Scuderi G, Romano M, Perdicchi A, Cascone N, Lograno M. Apraclonidine hydrochloride: pharmacology and clinical use. Expert Rev Ophthalmol 2008;3(2):149–53. https://doi.org/10.1586/17469899.3.2.149.

[50] Oh DJ, Chen JL, Vajaranant TS, Dikopf MS. Brimonidine tartrate for the treatment of glaucoma. Expert Opin Pharmacother 2019;20(1):115–22. https://doi.org/10.1080/14656566.2018.1544241.

[51] Loftsson T, Jansook P, Stefansson E. Topical drug delivery to the eye: dorzolamide. Acta Ophthalmol 2012;90(7):603–8. https://doi.org/10.1111/j.1755-3768.2011.02299.x.

[52] Brinzolamide IM. Expert Opin Pharmacother 2008;9(4):653–62. https://doi.org/10.1517/14656566.9.4.653.

[53] Guo Y, Jiang L. Drug transporters are altered in brain, liver and kidney of rats with chronic epilepsy induced by lithium-pilocarpine. Neurol Res 2010;32(1):106–12. https://doi.org/10.1179/174313209X408954.

[54] Matsou A, Anastasopoulos E. Investigational drugs targeting prostaglandin receptors for the treatment of glaucoma. Expert Opin Investig Drugs 2018;27(10):777–85. https://doi.org/10.1080/13543784.2018.1526279.

[55] Carta F, Supuran CT, Scozzafava A. Novel therapies for glaucoma: a patent review 2007–2011. Expert Opin Ther Pat 2012;22(1):79–88. https://doi.org/10.1517/13543776.2012.649006.

[56] Winkler NS, Fautsch MP. Effects of prostaglandin analogues on aqueous humor outflow pathways. J Ocul Pharmacol Ther 2014;30(2–3):102–9. https://doi.org/10.1089/jop.2013.0179.

[57] Marquis RE, Whitson JT. Management of glaucoma: focus on pharmacological therapy. Drugs Aging 2005;22(1):1–21. https://doi.org/10.2165/00002512-200522010-00001.

[58] Nocentini A, Ceruso M, Bua S, Lomelino CL, Andring JT, McKenna R, Lanzi C, Sgambellone S, Pecori R, Matucci R, Filippi L. Discovery of β-adrenergic receptors blocker-carbonic anhydrase inhibitor hybrids for multitargeted antiglaucoma therapy. J Med Chem 2018;61(12):5380–94. https://doi.org/10.1021/acs.jmedchem.8b00625.

[59] Arthur S, Cantor LB. Update on the role of alpha-agonists in glaucoma management. Exp Eye Res 2011;93(3):271–83. https://doi.org/10.1016/j.exer.2011.04.002.

[60] Sambhara D, Aref AA. Glaucoma management: relative value and place in therapy of available drug treatments. Ther Adv Chronic Dis 2014;5(1):30–43. https://doi.org/10.1177/2040622313511286.

[61] Masini E, Carta F, Scozzafava A, Supuran CT. Antiglaucoma carbonic anhydrase inhibitors: a patent review. Expert Opin Ther Pat 2013;23(6):705–16. https://doi.org/10.1517/13543776.2013.794788.

[62] Aslam S, Gupta V. Carbonic anhydrase inhibitors. StatPearls; 2022.

[63] Yadav KS, Rajpurohit R, Sharma S. Glaucoma: current treatment and impact of advanced drug delivery systems. Life Sci 2019;221:362–76. https://doi.org/10.1016/j.lfs.2019.02.029.

[64] Cheng YH, Ko YC, Chang YF, Huang SH, Liu CJ. Thermosensitive chitosan-gelatin-based hydrogel containing curcumin-loaded nanoparticles and latanoprost as a dual-drug delivery system for glaucoma treatment. Exp Eye Res 2019;179:179–87. https://doi.org/10.1016/j.exer.2018.11.017.

[65] Xu J, Ge Y, Bu R, Zhang A, Feng S, Wang J, Gou J, Yin T, He H, Zhang Y, Tang X. Co-delivery of latanoprost and timolol from micelles-laden contact lenses for the treatment of glaucoma. J Control Release 2019;305:18–28. https://doi.org/10.1016/j.jconrel.2019.05.025.

[66] Bessone CD, Akhlaghi SP, Tártara LI, Quinteros DA, Loh W, Allemandi DA. Latanoprost-loaded phytantriol cubosomes for the treatment of glaucoma. Eur J Pharm Sci 2021;160:105748. https://doi.org/10.1016/j.ejps.2021.105748.

[67] Giarmoukakis A, Labiris G, Sideroudi H, Tsimali Z, Koutsospyrou N, Avgoustakis K, Kozobolis V. Biodegradable nanoparticles for controlled subconjunctival delivery of latanoprost acid: in vitro and in vivo evaluation. Preliminary results. Exp Eye Res 2013;112:29–36. https://doi.org/10.1016/j.exer.2013.04.007.

[68] Fathalla D, Fouad EA, Soliman GM. Latanoprost niosomes as a sustained release ocular delivery system for the management of glaucoma. Drug Dev Ind Pharm 2020;46(5):806–13. https://doi.org/10.1080/03639045.2020.1755305.

[69] Kashiwagi K, Ito K, Haniuda H, Ohtsubo S, Takeoka S. Development of latanoprost-loaded biodegradable nanosheet as a new drug delivery system for glaucoma. Invest Ophthalmol Vis Sci 2013;54(8):5629–37. https://doi.org/10.1167/iovs.12-9513.

[70] Schnichels S, Hurst J, de Vries JW, Ullah S, Gruszka A, Kwak M, Löscher M, Dammeier S, Bartz-Schmidt KU, Spitzer MS, Herrmann A. Self-assembled DNA nanoparticles loaded with travoprost for glaucoma-treatment. Nanomedicine 2020;29:102260. https://doi.org/10.1016/j.nano.2020.102260.

[71] Wadetwar RN, Agrawal AR, Kanojiya PS. In situ gel containing bimatoprost solid lipid nanoparticles for ocular delivery: in-vitro and ex-vivo evaluation. J Drug Deliv Sci Technol 2020;56:101575. https://doi.org/10.1016/j.jddst.2020.101575.

[72] Franca JR, Foureaux G, Fuscaldi LL, Ribeiro TG, Rodrigues LB, Bravo R, Castilho RO, Yoshida MI, Cardoso VN, Fernandes SO, Cronemberger S. Bimatoprost-loaded ocular inserts as sustained release drug delivery systems for glaucoma treatment: in vitro and in vivo evaluation. PLoS One 2014;9(4):e95461. https://doi.org/10.1371/journal.pone.0095461.

[73] Yadav M, Guzman-Aranguez A, de Lara MJ, Singh M, Singh J, Kaur IP. Bimatoprost loaded nanovesicular long-acting sub-conjunctival in-situ gelling implant: in vitro and in vivo evaluation. Mater Sci Eng C 2019;103:109730. https://doi.org/10.1016/j.msec.2019.05.015.

[74] Xu W, Jiao W, Li S, Tao X, Mu G. Bimatoprost loaded microemulsion laden contact lens to treat glaucoma. J Drug Deliv Sci Technol 2019;54:101330. https://doi.org/10.1016/j.jddst.2019.101330.

[75] Fernández-Colino A, Quinteros DA, Allemandi DA, Girotti A, Palma SD, Arias FJ. Self-assembling elastin-like hydrogels for timolol delivery: development of an ophthalmic formulation against glaucoma. Mol Pharm 2017;14(12):4498–508. https://doi.org/10.1021/acs.molpharmaceut.7b00615.

[76] Ramadan AA, Eladawy SA, El-Enin AS, Hussein ZM. Development and investigation of timolol maleate niosomal formulations for the treatment of glaucoma. J Pharm Investig 2020;50(1):59–70. https://doi.org/10.1007/s40005-019-00427-1.

[77] Dubey A, Prabhu P. Formulation and evaluation of stimuli-sensitive hydrogels of timolol maleate and brimonidine tartrate for the treatment of glaucoma. Int J Pharm Investig 2014;4(3):112. https://www.ncbi.nlm.nih.gov/pmc/articles/PMC4131382/#.

[78] Tighsazzadeh M, Mitchell JC, Boateng JS. Development and evaluation of performance characteristics of timolol-loaded composite ocular films as potential delivery platforms for treatment of glaucoma. Int J Pharm 2019;566:111–25. https://doi.org/10.1016/j.ijpharm.2019.05.059.

[79] Hathout RM, Gad HA, Abdel-Hafez SM, Nasser N, Khalil N, Ateyya T, Amr A, Yasser N, Nasr S, Metwally AA. Gelatinized core liposomes: a new Trojan horse for the development of a novel timolol maleate glaucoma medication. Int J Pharm 2019;556:192–9. https://doi.org/10.1016/j.ijpharm.2018.12.015.

[80] Cuggino JC, Tártara LI, Gugliotta LM, Palma SD, Igarzabal CI. Mucoadhesive and responsive nanogels as carriers for sustainable delivery of timolol for glaucoma therapy. Mater Sci Eng C 2021;118:111383. https://doi.org/10.1016/j.msec.2020.111383.

[81] Allam A, Elsabahy M, El Badry M, Eleraky NE. Betaxolol-loaded niosomes integrated within pH-sensitive in situ forming gel for management of glaucoma. Int J Pharm 2021;598:120380. https://doi.org/10.1016/j.ijpharm.2021.120380.

[82] Li J, Tian S, Tao Q, Zhao Y, Gui R, Yang F, Zang L, Chen Y, Ping Q, Hou D. Montmorillonite/chitosan nanoparticles as a novel controlled-release topical ophthalmic delivery system for the treatment of glaucoma. Int J Nanomedicine 2018;13:3975. https://www.ncbi.nlm.nih.gov/pmc/articles/PMC6045908/#.

[83] Kumar N, Aggarwal R, Chauhan MK. Extended levobunolol release from Eudragit nanoparticle-laden contact lenses for glaucoma therapy. Future J Pharm Sci 2020;6(1):1–4. https://doi.org/10.1186/s43094-020-00128-9.

[84] Karataş A, Sonakin O, KiliÇarslan M, Baykara T. Poly(ε-caprolactone) microparticles containing levobunolol HCl prepared by a multiple emulsion (W/O/W) solvent evaporation technique: effects of some formulation parameters on microparticle characteristics. J Microencapsul 2009;26(1):63–74. https://doi.org/10.1080/02652040802141039.

[85] El-Kamel A, Al-Dosari H, Al-Jenoobi F. Environmentally responsive ophthalmic gel formulation of carteolol hydrochloride. Drug Deliv 2006;13(1):55–9. https://doi.org/10.1080/10717540500309073.

[86] Nagai N, Yamaoka S, Fukuoka Y, Ishii M, Otake H, Kanai K, Okamoto N, Shimomura Y. Enhancement in corneal permeability of dissolved carteolol by its combination with magnesium hydroxide nanoparticles. Int J Mol Sci 2018;19(1):282. https://doi.org/10.3390/ijms19010282.

[87] Alkholief M, Kalam MA, Almomen A, Alshememry A, Alshamsan A. Thermoresponsive sol-gel improves ocular bioavailability of dipivefrin hydrochloride and potentially reduces the elevated intraocular pressure in vivo. Saudi Pharm J 2020;28 (8):1019–29. https://doi.org/10.1016/j.jsps.2020.07.001.
[88] Maiti S, Paul S, Mondol R, Ray S, Sa B. Nanovesicular formulation of brimonidine tartrate for the management of glaucoma: in vitro and in vivo evaluation. AAPS PharmSciTech 2011;12(2):755–63. https://doi.org/10.1208/s12249-011-9643-9.
[89] Chiang B, Kim YC, Doty AC, Grossniklaus HE, Schwendeman SP, Prausnitz MR. Sustained reduction of intraocular pressure by supraciliary delivery of brimonidine-loaded poly(lactic acid) microspheres for the treatment of glaucoma. J Control Release 2016;228:48–57. https://doi.org/10.1016/j.jconrel.2016.02.041.
[90] Bhagav P, Upadhyay H, Chandran S. Brimonidine tartrate-eudragit long-acting nanoparticles: formulation, optimization, in vitro and in vivo evaluation. AAPS PharmSciTech 2011;12(4):1087–101. https://doi.org/10.1208/s12249-011-9675-1.
[91] Bhagav P, Trivedi V, Shah D, Chandran S. Sustained release ocular inserts of brimonidine tartrate for better treatment in open-angle glaucoma. Drug Deliv Transl Res 2011;1(2):161–74. https://doi.org/10.1007/s13346-011-0018-2.
[92] Zhang L, Zhang C, Dang H. Controlled brimonidine release from nanostructured lipid carriers-laden silicone contact lens to treat glaucoma. J Drug Deliv Sci Technol 2021;66:102753. https://doi.org/10.1016/j.jddst.2021.102753.
[93] Katiyar S, Pandit J, Mondal RS, Mishra AK, Chuttani K, Aqil M, Ali A, Sultana Y. In situ gelling dorzolamide loaded chitosan nanoparticles for the treatment of glaucoma. Carbohydr Polym 2014;102:117–24. https://doi.org/10.1016/j.carbpol.2013.10.079.
[94] Franca JR, Foureaux G, Fuscaldi LL, Ribeiro TG, Castilho RO, Yoshida MI, Cardoso VN, Fernandes SO, Cronemberger S, Nogueira JC, Ferreira AJ. Chitosan/hydroxyethyl cellulose inserts for sustained-release of dorzolamide for glaucoma treatment: in vitro and in vivo evaluation. Int J Pharm 2019;570:118662. https://doi.org/10.1016/j.ijpharm.2019.118662.
[95] Warsi MH, Anwar M, Garg V, Jain GK, Talegaonkar S, Ahmad FJ, Khar RK. Dorzolamide-loaded PLGA/vitamin E TPGS nanoparticles for glaucoma therapy: pharmacoscintigraphy study and evaluation of extended ocular hypotensive effect in rabbits. Colloids Surf B: Biointerfaces 2014;122:423–31. https://doi.org/10.1016/j.colsurfb.2014.07.004.
[96] Shinde U, Ahmed MH, Singh K. Development of dorzolamide loaded 6-o-carboxymethyl chitosan nanoparticles for open angle glaucoma. J Drug Deliv 2013;2013. https://doi.org/10.1155/2013/562727.
[97] Ammar HO, Salama HA, Ghorab M, Mahmoud AA. Nanoemulsion as a potential ophthalmic delivery system for dorzolamide hydrochloride. AAPS PharmSciTech 2009;10(3):808–19. https://doi.org/10.1208/s12249-009-9268-4.
[98] Dubey V, Mohan P, Dangi JS, Kesavan K. Brinzolamide loaded chitosan-pectin mucoadhesive nanocapsules for management of glaucoma: formulation, characterization and pharmacodynamic study. Int J Biol Macromol 2020;152:1224–32. https://doi.org/10.1016/j.ijbiomac.2019.10.219.
[99] Wang F, Bao X, Fang A, Li H, Zhou Y, Liu Y, Jiang C, Wu J, Song X. Nanoliposome-encapsulated brinzolamide-hydropropyl-β-cyclodextrin inclusion complex: a potential therapeutic ocular drug-delivery system. Front Pharmacol 2018;9:91. https://doi.org/10.3389/fphar.2018.00091.
[100] Jain N, Verma A, Jain N. Formulation and investigation of pilocarpine hydrochloride niosomal gels for the treatment of glaucoma: intraocular pressure measurement in white albino rabbits. Drug Deliv 2020;27(1):888–99. https://doi.org/10.1080/10717544.2020.1775726.

[101] Lai JY, Luo LJ. Chitosan-g-poly(N-isopropylacrylamide) copolymers as delivery carriers for intracameral pilocarpine administration. Eur J Pharm Biopharm 2017;(113): 140–8. https://doi.org/10.1016/j.ejpb.2016.11.038.

[102] Li J, Wu L, Wu W, Wang B, Wang Z, Xin H, Xu Q. A potential carrier based on liquid crystal nanoparticles for ophthalmic delivery of pilocarpine nitrate. Int J Pharm 2013;455 (1–2):75–84. https://doi.org/10.1016/j.ijpharm.2013.07.057.

[103] Moghassemi S, Hadjizadeh A. Nano-niosomes as nanoscale drug delivery systems: an illustrated review. J Control Release 2014;185:22–36. https://doi.org/10.1016/j. jconrel.2014.04.015.

[104] Lohumi A. A novel drug delivery system: niosomes review. J Drug Deliv Ther 2012;2(5). https://doi.org/10.22270/jddt.v2i5.274.

[105] Kaur D, Kumar S. Niosomes: present scenario and future aspects. J Drug Deliv Ther 2018;8(5):35–43. https://doi.org/10.22270/jddt.v8i5.1886.

[106] Patil YP, Jadhav S. Novel methods for liposome preparation. Chem Phys Lipids 2014;177:8–18. https://doi.org/10.1016/j.chemphyslip.2013.10.011.

[107] Dua JS, Rana AC, Bhandari AK. Liposome: methods of preparation and applications. Int J Pharm Stud Res 2012;3(2):14–20.

[108] Ahmed KS, Hussein SA, Ali AH, Korma SA, Lipeng Q, Jinghua C. Liposome: composition, characterisation, preparation, and recent innovation in clinical applications. J Drug Target 2019;27(7):742–61. https://doi.org/10.1080/1061186X.2018.1527337.

[109] Nagavarma BV, Yadav HK, Ayaz AV, Vasudha LS, Shivakumar HG. Different techniques for preparation of polymeric nanoparticles—a review. 2012;5(3):16–23.

[110] Pridgen EM, Alexis F, Farokhzad OC. Polymeric nanoparticle technologies for oral drug delivery. Clin Gastroenterol Hepatol 2014;12(10):1605–10. https://doi.org/10.1016/j. cgh.2014.06.018.

[111] Patel T, Zhou J, Piepmeier JM, Saltzman WM. Polymeric nanoparticles for drug delivery to the central nervous system. Adv Drug Deliv Rev 2012;64(7):701–5. https://doi.org/ 10.1016/j.addr.2011.12.006.

[112] Sultana F, Imran-Ul-Haque M, Arafat M, Sharmin S. An overview of nanogel drug delivery system. J Appl Pharm Sci 2013;3(8):S95–S105.

[113] Zhang H, Zhai Y, Wang J, Zhai G. New progress and prospects: the application of nanogel in drug delivery. Mater Sci Eng C 2016;60:560–8. https://doi.org/10.1016/j. msec.2015.11.041.

[114] Yin Y, Hu B, Yuan X, Cai L, Gao H, Yang Q. Nanogel: a versatile nano-delivery system for biomedical applications. Pharmaceutics 2020;12(3):290. https://doi.org/10.3390/ pharmaceutics12030290.

[115] Jaiswal M, Dudhe R, Sharma PK. Nanoemulsion: an advanced mode of drug delivery system. 3 Biotech 2015;5:123–7. https://doi.org/10.1007/s13205-014-0214-0.

[116] Sharma N, Bansal M, Visht S, Sharma PK, Kulkarni GT. Nanoemulsion: a new concept of delivery system. Chron Young Sci 2010;1(2):2–6.

[117] Patel RP, Joshi JR. An overview on nanoemulsion: a novel approach. Int J Pharm Sci Res 2012;3(12):4640.

[118] Yukuyama MN, Ghisleni DD, Pinto TD, Bou-Chacra NA. Nanoemulsion: process selection and application in cosmetics—a review. Int J Cosmet Sci 2016;38(1):13–24. https:// doi.org/10.1111/ics.12260.

[119] Rathore KS, Nema RK. Review on ocular inserts. Int J Pharmtech Res 2009;1(2): 164–9.

[120] Kumari A, Sharma PK, Garg VK, Garg G. Ocular inserts—advancement in therapy of eye diseases. J Adv Pharm Technol Res 2010;1(3):291. https://www.ncbi.nlm.nih.gov/ pmc/articles/PMC3255407/#.

Age-associated macular degeneration: Epidemiologic features, complications, and potential therapeutic approaches

Sumel Ashique[a], Shubneesh Kumar[b], Afzal Hussain[c], Arshad Farid[d], Neeraj Mishra[e], and Ashish Garg[f]

[a]Department of Pharmaceutics, Pandaveswar School of Pharmacy, Pandaveswar, West Bengal, India, [b]Department of Pharmaceutics, Bharat Institute of Technology, School of Pharmacy, Meerut, Uttar Pradesh, India, [c]Department of Pharmaceutics, College of Pharmacy, King Saud University, Riyadh, Saudi Arabia, [d]Gomal Center of Biochemistry and Biotechnology, Gomal University, Dera Ismail Khan, Khyber Pakhtunkhwa, Pakistan, [e]Amity Institute of Pharmacy, Amity University, Gwalior, Madhya Pradesh, India, [f]Department of Pharmaceutics, Guru Ramdas Khalsa Institute of Science and Technology (Pharmacy), Jabalpur, Madhya Pradesh, India

1 Introduction

Age-linked macular degeneration is a long-lasting optic disc ailment that gradually impairs vision, scheduled to momentous visual forfeiture and serving as a foremost source of global visual impairment. The advanced stages of the disease can result in what's more neovascular AMD otherwise geographic atrophy, causing central visual acuity loss and severe impairment. This has profound effects on a character's eminence of life expectancy, and functional liberation, and carries significant socio-economic implications. According to Wong et al. [1], it is projected that the global AMD population will reach nearly 20 crores by 2020 in addition to virtually 30 crores by 2040. AMD, also acknowledged as age-linked macular degeneration, primarily affects the macula, answerable for sharp central vision. It exhibits higher prevalence in lightly-pigmented and womanly populations, with its incidence progressively rising with age. As Friedman et al. [2] stated, it affects 2% of the populace at stage 40 and one in four individuals by stage 80. The disease affects multiple structures within the eye, including optical disc pigment or biochromes epithelial tissue, Bruch's crust, in addition, photoreceptors, and the choroid, extending beyond the macula. Despite its widespread pervasiveness, there are currently no FDA-approved rehabilitations for AMD, except for a slight fraction of culmination point cases [3]. AMD is written off as by the gradual fall of photoreceptors besides the retinal biochromes epithelial tissue (RBE),

Targeting Angiogenesis, Inflammation and Oxidative Stress in Chronic Diseases. https://doi.org/10.1016/B978-0-443-13587-3.00010-2

resulting in central vision loss. The accumulation of drusen, small extracellular deposits, in the retina is a key feature of AMD. Subretinal drusenoid deposits (SDD), another type of deposit, are also frequently detected in the subretinal space. AMD is a multifarious bug influenced by factors such as aging, genetics, and ecological jeopardy features. Adjustable jeopardy features like smoking besides regime consistently contribute to AMD, while genetic modifications in CFH then ARMS2-HTRA1 deliberate the peak peril of emergent ailment [4,5]. Clinical classification systems commonly use drusen size to evaluate the severity of the first points of AMD. It is alphabetized as "early" if intermediate-sized drusen are present plus "intermediate" if large drusen are observed [6]. AMD, a prevalent footing of visual fall in older individuals, carries the greatest menace of progressing to late-point AMD, characterized by geographic atrophy (GA) in addition/otherwise neovascular AMD, associated with bulky drusen moreover pigmentary vagaries. Unfortunately, there is no remedy for AMD, and handling possibilities viz. laser photocoagulation too photodynamic therapy provide only limited delay in visual function loss for a trivial subset of affected role by neovascularization [7]. AMD, a bug distressing the essential portion of the retina, has prompted considerable interest in prevention strategies. Vascular factors and cardiovascular disease menace influences have been inspected for their character in AMD development. Additionally, it is believed that the retina's susceptibility to highly reactive oxygen species generated during visual processing may contribute to AMD's onset. Recent advancements in molecular biology and genetics have opened up potential pathways to uncover the genetic influences that may subsidize to the underlying mechanisms of the bug. The objective of this evaluation is to mete out an all-embracing overview of the medical besides pathological facets of AMD, current investigative procedures, in addition to the existing otherwise emerging therapeutic approaches for mutually the atrophic in addition wet forms of the condition. A thorough literature search was conducted using relevant keywords such as "age-related macular degeneration," "risk factors," "VEGF," "prevention," "genetics," and "management," in various combinations. Projections suggest that approximately 28 crore sentities worldwide will be pretentious by AMD by the day 2040 [8].

2   Different types of AMD

Fig. 1 describes the various types of AMD. AMD, or age-associated macular deterioration, is an eye state that leads to the fall of the macula, the dominant fragment of the retina in control of detailed dominant vision. The condition can be categorized based on histopathological characteristics. Premature and in-between steps of AMD are well-thought-out by the attendance of protein, drusen, then phospholipid-rich quantities that accumulate flanked by the Bruch's skin along with retinal pigment epithelial tissue (RPE). Conversely, wet AMD is less common, affecting a slight section of AMD-affected roles, progressing rapidly, and potentially causing impaired vision if left untouched [9]. Dry AMD often heralds the spread of wet AMD and can be considered a precursor or risk factor for the latter. Drusen deposition occurs in the retina thru the asymptomatic premature points of AMD. Late-step dry AMD acknowledged

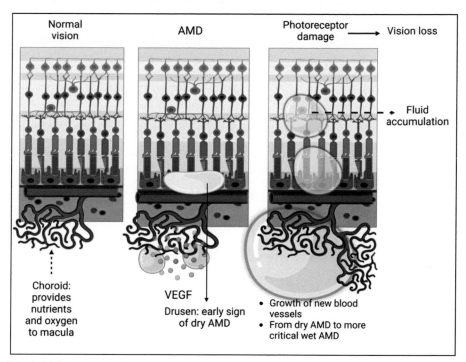

Fig. 1 Different types of AMD representation.

as geographic atrophy (GA), is well-thought-out by the relapse of RPE cells also the overlying retinal photoreceptors in scattered else confluent areas. In contrast, the late-step wet arrangement of AMD is marked by choroidal neovascularization (CNV), where undeveloped blood venules nurture the beginning of the underlying choroid toward the outer optics, ensuing in fluid seepage. Countless affected roles with in-between AMD sooner or later advancement to the progressive period. Progressive AMD encompasses two primary types: geographic atrophy (GA) in addition neovascular AMD (NVAMD), likewise acknowledged as wet else exudative AMD. The term "dry AMD" is used to refer to the premature besides in-between phases of the ailment, over and above GA. Despite the fact premature and in-between phases of AMD stereotypically origin mild visual acuity diminishing, advanced AMD is the chief origin of sightlessness wide-reaching. In some cases, both arrangements of progressive AMD may coexist in a similar eye due to shared risk factors. Intermediate AMD involves degenerative fluctuations in the Bruch's skin, retinal pigment epithelial tissue (RPE), besides choroid, which can lead to photoreceptor relapse in GA or else choroidal neovascularization trailed by photoreceptor relapse in NVAMD. However, the explicit pathological proceedings contributing to the headway from in-between to progressive AMD are so far under inquiry. Current research suggests that several factors, including genetics, age, smoking, diet, too oxidative anxiety, may subsidize the expansion in addition headway of AMD. Certain genetic variations, particularly those related to the counterpart structure, may also show a title role [10]. Although the wet system of AMD can be accomplished with intraocular athwart-VEGF shots to minimize neovascularization in addition to retinal deepening, there is at this time no operative handling for dry AMD. Consequently, there is increasing research interest in developing therapies for dry AMD, such as steroids, antioxidants, and neuroprotective representatives, which have remained inspected in vitro besides in vivo to postpone retinal relapse. Recent advances in the basic examination have shed light on the mechanisms of a deed of bioactive fragments for the handling of AMD. In this review, we present the up-to-the-minute and emergent contenders for AMD healing and highpoint their cytoprotective mechanisms in the framework of AMD [11].

3   Critical alterations during AMD

The bug procedure of age-linked macular degeneration (AMD) involves various ocular structures, including specialized neurons called photoreceptor cells, which reside in the superficial retina to drama a vital character in converting light into electrical signals. The photoreceptor cells rely on metabolic support provided by the retinal pigment epithelial tissue (RPE), a one-track sheet of polar cells. Situated among the RPE also the choroid is the Bruch skin, a multipart layer composed of elastic fibers collagen. Lastly, the inward choroid encompasses the choriocapillaris, a passageway cot responsible for meeting the metabolic desires of the superficial retina. These constructions are intricately connected metabolically, besides, there is a significant conversation of metabolites crosswise the Bruch crust to certify the health and optimal functioning of the visual system. Fig. 2 depicts the clinical features seen during AMD pathogenesis.

3.1   Structural alterations to the choroid during AMD

The choriocapillaris, located flanked by the superficial retina in addition to the Bruch's skin, is a layer characterized by a fenestrated sinusoidal complex without tight junctions. Its development is influenced in part by the outward componential countenance of VEGF from the retinal pigment epithelial tissue (RPE) [12,13]. Research by Ramrattan et al. [14] revealed the steadfastness of the choriocapillaris shrinkages with the time of life in healthy eyes lacking AMD. Additionally, studies using neoprene casts demonstrated that the choriocapillaris undergoes structural changes on or after a sinusoidal arrangement to a canular vascular arrangement as individuals age. In the advanced legs of AMD, there is a for feature besides thinning of the choriocapillaris [15]. It is worth noting that Sorsby fundus dystrophy, a monaural-genic set of symptoms, shares pathological features with AMD and leads to dominant visualization loss in the third or fourth time of lifecycle [16].

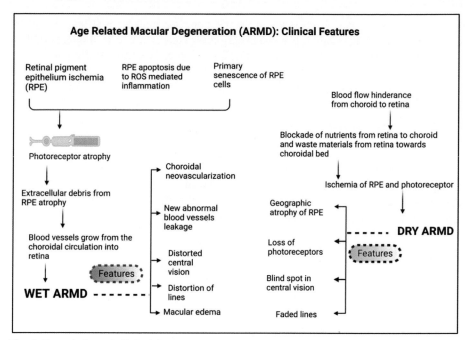

Fig. 2 Description of clinical features and alterations during AMD.

3.2 Structural vagaries to the Bruch skin in AMD

Through electron and light microscopy, studies have established a correlation flanked by aging to the thickness of the Bruch film [17]. However, one study examining Bruch membrane thickness among elderly individuals found considerable variation, suggesting that factors beyond oldness, as per inherited or else environmental provocations, may contribute to approximately half of the observed thickness changes [18]. A hypothesis proposes that the deepening of the Bruch crust results from the unfinished clearance of leftover material discharged superficial by the retinal biochromes epithelial tissue, leading to the accumulation of deposits. These deposits have been the subject of the investigation to understand their nature and potential impact on the job of the congealed Bruch skin. Discussions on the pathogenesis of pigment epithelial detachment (PED) have provided insights into the content of these deposits. AMD, characterized by a continuous peripheral drive of ions in addition to aquatic from the superficial retina in the direction of the choroid, may be affected by compact hydraulic conveyance in the Bruch skin, impeding water drive and resulting in fluid accumulation in the subRPE space. It is suggested that the Bruch membrane contains a high level of lipids, increasing fluid flow resistance. Histopathological, biochemical, biophysical, and clinical observations have been conducted to investigate these hypotheses, providing supportive evidence [19]. In cases where an individual is at risk of PED tearing, the Bruch membrane located in the eye may exhibit significant resistance to water flow. Such tearing occurs when peripheral strain in the disconnected muscles is strong enough to source falling out. Therefore, if one eye experiences a tear in the RPE, it may point to a high chance of a comparable event occurring in the other eye. Lipid trafficking within the Bruch membrane plays a significant role, and when lipids fail to pass through the tissue, they accumulate, scheduled to the deepening of the Bruch crust. Consequently, proteins within the Bruch membrane, particularly those associated with the immune system, are being investigated to determine if they are implicated in the thickening process. Vitronectin, present in high levels within the deepest fragment of the Bruch film, may protect against immune attacks [20]. The source of proteins tangled in the morbific of age-linked macular deterioration (AMD) is not fully perceived, but some constituents are stated in the retinal biochromes epithelial tissue and may originate after plasma [21]. The state of these proteins within the Bruch membrane remains unclear, but evidence suggests that certain proteins, such as CFH, may be oligomerized due to elevated levels of metallic ions, including zinc, in the Bruch membrane [22]. The points of bioavailable zinc in the Bruch membrane are significantly higher than what is necessary to induce CFH oligomerization in vitro [23]. The rigidification of the Bruch skin caused by the accumulation of these proteins can disrupt metabolic conversation besides fluid drive, contributing to the morbific of AMD. One potential beneficial tactic to slowing down the disease progression is to reduce the convenience of these integral proteins through the chronic practice of antiinflammatory representatives.

3.3 Structural vagaries to the RPE in AMD

The paragraph describes a research study that investigated the relationship between aging and the accretion of fluorescent remaining figures in the optical biochromes epithelial tissue. The study observed a quadratic relationship, indicating that both auto-fluorescence and residual body quantity increase with age. However, the study found that the accumulation of these bodies slows down in elderly individuals, which may be attributed to the decreasing population of photoreceptors in late life [24]. Despite the general trend, the study also discovered a wide variation in auto-fluorescence levels among the elderly, suggesting that aspects further than aging, let's say genetic or else ecological stimuli, contribute to this variance. Approximately 50% of the discrepancy in mutual auto-fluorescence in addition remaining figures could not be merely explained by aging [25]. This implies that individual variances illustrate a momentous character in the detected stages of auto-fluorescence and residual bodies. Geographic atrophy (GA) in the optical biochromese pithelial tissue is still not entirely understood, and Researchers have engaged in discussions regarding the molecular mechanisms that underlie this condition. Certain individuals have put forth the notion that the residual bodies found in the cytoplasm of the RPE might disrupt cellular metabolism as a result of their size. However, alternative theories propose that lipofuscin, a substance known for generating free radicals, could potentially cause damage to cells [26]. Additionally, specific constituents of lipofuscin, including A2-E, have demonstrated toxic properties. A2-E can modify the pH of lysosomes by regulating the impact of the ATP-reliant on the lysosomal proton pump.

3.4 Structural vagaries to the superficial retina in AMD

AMD is a prominent root of vision forfeiture in aging, principally distressing the neural retina. Although the photoreceptors play a crucial role in AMD, there is limited knowledge regarding the physical alterations that take place in this component of the neuronic retina. In the initial phases of AMD, early histological investigations have indicated a potential association between the gradual failure of photoreceptor cells, additionally, the impaired functioning of the retinal biochromes epithelium [27]. Recent imaging examinations have fetched valuable perceptions into the morbific of AMD. Specifically, these revisions have revealed that in regions of the fundus where the retina appears structurally normal except for the presence of drusen, a significant loss of photoreceptor cells is observed. This finding advocate that the forfeiture of photoreceptor cells might show a vivacious character in the functional impairments allied with AMD. The extent of photoreceptor cell loss in early-stage AMD can vary considerably among individuals, but emerging evidence indicates that it can be profound.

4 Diagnosis and symptoms

During routine eye examinations, age-linked macular degeneration (AMD) is frequently detected, as small regions of geographic atrophy may not immediately impact vision. However, individuals with neovascular AMD may experience symptoms such as blurred vision and central distortion in the affected eye. In certain instances, the condition may be discovered incidentally if it affects only one eye. AMD is commonly diagnosed through clinical examination and fundus photography, which allow for the detection of characteristic lesions. However, advancements in technology, such as spectral-realm optical rationality tomography in addition to fundus autofluorescence imaging, have enhanced the ability to perceive these grazes with greater precision. Additionally, fluorescein angiography remnants are a valuable tool for identifying choroidal neovascularization and confirming the presence and location of neovascular AMD by detecting dye leakage. These newer diagnostic modalities have brushed up the exactness and proficiency of AMD finding. Optical coherence imaging or tomography angiography (OCT-A) is an auspicious and noninvasive imaging practice that has fully-fledged implications in the finding of age-linked macular deterioration or (AMD). It empowers the recognition of choroidal neovascularization, a symbol feature of neovascular AMD, without the need for invasive dye injections. Although OCT-A cannot directly detect leakage, its ability to visualize abnormal blood vessel growth provides valuable information when combined with other imaging modalities. In the early stages, AMD may not present with noticeable symptoms, but it can cause mild central distortion and reduced reading ability. Therefore, the use of advanced imaging techniques like OCT-A, along with clinical evaluation, portraying a conclusive title role in identifying besides perceiving AMD, facilitating early intervention and management. Late AMD can ensue to quick worsening of essential vision in the neovascular arrangement, or a sluggish decay in the atrophic form over the passage of years or times [28]. Warning signs of AMD may embrace distorted vision, dark or gray coverings in the central vision, in addition, struggle to recognize faces. In cases where solitary one eye is affected, the warning sign may not be superficial until the good eye is shielded [29]. Choroidal neovascularization, which allusions to the construction of anomalous blood vessels in age-linked macular deterioration or (AMD), is a distinct characteristic of this ailment. It represents a complex lesion formation in the macule, the dominant fragment of the optic disc answerable for detailed vision. The complex presentation of age-linked macular degeneration (AMD) includes several characteristic lesions. These encompass the occurrence of fluid accumulation or retinal hemorrhage, which can be found in the inland the retina, beneath the optic nerve, or else below the retinal biochromes epithelial tissue. Additionally, subretinal fibrous scar tissue, detachments of the optical biochromes epithelial tissue, to the incidence of hard exudates are common features. Utilizing multimodal imaging techniques, particularly optical coherence tomography, is highly effective in visualizing and identifying these manifestations. It plays a crucial role in providing critical evidence about the position, extent, and extent of drusen (yellow deposits), on top of determining the commonness and movement of choroidal neovascularization [30].

5   Pathogenesis of age-linked macular deterioration

AMD is a communal root of visual waning amid entities grow older 50 besides overhead in developed countries. Despite extensive research, the precise pathophysiology of AMD remains incompletely understood. However, numerous studies have contributed valuable insights into the ailment and its vital mechanisms. The morbific of AMD is believed to encompass a multipart interplay of various influences, including metabolic, functional, genetic, and environmental influences. Studies have indicated that senescence, noticeable by the buildup of lipofuscin in optical biochromes epithelial tissue cells, accompanied by choroidal ischemia besides oxidative impairment, portrays a title role in the progress of AMD. These aspects subsidize the progressive fall of the macula and can in due course mode to vision loss [31], which are among the biological pathways implicated in AMD. The retina and its pigmented epithelium are constantly exposed to significant levels of light energy and oxygen, making them defenseless to the production of unhindered radicals. This exposure has commanded to the tender that oxidative anxiety, resulting from the cumulative effects of this oxidative environment throughout a person's lifetime, may serve as the initial trigger for the expansion of macular degeneration. The oxidative anxiety can induce offended to cellular gears and disrupt normal retinal function, contributing to the morbific of the bug [32,33]. Age-linked macular deterioration (AMD) is well-thought-out by the liberal deterioration of the dominant retina, scheduled to vision forfeiture. Despite ongoing research, the precise molecular mechanisms underlying the development of AMD remain incompletely elucidated. However, recent investigations are providing valuable insights into this intricate disease. Epidemiological reconsiderations have acknowledged several menace influences associated with AMD, such as ciggy smoking and prolonged exposure to sunlight over one's lifetime [34]. These verdicts highpoint the status of environmental factors in the morbific of AMD and contribute to our understanding of the disease's etiology. For example, populace-grounded learning in the European Union detected those entities with near-to-ground antioxidant points in their fluid besides giant communal days of sunshine contact had a dual amplified hazard of evolving late-stage AMD [35]. Indeed, recent research has uncovered significant genetic influences that subsidize the advancement of age-linked macular degeneration (AMD). Genetic studies have identified a specific inheritable factor that is allied with the clinical spectrum of AMD, shedding light on the complex mechanisms underlying the condition. By understanding the genetic factors involved in AMD, researchers hope to develop better diagnostic tools, risk assessment models, and potential targeted therapies for this complex eye condition. Age-linked macular deterioration (AMD) is widely considered to be the upshot of a multifactorial back-and-forth among hereditary then environmental influences. Extensive research has been conducted to better comprehend the underlying causes of AMD and the complex interactions involved in its development [36,37]. The passage labels the expansion of age-linked macular deterioration (AMD), a deteriorating eye illness viz. a foremost source of sightlessness midst elderly entities. Conferring to the passage, carrying manifold vulnerable alleles at counterpart loci in addition to engaging in such lifestyle activities, for instance, smoking, can upsurge the jeopardy of evolving AMD by as considerably as 80% [38]. The passage also proceedings that AMD as well as aging is mutually categorized by oxidative stress, which can source hurt to the optical biochromes epithelial tissue besides choriocapillaris. RPE hurt can provoke a provocative retort in the Bruch skin and choroid, resulting in the construction of an anomalous extracellular matrix (ECM) consequent from innumerable cells in the eye besides constituents in the systemic motion. The anomalous ECM can impair the retina, and choroid, besides RPE. In a count of, oxidative maltreatment to the choriocapillaris may also subsidize AMD morbific. The progress of choroidal neovascularization (CNV) in addition to geographic atrophy encompasses turbulences in RPE-choriocapillaris homeostasis. RPE demise may be answerable for choriocapillaris forfeiture in geographic atrophy [39]. Fig. 3 depicts the pathogenesis involved in AMD.

6 Prevalence, incidence, and natural history of AMD

There have been numerous epidemiologic revisions conducted on age-linked macular degeneration (AMD) over the former few decades. These reconsiderations have vacant treasured insights into the pervasiveness and distribution of AMD among different populations. The meta-analysis by Smith et al. [34] you mentioned focused on population-based studies in white individuals who grow older than 40 ages and grown-up. It estimated the pervasiveness of premature AMD to be 6.8% too late AMD to be 1.5% in this population. Additionally, epidemiological studies have examined the pervasiveness of AMD in other national assemblages as well. The Baltimore Eye Reconsideration, as cited in your statement, sense that late AMD was significantly more rampant in white participants equated to dark participants. The study reported a nine

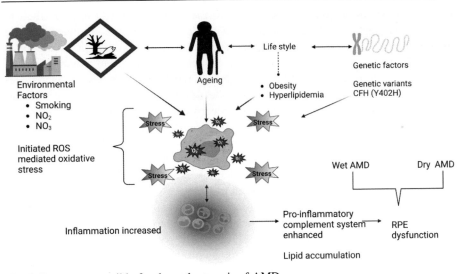

Fig. 3 Factors responsible for the pathogenesis of AMD.

to ten times higher pervasiveness of late AMD in white individuals equated to black individuals [40]. Nevertheless, an Asiatic meta-analysis [41] disclosed that the stage of development fixed commonness of late age-linked macular deterioration in Asiatic was mostly alike to that in Caucasian the social order. There is an incipient suggestion that sundry Asiatic affected roles thru neovascular age-linked macular deterioration have polypoidal distension of the choroidal vasculature, a modified labeled polypoidal choroidal vasculopathy (PCV). PCV can justification for 50% of neovascular age-linked macular deterioration specimens in Asiatic, but then again solitary 8%–13% in Caucasian individuals [42]. Another modification of age-linked macular deterioration is optical angiomatous propagation, which is justification intended for 12%–15% of neovascular age-linked macular deterioration [43]. These modifications might not retort as sound to distinctive supervision of neovascular age-linked macular deterioration. The predictable pervasiveness of late AMD in the three huge populace-grounded lessons was 0.2% (10 as of 4797 partakers) for the public growing older 55–64 ages, in addition, enlarged to 13.1% (68 as of 521) aimed at publics added than 85 years of oldness [34]. BMES disclosed the 15-cycle of year commonness was 22.7% (462 as of 2036) aimed at premature AMD besides 6.8% (165 as of 2421) aimed at late AMD [44]. AMD commonness was superior in women than in men for entirely time-of-life assemblages. Meta-scrutinized facts from 14 populace-grounded cohort revisions in the European Eye Epidemiology consortium [45] disclosed that inclusive commonness was 13.2% intended for premature AMD besides 3.0% intended for late AMD for populaces grow older than 70 ages or senior. This outline was analog to entities logged in BDES, in which the five-cycle of a year commonness of AMD was 60% lesser for individually succeeding peer groups, utilizing well-defined by the time of birth (1901–24, 1925–45, 1946–64, in addition, 1965–84) [46].

the variability in the headway of first macular degeneration to late indices and the impact of specific features on the menace of headway. The menace of headway in macular degeneration can diverge reliant on the harshness of the scope of premature bug landscapes. The AREDS(Age-Related Eye Disease Study) has offered treasured insights into the menace of headway in macular relapse. The learning quantified the risk based on specific features observed in early macular degeneration. According to the findings reported by Pauleikhoff [47], individuals with insignificant drusen in duple eyes have a very trivial menace of headway, ranging from 0.4% to 3.0% concluded 5 years. Nevertheless, if an individual has vast drusen besides pigmentary anomalies in duple eyes, the menace of progression surges significantly. The study reported a risk of approximately 47.3% over 5 years for individuals with these advanced features. Geographic atrophy typically begins as focal areas of depigmentation in the macula, which may eventually merge or else inflate to encompass the central macula. This liberal involvement of the macula leads to a gradual worsening of vision, often ensuing in legal sightlessness. The forfeiture of optical biochromes epithelial tissue cells correspondingly to the underlying photoreceptor cells in the macula contributes to vision forfeiture in geographic atrophy. However, neovascular complications in AMD present with a more acute commencement. The hasty advance of central blurring in addition to bias is often an upshot of the evolution of anomalous blood vessels beneath the retina. If left untreated, these abnormal blood vessels can continue to expand rapidly, causing leakage, bleeding, and the construction of a fibrous scratch in the macula. This can brutally distress foremost vision and lead to significant visual impairment. The meta-analysis conducted by Wong et al. [9] analyzed data from controlled clinical trials and also presented imperative intuitions into the natural progression of neovascular AMD. The analysis directed that in troika years of the inception of neovascularization, added than semi of untouched eyes would have a stage of vision of 20/200 or worse (Snellen 6/60). This level of vision impairment falls within the World Health Organization's elucidation of severe visual damage.

7 Risk influences and possible drugs targets for the administration of AMD

7.1 Drug targeting: Angiogenesis

The learning by Kersten et al. [48] extensively inspected the slice of angiogenic influences in AMD. According to the findings, angiogenic influences let's say transforming growth factor beta termed TGF-β, vascular endothelial growth factor termed VEGF, besides pigment epithelium-derivative factor (PEDF) have been acknowledged as important players in the morbific of AMD. VEGF, let's say, is a hypoxia-motivated indication that tempts the construction of newfangled blood vessels in addition is at the moment the primary bull's eye in the handling of nAMD. In disparity, PEDF is shaped by RPE cells too have athwart-angiogenic assets, conflicting with the assets of VEGF. Accompanying through PEDF has remained projected as latent handling for nAMD [49]. TGF-β, on the added hand, surges the appearance of VEGF and is consequently also caught up as an angiogenic aspect [50]. While cross-sectional revisions have fashioned disagreeing fallouts concerning whether the extents of systemic besides local angiogenic aspects diverge among AMD-affected roles besides controls, there are at this time no forthcoming cohorts that have restrained angiogenic aspects to foresee sickness evolution in AMD.

7.1.1   AntiVEGF therapies

Athwart-VEGF cure has transfigured the handling of neovascular AMD since the initial 2000s by introducing biologicals that hinder VEGF, a key intermediary of angiogenesis [51,52]. This dealing has given rise to a lessening of more than 50% in the frequency of lawful sightlessness from neovascular AMD in more than a few nations states [53]. There are some antiVEGF representatives reachable for the dealing of neovascular AMD, counting bevacizumab, plus ranibizumab, besides a flibercept. Ranibizumab is a cultivated monophonic clonal antibody portion that fixes with all VEGF isoarrangements, despite the fact aflibercept is a biogenetics blend protein that turns in place of a solvable decoy receptor. While aflibercept is a combative retarder of VEGF nevertheless then again fixes placental growing influences one also two, although the superfluous upshot of this is still under debate. Bevacizumab, however, is a head-to-toe cultivated monoclonal VEGF antibody. In 2019, brolucizumab, a cultivated soles hackle antibody bit, was added to the available antiVEGF biologicals, which allows for higher molar dosing compared to previous antiVEGF treatments [54]. AntiVEGF drugs are commonly cast-off to treat neovascular AMD by administering them through intravitreal injection. However, due to the need for frequently repeated treatments, the prime of which antiVEGF representative to practice is every so often determined by its cost as well as alleged variances in efficiency. Moreover, the diverse nation-state has varying convenience for these medications. While newfangled antiVEGF remedies whether may be extra operative or partake in a longer interval of deed are on the horizon, their dwelling in the handling of nAMD will become vibrant as per clinicians gain more experience using them (Table 1).

Ranibizumab

Ranibizumab is a monophonic clonal antibody bit that has been developed to target all known isoforms of VEGF and then inhibit their activity. Its linking affinity is almost 100-fold more advanced than that of bevacizumab, even nevertheless it only has one linking site. Contrasting bevacizumab, ranibizumab does not have an Fc segment and therefore has a shorter systemic half-life, better penetration into the retina, and potentially less risk of causing an inflammatory reaction. The effectiveness of ranibizumab has been established in two stages III randomized controlled hearings, where patients receiving monthly injections of ranibizumab showed superior improvements in visual acuity compared [65] to those receiving either sham injections or photodynamic therapy over 12 months [66,67]. In the stage III Marginally Definitive/Occult Experimental of the Athwart-VEGF Antibody Ranibizumab, researchers estimated the efficacy the safety of ranibizumab as a dealing for neovascular AMD with marginally definitive or else occult with no definitive CNV. Ranibizumab was legalized by the FDA in

Table 1 Up-to-date handling for exudative AMD then illustrations of hearings aiming exudative macular degeneration (AntiVEGF Handling) [55].

Handling tactics or medications	Principle or amalgamated	Establishment	Current evolution besides confines	Reference
Athwart-VEGFl65 protein stratagem	Polyethylene glycol athwart-VEGF aptamer (pegaptanib, promoted as Macugen)	Pfizer in addition Eyetech Pharmaceuticals	FDA in addition to EMEA permitted. Harmless besides operative in the brief tenure. No facts were accessible concerning enduring upshot.	[56]
Athwart-VEGF protein stratagem	Cultivated athwart-VEGF monophonicclonal antibody (bevacizumab, promoted as Avastin)	Roche also Genentech Inc.	Far-reaching use of the tag. NIH-subsidized randomized test proceeding	[57]
AntiVEGF protein stratagem	Greater-attraction athwart-VEGF Fab (ranibizumab, promoted as Lucentis)	Novartis also Genentech Inc.	USFDA was also EMEA-permitted in 2006. Harmless in addition operative in the brief tenure. Upgrading of visual perspicacity. No facts accessible concerning enduring upshot.	[58]
Athwart-VEGF tooPlGF proteins stratagem	VEGF-TRAP, a biogenetic protein encompassing The requisite provinces of VEGF-R1 in addition R2	Regeneron Pharmaceuticals Inc.	Further down medical test	[59]
Athwart-VEGF inheritable factor Countenance stratagem	VEGF-aiming siRNA (Cand5-bevasiranib)	Opko (Acuity Pharmaceuticals Inc.)	Further down medical test	[60]

Continued

Table 1 Continued

Handling tactics or medications	Principle or amalgamated	Establishment	Current evolution besides confines	Reference
Athwart-VEGF receptor Countenance stratagem	VEGF-R1aiming siRNA (siRNA-027)	Merck and Co. Inc.	Further down medical test	[61]
Athwart-angiogenic inheritable factor healing	Shot of adenovirus to prompthugepoints of PEDF	GenVec Inc.	Further down medical test	[62]
Antiangiogenic stratagem	IV Shot of squalamine (Evizon)	Genaera Corp.	Further down medical test	[63]
Angiostatic steroids	Anecortave acetate (Retaane)	Alcon	Further down medical test	[64]

2006, and subsequently received approval from other international agencies. Results from randomized controlled trials showed that over 90% of affected role attainment ranibizumab practiced damage of fewer than 15 letters of visualization on a caliber image diagram next 12 agenda calendar months, while only around 60% of affected role attainment sham injections experienced the same level of improvement.

Aflibercept

Two critical stages III randomized controlled hearings were conducted to judge the nonsubservience of aflibercept equated to ranibizumab in dealing with neovascular AMD. The fallouts of both trials directed that aflibercept was not substandard to once-a-month ranibizumab in rapports of the number of affected roles who vanished not as much of than 15 letters of vision in addition to the mean variation in best-amended visual acuity as of reference line. Following the FDA sanction in 2011, aflibercept was subsequently approved in other jurisdictions [68].

Bevacizumab

Bevacizumab is a drug that has remained cast off off-label for the dealing of neo-vascular AMD subsequently 2005 [69]. Studies have publicized that it is non-substandard to ranibizumab, another drug used for the same purpose [70,71]. Bevacizumab was primarily agreed upon via the FDA in 2004 for the handling of met-astatic colon malignancy [72] due to its antiangiogenic effects. Animal studies dem-onstrated that after circulating direction to cynomolgus monkeys, fluorescein-coupled bevacizumab trickled from laser-tempted CNV, which advised that bioavailable bevacizumab may perhaps, also seepage from CNV in affected role thru AMD also competitively inhibit extravascular VEGF [73]. The SANA learning was a forthcoming medical experimental premeditated to consider the wellbeing, worth, and robust-ness of bevacizumab for handling subfoveal choroidal neovascularization (CNV) in AMD. In this open-label study, 18 participants were administered a reference point drink of bevacizumab (5 mg/kg), and afterward ace otherwise binary added prescrip-tions at two calendar week pauses. After 24 calendar weeks, the systemic direction of bevacizumab was estimated well-endured, affected roles experienced a typical surge of 14 letters in best-corrected visual acuity shorten as BCVA also a drop of 112 μm in dominant retinal width as scaled by optical coherence tomography shorten as OCT [74]. Bevacizumab is a type of monaural clonal antibody that can impede the move-ment of VEGF, which is in the authority of the progress and preservation of anomalous blood vessels in quite a few eye ailments, for instance, diabetic macular edema (DME), age-linked macular deterioration, in addition, retinal vein occlusion (RVO). The shot of bevacizumab directly into the eye has been sketchily cast off as an off-label handling for these conditions due to its antiVEGF properties.

In 2005, a case report was published unfolding the first-ever testified illustration of bevacizumab given a shot into a human eye. The affected role was a 63-year-age adult female with subfoveal, predominantly definitive CNV. After four calendar weeks of attainment a single-intravitreal shot of bevacizumab (1 mg) [69], subretinal liquid on OCT had resolved, and no adverse effects were alleged. Since then, retrospective case series have shown that once-a-month intravitreal shots of bevacizumab (1.25 mg) are generally well-endured and accompanied by progress in visual perspicacity, declined retinal width, and abridged angiographic seepage in most affected roles. Many of these affected roles had previously been served with PDT besides/or else pegaptanib [75].

Brolucizumab

Brolucizumab has demonstrated nonsubservience to aflibercept about visual jobs at the two-year mark, according to a study by Dugel et al. [54]. The latent for a longer interval of action could cut the treatment load for the affected role with MNV, which is an auspicious outcome. Nevertheless, there has been intelligence of severe uveitis in addition to occlusive retinal vasculitis associated with brolucizumab, which entails supplementary inquiry [76]. One possible explanation is that the construction of immune developments due to delayed hypersensitivity with local antibodies could lead to vasculitis [77].

Pegaptanib

Pegaptanib, marketed as Macugen by Eyetech in Palm Beach Gardens, Florida, is a drug that selectively targets VEGF, a protein tangled in the construction of newfangled blood vessels. The drug has been made known to cut the likelihood of modest to stringent vision forfeiture in affected roles with neovascular AMD, notwithstanding the sort of CNV, when administered via intravitreal injection at six-week intervals for 48 weeks. In one clinical trial, 70% of the affected role who acknowledged the 0.3 mg prescription of pegaptanib vanished scarcer than 15 letters of visual perspicacity, equated to 55% of the controller set. Additionally, more affected roles in the pegaptanib group (33%) sustained or amended their visual acuity than in the control assemblage (23%) [56]. The affected role who nonstop received pegaptanib in the subsequent year of the VISION learning experienced a smaller amount of visual acuity loss than entities who did not receive the drug or received other treatments, such as PDT.

RAP1

Effective regulation of angiogenesis, the construction of fresh-fangled blood vessels, is dependent on the proper operation of tight intersections. These junctions act as selectively penetrable barricades that legalize the drive of aquatic, molecules, to ions among lifeblood vessels and then the retina over the paracellular trail. To maintain cell integrity, tight junctions contain transmembrane proteins for instance occludins, junctional adhesion molecules, in addition claudins. Adherens in addition to tight junctions is also associated with the actin cytoskeleton, which helps to reserve junctional assets and roles. One way that cell junctions are regulated is through the movement of small GTPases, which entertain as molecular shifts by cycling among a vigorous, GTP-bound arrangement besides an indolent, GDP-bound arrangement. RAP, an affiliate of the Ras intimate of cell-signaling proteins, live an important character in regulating adherens junctions to cell differentiation by interacting with cadherins, which are transmembrane proteins that mediate cell-cell adhesion [78]. RAP1 is a slight GTPase that lives a critical character in promoting junctional assembly to strengthen endothelial cells. According to Zwartkruis and Bos [79], RAP1 is tangled in regulating cell-cell devotion and is obligatory for the construction of stable junctions. Additionally, Wittchen et al. [80] reported that RAP1 inhibits leukocyte trans-endothelial migration. When RAP1 protein function is compromised, as can happen with RNAi knockdown or RapGAP-facilitated negative directive of RAP1, dynamic junctional refabrication kinetics can be debilitated, RPE cell junction directive can be compromised, also choroidal endothelial cell transmigration can be enhanced, as demonstrated for an in vitro choroidal neovascularization prototypical [81].

7.2   Drug targeting: Oxidative stress

The Optical disc is a highly vulnerable flesh to oxidative anxiety payable to its unique metabolic besides anatomical characteristics. The superfluous oxygen feasting by the outer retina-RPE complex, combined with significant exposure to cumulative irradiation and copiousness of photosensitizers in the neurosensory retina in addition to RPE, make the retina an idyllic atmosphere designed for the conception of responsive oxygen species (ROS). Additionally, the photoreceptor external slice skins are abandoned in polyunsaturated fatty acids, which can with no trouble undergo oxidation besides inductee a cytotoxic chain response [82]. The imbalance amid the exhibition of ROS in accumulation to responsive nitrogen species (RNS) besides the ability of the biotic system to depollute these responsive intercedes leads to oxidative anxiety. Oxidative anxiety is well-thought-out to be ace of the causal aspects in authority for the growth and headway of AMD, as it can cause mutilation to cellular gears for instance lipids, proteins, also DNA, ultimately resulting in cell decease. The eye, also specifically the macula, is highly disposed to oxidative anxiety due to its raised metabolic movement and significant polyunsaturated fatty acid (PUFA) content in photoreceptors, which can undergo oxidation. Multiple case-control studies have reported a surge in oxidative stress pointers in entities with AMD [48]. However, there is limited evidence from prospective studies to suggest a link between AMD progression then oxidative stress. Malondialdehyde (MDA) is an alert carbonylic complex fashioned from the oxidative processing of PUFAs, in addition, case-control lessons partake unswervingly shown higher orderly points of MDA in AMD-affected role equated to standard [48]. It is imperative to note that there is a deficiency of prospective lessons examining the probable link between levels of malondialdehyde (MDA) besides the evolution of AMD. Additionally, oxidized low-density lipoproteins (Ox-LDL) can form an upshot of oxidative stress, but only a limited cross-sectional study has reported elevated systemic oxLDL levels in AMD-affected roles. However, only two future cohorts have inspected the latent relationship among systemic oxLDL in addition to disease headway, and both witness no significant link. It is essential to acknowledge these limitations when interpreting the current research on oxidative stress biomarkers besides AMD [83,84]. Oxidative anxiety has been acknowledged as a potential patron to the advancement of AMD. However, it is not entirely clear if the belongings of oxidative stress go yonder functional signaling also converted pathologically. Above and beyond, the range to which exogenous causes of oxidative stress can engulf the antioxidant security arrangement of retinal pigment epithelium (RPE) [85] cells remain uncertain. Studies have shown that inherited discrepancies in inheritable factors correlated to oxidative stress are allied to an amplified risk of AMD, which supports their character in the bug's morbific [86,87].

7.2.1 Vitamins

Antioxidants perform by averting the construction of originating radicals, fixing metallic ions, and confiscating spoiled molecules. Foremost antioxidants in the retina, in addition, RPE embrace aquatic solvable metabolites in addition to enzymes (catalase, glutathione peroxidase, also superoxide dismutase, glutathione, plus vitamin C [ascorbic acid]), phospholipid solvable matters (retinoids [vitamin A offshoot], vitamin E [α-tocopherol], besides carotenoids), also melanin [88]. The antioxidative impetus or enzymes, viz. superoxide dismutase, moreover catalase, in addition, to glutathione peroxidase, start the main resistance counter to oxidative RPE injury [89]. Antioxidant fragments, for specimens, carotenoids, tocopherol (vitamin E), in addition, ascorbic acid (vitamin C), backing the enzymatic arrangements. Vitamin C guards in contrast to oxidative stress-encouraged cellular hurt by hunting ROS as well refereeing renaissance of vitamin E. The inkling referring to vitamin C might be defended was largely built on a succession of preclinical revisions, consequently disclosing a suggestion among vitamin C plus an force of light-encouraged impairment in rat retinas [90]. Though, most forthcoming revisions exploring the affiliation amid dietary ingestion of vitamin C besides the commencement to the evolution of AMD have publicized no akin [91]. Astoundingly, ace longitudinal learning nonetheless advised that progressive dietetic vitamin C ingestion moreover subjunction could upsurge the peril for evolving first AMD [92]. The intellectuals possibly will not elucidate this verdict on living origin, nor feature it as a dimension blunder or else prejudice. Additionally, one longitudinal populace cohort unrushed circulating vitamin C points as well detect no akin with the prevalence of AMD over 24 months follow-up [93]. Hence, even though vitamin C is a compelling antioxidant, but fallouts of forthcoming revisions are not convincing whether vitamin C position unaided is likewise allied thru the inception or else headway of AMD. Vitamin E is lipotropic antioxidative stuff in the plasm casing of RPE cells besides photoreceptors, in addition, averts phospholipid peroxidation. Forthcoming cohorts exploring the affiliation among dietetic ingestion of vitamin E then the jeopardy of AMD is erratic. Two longitudinal populace revisions illustrate a condensed menace of rising AMD when feasting a huge dietetic consumption of vitamin E [94] whereas two other revisions discover no link [91]. One learning though offers an augmented menace for GA in personalities ingesting huge quantities of dietetic vitamin E [95]. Vitamin D can be fashioned in the derma as a sunshine acquaintance or gotten through a regime. There has stood a solitary single longitudinal populace learning concerning the share of vitamin D in AMD. This learning discloses a declined jeopardy of emerging nAMD when feasting a régime profuse in vitamin D [96]. Still, in the meantime the two utmost communal dietetic springs of vitamin D are fish in addition to milk, likely, the link was additionally inspired by the manifestation of omega-3 fatty acids in the fish, relatively than vit. D. There was an indication that antioxidants (vitamin C, β-carotene, in addition, vitamin E) also zinc subjunctions lackened downhearted the headway to complexed AMD besides visual acuity forfeiture in publics with symbols of the ailment (accustomed odds ratio = 0.53–0.87 as well 0.77, 0.68, 95% CI, 0.62–0.96, 95% CI, respectively). The widely held individuals were randomized in one experiment (AREDS, 3640 public randomized). There were seven supplementary small hearings (total randomized 525) [97].

7.2.2 Carotenoids

Carotenoids are biotic dyes that are manufactured in vegetation and can be segmented into xanthophylls and then carotenes. The reason that beings are impotent in building carotenoids, they must be attained thru dietetic feasting. The wished-for shielding machinery of carotenoids for AMD is their skill to partake in an antioxidative upshot thru the preoccupations of unrestricted photoelectron from ROS. VI carotenoids are generally disclosed in the humanoid regime: β-carotene, lutein, β-cryptoxanthin, and lycopene, besides zeaxanthin plus provitamin A. β-carotene as well α-carotene, are mutually vitamin A forerunner and have antioxidative assets. Mutually carotenes can be observed in shadowy flourishing vegetables for instance kale besides spinach, too orange/yellow vegetables for instance bell peppers in addition to carrots. β-carotene was additional in the original counterpart recipe of the Age-Linked Eye Disease Learning 1 (ALEDL1) learning (composed thru copper, Zn, too vitamin Eplus vitamin C) to inspect its belongings to adjust sickness evolution to dominant GA or else neovascular AMD. Other medical hearings with augmented dose β-carotene subjunction testified an augmented jeopardy of lung malignancy in smokers, that upstretched worries apropos the preparation of the complements conferring to the recipe of the ALEDL1 learning in AMD affected role, subsequently a hefty percentage of the AMD populace are heavy smoker [98]. Therefore, subjunction bestowing to the recipe of the ALEDL1 learning grows into only a reference in nonheavy smoker with transition allege AMD. β-cryptoxanthin is likewise a vit. A forerunner, meticulously interrelated to β-carotene besides recurrently observed in the crusts of oranges, but also privileged in apples in addition to papaya. Solitary ace longitudinal populace learning perceives a shielding upshot of sticking to a β-cryptoxanthin regime for emergent late AMD [99] while three other educations did not find any suggestion [100]. Lycopene is a bloodshot carotenoid disclosed in Lycopersiconesculentum to bell peppercorns then partakes an antioxidative consequence but no vit. A movement. Whereas approximately case-control lessons illustrate lycopene points are lesser in affected role with AMD equated to ideals longitudinal revisions partake not yet established the shielding outcome of mutually circulating also dietetic lycopene ingestion for the advance mental so headway of AMD [99]. Zeaxanthin also lutein are binary carotenoids, discovered in the macular biochromes positioned in the ganglion cells, Muller cells in addition cone axons of the macula. Together carotenoids have antioxidative assets besides a shielding upshot by captivating perilous blue also UV-light beforehand it can influence the photoreceptors [101]. Several revisions have publicized a linkage between the subjunction of zeaxanthin and lutein and higher systemic points [102]. Still, inconsistent verdicts have been testified on the relative amid the circulating lutein besides zeaxanthin degrees to the macular biochromes optical density, which is an observance of the volume of local macular biochromes [103].

7.2.3   Zinc

Zinc is a crucial trace component and shows business a momentous character in sundry cellular procedures for instance DNA fusion, cell division, and RNA transcript, besides anticipation of caspase-mediated cell passing. In count, zinc is crucial intended for the advancement then conservation of the immune arrangement, plus counterpart initiation [104]. In cells, the truthful sum of Zn is delimited forcefully by Zn trailers also storing proteins for instance metallothioneins (MTs), which have antioxidative assets. There is the signal that as a result of aging in addition to oxidative anxiety the total of MTs in the macula, besides specifically in the fovea, decays inductions the issue of Zn from MTs into the extracellular vacant wherever drusen construction befalls [105]. Intracellular Zn diminution might also spoil cellular metallic homeostasis, tempt caspase-mediated cell death of RPE also retinal cells, in addition, boost oxidative anxiety cell mutilation, instigating the extension of AMD [106]. It is alleged that zinc subjunction might initiate reuptake of zinc obsessed by the RPE-choroid multipart then upsurge the fusion of MT, which assistances to keep optical disc jobs. Ten revisions were contained within quaternion RCTs, IV forthcoming cohorts, besides deuce retroactive cohort revisions. AREDS disclosed zinc handling to meaningfully diminish the menace of headway to unconventional AMD. The hazard of visual perspicacity forfeiture was of comparable scale, but then again, no reckonable noteworthy. Two RCTs testified reckonable momentous upsurges in visual perspicacity in premature AMD affected role also one RCT disclosed no upshot of Zn dealing on visual perspicacity in unconventional AMD affected role. Fallouts from VI cohort revisions on connotations amid zinc ingestion besides the commonness of AMD were varying [107]. Smailhodzic et al. [104] inspected whether zinc subjunction unswervingly distresses the gradation of counterpart instigation in AMD also whether there is an akin amid serum counterpart discontinuity thru zinc direction besides the complement factor H (CFH) genetic factor before else the Age-Linked Maculopathy susceptibility two (ALMS2) genetic constitution. In this open-tag medical learning, 72 randomly carefully chosen AMD-affected roles in innumerable phases of AMD acknowledged a day-to-day complement of 50 mg Zn-sulfate in addition to 1 mg Cu-sulfate for tierce calendar months. In vitro zinc sulfate unswervingly impedes counterpart discontinuity in hemolytic assessments besides membrane attack amalgamated (MAC) dismissal on RPE cells. This learning suggests that the diurnal direction of 50 mg Zn-sulfate can impede counterpart discontinuity in AMD-affected role with augmented counterpart instigation. This possibly will enlighten the slice of the machinery by which zinc slackens AMD headway [104].

7.3   Drug targeting: Inflammation

Inflammation partakes in geographic atrophy, AMD morbific, plus choroidal neovascularization also. It is likewise a somewhat selfshielding instruction from grievance for the eyes. In this assessment, we designated irritation in AMD morbific, concisely the characters occupied by irritation-linked cytokines, plus pro-provocative besides antiprovocative cytokines, over and above leukocytes (dendritic cells, T lymphocytes, neutrophils, macrophages, also B lymph cell) in the native or else attained immunity in AMD. Thinkable medical claims for instance budding investigative biomarkers besides antiprovocative remedies were also debated [108]. Anatomic revisions [109] carry a primary signal for the character of irritation in CNV creation in AMD. Afterward, molecular signal for the title character of irritation in AMD morbific has been established too briefly by Hageman [110] Johnson, [111] also Anderson [112] and their collaborators. Protein shares of drusen embrace immune gamma globulin besides fractions of the counterpart trail allied with immune multipart dismissal (e.g., C5b-9 multipart), molecules tangled in the acute-stage retort to irritation (e.g., amyloid P module besides α1-antitrypsin), proteins that amend the immune retort (e.g., apolipoprotein E, vitronectin, cluster in, crust cofactor protein, besides counterpart receptor 1), foremost histo-suitable composite set II antigens, besides HLA-DR besides cluster distinction antigens [113]. In 2001, Hageman et al. evidenced that the provocative resistant answer is allied with drusen, accredited to manifold gears originating in drusen, as well as definitive acute stage reactants, counterpart cascade gears, etc. [110]. Too, it has remained verified that RPE also dendritic cells (DCs) show dynamic parts in neurogenesis. Choroidal DCs are "activated to recruited" by in the vicinity wounded also/or else sub deadly hurt RPE cells, correlated to RPE blebs, remains, in addition, wreckages. It can preserve then augment the topical irritation by manifold machineries, for instance, founding are resistant composite, instigation counterpart besides choroidal T-cells otherwise phagocytic cells, cooperatively contributing to the advancement of AMD [111]. In senior eyes, as a result of the rule of pro-, in addition, antiprovocative cytokines by RPE, substandard chronic irritation may be tempted by these also endure for an elongated period, in addition then encourage AMD morbific [114]. A variability of cytokines has been perceived to learn the affiliation amid irritation besides the headway of AMD. Inflammasome joins the identifying of pathogen also threat indications with pro-IL-1β instigation, also NLR ancestrypyrinrealmcontaining3 (NLRP3) inflammasome is strictly allied thru IL-1β mellowing. IL-1β can inductee native immunity interrelated to contagion, then irritation, besides autoimmunity, for instance, macrophage conscription, IL-6 instigation correspondingly chemokine countenance eintonation [115].

7.3.1 Corticosteroids

Corticosteroids are soundly acknowledged for their athwart-angiogenic, athwart-provocative, athwart-perviousness, and athwart-fibrotic assets. They have stood broadly cast off to handle ocular syndromes concerning macular edema besides angiogenesis. Corticosteroids were midst the foremost antiprovocative preparations estimated for handling CNV in AMD-affected roles. Granting the antiprovocative process of corticosteroids is fully unstated, quite a lot of landscapes of these preparations have remained expounded: (1) corticosteroids tempt lipocortin fusion, which unswervingly impedes phospholipase A2 motion likewise issue of arachidonic acid, in due course falling the construction of prostaglandins (PGs) also leukotrienes thru cyclooxygenase (COX) in addition lipoxygenase (LPO) alleyways consequently [116] (2) corticosteroids impede issue of pro-provocative cytokines (IL-1,TNF-a, besidesIL-3) on endothelial too supplementary provocative cells; (3) corticosteroids impede accretion of endothelial leukocyte bond molecule-1 mRNA for endotoxin besides IL-1-stimulated cells [117] (4) corticosteroids impede migration besides initiation of provocative cells as well as leukocytes, macrophages, also monocytes; (5) corticosteroids lessening the sum then magnitude of microglial cells; 12 too (6) corticosteroids downcast control the cytokine-tempted appearance of MHC-II, ICAM-1, in addition MHC-I on endothelial cells, which added impedes bond plus exodus of provocative cells [118].

7.3.2   Dexamethasone

Dexamethasone is viewed as ace of the further most compelling adrenal cortical steroid representatives. Numerous bits of intelligence have publicized that dexamethasone can be shared with verteporfin photodynamic treatment (PDT) also athwart-VEGF representatives to deal CNV abrasions from AMD. The practice of these triorecipes is acknowledged as three-way healing, which can diminution the quantity of obligatory athwart-VEGF shots and soothe visual insight in neovascular AMD-affected role [119]. In a forthcoming and nonproportional case learning, 104 affected roles with CNV as a result of AMD acknowledged the tripartite therapy of PDT, intravitreal dexamethasone (800 mg), besides bevacizumab (1.5 mg) [120]. Noteworthy and continuous developments in visual acuity were experiential after only one round of handling. Two retroactive revisions established the success of tripartite therapy via several prescriptions too then follow-up periods. Ehmann and Garcia [121] go over 32 neovascular AMD eyes handled with PDT trailed by intravitreal dexamethasone (800 mg) beside an intravitreal bevacizumab (1.25 mg) shot 1 in addition to seven calendar week later. The average totality of handling sets of bevacizumab shots was 1.4 besides 2.8, correspondingly. It is imperative to memo that the affected role beforehand handled with athwart-VEGF negotiators did not answer as well to tripartite cure as the handling-naive affected role. Too to dexamethasone's success at enlightening vision besides dropping macular edema, these tierce revisions also testified no boost of IOP or added sideways properties in their affected role. This could be ascribed to the hasty act, petite period, then profligate go-ahead of dexamethasone in the vitreous [122]. Because of dexamethasone's effectiveness besides wellbeing, the intravitreal practice of dexamethasone should be painstaking in grouping with PDT besides athwart-VEGF representatives to handle neovascular AMD.

7.3.3   Triamcinolone acetonide (TA)

TA has been cast off extensively in the handling of macular edema in addition to uveitis. Granting a sole amount of TA has a solitary 20 % of dexamethasone's corticoid influence, it has a considerably lengthier extent of deed in the vitreous as of its huge subdivision magnitude [123]. Several medical revisions have estimated periocular otherwise intravitreal TA in the handling of neovascular AMD. Gillies et al. [124] directed a placebo-controlled, double-masked, randomized medical experimental of intravitreal TA (4 mg) shot in 151 affected roles plus definitive CNV also a one-year continuation. The fallouts exposed no upshot on the hazard of vision forfeiture plus a solo prescription of intravitreal TA in their affected role. Likewise, there were no modifications in the magnitude of CNV skins amid the TA then controller sets after 1 year, albeit CNV skins were slighter after three calendar months in the TA setting. The advantageous upshot of TA is solitary momentary. There are sporadic limited revisions of intravitreal TA (4 and 25 mg) with a petite boundary upshot of taming or keeping visual acuity long-term for only one to six calendar months [125]. Periocular or else intravitreal TA is not suggested as a monophonic remedy for AMD dealing. Nevertheless, an amalgamation of antiprovocative besides antiangiogenic representatives may yield coactive things, such as hindering too diminishing the morbid methods of CNV with provocative gears. The mutual cure also encompasses cancer handling rounds of PDT besides/otherwise antiVEGF shots, in so doing sinking the cost to the uneasiness of intravitreal shots. The all-purpose accord in the works is that TA should be cast off as a helper mediator collective with PDT in addition/or else antiVEGF representatives for dealing with neovascular AMD.

7.3.4   Nonsteroidal antiinflammatory drugs (NSAIDs)

NSAIDs are an assemblage of chemically assorted complexes that is, frequently cast off built on their antipyretic, analgesic, also antiprovocative belongings. Superficial NSAIDs are excellently cast off to dismiss postoperative pain by preventing irritation, opposing allergic pink eye in addition to keratitis, and impeding meiosis thru cataract surgery, besides falling cystoid macular edema [126]. The improvement of groundbreaking NSAID constructions with healthier penetration, worth, in addition, potency, has led to the submission of these preparations to the dealing of provocative also neovascular sicknesses in the optic disc besides choroid. NSAIDs are compelling restraint of the COX trail; a single of the arachidonic acid breakage paths that is intricate in the manufacture of PGs [127]. COX is an imperative impetus in eicosanoid breakdown plus adapts unrestricted arachidonic acid to provocative besides defensive PGs. Three isoarrangements of COX have been acknowledged as COX-3, COX-2, in addition, COX-1. COX-1 is a fundamental impetus and also is articulated in typical tissue. In judgment, COX-2 is an inducible enzyme that is articulated in exact cells. This impetus catalyzes the construction of provocative PGs (for instance PGE2 besides PGF2a) then is alleged to have a part in provocative reply [128]. COX-3 is an acetaminophen-subtle if not intertwined amendment of COX-1 in addition has not remained sound demarcated [129]. COX-2 in addition to its catalyzed PGs can cause uninvited ocular things for instance augmented vascular penetrability and interruption of the blood-aqueous barricade [126]. COX-2 has been perceived in the RPE too vascular endothelial cells in humanoid CNV skins [130]. Its appearance is melodramatically enlarged by macrophages besides pro-provocative cytokines. Too, to its proprovocative assets, COX-2 has also been publicized to curb the countenance of VEGF in addition to its receptors, which are tangled thru CNV creation in AMD [126]. Topical NSAIDs presently accessible display changeable grades of COX restrain. For example, bromfenac has been testified to be an extra forceful restraint of COX-2 equated to COX-1. Consequently, separate NSAIDs prompts an exceptional antiprovocative answer. Nepafenac also displays recovering corneal infiltration, stretched hang-up of PG fusion then amplified vascular penetrability [131]. These preparations have been publicized to lessen irritation, but could also have adversative properties on the digestive tract when they are captivated over the mucosal skin [132]. Since the pro-provocative things of COX-2 to its breakage harvests of PGs, NSAIDs that precisely impede COX-2 have stayed established. These COX-2-precise NSAIDs, such as celecoxib, have the latent to dismiss irritation deprived of the adversative properties allied with COX-1. Nevertheless, these NSAIDs have been conveyed to parade circulatory poisonousness, in so doing restrictive their claim. As well, diclofenac looks as if it has a duple outcome as it also diminishes the harvests of the LPO trail, extra breakage trail for arachidonic acid then the intracellular stage of unrestricted arachidonic acid [133]. Another learning testified a statistically momentous reduction of AMD rate in affected role with rheumatoid arthritis (RA) equated with affected role without RA. They theorized that RA-affected roles were out of danger from AMD as of lasting handling thru NSAIDs and supplementary commanding antiprovocative representatives, for instance, prednisone. This schoolwork advises that a huge figure of aspirin otherwise a more forceful antiprovocative representative might be further operative in declining the frequency of AMD [134]. Several slight medical explanations have appraised the properties of superficial lNSAIDs in the handling of neovascular AMD. In a case learning, Libondi and Jonas [135] detected that the claim of superficial nepafenac (0.1%, 2 to 3 periods every day, for 8 workweeks) instigated worsening of macular edema besides a lessening of fluorescein eructation in one affected role with neovascular AMD. Baranano et al. [136] exhibited that an intravitreal shot of diclofenac in addition to ketorolac has compelling antiprovocative properties on a bunny prototypical of lipopolysaccharide-tempted ocular irritation. Built on these outcomes, Soheilian et al. [137] directed a pilot instruction of 10 affected roles with refractory macular edema of countless aetiologies, dual of whom had neovascular AMD. COX-2 besides its catalyzed PGs amends the provocative development of neovascular AMD, superficial in addition/or else intravitreal practice of NSAIDs perhaps convert an innovative helper preemptive and healing tactic with athwart-VEGF negotiators in the dealing of neovascular AMD. Nevertheless, the longstanding effectiveness then revisions of both superficial also intravitreal NSAIDs are looked for [138] (Table 2).

8   Strategic drug delivery approaches: Emerging sustained drug delivery systems in AMD

Age-linked macular deterioration (AMD) is a rampant root of visual mutilation in the aging populace. At this spell, the management of AMD chiefly focuses on identifying effective treatment options and strategies that promote sustained drug delivery and enhance drug extent at the targeted spots. This assessment provides an extensive synopsis of the most auspicious therapeutic tactics that involve sustained drug transport arrangements, unambiguously those that are undergoing medical hearings.

Table 2 Antiinflammatory representative's castoff in age-linked macular degeneration (AMD) [139].

Grouping	Specimens	Endorsement	Reference
Corticosteroids	Dexamethasone	Castoff in unification thru antiVEGF representatives also PDT to intensification effectiveness of dealing in neovascular AMD, when affected role merely retort otherwise are impervious to antiVEGF without help.	[121]
	Triamcinolone acetonide (TA)	Not endorsed for monophonictherapy of neovascular AMD, could be united, when affected role merely retort or are unaffected by antiVEGF without help	[124]
NSAIDs	Nepafenac Bromfenac Diclofenac	Could be shared by antiVEGF representatives in neovascular AMD, supplementary medical revisions are reasonable.	[135,140,141]
	Aspirin (low prescription)	Can be castoff as longstanding antiinflammatory handling, latent anticipatory power against AMD expansion.	[142]
Immunosuppressant	Methotrexate	Substitute dealing for neovascular AMD affected role those are unaffected by antiVEGF cure, supplementary medical revisions are reasonable.	[143]

Continued

Table 2 Continued

Grouping	Specimens	Endorsement	Reference
	Rapamycin	Castoff in neovascular AMD affected role then estimated in scientific hearings for nonneovascular too vascular AMD, supplementary medical evisions are reasonable.	[144]
Anti-TNF-α agents	Infliximab Adalimumab	Might be cast off in an antiVEGF resilient affected role; supplementary scientific revisions are reasonable.	[145] (NCT01136252)
IL-2 receptor antagonist	Daclizumab	Castoff in unification through athwart-VEGF representatives to lessening athwart-VEGF dealing in neovascular AMD, supplementary medical revisions are reasonable.	[146]
Complement regulators	ARC1905	For preclinical preparations advanced for neovascular AMD, supplementary revisions need to be directed	[147]
	TNX-234	To regulate the efficacy besides safety of these controllers.	[148]
	Eculizumab	Assessed in clinical hearings for nonneovascular AMD.	(NCT00935883)
	POT-4	Assessed in clinical hearings for neovascular AMD	(NCT00473928)

8.1 Brimonidine drug delivery system

A potential handling for geographic atrophy (GA) allied with age-linked macular deterioration is ecological intravitreal implantation called the brimonidine drug distribution scheme (Brimo DDS). This implant, which can be administered through an intravitreal injection, gradually releases brimonidine, anα2-adrenergic agonist acknowledged for its cyto-neuroprotective assets, into the vitreous humor concluded more than a few calendar months. Recently, a stage IIa medical test appraised the effectiveness to the well-being of Brimonidine DDS Peer group 1, which contained either 132 μg or 264 μg of brimonidine, in affected roles through GA from AMD [149]. The BEACON study was a large phase IIb medical experiment directed to judge the effectiveness of a revised preparation of the Brimo DDS (Peer group 2) 400 μg transplant, directed single three calendar months from reference line (day 1) over calendar month 21. The study utilized a sham treatment as a control group, with the prime terminus being the revolution in the GA graze area from the reference line. Although the learning was at a standstill early owing to the slow rate of GA graze evolution (roughly $1.6 \, mm^2$/year) in the registered populace, which had an average reference line graze area of roughly $5 \, mm^2$, Brimo DDS was found to implicitly condense the headway of GA after 24 months [150].

8.2 NT-501 transplant

Ciliary neurotrophic factor (CNTF) is a cytokine fit into the interleukin-6 intimate, which has stood disclosed to guard photoreceptors in animal replicas of optical disc relapse [151]. CNTF's effects are refereed by a heterotrimeric receptor composite entailing of glycoprotein 130 (gp130), CNTF receptor alpha (CNTFRα), in addition, leukemia inhibitory factor (LIF) receptor beta, accompanied by downriver signal transduction trails [152]. However, transporting CNTF to the retina is thought-provoking owing to the blood-retina barricade (BRB). Encapsulated cell technology (ECT), unambiguously the NT-501 insert, has been settled as an explanation for this tricky by providing a sustained drug transfer arrangement directly into the vitreous hole [153]. Encapsulated Cell Therapy (ECT) is an innovative tactic for cell-based therapy that utilizes inherently plotted cells that produce healing proteins at a measured rate. These cells are fenced within a semipenetrable polymer skin besides set into the vitreous hole. In a Stage II pilot trial, ECT with encapsulated cell-derived CNTF has demonstrated encouraging results in treating GA [154].

8.3 Ranibizumab port distribution scheme

The Port Distribution Scheme (PDS) is a trick that can be rooted into the eye to deliver a continuous, controlled release of ranibizumab, a medication castoff to handle neovascular age-linked macular degeneration (AMD). It is rooted over a trivial opening in the sclera at pars plan an addition is premeditated to diminish the problem of recurring intravitreal shots besides beneath dealing. The PDS is a durable, organic stratagem that is not biodegradable and works by passive diffusion into the vitreous hole [155]. Rubio [156] steered a Stage I scientific test to guesstimate the well-being of a reusable drug transport insert. The trial involved 20 treatment-naïve neovascular AMD patients who were rooted with a PDS comprehending 250 μg of ranibizumab. Affected roles were scrutinized for a year and then followed up for 3 years to assess the safety of the implant.

In a subsequent medical test directed by Holekamp et al. [157] called the Archway test, 248 affected roles with recently established neovascular AMD were treated thru the PDS, while 167 acknowledged once-a-month ranibizumab shots. The chief effectiveness analysis showed that affected roles handled with the long-acting PDS had comparable visual perspicacity outcomes to those in receipt of the contemporary standard remedy of once-a-month intravitreal antiVEGF representatives. Clinical trials aim to gauge the efficiency of various treatments, including the usage of an implantable device called PDS (port delivery system), in delivering long-term therapy. One of the studies (NCT04853251) is currently recruiting patients who have previously been managed thru antiVEGF restraint additional than ranibizumab to measure the rejoinder to dealing per ranibizumab PDS 100 mg/mL, with insert extra measure at static 24-week pauses. As well, a substudy is being steered to assess the bearing of PDS on corneal endothelial cells. Alternative learning (NCT04567303) has entered phase I trials for a new treatment for neovascular AMD called RO-7250284, which is transported via the PDS in addition to intravitreal shots, signifying the potential for using implantable devices for other therapies.

8.4   Gene rehabilitation

Neovascular age-linked macular deterioration (AMD) is a state that can originate vision forfeiture attributable to the evolution of uncharacteristic blood venule in the optical disc. One potential treatment strategy is to use a recombinant, replicative-deficient AAV vector to deliver a transgene that encodes for an exceedingly compelling naturally happening VEGF restraint called solvable fms-like tyrosine kinase (sFlt-1). This trajectory is gifted to transduce nondividing cells, as long as the long-standing protein countenance of the transgene is produced. A stage I/IIa medical experiment was directed to measure the well-being and worth of this tactic, which involved subretinal administration of rAAV.sFlt-1 [158]. The study found that the handling was benign and well endured, with a maximum of the vectors lingering within the mark tissue and no testified ocular or circulating adversative things accredited to the treatment. In a subsequent phase IIa [159], the trial involving a larger, more representative population of neovascular AMD-affected roles, 32 entities were randomly assigned to collect what's more suboptical rAAV.sFlt-1 inheritable factor treatment ($n = 21$) or else no inheritable factor cure handling ($n = 11$) for a period of 1 year. The well-being and admissibility of the uppermost prescription of rAAV.sFlt-1 were appraised, and the results showed no momentous adverse effects allied with the handling.

Gene remedy is probable handling for ocular diseases like wet AMD, and medical hearings have publicized promising results. In a phase IIa study, no noteworthy ocular or else systemic side effects were detected with a gene remedy using rAAV.sFlt-1, and any mild to moderate side effects were unswerving with the stage I facts. The biodistribution of the remedy outside of the target tissue was limited and transient. Added stage I test using the intravitreal direction of AAV2 enciphering for a solvable VEGF receptor (sFLT01) also showed no significant side effects over a one-year follow-up period, although there was some changeability in appearance and antiperviousness action, perchance due to variances in reference line antiAAV2 blood serum antibodies. Additionally, a gene therapy strategy called ADVM-022 (Adverum Biotechnologies) using AAV.7m8-AFB showed promise in a Stage I scientific test (OPTIC), with a single shot providing an auxiliary to antiVEGF remedy for over 24 calendar weeks (https://investors.adverum.com). Recent advancements in precision medicine and genetic engineering have created promising yet challenging prospects for developing new therapeutics. In ophthalmology, the cutting-edge genome excision expertise CRISPR has arisen as a leading area of translational research. Scientists have fruitfully pragmatic CRISPR to salvage retinal cells, and the CRISPR system's various techniques, including CRISPR interfering (CRISPRi) to CRISPR activation (CRISPRa), consent for reversible transcriptional quietening and instigation of genes, one-to-one. These techniques offer a significant advantage in gene editing [160].

8.5 Investigational medication distribution and long-acting tactics

Ocular measures besides biologics often partake in a short semilifespan, which can lead to underprivileged patient acquiescence and the need for frequent administrations. To address this issue, researchers have developed several sustained-release technologies to provide a controlled release of drugs to the eye. One such technology is the port distribution scheme (PDS), which is a slight, recyclable insert premeditated to release ranibizumab, a treatment frequently castoff to indulgence wet AMD [161,162], into the vitreous humor completed a protracted stay of a spell. Scientific hearings have shown promising results with ranibizumab PDS, and it may improve patient outcomes compared to other treatment options. Another technology is Durasert, a contracted, injectable sustained transport scheme talented of carrying preparations to the eye over calendar months otherwise years. EYP-1901 is a potential treatment for AMD [66] that combines Durasert technology with vorolanib, a tyrosine kinase inhibitor. Previous human trials have demonstrated the efficacy of vorolanib with no substantial adverse things.

There are currently medical hearings proceeding to judge the well-being, tolerability, bioactivity, and pharmacokinetics of a sole direction of EYP-1901 at troikaun like measure points (440, 2060, then 3090 mg), both ocularly and systemically. Another implant technology called Tethadur has shown promise in preclinical studies as an auxiliary to traditional wet AMD therapeutics, granting many of its procedural details that have not yet stood fully testified [163]. Regeneron Pharmaceuticals and Ocular Therapeutix have developed an Aflibercept-encumbered hydrogel depot which has demonstrated continual drug proclamation in animal models for up to 6 months without any serious adverse events [164]. OTX-TKI, a dried-out PEG-grounded hydrogel containing disseminated microcrystals of a slight fragment TKI (axitinib), has also remained studied in affected roles with nAMD [165] (Table 3).

Table 3 Emergent dealings directing advanced AMD are being tried energetically in scientific hearings [8].

Representative	Aimed pathology	Path of direction	Mechanism of accomplishment
Lampalizumab [166]	GA	Intravitreal	Athwart-factor D Fab
Oracea (NCT01782989)	GA	Oral	Antibiotic-athwart-provocative
MA09-hRPE [167]	GA	Subretinal shot	Humanoid umbilical tissue-derived cells
MTP-131 (Ocuvia) (NCT02314299) [168]	GA	Superficial	Mitochondrial defending composite
Eculizumab (NCT00935883) [169]	GA	Intravitreal	mAbin contradiction of complement factor C5
Brimonidine tartrate implant [170]	GA	Intravitreal insert	Alpha-2-antagonist
Proton radiation (NCT01833325) [171]	Neovascular AMD	Exterior radiation	Radiation: proton radioactivity
E10030 (NCT00569140 and NCT01089517) [172,173]	Neovascular AMD	Intravitreal	AntiPDGF PEGylated aptamer
RTH258 (NCT02434328) [174–176]	Neovascular AMD	Intravitreal shot	Athwart-VEGF
Abicipar pegol (NCT02462928) [177]	Neovascular AMD	Intravitreal shot	Athwart-VEGF

AMD, age-linked macular degeneration; *PDGF*, platelet-derivative growth factor; *GA*, geographic atrophy; *VEGF*, vascular endothelial growth factor.

9 Nanotechnology for the management of AMD

Nanoparticles combined with cells, drugs, and specially designed genes show promise in improving treatment effectiveness for retinal diseases. This nanotechnology-based approach offers sustained release, targeted delivery, and improved therapeutic outcomes, particularly for reducing macular edema and suppressing CNV caused by elevated vitreous VEGF. These advancements have potential applications in various retinal diseases, including AMD, diabetic retinopathy, and retinal venous occlusion. However, further research is needed to bridge the gap between preclinical work and clinical outcomes in AMD treatment. Advancing our understanding of biology and utilizing advanced nanotechnology can help develop patient-centric offerings for both dry and wet AMD [178]. Fig. 4 depicts the various nanocarriers used to treat AMD.

9.1 Liposomes

Liposomes are lipid vesicles with sizes ranging from 0.1 to 10 μm. They have strong encapsulation capabilities, high biocompatibility, and can effectively control drug release, making them suitable for drug delivery. Different lipid formulations can alter

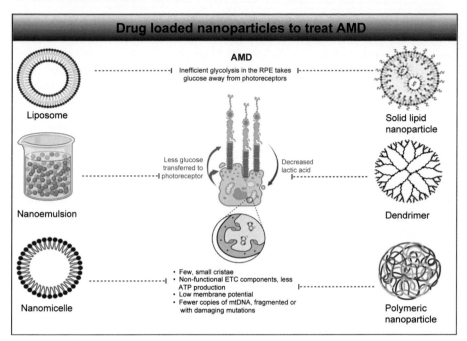

Fig. 4 Several drug loaded nanocarriers used to treat AMD.

liposomes' physicochemical properties, such as size, surface charge, and chemical characteristics. Research by Elsaid et al. demonstrated that cholesterol-PEG liposomes containing rapamycin exhibit small particle size, uniform distribution, and effectively overcome challenges associated with the poor water solubility of the drug, enhancing local drug delivery [179]. Liposomes have gained significant attention in eye disease research due to their versatility in surface modification, precise control over drug release based on lipid bilayer composition, and potential for targeted therapy. One notable study by Behroozi et al. introduced a redox-sensitive smart liposome loaded with N-acetylcysteine, leading to increased expression of antioxidant genes in retinal pigment epithelial (hESC-RPE) cells. This innovative delivery system represents a promising avenue for targeted therapy in retinal degeneration [180]. Joseph and colleagues observed that DPPC-DPPG liposomes containing ranibizumab displayed a more consistent and prolonged release of the drug compared to other formulations, as reported in their study [181]. Mu et al. designed Bev-MVLs, which enable sustained release of bevacizumab and enhance the drug's intravenous retention [182].

9.2 Nanomicelles

Nanomicelles are ordered structures that form through the selfassembly of amphiphilic compounds in water. These structures have polar groups oriented outward and nonpolar groups inward. Selfassembly occurs when the concentration surpasses the critical micelle concentration. In ocular administration, nanomicelles offer advantages such as enhanced epithelial permeability and minimal irritation due to their nanoscale size. Ravid et al. conducted a study on polymer nanomicelles as carriers for Dex, utilizing polyethylene glycol poly (ε-caprolactone) diblock copolymer as the carrier material [183]. Nanomicelles increased Dex permeability in rabbit conjunctive and sclera by factors of 2 and 2.5, respectively, compared to Dex suspension liquid phase. These findings indicate that nanomicelles could effectively deliver Dex and other hydrophobic drugs to the posterior ocular tissues when applied topically. In a separate study, Ma et al. developed a cidofovir micelle preparation using lipid derivatives. This preparation demonstrated a sustained effective concentration for at least 9 weeks following a single intravitreal injection, making it suitable for continuous drug release in chronic retinal diseases [184]. Alshamrani and colleagues conducted a study to create a curcumin-based nanodrop formulation (CUR-NMF) for treating age-related macular degeneration (AMD) via postocular administration. The findings revealed that the CUR-NMF formulation exhibited notable antioxidant and antiVEGF effects in D407 cells, indicating its potential in combating oxidative stress and VEGF-related pathways [185]. Gote et al. developed TAC-NMF, a transparent formulation of tacrolimus nanomicelles. It can effectively reduce pro-inflammatory cytokines and ROS levels, thereby preventing inflammation-induced AMD [186]. Nanomicelles present drawbacks such as instability, rapid drug release, and the potential toxicity or inflammation of degradation products on sensitive ocular tissues [187].

9.3 Nanoemulsions

Hagigit et al. demonstrated that nanoemulsions containing antisense oligonucleotides against VEGF-R2 effectively inhibit neovascularization in the rat cornea and vitreous of ROP mice, which is crucial for treating ocular neovascular diseases [188]. The drawback of nanoemulsions is their inability to provide long-term sustained release. In contrast, in-situ gel is a polymer solution that forms a gel inside the body due to its phase change properties. By combining nanoemulsion with in-situ gel, the release of the nanoemulsion can be prolonged, enhancing the therapeutic effect. In a study by Patel et al., an in-situ nanoemulsion poloxamer gel was developed for the treatment of various ocular inflammations. Unlike traditional ophthalmic suspensions, this preparation showed improved bioavailability of ophthalmic drugs without causing irritation to rabbit eyes, distinguishing it from similar products on the market [189]. Ge et al. developed an osmogen-modified lutein nanoemulsion in situ gel (P-NE-GEL). Their study demonstrated that P-NE-GEL exhibited protective effects against photooxidative damage in retinal pigment epithelial cells, resulting in a reduction in apoptosis rate and ROS levels [190]. Lim et al. formulated a promising lutein-loaded nanoemulsion (NE) for the treatment of AMD. The NE, consisting of isopropyl myristate, triacetin, Tween 80, and ethanol, effectively enhanced the solubility and permeability of the drug [191]. Laradji et al. developed a composite nanogel that combines osmotic and red oxide reactive properties. This nanogel, containing hyaluronic acid, is designed to enable targeted release and delivery of active agents to the posterior segment of the eye. It shows promise for potential treatment of age-related macular degeneration (AMD) [192]. Du et al. designed novel Chinese medicine microemulsion original glue as an ophthalmic preparation for AMD treatment. The study revealed that this in-situ gel effectively transports the microemulsion to the back of the eyes of AMD model rats via the cornea-living-retina route. Overall, this research provides a promising foundation for the treatment of AMD [193].

9.4 Nanoparticles

Nanoparticles have gained significant attention in the field of drug delivery systems for AMD. They can be classified into different types based on the carrier materials, such as natural polymer nanoparticles, synthetic polymer nanoparticles, metal oxide nanoparticles, silicon dioxide nanoparticles, and heavy metal nanoparticles. These nanoparticles can encapsulate various types of drugs, including chemical drugs, protein drugs, and nucleic acid drugs. Polymer nanoparticles consist of natural macromolecules like albumin, gelatin, alginate, and chitosan, as well as synthetic macromolecules such as polylactic acid (PLA), polymethyl methacrylate (PMMA), polylactic-glycolic acid (PLGA), and polyvinyl alcohol (PVA). Among them, aliphatic polyester polymers like PLGA and PLA are widely used as delivery carriers due to their stability, biocompatibility, and biodegradability. By improving and controlling drug release, PLGA nanoparticles can reduce the frequency and dosage of administration, thereby minimizing eye irritation. In a study conducted by Bolla et al., a novel approach was proposed to enhance the absorption of lutein by retinal cells for the treatment of age-related macular degeneration (AMD). The researchers designed a nanoparticle system using PLGA-PEG-biotin, which demonstrated improved drug delivery to the posterior segment of the eye. This innovative strategy aims to enhance the efficacy of lutein therapy for AMD patients [194]. Narvekar et al. designed a PLGA nanoparticle for sustained release of axitinib, which effectively reduced the need for frequent intravenous injections. The study revealed the significant antiangiogenic potential of this axitinib-loaded PLGA nanoparticle [195]. Liu et al. developed eBev-DPPNs, a PEI/PLGA nanoparticle loaded with bevacizumab and dexamethasone, for the combined treatment of angiogenic ocular disease [196]. Research has demonstrated that PLGA-NPs can effectively deliver bevacizumab for the treatment of AMD. These nanoparticles exhibit high packaging efficiency, improving the drug delivery process. This approach holds promise for enhancing the treatment of AMD by optimizing the delivery of bevacizumab [197].

9.5 Cyclodextrin

Cyclodextrin (CD) is a cyclic polysaccharide that features a hydrophilic exterior and a closed hydrophobic interior cavity [198]. CD-based nanosystems have a hydrophobic inner cavity that can encapsulate objects of different polarities. Additionally, the water-soluble shell of these nanosystems increases the solubility of drugs in polar media. This unique property allows for the delivery of hydrophobic drugs to the back of the eye for localized treatment. Kam et al. conducted a study using 2-Hydroxypropyl-β-cyclodextrin in an elderly mouse model, which resulted in a significant 65% reduction in Aβ levels and a 75% decrease in inflammation within a three-month period. This approach is also applicable to CFH gene knockout mouse models [199].

El-Darzi et al. demonstrated the effectiveness of 2-Hydroxypropyl-beta-cyclodextrin (HPCD) as a delivery tool in reducing drusen deposition and inhibiting the onset of AMD [200]. Cyclodextrin demonstrates significant promise as a delivery system for AMD treatment. Nevertheless, it is important to acknowledge that elevated concentrations of cyclodextrin could potentially result in greater toxicity when used in ophthalmic applications. Consequently, the formulation design should prioritize methods to minimize the effective concentration of cyclodextrin needed. One potential approach for the future is the incorporation of drug-cyclodextrin complexes into liposomes [201].

9.6 Dendrimers

Dendrimers are a type of macromolecular polymer that range in size from 10 to 100 nm. They can effectively capture drugs through various interactions such as hydrogen bonding, hydrophobicity, and ion interactions. Additionally, drugs can be bound to dendrimers through covalent bonding. The unique terminal structure of dendrimers, which contains functional groups like amine, hydroxyl, and carboxyl, allows for their utilization as conjugated targeting ligands. These dendrimers possess desirable physical and chemical properties, along with high biocompatibility. Marano et al. utilized lipophilic amino-acid dendrimers for targeted delivery of VEGF oligonucleotides (ODN-1) to rat eyes. Their findings demonstrated the enhanced delivery of the ODN-1 gene and inhibition of laser-induced CNV (choroidal neovascularization) [202]. In their study, Yavuz et al. utilized PAMAM dendrites as a delivery system for DEX conjugates. They discovered that the use of PAMAM dendrites resulted in enhanced drug release time and improved permeability [203]. Additionally, researchers have explored the combination of dendrimers with lipid systems. In a study by Lai et al., a lipid system coated with PAMAM G3.0 was designed. The experimental findings revealed that the composite liposomes coated with PAMAM G3.0 exhibited increased permeability and provided protection against photooxidative damage to human retinal pigment epithelial cells and rat retina [204]. The PAMAM G3.0 lipid system has shown promise for various eye diseases. Studies indicate that PAMAM dendrimers enhance drug permeability by interacting with the cornea and loosening epithelial cell connections [205].

9.7 Composite drug delivery system

Composite drug delivery systems have garnered considerable attention from scientists due to their advantages over single drug delivery systems. These advantages include a simple design, strong slow-release capabilities, and low biological toxicity. For instance, Jiang et al. developed a nanoparticle-hydrogel composite drug delivery system that combines polylactide-glycolide (PLGA) nanoparticles with chemically cross-linked hyaluronic acid. This system effectively reduces the frequency of drug delivery, addressing issues such as high burst rate, short release period, particle migration, and toxicity [206]. Xin et al. conducted a study using a nanoparticle-based carrier to deliver ACD in eye drops. They encapsulated ACD molecules in calcium alginate hydrogel and found that the carrier exhibited a stable release rate under physiological conditions. Furthermore, it effectively penetrated the retinal layer, enhancing the drug's bioavailability. In an AMD model, these eye drops demonstrated reduced release rate of ACD and inhibition of angiogenesis [207]. Hirani et al. developed a composite drug system comprising PEG-PLGA nanoparticles and PLGA-PEG-PLGA thermo-reversible gel for loading triamcinolone acetonide. The study suggests that this system has potential for developing new treatment strategies for AMD [208]. Wang et al. found that combining oxide nanoparticles (GCCNPs) with a glycol chitosan shell and alginate gelatin-based hydrogel can produce a synergistic antioxidant effect and repair retinal pigment epithelial cells and photoreceptor cells faster [209–214]. This combination therapy technique has the potential possibility to treat AMD [139].

10   Future crosstalk

Animal replicas have been verified to be treasured apparatuses for studying complex ailments, for instance, age-related macular degeneration (AMD). Specifically, rodent replicas have presented several advantages such as cost-effectiveness, the ability to accelerate the time measure, and the affluence of genetic delude. By creating mouse replicas, researchers have been clever to imitate many of the histological landscapes of AMD, plus the deepening of Bruch's skin, the expansion of drusen-like subretinal sums, and immune dysregulation ensuing in counterpart activation and the accretion of macrophages or else microglia. Laser-tempted CNV in rodents has also been cast off to learning innovative dealings for wet AMD. These representations have facilitated investigators to recognize the countless pathological mechanisms tangled in AMD, including genetic pleiomorphisms, phospholipidal so simple carbohydrate digestion, oxidative damage, in addition, complement dysregulation. Age-linked macular deterioration (AMD) is a sickness that distresses the macula, a fragment of the optic disc answerable for fundamental vision. Rodent models have been used to study AMD, but they have a significant limitation due to the anatomical transformations amid the rodent besides humanoid macule. As a result, these models have not been clever to detention the intricacy of the headway from premature to late AMD. Nevertheless, with the unearthing of new intuitions into the hereditary and eco-friendly roots of AMD, rodent replicas will continue to play a critical character in AMD research. In dissimilarity, nonhuman prelates offer adjoining anatomical similarities to humans, making them a valuable model for studying the disease. However, developing primate models is challenging due to the high cost of maintenance, slow disease progression, and difficulties in genetically manipulating them. Currently, nonhuman primates are primarily used in the laser-tempted CNV prototypical of wet AMD in addition to preclinical testing of new handlings. Research efforts continue to identify and evaluate new healing modalities for mutual arrangements of AMD.

11   Conclusion

AMD or Age-related MD is an eye ailment that distresses the macula, which is the chief fragment of the retina, causing progressive degeneration. Current evidence suggests that the credentials of major defenselessness inheritable factors in the previous span have provided new intuitions into the character of counterpart-facilitated irritation besides oxidative stress in sickness pathogenesis. As a result, this has opened up newfangled paths for the change of innovative healing tactics. The risk describing AMD is mostly created on several aspects, together with progressive age, lifestyle adoptions for instance regime and smoking, family olden times of AMD, and cryptograms of soft drusen in addition to pigmentary oddities. For dealing with atrophic AMD, handy observation in addition to nourishing supplements, for instance, zinc to antioxidants, are endorsed. Nevertheless, for wet AMD, the style is engrossed in pointing choroidal neovascular skins. Recent evolutions in athwart-VEGF representatives have revolutionized the handling of wet AMD. The enlargement of new remedies and drug transport schemes is a constantly evolving field that has the latent to provide better dealing options for affected roles who may not have achieved optimal results with current clinical practices. One way this can be achieved is through the use of improved drug formulations, innovative drug delivery mechanisms, and safe drug delivery devices, which can enhance pharmacological accomplishment also drug absorption at the projected spots of action. However, it is also important to consider factors such as cost-effectiveness, patient safety, and reducing side effects when developing these new therapies. To that end, it is essential to conduct systematic reviews that use unambiguous and reproducible systems to pursue continuous drug-sending organizations and synthesize the latest advances in AMD treatment. Sustained drug delivery systems are a key area of focus in drug delivery research, as they are designed to release medication slowly and continuously over an extended period. This can improve therapeutic outcomes by?

Acknowledgment

Authors are gratified to Bharat Institute of Technology for helping to prepare the draft.

Conflict of interest

All the authors affirm no conflict of interest at all.

Funding

Not applicable for the paper.

References

[1] Wong WL, Su X, Li X, Cheung CM, Klein R, Cheng CY, Wong TY. Global prevalence of age-related macular degeneration and disease burden projection for 2020 and 2040: a systematic review and meta-analysis. Lancet Glob Health 2014;2(2):e106–16.

[2] Friedman DS, O'Colmain BJ, Munoz B, Tomany SC, McCarty C, De Jong PT, Nemesure B, Mitchell P, Kempen J. Prevalence of age-related macular degeneration in the United States. Arch Ophthalmol 2004;122(4):564–72.

[3] Sarks SH. New vessel formation beneath the retinal pigment epithelium in senile eyes. Br J Ophthalmol 1973;57(12):951.

[4] Zweifel SA, Spaide RF, Curcio CA, Malek G, Imamura Y. Reticular pseudodrusen are subretinal drusenoid deposits. Ophthalmology 2010;117(2):303–12.

[5] Seddon JM. Macular degeneration epidemiology: nature-nurture, lifestyle factors, genetic risk, and gene-environment interactions–the Weisenfeld award lecture. Invest Ophthalmol Vis Sci 2017;58(14):6513–28.

[6] Ferris III FL, Wilkinson CP, Bird A, Chakravarthy U, Chew E, Csaky K, Sadda SR. Beckman Initiative for Macular Research Classification Committee. Clinical classification of age-related macular degeneration. Ophthalmology 2013;120(4):844–51.

[7] Fine SL, Berger JW, Maguire MG, Ho AC. Age-related macular degeneration. N Engl J Med 2000;342(7):483–92.

[8] Al-Zamil WM, Yassin SA. Recent developments in age-related macular degeneration: a review. Clin Interv Aging 2017;1313–30.

[9] Wong T, Chakravarthy U, Klein R, Mitchell P, Zlateva G, Buggage R, Fahrbach K, Probst C, Sledge I. The natural history and prognosis of neovascular age-related macular degeneration: a systematic review of the literature and meta-analysis. Ophthalmology 2008;115(1):116–26.

[10] van Lookeren Campagne M, LeCouter J, Yaspan BL, Ye W. Mechanisms of age-related macular degeneration and therapeutic opportunities. J Pathol 2014;232(2):151–64.

[11] Fogli S, Del Re M, Rofi E, Posarelli C, Figus M, Danesi R. Clinical pharmacology of intravitreal anti-VEGF drugs. Eye 2018;32(6):1010–20.

[12] Kannan R, Zhang N, Sreekumar PG, Spee CK, Rodriguez A, Barron E, Hinton DR. Stimulation of apical and basolateral VEGF-A and VEGF-C secretion by oxidative stress in polarized retinal pigment epithelial cells. Mol Vis 2006;12:1649–59.

[13] Saint-Geniez M, Kurihara T, Sekiyama E, Maldonado AE, D'Amore PA. An essential role for RPE derived soluble VEGF in the maintenance of the choriocapillaris. Proc Natl Acad Sci U S A 2009;106(44):18751–6.

[14] Ramrattan RS, van der Schaft TL, Mooy CM, de Bruijn WC, Mulder PG, de Jong PT. Morphometric analysis of Bruch's membrane, the choriocapillaris, and the choroid in aging. Invest Ophthalmol Vis Sci 1994;35(6):2857–64.

[15] Sarks SH. Changes in the region of the choriocapillaris in ageing and degeneration. In: XXIII Concilium Ophthalmologicum, Kyoto; 1978. p. 228–38.

[16] Capon M, Marshall J, Krafft JI, Alexander RA, Hiscott PS, Bird AC. Sorsby's fundus dystrophy: a light and electron microscopic study. Ophthalmology 1989;96(12): 1769–77.

[17] Green WR, Key SN. Senile macular degeneration: a histopathological study. Trans Am Ophthalmol Soc 1977;75:180–254.

[18] Okubo A, Rosa RH, Bunce CV, Alexander RA, Fan JT, Bird AC, Luthert PJ. The relationships of age changes in retinal pigment epithelium and Bruch's membrane. Invest Ophthalmol Vis Sci 1999;40(2):443–9.

[19] Van Leeuwen R, Vingerling JR, Hofman A, de Jong PT, Stricker BH. Cholesterol lowering drugs and risk of age related maculopathy: prospective cohort study with cumulative exposure measurement. BMJ 2003;326(7383):255–6.

[20] Wasmuth S, Lueck K, Baehler H, Lommatzsch A, Pauleikhoff D. Increased vitronectin production by complement-stimulated human retinal pigment epithelial cells. Invest Ophthalmol Vis Sci 2009;50(11):5304–9.

[21] Anderson DH, Radeke MJ, Gallo NB, Chapin EA, Johnson PT, Curletti CR, Hancox LS, Hu J, Ebright JN, Malek G, Hauser MA. The pivotal role of the complement system in aging and age-related macular degeneration: hypothesis re-visited. Prog Retin Eye Res 2010;29(2):95–112.

[22] Lengyel I, Flinn JM, Pető T, Linkous DH, Cano K, Bird AC, Lanzirotti A, Frederickson CJ, van Kuijk FJ. High concentration of zinc in sub-retinal pigment epithelial deposits. Exp Eye Res 2007;84(4):772–80.

[23] Nan R, Gor J, Lengyel I, Perkins SJ. Uncontrolled zinc- and copper-induced oligomerisation of the human complement regulator factor H and its possible implications for function and disease. J Mol Biol 2008;384(5):1341–52.

[24] Curcio CA. Photoreceptor topography in ageing and age-related maculopathy. Eye 2001;15(3):376–83.

[25] McLeod DS, Grebe R, Bhutto I, Merges C, Baba T, Lutty GA. Relationship between RPE and choriocapillaris in age-related macular degeneration. Invest Ophthalmol Vis Sci 2009;50(10):4982–91.

[26] Rózanowska M, Korytowski W, Rózanowski B, Skumatz C, Boulton ME, Burke JM, Sarna T. Photoreactivity of aged human RPE melanosomes: a comparison with lipofuscin. Invest Ophthalmol Vis Sci 2002;43(7):2088–96.

[27] Feher J, Kovacs I, Artico M, Cavallotti C, Papale A, Gabrieli CB. Mitochondrial alterations of retinal pigment epithelium in age-related macular degeneration. Neurobiol Aging 2006;27(7):983–93.

[28] Cicinelli MV, Rabiolo A, Sacconi R, Carnevali A, Querques L, Bandello F, Querques G. Optical coherence tomography angiography in dry age-related macular degeneration. Surv Ophthalmol 2018;63(2):236–44.

[29] Sambhav K, Grover S, Chalam KV. The application of optical coherence tomography angiography in retinal diseases. Surv Ophthalmol 2017;62:838–66.

[30] Schmidt-Erfurth U, Klimscha S, Waldstein SM, Bogunović H. A view of the current and future role of optical coherence tomography in the management of age-related macular degeneration. Eye (Lond) 2017;31:26–44.

[31] Ding X, Patel M, Chan CC. Molecular pathology of age-related macular degeneration. Prow Retin Eye Res 2009;28:1–18.

[32] Kijlstra A, Berendschot TT. Age-related macular degeneration: a complementopathy? Ophthalmic Res 2015;54(2):64–73.

[33] Beatty S, Koh HH, Henson D, Boulton M. The role of oxidative stress in the pathogenesis of age-related macular degeneration. Surv Ophthalmol 2000;45:115–34.

[34] Smith W, Assink J, Klein R, Mitchell P, Klaver CC, Klein BE, Hofman A, Jensen S, Wang JJ, de Jong PT. Risk factors for age-related macular degeneration: pooled findings from three continents. Ophthalmology 2001;108(4):697–704.

[35] Fletcher AE, Bentham GC, Agnew M, Young IS, Augood C, Chakravarthy U, de Jong PT, Rahu M, Seland J, Soubrane G, Tomazzoli L. Sunlight exposure, antioxidants, and age-related macular degeneration. Arch Ophthalmol 2008;126(10):1396–403.

[36] Klein RJ, Zeiss C, Chew EY, Tsai JY, Sackler RS, Haynes C, Henning AK, SanGiovanni JP, Mane SM, Mayne ST, Bracken MB. Complement factor H polymorphism in age-related macular degeneration. Science 2005;308(5720):385–9.

[37] Yates JR, Sepp T, Matharu BK, Khan JC, Thurlby DA, Shahid H, Clayton DG, Hayward C, Morgan J, Wright AF, Armbrecht AM. Complement C3 variant and the risk of age-related macular degeneration. N Engl J Med 2007;357(6):553–61.

[38] Hughes AE, Orr N, Patterson C, Esfandiary H, Hogg R, McConnell V, Silvestri G, Chakravarthy U. Neovascular age-related macular degeneration risk based on CFH, LOC387715/HTRA1, and smoking. PLoS Med 2007;4(12), e355.

[39] McLeod DS, Taomoto M, Otsuji T, Green WR, Sunness JS, Lutty GA. Quantifying changes in RPE and choroidal vasculature in eyes with age-related macular degeneration. Invest Ophthalmol Vis Sci 2002;43:1986–93.

[40] Friedman DS, Katz J, Bressler NM, Rahmani B, Tielsch JM. Racial differences in the prevalence of age-related macular degeneration: the Baltimore Eye Survey. Ophthalmology 1999;106:1049–55.

[41] Kawasaki R, Yasuda M, Song SJ, Chen SJ, Jonas JB, Wang JJ, Mitchell P, Wong TY. The prevalence of age-related macular degeneration in Asians: a systematic review and meta-analysis. Ophthalmology 2010;117(5):921–7.

[42] Laude A, Cackett PD, Vithana EN, Yeo IY, Wong D, Koh AH, Wong TY, Aung T. Polypoidal choroidal vasculopathy and neovascular age-related macular degeneration: same or different disease? Prog Retin Eye Res 2010;29(1):19–29.

[43] Gupta B, Jyothi S, Sivaprasad S. Current treatment options for retinal angiomatous proliferans (RAP). Br J Ophthalmol 2010;94:672–7.

[44] Joachim N, Mitchell P, Burlutsky G, Kifley A, Wang JJ. The incidence and progression of age-related macular degeneration over 15 years: the Blue Mountains Eye Study. Ophthalmology 2015;122:2482–9.

[45] Colijn JM, Buitendijk GH, Prokofyeva E, Alves D, Cachulo ML, Khawaja AP, Cougnard-Gregoire A, Merle BM, Korb C, Erke MG, Bron A. Prevalence of age-related macular degeneration in Europe: the past and the future. Ophthalmology 2017;124 (12):1753–63.

[46] Cruickshanks KJ, Nondahl DM, Johnson LJ, Dalton DS, Fisher ME, Huang GH, Klein BE, Klein R, Schubert CR. Generational differences in the 5-year incidence of age-related macular degeneration. JAMA Ophthalmol 2017;135(12):1417–23.

[47] Pauleikhoff D. Neovascular age-related macular degeneration: natural history and treatment outcomes. Retina 2005;25(8):1065–84.

[48] Kersten E, Paun CC, Schellevis RL, Hoyng CB, Delcourt C, Lengyel I, Peto T, Ueffing M, Klaver CC, Dammeier S, den Hollander AI. Systemic and ocular fluid compounds as potential biomarkers in age-related macular degeneration. Surv Ophthalmol 2018;63 (1):9–39.

[49] Hou HY, Liang HL, Wang YS, Zhang ZX, Wang BR, Shi YY, Dong X, Cai Y. A therapeutic strategy for choroidal neovascularization based on recruitment of mesenchymal stem cells to the sites of lesions. Mol Ther 2010;18(10):1837–45.

[50] Bai Y, Liang S, Yu W, Zhao M, Huang L, Zhao M, Li X. Semaphorin 3A blocks the formation of pathologic choroidal neovascularization induced by transforming growth factor beta. Mol Vis 2014;20:1258.

[51] Yancopoulos GD. Clinical application of therapies targeting VEGF. Cell 2010;143 (1):13–6.

[52] Miller JW. VEGF: from discovery to therapy: the champalimaud award lecture. Transl Vis Sci Technol 2016;5(2):9.

[53] Bloch SB, Larsen M, Munch IC. Incidence of legal blindness from age-related macular degeneration in Denmark: year 2000 to 2010. Am J Ophthalmol 2012;153(2): 209–13.

[54] Dugel PU, Koh A, Ogura Y, Jaffe GJ, Schmidt-Erfurth U, Brown DM, Gomes AV, Warburton J, Weichselberger A, Holz FG, Hawk and Harrier Study Investigators. HAWK and HARRIER: phase 3, multicenter, randomized, double-masked trials of brolucizumab for neovascular age-related macular degeneration. Ophthalmology 2020;127(1):72–84.

[55] Noël A, Jost M, Lambert V, Lecomte J, Rakic JM. Anti-angiogenic therapy of exudative age-related macular degeneration: current progress and emerging concepts. Trends Mol Med 2007;13(8):345–52.

[56] Gragoudas ES, Adamis AP, Cunningham Jr ET, Feinsod M, Guyer DR. Pegaptanib for neovascular age-related macular degeneration. N Engl J Med 2004;351(27):2805–16.

[57] Michels S, Rosenfeld PJ, Puliafito CA, Marcus EN, Venkatraman AS. Systemic bevacizumab (Avastin) therapy for neovascular age-related macular degeneration: twelve-week results of an uncontrolled open-label clinical study. Ophthalmology 2005;112(6):1035–47.

[58] Steinbrook R. The price of sight—ranibizumab, bevacizumab, and the treatment of macular degeneration. N Engl J Med 2006;355(14):1409–12.

[59] Nguyen QD, Shah SM, Hafiz G, Quinlan E, Sung J, Chu K, Cedarbaum JM, Campochiaro PA, CLEAR-AMD 1 Study Group. A phase I trial of an IV-administered vascular endothelial growth factor trap for treatment in patients with choroidal neovascularization due to age-related macular degeneration. Ophthalmology 2006;113(9):1522–e1.

[60] Reich SJ, Fosnot J, Kuroki A, Tang W, Yang X, Maguire AM, Bennett J, Tolentino MJ. Small interfering RNA (siRNA) targeting VEGF effectively inhibits ocular neovascularization in a mouse model. Mol Vis 2003;9(5):210–6.

[61] Shen J, Samul R, Silva RL, Akiyama H, Liu H, Saishin Y, Hackett SF, Zinnen S, Kossen K, Fosnaugh K, Vargeese C. Suppression of ocular neovascularization with siRNA targeting VEGF receptor 1. Gene Ther 2006;13(3):225–34.

[62] Campochiaro PA, Nguyen QD, Shah SM, Klein ML, Holz E, Frank RN, Saperstein DA, Gupta A, Stout JT, Macko J, DiBartolomeo R. Adenoviral vector-delivered pigment epithelium-derived factor for neovascular age-related macular degeneration: results of a phase I clinical trial. Hum Gene Ther 2006;17(2):167–76.

[63] Connolly B, Desai A, Garcia CA, Thomas E, Gast MJ. Squalamine lactate for exudative age-related macular degeneration. Ophthalmol Clin North Am 2006;19(3):381–91.

[64] Slakter JS, Bochow TW, D'Amico DJ, Marks B, Jerdan J, Sullivan EK, Robertson SM, Slakter JS, Sullins G, Zilliox P, Anecortave Acetate Clinical Study Group. Anecortave acetate (15 milligrams) versus photodynamic therapy for treatment of subfoveal neovascularization in age-related macular degeneration. Ophthalmology 2006;113 (1):3–13.

[65] Ferrara N, Damico L, Shams N, Lowman H, Kim R. Development of ranibizumab, an anti–vascular endothelial growth factor antigen binding fragment, as therapy for neovascular age-related macular degeneration. Retina 2006;26(8):859–70.

[66] Rosenfeld PJ, Brown DM, Heier JS, Boyer DS, Kaiser PK, Chung CY, Kim RY. Ranibizumab for neovascular age-related macular degeneration. N Engl J Med 2006;355(14):1419–31.

[67] Brown DM, Kaiser PK, Michels M, Soubrane G, Heier JS, Kim RY, Sy JP, Schneider S. Ranibizumab versus verteporfin for neovascular age-related macular degeneration. N Engl J Med 2006;355(14):1432–44.

[68] Heier JS, Brown DM, Chong V, Korobelnik JF, Kaiser PK, Nguyen QD, Kirchhof B, Ho A, Ogura Y, Yancopoulos GD, Stahl N. Intravitreal aflibercept (VEGF trap-eye) in wet age-related macular degeneration. Ophthalmology 2012;119(12):2537–48.

[69] Rosenfeld PJ, Moshfeghi AA, Puliafito CA. Optical coherence tomography findings after an intravitreal injection of bevacizumab (Avastin®) for neovascular age-related macular degeneration. Ophthalm Surgery Lasers Imaging Retina 2005;36(4):331–5.

[70] Chakravarthy U, Harding SP, Rogers CA, Downes SM, Lotery AJ, Wordsworth S, Reeves BC, IVAN Study Investigators. Ranibizumab versus bevacizumab to treat neovascular age-related macular degeneration: one-year findings from the IVAN randomized trial. Ophthalmology 2012;119(7):1399–411.

[71] Martin DF, Maguire MG, Ying GA, Grunwald JE, Fine SL, Jaffe GJ. Ranibizumab and bevacizumab for neovascular age-related macular degeneration. N Engl J Med 2011;364(20):1897–908.

[72] Ferrara N, Hillan KJ, Gerber HP, Novotny W. Discovery and development of bevacizumab, an anti-VEGF antibody for treating cancer. Nat Rev Drug Discov 2004;3(5):391–400.

[73] Tolentino MJ, Husain D, Theodosiadis P, Gragoudas ES, Connolly E, Kahn J, Cleland J, Adamis AP, Cuthbertson A, Miller JW. Angiography of fluoresceinated anti–vascular endothelial growth factor antibody and dextrans in experimental choroidal neovascularization. Arch Ophthalmol 2000;118(1):78–84.

[74] Moshfeghi AA, Rosenfeld PJ, Puliafito CA, Michels S, Marcus EN, Lenchus JD, Venkatraman AS. Systemic bevacizumab (Avastin) therapy for neovascular age-related macular degeneration: twenty-four–week results of an uncontrolled open-label clinical study. Ophthalmology 2006;113(11):2002–11.

[75] Avery RL, Pieramici DJ, Rabena MD, Castellarin AA, Ma'an AN, Giust MJ. Intravitreal bevacizumab (Avastin) for neovascular age-related macular degeneration. Ophthalmology 2006;113(3):363–72.

[76] Rosenfeld PJ, Browning DJ. Is this a 737 Max moment for brolucizumab? Am J Ophthalmol 2020;216:A7–8.

[77] Baumal CR, Spaide RF, Vajzovic L, Freund KB, Walter SD, John V, Rich R, Chaudhry N, Lakhanpal RR, Oellers PR, Leveque TK. Retinal vasculitis and intraocular inflammation after intravitreal injection of brolucizumab. Ophthalmology 2020;127(10):1345–59.

[78] Bokoch GM. Biology of the rap proteins, members of the ras superfamily of GTP-binding proteins. Biochem J 1993;289:17–24.

[79] Zwartkruis FJ, Bos JL. Ras and Rap1: two highly related small GTPases with distinct function. Exp Cell Res 1999;253:157–65.

[80] Wittchen ES, Worthylake RA, Kelly P, Casey PJ, Quilliam LA, Burridge K. Rap1 GTPase inhibits leukocyte transmigration by promoting endothelial barrier function. J Biol Chem 2005;280:11675–82.

[81] Wittchen ES, Hartnett ME. The small GTPase Rap1 is a novel regulator of RPE cell barrier function. Investig Ophthalmol Vis Sci 2011;52:7455–63.

[82] Jabbehdari S, Handa JT. Oxidative stress as a therapeutic target for the prevention and treatment of early age-related macular degeneration. Surv Ophthalmol 2021;66(3):423–40.

[83] Klein R, Lee KE, Tsai MY, Cruickshanks KJ, Gangnon RE, Klein BEK. Oxidized low-density lipoprotein and the incidence of age-related macular degeneration. Ophthalmology 2019;126:752–8.

[84] Klein R, Klein BE, Knudtson MD, et al. Subclinical atherosclerotic cardiovascular disease and early age-related macular degeneration in a multiracial cohort: the Multiethnic Study of Atherosclerosis. Arch Ophthalmol 2007;125:534–43.

[85] Datta S, Cano M, Ebrahimi K, Wang L, Handa JT. The impact of oxidative stress and inflammation on RPE degeneration in non-neovascular AMD. Prog Retin Eye Res 2017;60:201–18.

[86] SanGiovanni JP, Arking DE, Iyengar SK, Elashoff M, Clemons TE, Reed GF, Henning AK, Sivakumaran TA, Xu X, DeWan A, Agrón E. Mitochondrial DNA variants of respiratory complex I that uniquely characterize haplogroup T2 are associated with increased risk of age-related macular degeneration. PloS One 2009;4(5), e5508.

[87] Canter JA, Olson LM, Spencer K, Schnetz-Boutaud N, Anderson B, Hauser MA, Schmidt S, Postel EA, Agarwal A, Pericak-Vance MA, Sternberg Jr P. Mitochondrial DNA polymorphism A4917G is independently associated with age-related macular degeneration. PloS One 2008;3(5), e2091.

[88] Cai J, Nelson KC, Wu M, Sternberg P, Jones DP. Oxidative damage and protection of the RPE. Prog Retin Eye Res 2000;19:205–22.

[89] Cai J, Wu M, Nelson K, Jones DP, Sternberg Jr P. Oxidant induced apoptosis in cultured human retinal pigment epithelial cells. Invest Ophthalmol Vis Sci 1999;40:959–66.

[90] Li ZY, Tso MO, Wang HM, Organisciak DT. Amelioration of photic injury in rat retina by ascorbic acid: a histopathologic study. Invest Ophthalmol Vis Sci 1985;26:1589–98.

[91] Cho E, Hankinson SE, Rosner B, Willett WC, Colditz GA. Prospective study of lutein/zeaxanthin intake and risk of age-related macular degeneration. Am J Clin Nutr 2008;87:1837–43.

[92] Flood V, Smith W, Wang JJ, Manzi F, Webb K, Mitchell P. Dietary antioxidant intake and incidence of early age-related maculopathy: the Blue Mountains Eye Study. Ophthalmology 2002;109:2272–8.

[93] West S, Vitale S, Hallfrisch J, et al. Are antioxidants or supplements protective for age-related macular degeneration? Arch Ophthalmol 1994;112:222–7.

[94] VandenLangenberg GM, Mares-Perlman JA, Klein R, Klein BEK, Brady WE, Palta M. Associations between antioxidant and zinc intake and the 5-year incidence of early age related maculopathy in the Beaver Dam Eye Study. Am J Epidemiol 1998;148:204–14.

[95] Tan JS, Wang JJ, Flood V, Rochtchina E, Smith W, Mitchell P. Dietary antioxidants and the long-term incidence of age-related macular degeneration: the Blue Mountains Eye Study. Ophthalmology 2008;115:334–41.

[96] Merle BMJ, Silver RE, Rosner B, Seddon JM. Associations between vitamin D intake and progression to incident advanced age-related macular degeneration. Invest Ophthalmol Vis Sci 2017;58:4569–78.

[97] Evans J. Antioxidant supplements to prevent or slow down the progression of AMD: a systematic review and meta-analysis. Eye 2008;22(6):751–60.

[98] Albanes D, Heinonen OP, Taylor PR, Virtamo J, Edwards BK, Rautalahti M, Hartman AM, Palmgren J, Freedman LS, Haapakoski J, Barrett MJ. Tocopherol and beta-carotene supplements and lung cancer incidence in the alpha-tocopherol, beta-carotene cancer prevention study: effects of base-line characteristics and study compliance. JNCI J Nat Cancer Inst 1996;88(21):1560–70.

[99] Wu J, Cho E, Willett WC, Sastry SM, Schaumberg DA. Intakes of lutein, zeaxanthin, and other carotenoids and age related macular degeneration during 2 decades of prospective follow-up. JAMA Ophthalmol 2015;133:1415–24.

[100] Ouyang Y, Heussen FM, Hariri A, Keane PA, Sadda SR. Optical coherence tomography-based observation of the natural history of drusenoid lesion in eyes with dry age-related macular degeneration. Ophthalmology 2013;120:2656–65.

[101] Krinsky NI, Landrum JT, Bone RA. Biologic mechanisms of the protective role of lutein and zeaxanthin in the eye. Annu Rev Nutr 2003;23:171–201.

[102] Korobelnik JF, Rougier MB, Delyfer MN, Bron A, Merle BM, Savel H, Chêne G, Delcourt C, Creuzot-Garcher C. Effect of dietary supplementation with lutein, zeaxanthin, and ω-3 on macular pigment: a randomized clinical trial. JAMA Ophthalmol 2017;135(11):1259–66.

[103] Conrady CD, Bell JP, Besch BM, Gorusupudi A, Farnsworth K, Ermakov I, Sharifzadeh M, Ermakova M, Gellermann W, Bernstein PS. Correlations between macular, skin, and serum carotenoids. Invest Ophthalmol Vis Sci 2017;58(9):3616–27.

[104] Smailhodzic D, van Asten F, Blom AM, Mohlin FC, den Hollander AI, van de Ven JP, van Huet RA, Groenewoud JM, Tian Y, Berendschot TT, Lechanteur YT. Zinc supplementation inhibits complement activation in age-related macular degeneration. PloS One 2014;9(11), e112682.

[105] Gonzalez-Iglesias H, Alvarez L, Garcia M, Petrash C, Sanz Medel A, Coca-Prados M. Metallothioneins (MTs) in the human eye: a perspective article on the zinc-MT redox cycle. Metallomics 2014;6:201–8.

[106] Ugarte M, Osborne NN. Recent advances in the understanding of the role of zinc in ocular tissues. Metallomics 2014;6:189–200.

[107] Vishwanathan R, Chung M, Johnson EJ. A systematic review on zinc for the prevention and treatment of age-related macular degeneration. Invest Ophthalmol Vis Sci 2013;54 (6):3985–98.

[108] Tan W, Zou J, Yoshida S, Jiang B, Zhou Y. The role of inflammation in age-related macular degeneration. Int J Biol Sci 2020;16(15):2989.

[109] Penfold PL, Provis JM, Billson FA. Age-related macular degeneration: ultra-structural studies of the relationship of leucocytes to angiogenesis. Graefes Arch Clin Exp Ophthalmol 1987;225:70–6.

[110] Hageman GS, Luthert PJ, Chong VNH, Johnson LV, Anderson DH, Mullins RF. An integrated hypothesis that considers drusen as biomarkers of immune mediated processes at the RPE–Bruch's membrane interface in aging and age related macular degeneration. Prog Retin Eye Res 2001;20:705–32.

[111] Johnson LV, Leitner WP, Staples MK, Anderson DH. Complement activation and inflammatory processes in drusen formation and age-related macular degeneration. Exp Eye Res 2001;73:887–96.

[112] Anderson DH, Mullins RF, Hageman GS, Johnson LV. A role for local inflammation in the formation of drusen in the aging eye. Am J Ophthalmol 2002;134:411–31.

[113] Johnson L, Ozaki S, Staples M, Erickson P, Anderson D. A potential role for immune complex pathogenesis in drusen formation. Exp Eye Res 2000;70:441–9.

[114] Spindler J, Zandi S, Pfister IB, et al. Cytokine profiles in the aqueous humor and serum of patients with dry and treated wet age-related macular degeneration. PloS One 2018;13, e0203337.

[115] Zhao M, Bai Y, Xie W, et al. Interleukin-1beta level is increased in vitreous of patients with neovascular age-related macular degeneration (nAMD) and polypoidal choroidal vasculopathy (PCV). PloS One 2015;10, e0125150.

[116] Jermak CM, Dellacroce JT, Heffez J, Peyman GA. Triamcinolone acetonide in ocular therapeutics. Surv Ophthalmol 2007;52:503–22.

[117] Cronstein BN, Kimmel SC, Levin RI, Martiniuk F, Weissmann G. A mechanism for the anti inflammatory effects of corticosteroids: the glucocorticoid receptor regulates leukocyte adhesion to endothelial cells and expression of endothelial-leukocyte adhesion molecule 1 and intercellular adhesion molecule 1. Proc Natl Acad Sci U S A 1992;89:9991–5.

[118] Penfold PL, Wong JG, Gyory J, Billson FA. Effects of triamcinolone acetonide on microglial morphology and quantitative expression of MHC-II in exudative age-related macular degeneration. Clin Exp Ophthalmol 2001;29:188–92.

[119] Bakri SJ, Couch SM, McCannel CA, Edwards AO. Same day triple therapy with photodynamic therapy, intravitreal dexamethasone, and bevacizumab in wet age-related macular degeneration. Retina 2009;29:573–8.

[120] Augustin AJ, Puls S, Offermann I. Triple therapy for choroidal neovascularization due to age-related macular degeneration: verteporfin PDT, bevacizumab, and dexamethasone. Retina 2007;27:133–40.

[121] Ehmann D, Garcia R. Triple therapy for neovascular age-related macular degeneration (verteporfin photodynamic therapy, intravitreal dexamethasone, and intravitreal bevacizumab). Can J Ophthalmol 2010;45:36–40.

[122] Graham RO, Peyman GA. Intravitreal injection of dexamethasone. Treatment of experimentally induced endophthalmitis. Arch Ophthalmol 1974;92:149–54.

[123] Francis BA, Chang EL, Haik BG. Particle size and drug interactions of injectable corticosteroids used in ophthalmic practice. Ophthalmology 1996;103:1884–8.

[124] Gillies MC, Simpson JM, Luo W, Penfold P, Hunyor AB, Chua W, Mitchell P, Billson F. A randomized clinical trial of a single dose of intravitreal triamcinolone acetonide for neovascular age-related macular degeneration: one-year results. Arch Ophthalmol 2003;121(5):667–73.

[125] Danis RP, Ciulla TA, Pratt LM, Anliker W. Intravitreal triamcinolone acetonide in exudative age-related macular degeneration. Retina 2000;20:244–50.

[126] Kim SJ, Flach AJ, Jampol LM. Nonsteroidal anti-inflammatory drugs in ophthalmology. Surv Ophthalmol 2010;55(2):108–33.

[127] Luan M, Wang H, Wang J, Zhang X, Zhao F, Liu Z, Meng Q. Advances in anti-inflammatory activity, mechanism and therapeutic application of ursolic acid. Mini Rev Med Chem 2022;22(3):422–36.

[128] Vane JR, Bakhle YS, Botting RM. Cyclooxygenases 1 and 2. Annu Rev Pharmacol Toxicol 1998;38:97–120.

[129] Botting RM. Vane's discovery of the mechanism of action of aspirin changed our understanding of its clinical pharmacology. Pharmacol Rep 2010;62:518–25.

[130] Maloney SC, Fernandes BF, Castiglione E, Antecka E, Martins C, Marshall JC, Di Cesare S, Logan P, BurnierJr MN. Expression of cyclooxygenase-2 in choroidal neovascular membranes from age-related macular degeneration patients. Retina 2009;29(2):176–80.

[131] Gamache DA, Graff G, Brady MT, Spellman JM, Yanni JM. Nepafenac, a unique nonsteroidal prodrug with potential utility in the treatment of trauma-induced ocular inflammation: I. Assessment of anti-inflammatory efficacy. Inflammation 2000;24:357–70.

[132] Chakraborti AK, Garg SK, Kumar R, Motiwala HF, Jadhavar PS. Progress in COX-2 inhibitors: a journey so far. Curr Med Chem 2010;17(15):1563–93.

[133] Ku EC, Lee W, Kothari HV, Scholer DW. Effect of diclofenac sodium on the arachidonic acid cascade. Am J Med 1986;80:18–23.

[134] McGeer PL, Sibley J. Sparing of age-related macular degeneration in rheumatoid arthritis. Neurobiol Aging 2005;26:1199–203.

[135] Libondi T, Jonas JB. Topical nepafenac for treatment of exudative age-related macular degeneration. Acta Ophthalmol 2010;88:e32–3.

[136] Baranano DE, Kim SJ, Edelhauser HF, Durairaj C, Kompella UB, Handa JT. Efficacy and pharmacokinetics of intravitreal non-steroidal anti-inflammatory drugs for intraocular inflammation. Br J Ophthalmol 2009;93:1387–90.

[137] Soheilian M, Karimi S, Ramezani A, Peyman GA. Pilot study of intravitreal injection of diclofenac for treatment of macular edema of various etiologies. Retina 2010;30:509–15.

[138] Grimes KR, Aloney A, Skondra D, Chhablani J. Effects of systemic drugs on the development and progression of age-related macular degeneration. Surv Ophthalmol 2023.

[139] Wang K, Zheng M, Lester KL, Han Z. Light-induced Nrf2−/− mice as atrophic age-related macular degeneration model and treatment with nanoceria laden injectable hydrogel. Sci Rep 2019;9(1):14573.

[140] Zweifel SA, Engelbert M, Khan S, Freund KB. Retrospective review of the efficacy of topical bromfenac (0.09%) as an adjunctive therapy for patients with neovascular age-related macular degeneration. Retina 2009;29:1527–31.

[141] Boyer DS, Beer PM, Joffe L, Koester JM, Marx JL, Weisberger A, Yoser SL. Effect of adjunctive diclofenac with verteporfin therapy to treat choroidal neovascularization due to age-related macular degeneration: phase II study. Retina (Philadelphia, Pa) 2007;27 (6):693–700.

[142] Christen WG, Glynn RJ, Chew EY, Buring JE. Low-dose aspirin and medical record-confirmed age-related macular degeneration in a randomized trial of women. Ophthalmology 2009;116:2386–92.

[143] Kurup SK, Gee C, Greven CM. Intravitreal methotrexate in therapeutically resistant exudative age-related macular degeneration. Acta Ophthalmol 2010;88:e145–6.

[144] Zohlnhofer D, Nuhrenberg TG, Neumann FJ, Richter T, May AE, Schmidt R, et al. Rapamycin effects transcriptional programs in smooth muscle cells controlling proliferative and inflammatory properties. Mol Pharmacol 2004;65:880–9.

[145] Theodossiadis PG, Liarakos VS, Sfikakis PP, Vergados IA, Theodossiadis GP. Intravitreal administration of the anti-tumor necrosis factor agent infliximab for neovascular age-related macular degeneration. Am J Ophthalmol 2009;147:825–30. e821.

[146] Goebel J, Stevens E, Forrest K, Roszman TL. Daclizumab (Zenapax) inhibits early interleukin-2 receptor signal transduction events. Transpl Immunol 2000;8:153–9.

[147] Gehrs KM, Jackson JR, Brown EN, Allikmets R, Hageman GS. Complement, age-related macular degeneration and a vision of the future. Arch Ophthalmol 2010;128:349–58.

[148] Ricklin D, Lambris JD. Complement-targeted therapeutics. Nat Biotechnol 2007;25 (11):1265–75.

[149] Kuppermann BD, Patel SS, Boyer DS, Augustin AJ, Freeman WR, Kerr KJ, Guo Q, Schneider S, López FJ. Phase 2 study of the safety and efficacy of brimonidine drug delivery system (Brimo DDS) generation 1 in patients with geographic atrophy secondary to age-related macular degeneration. Retina 2021;41(1):144–55.

[150] Freeman WR, Bandello F, Souied E, Guymer RH, Garg SJ, Chen FK, Rich R, Holz FG, Patel SS, Kim K, López FJ. Randomized phase IIb study of brimonidine drug delivery system generation 2 for geographic atrophy in age-related macular degeneration. Ophthalmol Retina 2023;7(7):573–85.

[151] LaVail MM, Yasumura D, Matthes MT, Lau-Villacorta C, Unoki K, Sung CH, Steinberg RH. Protection of mouse photoreceptors by survival factors in retinal degenerations. Investig Ophthalmol Vis Sci 1998;39:592–602.

[152] Stahl N, Yancopoulos GD. The tripartite CNTF receptor complex: activation and signaling involves components shared with other cytokines. J Neurobiol 1994;25:1454–66.

[153] Tao W. Application of encapsulated cell technology for retinal degenerative diseases. Expert Opin Biol Ther 2006;6:717–26.

[154] Zhang K, Hopkins JJ, Heier JS, Birch DG, Halperin LS, Albini TA, Brown DM, Jaffe GJ, Tao W, Williams GA. Ciliary neurotrophic factor delivered by encapsulated cell

intraocular implants for treatment of geographic atrophy in age-related macular degeneration. Proc Natl Acad Sci U S A 2011;108:6241–5.

[155] Campochiaro PA, Marcus DM, Awh CC, Regillo C, Adamis AP, Bantseev V, Chiang Y, Ehrlich JS, Erickson S, Hanley WD, Horvath J. The port delivery system with ranibizumab for neovascular age-related macular degeneration: results from the randomized phase 2 ladder clinical trial. Ophthalmology 2019;126(8):1141–54.

[156] Rubio RG. Long-acting anti-VEGF delivery. Retina Today 2014;2014:78–80.

[157] Holekamp NM, Campochiaro PA, Chang MA, Miller D, Pieramici D, Adamis AP, Brittain C, Evans E, Kaufman D, Maass KF, Patel S. Archway randomized phase 3 trial of the port delivery system with ranibizumab for neovascular age-related macular degeneration. Ophthalmology 2022;129(3):295–307.

[158] Rakoczy EP, Lai CM, Magno AL, Wikstrom ME, French MA, Pierce CM, Schwartz SD, Blumenkranz MS, Chalberg TW, Degli-Esposti MA, Constable IJ. Gene therapy with recombinant adeno-associated vectors for neovascular age-related macular degeneration: 1 year follow-up of a phase 1 randomised clinical trial. Lancet 2015;386(10011):2395–403.

[159] Constable IJ, Pierce CM, Lai CM, Magno AL, Degli-Esposti MA, French MA, McAllister IL, Butler S, Barone SB, Schwartz SD, Blumenkranz MS. Phase 2a randomized clinical trial: safety and post hoc analysis of subretinal rAAV. sFLT-1 for wet age-related macular degeneration. EBioMedicine 2016;14:168–75.

[160] Nashine S. Potential therapeutic candidates for age-related macular degeneration (AMD). Cell 2021;10(9):2483.

[161] Schlottmann PG, Alezzandrini AA, Zas M, Rodriguez FJ, Luna JD, Wu L. New treatment modalities for neovascular age-related macular degeneration. Asia-Pacific J Ophthalmol 2017;6(6):514–9.

[162] Chen ER, Kaiser PK. Therapeutic potential of the ranibizumab port delivery system in the treatment of AMD: evidence to date. Clin Ophthalmol 2020;14:1349–55.

[163] Stevenson CL, SantiniJr JT, Langer R. Reservoir-based drug delivery systems utilizing microtechnology. Adv Drug Deliv Rev 2012;64(14):1590–602.

[164] Kim S, Kang-Mieler JJ, Liu W, Wang Z, Yiu G, Teixeira LB, Mieler WF, Thomasy SM. Safety and biocompatibility of aflibercept-loaded microsphere thermo-responsive hydrogel drug delivery system in a nonhuman primate model. Transl Vis Sci Technol 2020;9(3):30.

[165] Seah I, Zhao X, Lin Q, Liu Z, Su SZ, Yuen YS, Hunziker W, Lingam G, Loh XJ, Su X. Use of biomaterials for sustained delivery of anti-VEGF to treat retinal diseases. Eye 2020;34(8):1341–56.

[166] Geerlings MJ, de Jong EK, den Hollander AI. The complement system in age-related macular degeneration: a review of rare genetic variants and implications for personalized treatment. Mol Immunol 2017;84:65–76.

[167] Schwartz SD, Regillo CD, Lam BL, Eliott D, Rosenfeld PJ, Gregori NZ, Hubschman JP, Davis JL, Heilwell G, Spirn M, Maguire J. Human embryonic stem cell-derived retinal pigment epithelium in patients with age-related macular degeneration and Stargardt's macular dystrophy: follow-up of two open-label phase 1/2 studies. Lancet 2015;385(9967):509–16.

[168] Clinical Trial. Study of RO7250284 in participants with neovascular age-related macular degeneration. Identifier NCT04567303, 2023. Available online: https://clinicaltrials.gov/ct2/show/NCT04567303. [Accessed 18 April 2023].

[169] Heier JS, Kherani S, Desai S, Dugel P, Kaushal S, Cheng SH, Delacono C, Purvis A, Richards S, Le-Halpere A, Connelly J. Intravitreous injection of AAV2-sFLT01 in

patients with advanced neovascular age-related macular degeneration: a phase 1, open-label trial. Lancet 2017;390(10089):50–61.

[170] Tsai JC. Canadian Journal of Ophthalmology Lecture: translational research advances in glaucoma neuroprotection. Can J Ophthalmol 2013;48(3):141–5.

[171] ADVERUM. Adverum Biotechnologies Presents Long-term Data through March 10, 2021 from the OPTIC Trial of ADVM-022 Intravitreal Gene Therapy in Treatment-experienced Wet AMD Patients at ARVO 2021., 2021, https://investors.adverum.com/news/news-details/2021/Adverum-Biotechnologies-Presents-Long-term-Data-through-March-10-2021-from-the-OPTIC-Trial-of-ADVM-022-Intravitreal-Gene-Therapy-in-Treatment-experienced-Wet-AMD-Patients-at-ARVO-2021-2021-5-1-2021-5-1/default.aspx. [Accessed 18 April 2023].

[172] ClinicalTrials.gov. Complement inhibition with eculizumab for the treatment of non-exudative macular degeneration (AMD) (COMPLETE). NLM identifier: NCT00935883, 2023. Available from: https://clinicaltrials.gov/ct2/show/NCT00935883. [Accessed 28 July 2017].

[173] ClinicalTrials.gov. A phase 1, safety, tolerability and pharmacokinetic profile of intra-vitreous injections of E10030 (anti-PDGF pegylated aptamer) in subjects with neo-vascular age-related macular degeneration. NLM identifier: NCT00569140, 2023. Available from: https://clinicaltrials.gov/show/NCT00569140. [Accessed 28 July 2017].

[174] ClinicalTrials.gov. A safety and efficacy study of E10030 (anti-PDGF pegylated aptamer) plus lucentis for neovascular age-related macular degeneration. NLM identi-fier: NCT01089517, 2023. Available from: https://clinicaltrials.gov/ct2/show/results/NCT01089517?term=E10030&rank=3. [Accessed 28 July 2017].

[175] ClinicalTrials.gov. Proton radiation therapy for macular degeneration. NLM identi-fier: NCT01833325, 2023. Available from: https://clinicaltrials.gov/ct2/show/NCT01833325?term=proton+radiation&recrs=d&cond=AMD&rank=1. [Accessed 28 July 2017].

[176] ClinicalTrials.gov. A safety and efficacy study of abicipar pegol in patients with neo-vascular age-related macular degeneration (CDER). NLM identifier: NCT02462928, 2023. Available from: https://clinicaltrials.gov/ct2/show/NCT02462928?term=Abicipar+pegol&recrs=d&cond=AMD&rank=3. [Accessed 28 July 2017].

[177] ClinicalTrials.gov. Efficacy and safety of RTH258 versus aflibercept – study 2. NLM identifier: NCT02434328, 2023. Available from: https://clinicaltrials.gov/ct2/show/NCT02434328?term=RTH258&recrs=d&cond=AMD&rank=1. [Accessed 28 July 2017].

[178] Ponnusamy C, Ayarivan P, Selvamuthu P, Natesan S. Age-related macular degeneration-therapies and their delivery. Curr Drug Deliv 2023.

[179] Elsaid N, Somavarapu S, Jackson TL. Cholesterol-poly (ethylene) glycol nanocarriers for the transscleral delivery of sirolimus. Exp Eye Res 2014;121:121–9.

[180] Behroozi F, Abdkhodaie MJ, Abandansari HS, Satarian L, Ashtiani MK, Jaafari MR, Baharvand H. Smart liposomal drug delivery for treatment of oxidative stress model in human embryonic stem cell-derived retinal pigment epithelial cells. Int J Pharm 2018;548(1):62–72.

[181] Joseph RR, Tan DW, Ramon MR, Natarajan JV, Agrawal R, Wong TT, Venkatraman SS. Characterization of liposomal carriers for the trans-scleral transport of ranibizumab. Sci Rep 2017;7(1):16803.

[182] Mu H, Wang Y, Chu Y, Jiang Y, Hua H, Chu L, Wang K, Wang A, Liu W, Li Y, Fu F. Multivesicular liposomes for sustained release of bevacizumab in treating laser-induced choroidal neovascularization. Drug Deliv 2018;25(1):1372–83.

[183] Vaishya RD, Gokulgandhi M, Patel S, Minocha M, Mitra AK. Novel dexamethasone-loaded nanomicelles for the intermediate and posterior segment uveitis. AAPS PharmSciTech 2014;15:1238–51.

[184] Ma F, Nan K, Lee S, Beadle JR, Hou H, Freeman WR, Hostetler KY, Cheng L. Micelle formulation of hexadecyloxypropyl-cidofovir (HDP-CDV) as an intravitreal long-lasting delivery system. Eur J Pharm Biopharm 2015;89:271–9.

[185] Alshamrani M, Sikder S, Coulibaly F, Mandal A, Pal D, Mitra AK. Self-assembling topical nanomicellar formulation to improve curcumin absorption across ocular tissues. Aaps Pharmscitech 2019;20:1–6.

[186] Gote V, Mandal A, Alshamrani M, Pal D. Self-assembling tacrolimus nanomicelles for retinal drug delivery. Pharmaceutics 2020;12(11):1072.

[187] Vaishya RD, Khurana V, Patel S, Mitra AK. Controlled ocular drug delivery with nanomicelles. Wiley Interdiscip Rev Nanomed Nanobiotechnol 2014;6(5):422–37.

[188] Hagigit T, Abdulrazik M, Valamanesh F, Behar-Cohen F, Benita S. Ocular antisense oligonucleotide delivery by cationic nanoemulsion for improved treatment of ocular neovascularization: an in-vivo study in rats and mice. J Control Release 2012;160 (2):225–31.

[189] Patel N, Nakrani H, Raval M, Sheth N. Development of loteprednoletabonate-loaded cationic nanoemulsified in-situ ophthalmic gel for sustained delivery and enhanced ocular bioavailability. Drug Deliv 2016;23(9):3712–23.

[190] Ge Y, Zhang A, Sun R, Xu J, Yin T, He H, Gou J, Kong J, Zhang Y, Tang X. Penetratin-modified lutein nanoemulsion in-situ gel for the treatment of age-related macular degeneration. Expert Opin Drug Deliv 2020;17(4):603–19.

[191] Lim C, Kim DW, Sim T, Hoang NH, Lee JW, Lee ES, Youn YS, Oh KT. Preparation and characterization of a lutein loading nanoemulsion system for ophthalmic eye drops. J Drug Deliv Sci Technol 2016;36:168–74.

[192] Laradji AM, Kolesnikov AV, Karakocak BB, Kefalov VJ, Ravi N. Redox-responsive hyaluronic acid-based nanogels for the topical delivery of the visual chromophore to retinal photoreceptors. ACS Omega 2021;6(9):6172–84.

[193] Du M, Shen S, Liang L, Xu K, He A, Yao Y, Liu S. Evaluations of the Chuanqi ophthalmic microemulsion in situ gel on dry age-related macular degeneration treatment. Evid Based Complement Alternat Med 2020;2020:1–4.

[194] Bolla PK, Gote V, Singh M, Patel M, Clark BA, Renukuntla J. Lutein-loaded, biotin-decorated polymeric nanoparticles enhance lutein uptake in retinal cells. Pharmaceutics 2020;12(9):798.

[195] Narvekar P, Bhatt P, Fnu G, Sutariya V. Axitinib-loaded poly (lactic-co-glycolic acid) nanoparticles for age-related macular degeneration: formulation development and in vitro characterization. Assay Drug Dev Technol 2019;17(4):167–77.

[196] Liu J, Zhang X, Li G, Xu F, Li S, Teng L, Li Y, Sun F. Anti-angiogenic activity of bevacizumab-bearing dexamethasone-loaded PLGA nanoparticles for potential intravitreal applications. Int J Nanomedicine 2019;88:19–34.

[197] Zhang L, Si T, Fischer AJ, Letson A, Yuan S, Roberts CJ, Xu RX. Coaxial electrospray of ranibizumab-loaded microparticles for sustained release of anti-VEGF therapies. PloS One 2015;10(8), e0135608.

[198] Loftsson T, Duchene D. Cyclodextrins and their pharmaceutical applications. Int J Pharm 2007;329(1–2):1.

[199] Kam JH, Lynch A, Begum R, Cunea A, Jeffery G. Topical cyclodextrin reduces amyloid beta and inflammation improving retinal function in ageing mice. Exp Eye Res 2015;135:59–66.

[200] El-Darzi N, Mast N, Petrov AM, Pikuleva IA. 2-Hydroxypropyl-β-cyclodextrin reduces retinal cholesterol in wild-type and Cyp27a1−/− Cyp46a1−/− mice with deficiency in the oxysterol production. Br J Pharmacol 2021;178(16):3220–34.

[201] Kaur IP, Chhabra S, Aggarwal D. Role of cyclodextrins in ophthalmics. Curr Drug Deliv 2004;1(4):351–60.

[202] Marano RJ, Toth I, Wimmer N, Brankov M, Rakoczy PE. Dendrimer delivery of an anti-VEGF oligonucleotide into the eye: a long-term study into inhibition of laser-induced CNV, distribution, uptake and toxicity. Gene Ther 2005;12(21):1544–50.

[203] Yavuz B, Bozdağ Pehlivan S, Sümer Bolu B, Nomak Sanyal R, Vural İ, Ünlü N. Dexamethasone–PAMAM dendrimer conjugates for retinal delivery: preparation, characterization and in vivo evaluation. J Pharm Pharmacol 2016;68(8):1010–20.

[204] Lai S, Wei Y, Wu Q, Zhou K, Liu T, Zhang Y, Jiang N, Xiao W, Chen J, Liu Q, Yu Y. Liposomes for effective drug delivery to the ocular posterior chamber. J Nanobiotechnol 2019;17:1–2.

[205] Yao WJ, Sun KX, Liu Y, Liang N, Mu HJ, Yao C, Liang RC, Wang AP. Effect of poly (amidoamine) dendrimers on corneal penetration of puerarin. Biol Pharm Bull 2010;33 (8):1371–7.

[206] Jiang Y, Krishnan N, Heo J, Fang RH, Zhang L. Nanoparticle–hydrogel superstructures for biomedical applications. J Control Release 2020;324:505–21.

[207] Xin G, Zhang M, Zhong Z, Tang L, Feng Y, Wei Z, Li S, Li Y, Zhang J, Zhang B, Zhang M. Ophthalmic drops with nanoparticles derived from a natural product for treating age-related macular degeneration. ACS Appl Mater Interfaces 2020;12(52):57710–20.

[208] Hirani A, Grover A, Lee YW, Pathak Y, Sutariya V. Triamcinolone acetonide nanoparticles incorporated in thermoreversible gels for age-related macular degeneration. Pharm Dev Technol 2016;21(1):61–7.

[209] Velilla S, García-Medina JJ, García-Layana A, Dolz-Marco R, Pons-Vázquez S, Pinazo-Durán MD, Gómez-Ulla F, Arévalo JF, Díaz-Llopis M, Gallego-Pinazo R. Smoking and age-related macular degeneration: review and update. J Ophthalmol 2013;2013.

[210] Lee S, Song SJ, Yu HG. Current smoking is associated with a poor visual acuity improvement after intravitreal ranibizumab therapy in patients with exudative age-related macular degeneration. J Korean Med Sci 2013;28:769–74.

[211] ClinicalTrials.gov. Intravitreal adalimumab in patients with choroidal neovascularization secondary to age-related macular degeneration. Identifier: NCT01136252, 2010. Available at http://www.clinicaltrial.gov/ct2/show/NCT01136252. [Accessed 9 October 2010].

[212] ClinicalTrials.gov. Complement inhibition with eculizumab for the treatment of non-exudative macular degeneration (AMD) (COMPLETE). Identifier: NCT00935883, 2009. Available at: http://www.clinicaltrial.gov/ct2/show/NCT00935883. [Accessed 9 October 2010].

[213] ClinicalTrials.gov. Safety of intravitreal POT-4 therapy for patients with neovascular age-related macular degeneration (AMD) (ASaP). Identifier: NCT00473928, 2010. Available at http://www.clinicaltrial.gov/ct2/show/NCT00473928. [Accessed 9 October 2010].

[214] Clinical Trial. A study of the response to treatment after transition to the port delivery system with ranibizumab [susvimo (ranibizumab injection)] in patients with neovascular age-related macular degeneration previously treated with intravitreal agents other than ranibizumab (belvedere). Identifier NTC04853251, 2023. Available online: https://clinicaltrials.gov/ct2/show/NCT04853251. [Accessed 18 April 2023].

Exploring the potential role of nanotechnology as cutting-edge for management of hirsutism and gynecomastia: A paradigm in therapeutics

17

Neelam Sharma[a], Sonam Grewal[a], Sukhbir Singh[a], Sumeet Gupta[b], Tapan Behl[c], and Ishrat Zahoor[a]
[a]Department of Pharmaceutics, MM College of Pharmacy, Maharishi Markandeshwar (Deemed to be University), Mullana-Ambala, Haryana, India, [b]Department of Pharmacology, MM College of Pharmacy, Maharishi Markandeshwar (Deemed to be University), Mullana-Ambala, Haryana, India, [c]Amity School of Pharmaceutical Sciences, Amity University, Mohali, Punjab, India

1 Introduction

Women experience a wide range of hair development variations therefore it is critical to distinguish between normal variability and hirsutism. Though Mediterranean women typically have somewhat heavy body hair, most Asian and Native American women tend to have very little [1]. Excessive terminal hair growth in androgen-dependent parts of a woman's body is known as hirsutism. Particularly, the lip, sideburn region, chin, and chest are places where hair grows [2]. It is important to distinguish hirsutism from hypertrichosis, a broad excessive hair growth disorder not brought on by an excess of androgen. Hypertrichosis can be inherited or brought on by metabolic conditions such as thyroid problems, anorexia nervosa, and porphyria [3]. Because there is such a wide difference in attitudes toward facial hair and body for personal and societal reasons, very few hirsute women will seek medical assistance. Consider the actuality that more than 95% of women who have hirsutism have a relatively benign condition like PCOS or idiopathic hirsutism, which may also be caused by more serious conditions like congenital adrenal hyperplasia, Cushing's syndrome, and benign and malignant androgen-secreting adrenal or ovarian tumors [4]. Between 5% and 15% of the women surveyed have hirsutism, which is the presence of terminal (coarse) hairs in a male-like pattern and their presence causes patients great distress and has a negative impact on their psychosocial development [5]. The most prevalent benign breast ailment in men is gynecomastia, which is defined as a generalized expansion of the male breast. The typical male breast has a few ducts imbedded in loose connective tissue and resembles the prepubertal female breast histologically. Gynecomastia is a condition that affects

Targeting Angiogenesis, Inflammation and Oxidative Stress in Chronic Diseases. https://doi.org/10.1016/B978-0-443-13587-3.00015-1

40%–65% of adult men and is characterized by an enlarged male breast that is sensitive and symmetrical [6]. The development of the breast ducts in both sexes is directly influenced by estrogen, whereas testosterone is a strong inhibitor of breast growth. Gynecomastia typically results from either an excess of estrogens or estrogen precursors, a decrease in androgens, or impairment of their action [7]. Gynecomastia has a variety of conditions connected with it, but in around 50% of instances, the condition is either idiopathic or persistent pubertal gynecomastia without a significant secondary reason [8]. Gynecomastia that develops naturally occurs during puberty, aging, and childhood. Gynecomastia must be recognized from lipomastia by comparing the findings of palpating the subareolar tissue to the surrounding subcutaneous adipose tissue in the anterior axillary fold or other regions on the chest wall also known as fatty breasts or pseudogynecomastia [9]. The number of hirsute premenopausal women in the United States is thought to be over 4 million and they are likely responsible for about 1.5 billion dollars in annual hair removal products and procedures [10]. In general, hirsutism prevalence grew up with age. Women with irregular monthly cycles (amenorrhea/oligomenorrhea) exhibited a higher prevalence of hirsutism (17.34%) compared to those with normal cycles (7.83%). In North India, hirsutism was more common among obese women than among nonobese women and women with a family history of the condition [4]. In general, 32%–40% of men experience gynecomastia, and the maximum prevalence is found in old age and up to 65% in men [11]. Asymptomatic gynecomastia is much more common than symptomatic gynecomastia, with prevalence rates ranging from 60% to 90% in newborns, 50% to 60% in adolescents, and up to 70% in men between the ages of 50 and 69 [12]. Gynecomastia incidence has recently peaked in men between the ages of 50 and 85, with a prevalence of up to 70%. Bilateral gynecomastia occurs more frequently than unilateral gynecomastia. It has been estimated that 35%–45% of men have unilateral gynecomastia [13]. Angiogenesis may affect the surrounding skin's health and is also involved in skin diseases. Both the papilla and the outer root sheath keratinocytes produce angiogenic substances such as VEGF. In hirsutism, the anagen growth phase is prolonged and the size of the hair follicles changes from vellus to terminal hair production [14]. Polycystic ovarian syndrome (PCOS) is the primary cause of hirsutism, although only 5% of patients had unique endocrine problems detected. It is possible to think of PCOS as a multiorgan disease that affects the release of growth hormone, gonadotropins, and adrenocorticotrophic hormone (ACTH) from the pituitary in addition to elevated levels of adrenal and ovarian sex hormones. An enhanced inflammatory state, abdominal obesity, and an increased release of interleukins, chemokines, and adipokines are characteristics of PCOS and sufferers have insulin resistance [15]. MDA levels are likely to rise in response to insulin resistance, hyperandrogenism, dyslipidemia, and obesity caused by PCOS [16]. Male patients represent a small but important portion of those who require help for breast issues. Breast growth, lumps, or soreness are the most common complaints, whereas nipple discharge and inflammation are extremely uncommon appearances [17]. Melatonin is a powerful antioxidant since it may help to stop the creation of free radicals. It also appears to have an anticancer impact and can improve immune system performance. Melatonin therapy may also cause other negative effects, including breast enlargement in men, disorientation, headaches, and stomach cramps (called gynecomastia).

2 Pathophysiology of hirsutism

Androgens in men mostly come from adrenal glands and the testes, whereas in women they are primarily produced by the ovaries and adrenal glands. The ovaries and adrenals produce 40%–50% of the testosterone found in females, whereas 50%–60% comes through peripheral conversion of androgen precursors such as androstenedione [18]. Women's ovaries and adrenal glands both produce androstenedione. It has two possible biologic endpoints, which is significant. The enzyme 17-hydroxysteroid dehydrogenase in the cell or the enzyme aromatase in the granulosa cell can either convert it to testosterone or estrogen. Although comparatively small compared to the more prevalent testosterone, androstenedione in women can be converted to testosterone and then to dihydrotestosterone, which normally promotes the growth of sexually dependent hair in the axillary and pubic regions. The Leydig cells of the testicles are where more than 95% of testosterone in males is produced [19]. Numerous local and systemic variables, growth factors, cytokines, and sex hormones all influence hair development. It has also been demonstrated that thyroid and growth hormones can change how hair grows. Androgens in particular, which are found in sex steroids, have a significant impact on the type of hair that develops and how it is dispersed throughout the human body. During puberty, vellus follicles in particular places grow into terminal hair as levels of androgen rise. Additionally, androgens promote sebum production, which results in oilier hair and skin. Similarly, androgens lengthen the anagen phase of body hair while prolonging the anagen phase of hair growth [20,21]. Hirsutism is caused by an interaction between androgen levels and the hair follicle's sensitivity to androgens. The amount and length of androgen exposure, the activity of the local 5-alpha-reductase enzyme, and the inherent sensitivity of the hair follicle to androgen action all play a role in whether vellus hair develops into terminal hair [22]. Men who are deficient in 5-alpha reductase have thin body hair and do not suffer from androgenetic alopecia. Androgenetic alopecia is the outcome of androgens' ability to miniaturize hair follicles on the scalp. In the axillary, pubic, and beard regions, androgens lengthen the anagen phase of the hair development cycle as well as the size, diameter, and duration of the hair fibers. Sebaceous glands also have androgen receptors and when stimulated, they expand and generate more sebum [23]. Hirsutism is caused by either an exogenous or endogenous rise in circulating androgens or by an increase in the hair follicle's sensitivity to normal serum androgen levels. The predicted amounts of circulating androgens do not usually correlate well with the clinical severity of hirsutism. According to theory, various people's androgen-dependent follicles respond in a variety of ways [21].

3 Angiogenesis, oxidative stress, and inflammation in hirsutism

Significant vascularization around the hair follicle occurs during the anagen phase; in the catagen and telogen phases, blood vessels degenerate and disappear. Hence, angiogenesis is induced along with active hair development to accommodate the increased demand for blood flow, and other substances such as nutrients, which are essential for the quick proliferation of follicular keratinocytes and the ensuing elongation and thickness of the hair shaft. It has been demonstrated that increased vascularization of the hair follicle fosters hair development and expands the width of hair follicles and hair shafts [24]. Melatonin's lower concentration in the ovarian follicles has been used as an explanation for elevated evening melatonin levels in PCOS. Women with PCOS who experience high levels of oxidative stress, therefore, produce more melatonin, likely in an effort to get rid of additional free radicals. High levels of melatonin in the ovarian follicle fluid are essential for ovulation, follicular growth, and oocyte quality, whereas low levels of melatonin in the follicular fluid may be responsible for these women who had decreased oocyte quality and ovulation. However, it has not yet been determined whether melatonin affects these patients' hirsutism, hormonal profiles, and indicators of inflammation or oxidative stress directly [25].

4   Pathophysiology of gynecomastia

The primary cause of gynecomastia appears to be an imbalance between the effects of androgenic and estrogenic influences on breast tissue. Normal male breast tissue has receptors for both estrogen and androgen hormones; estrogens promote breast tissue development while androgens hinder it. The development of male breasts is consequently brought by either an absolute or relative shortage of androgens, a deficiency in androgen action, or a rise in estrogen levels or estrogenic activity [26,27]. Absolute estrogen refers to serum or tissue estrogen concentrations above the usual range seen in healthy young adult women, whereas absolute androgen deficiency refers to serum or tissue estrogen concentrations below the normal range seen in healthy young adult men. A relative androgen deficit or relative estrogen excess is defined as the presence of both androgen and estrogen levels that are within the normal range but have an abnormal androgen-to-estrogen ratio [28]. More frequently than estrogen-secreting tumors or their precursors, tissue aromatase-increased extragonadal conversion of androgens to estrogens accounts for raised serum estrogen levels (such as Leydig or Sertoli cell tumors, hCG-producing tumors, and adrenocortical tumors). Patients with primary (Klinefelter syndrome, mumps orchitis, castration) or secondary (gonadal failure) gonadal failure have lower levels of free blood testosterone (hypothalamic and pituitary disease). Gynecomastia may also be connected to androgen resistance syndromes, which are brought on by decreased activity of the enzymes needed to produce testosterone [12]. Adiposity rises with age, and some research indicates that as men age, the expression of the aromatase enzyme in adipose tissue may also rise. It has also been proposed that increased levels of inflammatory cytokines cause the expression of aromatase to rise in adipose tissue with age and in certain diseases. Gynecomastia has been linked to an increased prevalence by polymorphic variations in aromatase that lead to higher enzymatic activity [9]. Local variables in breast tissue may also be crucial in gynecomastia in addition to systemic hormone levels. Gynecomastia may be brought on by alterations in the number and/or activity of androgen or estrogen receptors in nearby breast tissue as well as alterations in the production of estrogens or androgens locally, a reduction in the inactivation of estrogen, or an increase in the local production of estrogen [29]. Estrogen enhances IGF-1R expression in breast cancer cells, which may improve IGF-1 activity in healthy breast tissue. It is possible that progesterone and IGF-1 work in concert to promote breast ductular growth and development [30]. In addition, it has been proposed that particular mutations in the leptin receptor gene predispose boys to develop pubertal gynecomastia. Leptin may help to cause gynecomastia by increasing the amount of aromatase produced in breast and fat tissue [29,31].

5 Drugs used for the treatment of hirsutism

To improve patient quality of life, medical treatment for hirsutism seeks to rectify hormonal abnormalities. The manner of therapy depends on the underlying cause, the location and amount of excessive hair growth, patient preferences, and the accessibility and cost of the products that are already on the market. Oral contraceptives with antiandrogenic activity are the first-line treatment for hirsutism for the majority of premenopausal women. If a patient does not exhibit any clinical improvement, it is recommended that they have a combined therapy that includes oral contraceptives and antiandrogens. It is advisable to continue pharmacologic hirsutism treatment for 6–9 months before changing the dosage or drug class. Insulin sensitizers also reduce hirsutism in females who also have hyperandrogenism and insulin resistance [32]. The classification of drugs used in the medical treatment of hirsutism is depicted in Fig. 1 and their mechanism of action is given in Fig. 2 [33]. Table 1 describes the pharmacokinetics of these drugs.

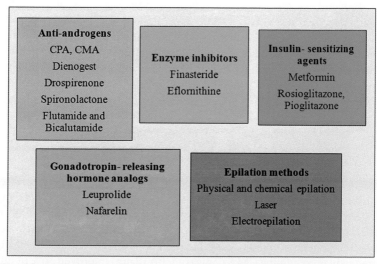

Fig. 1 Classification of drugs used in the medical treatment of hirsutism.

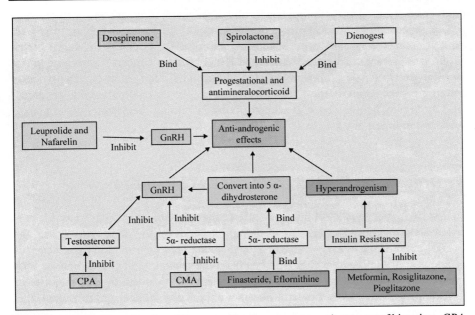

Fig. 2 The mechanism of action of drugs used for the management/treatment of hirsutism. *CPA*, cyproterone acetate; *CMA*, chlormadinone acetate; and *GnRH*, gonadotropin-releasing hormone.

5.1 Antiandrogens

5.1.1 CPA: Cyproterone acetate

Cyproterone acetate appears to be as effective as other treatments for women with hirsutism brought on by excessive androgen production by the ovaries. The antiandrogen cyproterone acetate inhibits five alpha-reductase activities in hirsute women's skin, increases SHBG levels, and has a potent antigonadotrophins action when paired with ethinyl estradiol [54]. It reduces androgen activity in hirsute women by blocking receptors and causing hepatic T clearance. Treatment of hirsutism has proven successful when 2 mg CPA and 35 mg ethyl estradiol are taken daily. In cases of moderate to severe hirsutism or when a quicker response is needed, higher doses of CPA are advised [55].

5.1.2 CMA: Chlormadinone acetate

Compared to CPA, the antiandrogenic potential of CMA is regarded as lesser [33]. The combined monophasic contraceptive pill EE/CMA 0.03 mg/2 mg has antiandrogenic qualities. Acne and hirsutism can develop in women with normal serum levels of testosterone while in some women with elevated circulating androgen levels (hyperandrogenemia) [56]. A combination of ethinyl estradiol (30–35 mg) plus oral contraceptives containing CMA (2 mg) or CPA (2 mg) improved 36% of hirsutism patients [57].

Table 1 The pharmacokinetics of drugs used for the treatment of hirsutism.

Drugs	Absorption	Distribution	Metabolism	Excretion	Marketed drug	References
Antiandrogens						
CPA	Oral: 68%–100%	3 h	Hepatic (CYP3A4)	Feces: 70% Urine: 30%	Diane-35, Cyestra-35	[34–36]
CMA	Mouth: 100%	96.6%–99.4% (to albumin not to SHBG or CBG)	Liver	Urine: 33%–45% Feces: 24%–41%	Belara, Gynorelle	[37]
Dienogest	90% (p.o.)	Albumin: 90% Free: 10%	Liver	Urine	Visanne	[38,39]
Droespirone	66% and 85% (p. o.)	1.6–2 h	CYP3A4	Between 25 and 33h	Slynd	[36]
Spironolactone Bicalutamide	60%–90% Oral (6 days)	88% to albumin Plasma protein binding mainly to albumin	Liver Liver	Urine and bile Urine and bile	Aldactone	[40,41] [42,43]
Flutamide	Oral (within 2 h)	Plasma protein binding (94%–96%)	Liver	Urine and feces	Eulexin	[44]
Enzyme inhibitors						
Finasteride	p.o. (65%)	Plasma protein binding (90%)	Liver (CYP3A4)	Urine (57%) and feces (40%)	Proscar and Propecia	[43]
Eflornithine (Topical cream)	Twice a day	Steady-state plasma $t_{1/2}$ is 8h	Not known to be metabolized	Primarily excreted unchanged in the urine	VANIQA	[45]

Continued

Table 1 Continued

Drugs	Absorption	Distribution	Metabolism	Excretion	Marketed drug	References
Insulin sensitizing agents						
Metformin	p.o. (40%–60%)	Rapidly distributed and not bound to PPB	Not metabolized in the liver	Between 4.0 and 8.7 h	Glucophage	[46,47]
Rosiglitazone	Oral (99%)	Plasma binding	Liver	Urine (approximately 65%)	Avandia	[48,49]
Pioglitazone	i.v. (83% BA)	Plasma binding especially to albumin	Rapidly metabolized by cytochrome 450	Urine	Actos, Takeda	[50,51]
GnRH analogs						
Leuprolide	SC (50.60%)	487.40 mL		514.46 mL/h	Lucrin depot	[52]
Nafarelin	Intranasal and SC (within 5–60 min)	In extravascular organs and rapid distribution in 1 h	Liver	Urine and feces		[53]

5.1.3 Dienogest

Only the nortestosterone derivative dienogest has antiandrogenic potential. Its antiandrogenic efficacy is equivalent to roughly 30% of that of cyproterone acetate. Only 10% of dienogest is bonded to SHBG or CBG, allowing for high serum levels. In addition, dienogest is regarded as a tried-and-true method of treating endometriosis because of its acceptable profile, few side effects, safety during long-term use, and decreased chance of recurrence [58]. Dienogest, a selective progestin with a high progestogenic activity at the endometrium, combines the pharmacologic features of progesterone derivatives and 19-norprogestins [59]. It has a strong impact on endometrial tissue, causing endometriotic cells to stop proliferating and exhibiting antiinflammatory and antiangiogenic properties. Due to the high amounts of the unbound molecule in circulation, dienogest has a strong progestogenic action after binding to the progesterone receptor with great specificity [60].

5.1.4 Drospirenone

The first synthetic progestin with progesterone-like antimineralocorticoid action is drospirenone. A screening program for a progesterone antagonist that included powerful spironolactone analogs led to the discovery of DRSP [61]. Drospirenone, a progestin used in many oral contraceptives, has only marginal antiandrogenic properties. 3 mg is equivalent to 25 mg of spironolactone or 1 mg of CPA (in oral contraceptives). In a nonrandomized research, ethinyl estradiol (30 mg) and the oral contraceptive drospirenone (3 mg) together reduced the clinical symptoms of hirsutism through their antiandrogenic and antimineralocorticoid effects [62].

5.1.5 Spironolactone

The oral contraceptive pill drospirenone, which contains ethinyl estradiol and the spironolactone analog progestin, seems promising because it also has additional antimineralocorticoid and antiandrogenic properties. In addition to oral contraceptives, antiandrogens can be used to reduce the issue of hirsutism. Since it blocks testosterone's impact at the receptor level, the aldosterone antagonist spironolactone has been researched as a potential treatment for hirsutism associated with PCOS. Although few trials have examined the effectiveness of spironolactone in treating PCOS, it is effective in reducing both subjective and objective assessments of hirsutism [63].

5.1.6 Flutamide and bicalutamide

Flutamide, a nonsteroidal substance that functions as the androgen receptor site, is regarded as a pure antiandrogen. Flutamide is used at doses ranging from 62.5% to 500% daily. Serum transaminases should be monitored periodically because liver damage is possible. Other side effects that have been documented include dry skin, diarrhea, nausea, and vomiting [33]. Bicalutamide is a brand-new, powerful, and well-tolerated nonsteroidal pure antiandrogen that comes in a 25 mg/d dosage. The 50 mg/d dosage was designed to treat prostate cancer. It has been demonstrated to be beneficial in treating patients with PCOS-induced and idiopathic hirsutism without causing noticeably negative side effects [64].

5.2 Enzyme inhibitors

5.2.1 Finasteride and eflornithine

Due to its ability to inhibit the enzyme 5α-reductase, which prevents the conversion of testosterone into 5a-dihydrotestosterone, Finasteride is commonly thought of as an antiandrogen. Although it decreases the number of hormones available to interact with the androgen receptor, it does not affect how much androgen is secreted by the ovary or the adrenal glands. Women with hirsutism are treated with Finasteride (in doses ranging from 1 to 5 mg/d), and various clinical trials have examined the effectiveness of this medication. Finasteride can lower hirsutism scores by as much as 60% while also reducing the average hair diameter [65]. Eflornithine prevents the formation of new hair by permanently inhibiting the ornithine decarboxylase enzyme, which lowers the amount of polyamines, which are essential components of rapidly dividing tissue-like hair. Food and Drug Administration has authorized topical eflornithine cream for decreasing female facial hair. When eflornithine cream was used in conjunction with laser treatment for the upper lip and chin as opposed to a placebo or eflornithine cream alone, a more obvious decrease in hair count was observed [66].

5.3 Insulin-sensitizing agents

5.3.1 Metformin, rosioglitazone, and pioglitazone

Increased insulin sensitivity is a typical side effect of the drug metformin, which is used to treat type 2 diabetes mellitus. For diabetes, they reduce elevated sugar levels, for nondiabetes, they reduce insulin levels, with no impact on blood sugar and to improve insulin action, they raise insulin sensitivity. While metformin reduces the production of hepatic glucose and raises insulin levels, thiazolidinediones improve the action of insulin in the liver, skeletal muscle, and adipose tissue. Both may also increase sex hormone-binding globulin levels, decrease adrenal and ovarian androgen production, and improve gonadotropin secretion [67]. Insulin-sensitizing drugs may be used to treat hirsutism by reducing insulin levels and, consequently, the quantity of free and physiologically active androgens in the blood. However, Pioglitazone was taken off the market in 2011 and rosiglitazone was removed from the European and Swiss markets in 2010 as a result of reports associating these medications with increased cancer risk or cardiovascular risk [33].

5.4 Gonadotropin-releasing hormone analogs

5.4.1 Leuprolide and nafarelin

The natural GnRH hormone released from the arcuate nucleus of the hypothalamus is structurally identical to synthetic peptides called gonadotropin-releasing hormone (GnRH) agonists. GnRH agonists work by interacting with GnRH receptors to increase gonadotropin release initially. This is followed by desensitization, which causes the release of gonadotropins to decrease, simulating menopause. In addition to being utilized in assisted reproduction, hirsutism, abnormal uterine bleeding, and the treatment of premenstrual syndrome in women, GnRH agonists are also used to treat endometriosis [68]. Leuprolide may be administered as a daily dose of 0.5–2 mg, a monthly depot dose of 3.75 mg, or even three-monthly doses totaling 11.25 mg. It can be injected intramuscularly or subcutaneously. Depending on the intended application, Nafarelin is given orally once or twice a day in doses ranging from 200 to 400 µg [69].

6 Drugs used for the management/treatment of gynecomastia

The presence of subareolar hard glandular tissue with a diameter of less than 2 cm allows for the clinical diagnosis of gynecomastia. Breast cancer and lipomastia are two differential diagnoses. Lipomastia is an accumulation of soft fatty tissue, but clinical indications of breast cancer are a hard, asymmetrical, eccentrically positioned lump that may be felt, as well as probable alterations to the skin and nipple. Mammography, ultrasonography, and biopsy are recommended if malignancy is suspected [53]. Fig. 3 describes the surgical and nonsurgical approaches to the management of gynecomastia.

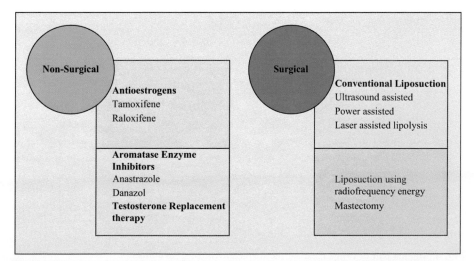

Fig. 3 The surgical and nonsurgical approaches for the management of gynecomastia.

6.1 Nonsurgical treatment of gynecomastia

The use of medical treatment is probably helpful if done during the early proliferative phase. Once gynecomastia has been present for more than a year, the glandular structure is replaced by stromal hyalinization and fibrosis. Breast tissue is less likely to respond to treatment as a result [70]. The mechanism of the various drugs used for the management of gynecomastia is given in Fig. 4 and their pharmacokinetics is given in Table 2.

6.1.1 Antiestrogens (tamoxifen and raloxifene)

Suppress transcription of growth genes and compete with estrogen at breast tissue receptor sites. In individuals with gynecomastia that are recently grown, this seems to be helpful [75].

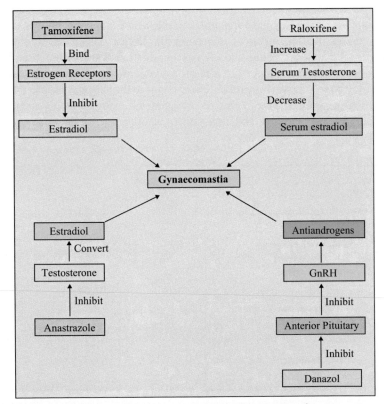

Fig. 4 Mechanism of drugs used for the management of gynecomastia.

Table 2 Pharmacokinetics of various drugs used for the management of gynecomastia.

Drug	Absorption	Distribution	Metabolism	Excretion	Marketed	References
Antiestrogens						
Tamoxifen	Rapid absorption from the intestine	PPB (99%)	Liver (CYP3A4)	Feces: 65% Urine: 9%	Nolvadex	[71]
Raloxifene	Rapid absorption from the intestine	PPB (95%)	Liver and intestine	Feces	Evista	[71]
Aromatase enzyme inhibitors						
Anastrazole	Unknown (p.o.)	PPB (40%)	Liver (85%)	Urine (11%)	Arimidex	[72]
Danazol	Low BA	PPB to albumin	Liver (CYP3A4)	Urine and feces	Danocrine	[73,74]

6.1.2 Aromatase enzyme inhibitors such as Anastrozole and Danazol

Men with hypogonadism and gynecomastia have used this medication because it will stop the peripheral conversion of testosterone into estradiol [76].

6.1.3 Testosterone replacement therapy

Testosterone replacement therapy leads to the disappearance of gynecomastia in many hypogonadal men. However, as testosterone can be converted to estradiol by aromatization, testosterone therapy has the potential to exacerbate gynecomastia and, in some situations, even trigger its recurrence. This is why nonaromatizable androgens like dihydrotestosterone (DHT) are utilized [76].

6.2 Surgical treatment of gynecomastia

The flattening of the thoracic region, the elimination of extra skin, and the symmetry of the two sides, among other things, are typically the goals of surgical therapy.

6.2.1 Conventional liposuction

By introducing an aspiration cannula through a small skin incision and sucking the fat out, liposuction is a surgical treatment that eliminates subcutaneous fat. In literature, terms like "fat suction," "blunt suction lipectomy," and "liposculpture" are interchangeable with "suction-assisted lipectomy," "suction lipoplasty," and "lipo-suction surgery" [77]. One of the top four regions for liposuction in guys is the male breasts (after love handles, abdomen, and submental fat). The primary cause is the cosmetic discomfort and confidence decline brought on by a feminine self-image [78].

Ultrasound-assisted liposuction

The higher-density fibro connective tissues are largely unharmed since the fat is more selectively emulsified. This method, which has proven to be quite effective in the treatment of gynecomastia successfully, removes the fibrous parenchymal framework of the male breasts [79].

Power-assisted liposuction

The cannula oscillating movement effortlessly simulates a surgeon's job during a conventional suction-assisted method, the vibration method easily penetrated fibrous fat while generating no thermal energy and providing no risk of skin burns [80].

6.2.2 Liposuction using radiofrequency energy

This method causes coagulation of adipose, vascular, and fibrous tissue by aspirating the coagulated tissue while using a blunt-nose suction cannula [81].

6.2.3 Laser-assisted lipolysis

Benefits of this include rapid recovery times, high patient toleration, and the additional benefit of dermal tightening [82].

6.2.4 Mastectomy

The subcutaneous mastectomy has been performed using a variety of incisions and methods. Webster's intraareolar, circumareolar, and periareolar incisions are the most often used types [83]. Treatment for fibrous gynecomastia has been described using the endoscope. With reduced scarring, a shorter recovery period, fewer problems, and better cosmetic outcomes, this procedure has enhanced the ultimate result [84].

7 Nanotechnological-based approaches for the management of hirsutism and gynecomastia

Several nanotechnology-based approaches have been investigated for the management of hirsutism and gynecomastia in recent years. Tables 3 and 4 provide a detailed description of the outcomes and clinical significance of nanocarriers explored for hirsutism and gynecomastia, respectively.

7.1 Strategic drug delivery system for hirsutism

Table 3 Recapitulation of nanocarriers investigated for the management of hirsutism.

Techniques	Dosage form	Excipients	Outcomes	References
Cyproterone acetate				
Solvent diffusion evaporation	NLCs	Cholesterol, triolein, and stearic acid	Enhanced the low solubility of the drug and effective for skin disorders	[85]
Thin film hydration technique	Liposomes	Egg phosphatidylcholine, butylated hydroxyl toluene, and cholesterol	Increased the permeability of the drug and suitable carriers for controlled release	[86]
	NLCs	Oleic acid, Miglyol, and Poloxamer 188	Effective in the increase of 2–3-fold absorption rate for the therapeutic option of acne and other diseases of the pilosebaceous unit	[87]
Chlormadinone acetate				
	SMEDDS	Ethyl oleate, Tween-80, Transcutol P, and PEG400	C_{max} and AUC of CMA-SMEDDS were significantly higher by 1.98-fold which could potentially be useful in enhancing oral bioavailability and reducing the clinical dose of CMA	[88]
Dienogest				
Milling method	Micronized tablet	Lactose monohydrate, microcrystalline cellulose, maize starch, povidone, talc magnesium stearate	Improved the dissolution rate of the drug which ultimately increases the BA	[89]

Continued

Table 3 Continued

Techniques	Dosage form	Excipients	Outcomes	References
Drospirenone				
Solvent evaporation method	Polymeric nanoparticles	PVA, povidone, *n*-butyl cyanoacrylate, polysorbate 80 tween 80, dichloromethane, triton X-100, potassium dihydrogen phosphate, sodium azide, and hydroxypropyl-b-cyclodextrin	Study showed more effect for prolonged and uniform release of the drug	[90]
Spironolactone				
Probe ultrasonication method	NLCs	Oleic acid, Span 80, and Tween 80	Showed about a 5.1- and 7.2-fold increase in solubility and more effective than SLNs	[91]
Ethanol injection method	Hyaluronic acid-enriched cerosomes	Phosphatidyl choline and ceramide III	Effective for the treatment of gynecomastia	[92]
Finasteride				
Coacervation phase separation	Proniosomes	Cholesterol, span 60	Improve stability and increase the encapsulation efficiency of the drug	[93]
Hot melt homogenization method	SLNs	Span 80, Precirol ATO-5, and Poloxamer	Become promised carriers for the treatment of over-hair growth	[94]
Eflornithine				
	Cream	Potassium phosphate monobasic and Honey wax	Enhanced the ability to inhibit extra hair growth	[95]
Emulsion solvent evaporation method	SLNs	Tween 20, PEG-20 sorbitan monolaurate, potassium dihydrogen phosphate	Promised therapy for the permeation and sustained release of drugs	[96]

NLCs, nanostructured lipid carriers; *SLNs*, solid lipid nanoparticles.

7.2 Strategic drug delivery system for gynecomastia

Table 4 Recapitulation of nanocarriers investigated for the management of hirsutism.

Techniques	Dosage form	Excipients	Outcomes	References
Tamoxifen				
Multiple-emulsion solvent evaporation	Polymeric nanoparticles	PVA, hydroxypropyl-β-cyclodextrin, fluorescein isothiocyanate, fetal bovine serum, and tetrazolium dye	Effective in the controlled release of the drug	[97]
High-pressure homogenization	SLNs	Phospholipon 90H, Hydrogenated palm oil Softisan 154, and CH3OH	Effective for the antiestrogen effects	[98]
Microemulsion and precipitation methods	SLNs	Taurocholate sodium salt and palmitic acid	Effective in the prolonged release of drugs in the i.v. route	[99]
Emulsification and high-pressure homogenization technique	SLNs	Sodium tauroglycocholate, glycerol behenate	Enhanced the entrapment efficiency of the drug	[100]
Solvent injection method	SLNs	Glycerol monostearate, stearic acid (SA), Poloxamer 188, and Tween 80	Enhanced the oral BA of the drug	[101]
High-pressure homogenization	SLNs	Polytetrafluoroethylene, Thimerosal, and Sorbitol	Effective for prolonged release and decrease in the hepatotoxic effects	[102]
Microemulsion technique	SLNs	Glycerol monostearate, Tristearin Soya lecithin, and Tween 80	Enhanced the permeation rate of drug	[103]
Ultrasonic homogenizer method	Liposomes	Distearoyl phosphatidylcholine, stearyl amine	Effective against the MCF-7 cells	[104]
Lipid film hydration	Liposomes	Dioleyl-*sn*-glycero-3-phosphoethanolamine, octadecyl-(1,1-dimethyl-piperidino-4-yl)-phosphate	Improved the therapeutic efficacy in several antiestrogen-resistant xenografts	[105]
Film hydration techniques	Liposomes	Phosphatidylcholine, soy lecithin, dicetyl phosphate, and stearylamine	Enhanced the skin permeation rate of the drug	[106]

Continued

Table 4 Continued

Techniques	Dosage form	Excipients	Outcomes	References
Raloxifene				
Reverse-phase evaporation	Liposomes	1,2-Dipalmitoyl-*sn*-glycero-3-phosphocholine, Sodium Taurocholate, Dimethyl-β-cyclodextrin, chitosan, and (3-(4,5-dimethylthiazol-2-yl)-2,5-diphenyltetrazolium bromide	Reduced the effect of tumor cells and increased by 3.5-fold in permeability	[107]
Solvent evaporation techniques	Chitosan NPs	Sulfobutylether-β-cyclodextrin	Effective to increase the low oral BA of the drug	[108]
Emulsion-diffusion-evaporation technique	Polymeric nanoparticles	PEG and Tween-80	Approximately 4.87 times more bioavailability than the free drug	[109]

7.3 Description of nanocarriers: As a boon for the management of hirsutism and gynecomastia

7.3.1 Polymeric nanoparticles

The drug is dissolved, entrapped, encapsulated, and linked to a nanoparticle matrix in the form of polymeric nanoparticles which are made of biocompatible and biodegradable polymers with sizes ranging from 10 to 1000 nm. Nanoparticles, nanospheres, or nanocapsules can be produced depending on the technique of preparation. Unlike nanospheres, which are matrix systems in which the drug is physically and uniformly spread, nanocapsules contain the drug inside a chamber that is enclosed by a special polymer membrane [25,110]. Several techniques have been used to create polymeric nanoparticles depending on the application and type of medication to be contained. These nanoparticles are widely employed to create nanomedicine by nanoencapsulating several beneficial bioactive chemicals and pharmaceuticals. Polymeric nanoparticles that degrade naturally are highly preferred because they have potential as a medication delivery mechanism [111]. In addition, these nanomedicines are biodegradable, avoid the reticuloendothelial system, are nontoxic, nonthrombogenic, nonimmunogenic, noninflammatory, do not activate neutrophils, and are stable in blood. They are also applicable to a variety of molecules, including drugs, proteins, peptides, and nucleic acids [112]. Dispersion of processed polymers and polymerization of monomers are the two processes used to create polymeric nanoparticles. Dialysis, nanoprecipitation, solvent evaporation, supercritical fluid technology (SCF), emulsification/solvent diffusion, and salting out are the methods utilized to disperse the performed polymers. Another method for making polymeric nanoparticles is through the polymerization of monomers using techniques including miniemulsion, microemulsion, emulsion, controlled/living radical polymerization, and interfacial polymerization [110,113].

7.3.2 Solid lipid nanoparticles

SLNs are introduced as an effective carrier approach for correcting dynamic medicines and water-soluble medicines. Colloidal particles between 10 and 1000nm in size are considered nanoparticles. They are composed of synthetic, unique polymers with improved drug delivery and decreased lethality as their main goals [114]. SLNs are colloidal carrier systems made up of an aqueous surfactant on top of a solid core of high melting point lipids. BCS classes II and IV are the categories of drugs utilized in SLNs. SLNs vary from other colloidal carriers in that they employ solid lipids rather than liquid lipids. Lipids include triglycerides, partial glycerides, fatty acids, hard fats, and waxes in a broad sense. A clear advantage of SLN is the fact that the lipid matrix is made up of physiological lipids, which lowers the risk of both acute and long-term toxicity [115,116]. The following techniques are used to create solid lipid nanoparticles: solvent emulsification-evaporation, high-pressure homogenization (cold and hot homogenization), solvent emulsification-diffusion, solvent injection, high shear homogenization, double emulsion (w/o/w), and ultrasound dispersion [117,118].

7.3.3 Liposomes

Because of their many varieties, liposomes have been studied more than other carrier systems. By adding phospholipids to a water solution, phospholipid bilayer membranes can create liposomes, which are sphere-shaped structures with internal hydrophilic compartments [119]. Lipids are amphipathic molecules with portions that love and hate water. Liposomes are made up of one or more lipid bilayers that interact with the aqueous phase in both hydrophilic and hydrophobic ways to form. Water molecules reject the hydrophobic tails of liposomes, which causes liposome self-assembly [120]. The therapeutic index of encapsulated pharmaceuticals like doxorubicin and amphotericin was improved by liposomal delivery, according to research on the clinical potential of conventional liposomes that was conducted in the 1980s. Traditional liposomal formulations improved medicine delivery to diseased tissue in comparison to free drugs, changing pharmacokinetics and biodistribution to lower chemical toxicity in vivo. The therapeutic effectiveness was however constrained by the delivery system's susceptibility to quick removal from the bloodstream [121]. Liposomes are prepared by following methods such as lipid-film hydration via hand and nonhand shaking, freeze-drying, sonication, French pressure cell, membrane extrusion, freeze-thawed liposome, microemulsification, and dried reconstituted vesicles [122].

7.3.4 Nanostructured lipid carriers

Lipid nanoformulations may help to overcome the difficulties in producing solubilized phases from which drug absorption occurs quickly, which are limitations of the moderately water-soluble medications, such as Biopharmaceutics classification

System (BCS) class II, that dissolve slowly and poorly [123]. The bioavailability and solubility of insoluble drugs are the two main factors that could be enhanced. NLCs can be categorized into three groups: imperfect, amorphous, and numerous structures, depending on their lipid content and formulation factors. NLCs are created using a variety of techniques, including high-pressure homogenization, solvent emulsification/evaporation, microemulsification, ultrasonification or high-speed homogenization, spray drying, and microfluidics technology [124,125]. NLCs are a kind of formulation that can deliver concentrated dispersions together with improved loading and stability. The key component of NLC that affects the formulations' stability, sustained release behavior, and drug loading capability is the lipid itself. The type of surfactant used has a big impact on the properties of NLCs, which offer magnificent characteristics and attributes that can enhance the presentation of a variety of integrated drug forms [126].

7.3.5   Self-emulsifying drug delivery system

SMEDDS, or self-microemulsifying drug delivery systems, have drawn a lot of interest as a potential treatment for both naturally occurring compounds and medications with low solubility. The main elements of lipid-based formulations include solid-lipid nanoparticles (SLN), macroemulsions (coarse emulsion), microemulsions, liposomes, and lipoplexes [127]. SMEDDS are isotropic mixtures made up of oil, a solubilizer (cosurfactant), a surfactant, and a medication. This system's ability to create excellent oil-in-water (o/w) microemulsion after aqueous phases has diluted its main selling point. The digestive motility of the stomach and intestine produces the agitation required for self-emulsification [128]. Several methods, such as spray drying, melt extrusion, adsorption to a solid carrier, extrusion spheronization, encapsulation of solid and semisolid SEDDS, etc., are used to turn liquid SMEDDS into solid SMEDDS [129]. As a result of this spontaneous emulsion formation in the gastrointestinal tract, the medication is supplied in a solubilized state, and the droplet's small size creates a large interfacial surface area for drug absorption. In addition to solubilization, the addition of fat to the formulation increases bioavailability by affecting medication absorption. The phase diagram area of the self-emulsifying zone, the distribution of the emulsion's droplet sizes, and the assessment of the drug's solubility in different components are all factors that are taken into consideration when choosing an acceptable self-emulsifying formulation [130].

8   Conclusion and future perspectives

For the therapy of hirsutism, a deep clinical evaluation and study are needed. For the treatment of hirsutism, antiandrogen and hair removal are part of the treatment. In some cases, treatment for underlying illnesses, stopping the use of medications that could lead to gynecomastia, medical therapy, and surgery may be used for gynecomastia who have symptoms or who have developed it recently. Nanotechnology due to its small particle size, high specific surface area, many active centers, higher surface reactivity, and superior adsorption capacity is used to improve efficacy and bioavailability. By enabling the precise delivery of nanodrugs into target cells, it avoids interfering with the physiological processes in other organs. The problem of enhancing the bioavailability and efficiency of further drugs whose nanoformulations are not formed awaits a solution.

Acknowledgments

The authors would like to thank the Department of Pharmaceutics, MM College of Pharmacy, Maharishi Markandeshwar (Deemed to be University), Mullana-Ambala, Haryana, India 133207.

References

[1] Mofid A, Seyyed Alinaghi SA, Zandieh S, Yazdani T. Hirsutism. Int J Clin Pract 2008; 62(3):433–43. https://doi.org/10.1111/j.1742-1241.2007.01621.x.

[2] Brodell LA, Mercurio MG. Hirsutism: diagnosis and management. Gend Med 2010;7(2): 79–87. https://doi.org/10.1016/j.genm.2010.04.002.

[3] Bode DV, Seehusen D, Baird D. Hirsutism in women. Am Fam Physician 2012;85(4): 373–80.

[4] Zargar AH, Wani AI, Masoodi SR, Laway BA, Bashir MI, Salahuddin M. Epidemiologic and etiologic aspects of hirsutism in Kashmiri women in the Indian subcontinent. Fertil Steril 2002;77(4):674–8. https://doi.org/10.1016/S0015-0282(01)03241-1.

[5] Azziz R. The evaluation and management of hirsutism. Obstet Gynecol 2003;101(5): 995–1007. https://doi.org/10.1016/S0029-7844(02)02725-4.

[6] Daniels IR, Layer GT. Gynaecomastia. Eur J Surg 2001;167(12):885–92. https://doi.org/ 10.1080/110241501753361550.

[7] Thiruchelvam P, Walker JN, Rose K, Lewis J, Al-Mufti R. Gynaecomastia. BMJ 2016;354. https://doi.org/10.1136/bmj.i4833.

[8] Ersöz HÖ, Önde ME, Terekeci H, Kurtoglu S, Tor H. Causes of gynaecomastia in young adult males and factors associated with idiopathic gynaecomastia. Int J Androl 2002;25 (5):312–6. https://doi.org/10.1046/j.1365-2605.2002.00374.x.

[9] Narula HS, Carlson HE. Gynaecomastia—pathophysiology, diagnosis and treatment. Nat Rev Endocrinol 2014;10(11):684–98.

[10] Nourbala M, Kefaei P. The prevalence of hirsutism in adolescent girls in Yazd, Central Iran. Iran Red Cresecent Med J 2010;12(2):111–7.

[11] Ahmad M. Prevalence of gynaecomastia in male Pakistani population. World J Plast Surg 2017;6(1):114.

[12] Johnson RE, Murad MH. Gynecomastia: pathophysiology, evaluation, and management. Mayo Clin Proc 2009;84(11):1010–5. Elsevier https://doi.org/10.1016/S0025-6196(11) 60671-X.

[13] Deepinder F, Braunstein GD. Gynecomastia: incidence, causes and treatment. Expert Rev Endocrinol Metab 2011;6(5):723–30. https://doi.org/10.1586/eem.11.57.

[14] Yu M, Finner A, Shapiro J, Lo B, Barekatain A, McElwee KJ. Hair follicles and their role in skin health. Expert Rev Dermatol 2006;1(6):855–71.

[15] Glintborg D, Andersen M. An update on the pathogenesis, inflammation, and metabolism in hirsutism and polycystic ovary syndrome. Gynecol Endocrinol 2010;26(4): 281–96. https://doi.org/10.3109/09513590903247873.

[16] Uçkan K, Demir H, Turan K, Sarıkaya E, Demir C. Role of oxidative stress in obese and nonobese PCOS patients. Int J Clin Pract 2022;2022. https://doi.org/10.1155/2022/4579831.

[17] Lanitis S, Dimopoulos N, Sivakumar S, Read J, Starren E, Al Mufti R, Hadjiminas DJ. Breast problems in male population; a nine-year single institution experience. Hell J Surg 2010;82(3):176–83. https://doi.org/10.1007/s13126-010-0030-x.

[18] Somani N, Harrison S, Bergfeld WF. The clinical evaluation of hirsutism. Dermatol Ther 2008;21(5):376–91. https://doi.org/10.1111/j.1529-8019.2008.00219.x.

[19] Gardner DG, Shoback DM. Greenspan's basic and clinical endocrinology. McGraw-Hill Education; 2017.

[20] Azziz R, Carmina E, Sawaya ME. Idiopathic hirsutism. Endocr Rev 2000;21(4):347–62.

[21] Rosenfield RL. Hirsutism and the variable response of the pilosebaceous unit to androgen. J Investig Dermatol Symp Proc 2005;10(3):205–8 [Elsevier] https://doi.org/10.1111/j.1087-0024.2005.10106.x.

[22] Hawryluk EB, English III JC. Female adolescent hair disorders. J Pediatr Adolesc Gynecol 2009;22(4):271–81. https://doi.org/10.1016/j.jpag.2009.03.007.

[23] Deplewski D, Rosenfield RL. Role of hormones in pilosebaceous unit development. Endocr Rev 2000;21(4):363–92.

[24] Jamilian M, Foroozanfard F, Mirhosseini N, Kavossian E, Aghadavod E, Bahmani F, Ostadmohammadi V, Kia M, Eftekhar T, Ayati E, Mahdavinia M. Effects of melatonin supplementation on hormonal, inflammatory, genetic, and oxidative stress parameters in women with polycystic ovary syndrome. Front Endocrinol 2019;10:273.

[25] Schmid G, editor. Nanoparticles: from theory to application. John Wiley & Sons; 2011.

[26] Dimitrakakis C, Zhou J, Bondy CA. Androgens and mammary growth and neoplasia. Fertil Steril 2002;77:26–33. https://doi.org/10.1016/S0015-0282(02)02979-5.

[27] Kanhai RC, Hage JJ, Van Diest PJ, Bloemena E, Mulder JW. Short-term and long-term histologic effects of castration and estrogen treatment on breast tissue of 14 male-to-female transsexuals in comparison with two chemically castrated men. Am J Surg Pathol 2000;24(1):74.

[28] Dejager S, Bry-Gauillard H, Bruckert E, Eymard B, Salachas F, LeGuern E, Tardieu S, Chadarevian R, Giral P, Turpin G. A comprehensive endocrine description of Kennedy's disease revealing androgen insensitivity linked to CAG repeat length. J Clin Endocrinol Metab 2002;87(8):3893–901. https://doi.org/10.1210/jcem.87.8.8780.

[29] Eren E, Edgunlu T, Korkmaz HA, Cakir ED, Demir K, Cetin ES, Celik SK. Genetic variants of estrogen beta and leptin receptors may cause gynecomastia in adolescent. Gene 2014;541(2):101–6. https://doi.org/10.1016/j.gene.2014.03.013.

[30] Ruan W, Monaco ME, Kleinberg DL. Progesterone stimulates mammary gland ductal morphogenesis by synergizing with and enhancing insulin-like growth factor-I action. Endocrinology 2005;146(3):1170–8. https://doi.org/10.1210/en.2004-1360.

[31] Dieudonné MN, Sammari A, Dos Santos E, Leneveu MC, Giudicelli Y, Pecquery R. Sex steroids and leptin regulate 11β-hydroxysteroid dehydrogenase I and P450 aromatase expressions in human preadipocytes: sex specificities. J Steroid Biochem Mol Biol 2006;99(4–5):189–96. https://doi.org/10.1016/j.jsbmb.2006.01.007.

[32] Blume-Peytavi U, Hahn S. Medical treatment of hirsutism. Dermatol Ther 2008;21 (5):329–39. https://doi.org/10.1111/j.1529-8019.2008.00215.x.

[33] Blume-Peytavi U. How to diagnose and treat medically women with excessive hair. Dermatol Clin 2013;31(1):57–65. https://doi.org/10.1016/j.det.2012.08.009.

[34] Miller JA, Jacobs HS. 11 treatment of hirsutism and acne with cyproterone acetate. Clin Endocrinol Metab 1986;15(2):373–89. https://doi.org/10.1016/s0300-595x(86)80031-7.

[35] Weber GF. Molecular therapies of cancer. Springer; 2015.

[36] Kromm J, Jeerakathil T. Cyproterone acetate–ethinyl estradiol use in a 23-year-old woman with stroke. CMAJ 2014;186(9):690–3. https://doi.org/10.1503/cmaj.130579.

[37] Kuhl H. Pharmacology of estrogens and progestogens: influence of different routes of administration. Climacteric 2005;8(Suppl):3–63. https://doi.org/10.1080/1369713 0500148875.

[38] Bińkowska M, Woroń J. Progestogens in menopausal hormone therapy. Prz Menopauzalny 2015;14(2):134–43. https://doi.org/10.5114/pm.2015.52154.

[39] Bizzarri N, Remorgida V, Leone Roberti Maggiore U, Scala C, Tafi E, Ghirardi V, Salvatore S, Candiani M, Venturini PL, Ferrero S. Dienogest in the treatment of endometriosis. Expert Opin Pharmacother 2014;15(13):1889–902. https://doi.org/10.1517/14656566.2014.943734.

[40] Sica DA. Pharmacokinetics and pharmacodynamics of mineralocorticoid blocking agents and their effects on potassium homeostasis. Heart Fail Rev 2005;10(1):23–9. https://doi.org/10.1007/s10741-005-2345-1.

[41] Takamura N, Maruyama T, Ahmed S, Suenaga A, Otagiri M. Interactions of aldosterone antagonist diuretics with human serum proteins. Pharm Res 1997;14(4):522–6. https://doi.org/10.1023/A:1012168020545.

[42] Cockshott ID. Bicalutamide. Clin Pharmacokinet 2004;43(13):855–78. https://doi.org/10.2165/00003088-200443130-00003.

[43] Foye WO, Lemke TL. Foye's principles of medicinal chemistry. Lippincott Williams and Wilkins; 2008.

[44] Neri R. Pharmacology and pharmacokinetics of flutamide. Urology 1989;34(4):19–21. https://doi.org/10.1016/0090-4295(89)90230-6.

[45] Shapiro J, Lui H. Vaniqa—eflornithine 13.9% cream. Skin Ther Lett 2001;6(7):1–3.

[46] Scheen AJ. Clinical pharmacokinetics of metformin. Clin Pharmacokinet 1996;30 (5):359–71. https://doi.org/10.2165/00003088-199630050-00003.

[47] Cosentino F, Grant PJ, Aboyans V, Bailey CJ, Ceriello A, Delgado V, Federici M, Filippatos G, Grobbee DE, Hansen TB, Huikuri HV. 2019 ESC guidelines on diabetes, pre-diabetes, and cardiovascular diseases developed in collaboration with the EASD. Eur Heart J 2020;41(2). https://doi.org/10.1093/eurheartj/ehz486.

[48] Cox PJ, Ryan DA, Hollis FJ, Harris AM, Miller AK, Vousden M, Cowley H. Absorption, disposition, and metabolism of rosiglitazone, a potent thiazolidinedione insulin sensitizer, in humans. Drug Metab Dispos 2000;28(7):772–80.

[49] Nissen SE, Wolski K. Effect of rosiglitazone on the risk of myocardial infarction and death from cardiovascular causes. N Engl J Med 2007;356(24):2457–71. https://doi.org/10.1056/NEJMoa072761.

[50] Eckland DA, Danhof M. Clinical pharmacokinetics of pioglitazone. Exp Clin Endocrinol Diabetes 2000;108(Suppl 2):234–42.

[51] Lee DS, Kim SJ, Choi GW, Lee YB, Cho HY. Pharmacokinetic–pharmacodynamic model for the testosterone-suppressive effect of leuprolide in normal and prostate cancer rats. Molecules 2018;23(4):909. https://doi.org/10.3390/molecules23040909.

[52] Chrisp P, Goa KL. Nafarelin. Drugs 1990;39(4):523–51. https://doi.org/10.2165/00003495-199039040-00005.

[53] Rahmani S, Turton P, Shaaban A, Dall B. Overview of gynecomastia in the modern era and the Leeds Gynaecomastia Investigation algorithm. Breast J 2011;17(3):246–55. https://doi.org/10.1111/j.1524-4741.2011.01080.x.

[54] van der Spuy ZM, Le Roux PA, Matjila MJ. Cyproterone acetate for hirsutism. Cochrane Database Syst Rev 2003;(4). https://doi.org/10.1002/14651858.CD001125.

[55] Batukan C, Muderris II, Ozcelik B, Ozturk A. Comparison of two oral contraceptives containing either drospirenone or cyproterone acetate in the treatment of hirsutism. Gynecol Endocrinol 2007;23(1):38–44. https://doi.org/10.1080/09637480601137066.

[56] Guerra-Tapia A, Pérez BS. Ethinylestradiol/chlormadinone acetate. Am J Clin Dermatol 2011;12(1):3–11.

[57] Raudrant D, Rabe T. Progestogens with antiandrogenic properties. Drugs 2003;63(5): 463–92.

[58] Chandra A, Rho AM, Jeong K, Yu T, Jeon JH, Park SY, Lee SR, Moon HS, Chung HW. Clinical experience of long-term use of dienogest after surgery for ovarian endometrioma. Obstet Gynecol Sci 2018;61(1):111–7. https://doi.org/10.5468/ogs.2018.61.1.111.

[59] Petraglia F, Hornung D, Seitz C, Faustmann T, Gerlinger C, Luisi S, Lazzeri L, Strowitzki T. Reduced pelvic pain in women with endometriosis: efficacy of long-term dienogest treatment. Arch Gynecol Obstet 2012;285(1):167–73. https://doi.org/10.1007/s00404-011-1941-7.

[60] Nadkarni P. Endometriosis managed using dienogest (oral progestin): a tailored individual management approach. Medicine 2019;26(10):17.

[61] Elger W, Beier S, Pollow K, Garfield R, Shi SQ, Hillisch A. Conception and pharmacodynamic profile of drospirenone. Steroids 2003;68(10–13):891–905. https://doi.org/10.1016/j.steroids.2003.08.008.

[62] Gregoriou O, Papadias K, Konidaris S, Bakalianou K, Salakos N, Vrachnis N, Creatsas G. Treatment of hirsutism with combined pill containing drospirenone. Gynecol Endocrinol 2008;24(4):220–3. https://doi.org/10.1080/09513590801948309.

[63] Christy NA, Franks AS, Cross LB. Spironolactone for hirsutism in polycystic ovary syndrome. Ann Pharmacother 2005;39(9):1517–21. https://doi.org/10.1345/aph.1G025.

[64] Müderris II, Bayram F, Özçelik B, Güven M. New alternative treatment in hirsutism: bicalutamide 25 mg/day. Gynecol Endocrinol 2002;16(1):63–6. https://doi.org/10.1080/gye.16.1.63.66.

[65] Lakryc EM, Motta EL, Soares JM, Haidar MA, Rodrigues de Lima G, Baracat EC. The benefits of finasteride for hirsute women with polycystic ovary syndrome or idiopathic hirsutism. Gynecol Endocrinol 2003;17(1):57–63. https://doi.org/10.1080/gye.17.1.57.63.

[66] Hamzavi I, Tan E, Shapiro J, Lui H. A randomized bilateral vehicle-controlled study of eflornithine cream combined with laser treatment versus laser treatment alone for facial hirsutism in women. J Am Acad Dermatol 2007;57(1):54–9. https://doi.org/10.1016/j.jaad.2006.09.025.

[67] Tang T, Norman RJ, Balen AH, Lord JM. Insulin-sensitising drugs (metformin, troglitazone, rosiglitazone, pioglitazone, D-chiro-inositol) for polycystic ovary syndrome. Cochrane Database Syst Rev 2003;(2). https://doi.org/10.1002/14651858.CD003053.

[68] Hodgson R, Chittawar PB, Farquhar C. GnRH agonists for uterine fibroids. Cochrane Database Syst Rev 2017;2017(10). https://doi.org/10.1002/14651858.CD012846.

[69] Magon N. Gonadotropin releasing hormone agonists: expanding vistas. Indian J Endocrinol Metab 2011;15(4):261. https://doi.org/10.4103/2230-8210.85575.

[70] Barros AC, Sampaio MD. Gynecomastia: physiopathology, evaluation and treatment. Sao Paulo Med J 2012;130:187–97. https://doi.org/10.1590/S1516-31802012000300009.

[71] Morello KC, Wurz GT, DeGregorio MW. Pharmacokinetics of selective estrogen receptor modulators. Clin Pharmacokinet 2003;42(4):361–72. https://doi.org/10.2165/00003088-200342040-00004.

[72] Lønning P, Pfister C, Martoni A, Zamagni C. Pharmacokinetics of third-generation aromatase inhibitors. Semin Oncol 2003;30:23–32 [WB Saunders] https://doi.org/10.1016/S0093-7754(03)00305-1.

[73] Brayfield A. Martindale: the complete drug reference. PHP; 2014.

[74] Dörwald FZ. Lead optimization for medicinal chemists: pharmacokinetic properties of functional groups and organic compounds. John Wiley & Sons; 2012.

[75] Braunstein GD. Gynecomastia. N Engl J Med 2007;357(12):1229–37. https://doi.org/10.1590/S1516-3180201200030000.

[76] Cuhaci N, Polat SB, Evranos B, Ersoy R, Cakir B. Gynecomastia: clinical evaluation and management. Indian J Endocrinol Metab 2014;18(2):150. https://doi.org/10.4103/2230-8210.129104.

[77] Coleman III WP, Glogau RG, Klein JA, Moy RL, Narins RS, Chuang TY, Farmer ER, Lewis CW, Lowery BJ. Guidelines of care for liposuction. J Am Acad Dermatol 2001; 45(3):438–47. https://doi.org/10.1067/mjd.2001.117045.

[78] Giuseppe A. Ultrasound-assisted liposuction for gynecomastia. In: Shiffman MA, Di Giuseppe A, editors. Liposuction principles and practice. Berlin; Heidelberg: Springer; 2006. p. 474–9. https://doi.org/10.1007/3-540-28043-X_70.

[79] Esme DL, Beekman WH, Hage JJ, Nipshagen MD. Combined use of ultrasonic-assisted liposuction and semicircular periareolar incision for the treatment of gynecomastia. Ann Plast Surg 2007;59:629–34. https://doi.org/10.1097/SAP.0b013e318038f762.

[80] Codazzi D, Bruschi S, Robotti E, Bocchiotti MA. Power-assisted liposuction (PAL) fat harvesting for lipofilling: the trap device. World J Plast Surg 2015;4:17.

[81] Blugerman G, Schalvezon D, Mulholland RS, Soto JA, Siguen M. Gynecomastia treatment using radiofrequency-assisted liposuction (RFAL). Eur J Plast Surg 2013;36:231–6. https://doi.org/10.1007/s00238-012-0772-5.

[82] Trelles MA, Mordon SR, Bonanad E, Moraga JM, Heckmann A, Unglaub F, et al. Laser-assisted lipolysis in the treatment of gynecomastia: a prospective study in 28 patients. Lasers Med Sci 2013;28:375–82. https://doi.org/10.1007/s10103-011-1043-6.

[83] Çelebioglu S, Ertaş NM, Özdil K, Öktem F. Gynecomastia treatment with subareolar glandular pedicle. Aesthetic Plast Surg 2004;28:281–6. https://doi.org/10.1007/s00266-004-1300-1.

[84] Jarrar G, Peel A, Fahmy R, Deol H, Salih V, Mostafa A. Single incision endoscopic surgery for gynaecomastia. J Plast Reconstr Aesthet Surg 2011;64:e231-6. https://doi.org/10.1016/j.bjps.2011.04.016.

[85] Ghasemiyeh P, Azadi A, Daneshamouz S, Mohammadi Samani S. Cyproterone acetate-loaded solid lipid nanoparticles (SLNs) and nanostructured lipid carriers (NLCs): preparation and optimization. Trends Pharm Sci 2017;3(4):275–86.

[86] Mohammadi-Samani S, Montaseri H, Jamshidnejad M. Preparation and evaluation of cyproterone acetate liposome for topical drug delivery. Iran J Pharm Sci 2009;5(4):199–204.

[87] Štecová J, Mehnert W, Blaschke T, Kleuser B, Sivaramakrishnan R, Zouboulis CC, Seltmann H, Korting HC, Kramer KD, Schäfer-Korting M. Cyproterone acetate loading to lipid nanoparticles for topical acne treatment: particle characterisation and skin uptake. Pharm Res 2007;24(5):991–1000. https://doi.org/10.1007/s11095-006-9225-9.

[88] Zeng J, Chen J, Chen L, Zheng W, Cao Y, Huang T. Enhanced oral bioavailability of chlormadinone acetate through a self-microemulsifying drug delivery system for a potential dose reduction. AAPS PharmSciTech 2018;19(8):3850–8. https://doi.org/10.1208/s12249-018-1193-y.

[89] Pankaj P, Kailash B, Rao RT, Kumud P, Ajit S, Singh KP. Micronization: an efficient tool for dissolution enhancement of Dienogest. Int J Drug Dev Res 2011;3:329–33.

[90] Nippe S, General S. Combination of injectable ethinyl estradiol and drospirenone drug-delivery systems and characterization of their in vitro release. Eur J Pharm Sci 2012;47(4):790–800. https://doi.org/10.1016/j.ejps.2012.08.009.

[91] Kelidari HR, Saeedi M, Akbari J, Morteza-Semnani K, Valizadeh H, Maniruzzaman M, Farmoudeh A, Nokhodchi A. Development and optimisation of spironolactone nanoparticles for enhanced dissolution rates and stability. AAPS PharmSciTech 2017;18(5):1469–74.

[92] Albash R, Fahmy AM, Hamed MI, Darwish KM, El-Dahmy RM. Spironolactone hyaluronic acid enriched cerosomes (HAECs) for topical management of hirsutism: in silico studies, statistical optimization, ex vivo, and in vivo studies. Drug Deliv 2021;28(1):2289–300. https://doi.org/10.1080/10717544.2021.1989089.

[93] Rungseevijitprapa W, Wichayapreechar P, Sivamaruthi BS, Jinarat D, Chaiyasut C. Optimization and transfollicular delivery of finasteride-loaded proniosomes for hair growth stimulation in C57BL/6Mlac mice. Pharmaceutics 2021;13(12):2177.

[94] Hamishehkar H, Ghanbarzadeh S, Sepehran S, Javadzadeh Y, Adib ZM, Kouhsoltani M. Histological assessment of follicular delivery of flutamide by solid lipid nanoparticles: potential tool for the treatment of androgenic alopecia. Drug Dev Ind Pharm 2016;42(6):846–53.

[95] Kumar A, Naguib YW, Shi YC, Cui Z. A method to improve the efficacy of topical eflornithine hydrochloride cream. Drug Deliv 2016;23(5):1495–501.

[96] Grewal IK, Singh S, Arora S, Sharma N. Application of central composite design for development and optimization of eflornithine hydrochloride-loaded sustained release solid lipid microparticles. Biointerface Res Appl Chem 2021;112:618–37.

[97] Maji R, Dey NS, Satapathy BS, Mukherjee B, Mondal S. Preparation and characterization of tamoxifen citrate loaded nanoparticles for breast cancer therapy. Int J Nanomedicine 2014;9:3107. https://doi.org/10.2147/IJN.S63535.

[98] Al Haj NA, Abdullah R, Ibrahim S, Bustamam A. Tamoxifen drug loading solid lipid nanoparticles prepared by hot high pressure homogenization techniques. Am J Pharmacol Toxicol 2008;3(3):219–24. https://doi.org/10.3844/ajptsp.2008.219.224.

[99] Fontana G, Maniscalco L, Schillaci D, Cavallaro G, Giammona G. Solid lipid nanoparticles containing tamoxifen characterization and in vitro antitumoral activity. Drug Deliv 2005;12(6):385–92. https://doi.org/10.1080/10717540590968855.

[100] Harivardhan Reddy L, Vivek K, Bakshi N, Murthy RS. Tamoxifen citrate loaded solid lipid nanoparticles (SLN™): preparation, characterization, in vitro drug release, and pharmacokinetic evaluation. Pharm Dev Technol 2006;11(2):167–77. https://doi.org/10.1080/10837450600561265.

[101] Hashem FM, Nasr M, Khairy A. In vitro cytotoxicity and bioavailability of solid lipid nanoparticles containing tamoxifen citrate. Pharm Dev Technol 2014;19(7):824–32. https://doi.org/10.3109/10837450.2013.836218.

[102] Abbasalipourkabir R, Salehzadeh A, Abdullah R. Tamoxifen-loaded solid lipid nanoparticles-induced apoptosis in breast cancer cell lines. J Exp Nanosci 2016;11(3):161–74. https://doi.org/10.1080/17458080.2015.1038660.

[103] Sudarshan B, Vikas S, Viveknand C, Vilas S, Suresh R, Ganesh D. Tamoxifen citrate loaded solid lipid nanoparticles—a novel approach in the treatment of ER+ breast cancer. Res J Pharm Dosage Forms Technol 2009;1(2):143–9.

[104] Shafaa M. Preparation, characterization and evaluation of cytotoxic activity of tamoxifen bound liposomes against breast cancer cell line. Egypt J Biomed Eng Biophys 2020;21(1):19–31.

[105] Zeisig R, Rückerl D, Fichtner I. Reduction of tamoxifen resistance in human breast carcinomas by tamoxifen-containing liposomes in vivo. Anticancer Drugs 2004;15(7):707–14. https://doi.org/10.1097/01.cad.0000136885.65293.e9.

[106] Bhatia A, Kumar R, Katare OP. Tamoxifen in topical liposomes: development, characterization and in-vitro evaluation. J Pharm Pharm Sci 2004;7(2):252–9.

[107] Ağardan NM, Değim Z, Yılmaz Ş, Altıntaş L, Topal T. Tamoxifen/raloxifene loaded liposomes for oral treatment of breast cancer. J Drug Deliv Sci Technol 2020;57:101612.

[108] Wang Z, Li Y. Raloxifene/SBE-β-CD inclusion complexes formulated into nanoparticles with chitosan to overcome the absorption barrier for bioavailability enhancement. Pharmaceutics 2018;10(3):76.

[109] Kala SG, Chinni S. Development of raloxifene hydrochloride loaded mPEG-PLA nanoparticles for oral delivery. Indian J Pharm Educ Res 2021;55(1):S135–48.

[110] Nagavarma BV, Yadav HK, Ayaz AV, Vasudha LS, Shivakumar HG. Different techniques for preparation of polymeric nanoparticles-a review. Asian J Pharm Clin Res 2012;5(3):16–23.

[111] Panyam J, Labhasetwar V. Biodegradable nanoparticles for drug and gene delivery to cells and tissue. Adv Drug Deliv Rev 2003;55(3):329–47.

[112] des Rieux A, Fievez V, Garinot M, Schneider YJ, Préat V. Nanoparticles as potential oral delivery systems of proteins and vaccines: a mechanistic approach. J Control Release 2006;116(1):1–27.

[113] Rao JP, Geckeler KE. Polymer nanoparticles: preparation techniques and size-control parameters. Prog Polym Sci 2011;36(7):887–913.

[114] Pottoo FH, Sharma S, Javed MN, Barkat MA, Harshita, Alam MS, Naim MJ, Alam O, Ansari MA, Barreto GE, Ashraf GM. Lipid-based nanoformulations in the treatment of neurological disorders. Drug Metab Rev 2020;52(1):185–204.

[115] Mishra V, Bansal KK, Verma A, Yadav N, Thakur S, Sudhakar K, Rosenholm JM. Solid lipid nanoparticles: emerging colloidal nano drug delivery systems. Pharmaceutics 2018;10(4):191.

[116] Muchow M, Maincent P, Müller RH. Lipid nanoparticles with a solid matrix (SLN®, NLC®, LDC®) for oral drug delivery. Drug Dev Ind Pharm 2008;34(12):1394–405.

[117] Parhi R, Suresh P. Production of solid lipid nanoparticles-drug loading and release mechanism. J Chem Pharm Res 2010;2(1):211–7.

[118] Garud A, Singh D, Garud N. Solid lipid nanoparticles (SLN): method, characterization and applications. Int Curr Pharm J 2012;1(11):384–93.

[119] Allen TM, Cullis PR. Liposomal drug delivery systems: from concept to clinical applications. Adv Drug Deliv Rev 2013;65(1):36–48.

[120] Valenzuela SM. Liposome techniques for synthesis of biomimetic lipid membranes. In: Nanobiotechnology of biomimetic membranes. Springer; 2007. p. 75–87.

[121] Sercombe L, Veerati T, Moheimani F, Wu SY, Sood AK, Hua S. Advances and challenges of liposome assisted drug delivery. Front Pharmacol 2015;6:286.

[122] Allahou LW, Madani SY, Seifalian A. Investigating the application of liposomes as drug delivery systems for the diagnosis and treatment of cancer. Int J Biomater 2021; 2021:3041969.

[123] Kumar S, Dilbaghi N, Saharan R, Bhanjana G. Nanotechnology as emerging tool for enhancing solubility of poorly water-soluble drugs. Bionanoscience 2012;2:227–50.

[124] Bhise K, Kashaw SK, Sau S, Iyer AK. Nanostructured lipid carriers employing polyphenols as promising anticancer agents: quality by design (QbD) approach. Int J Pharm 2017;526(1–2):506–15.

[125] Chauhan I, Yasir M, Verma M, Singh AP. Nanostructured lipid carriers: a groundbreaking approach for transdermal drug delivery. Adv Pharm Bull 2020;10(2):150.

[126] Subramaniam B, Siddik ZH, Nagoor NH. Optimization of nanostructured lipid carriers: understanding the types, designs, and parameters in the process of formulations. J Nanopart Res 2020;22:1–29.

[127] Krstić M, Medarević Đ, Đuriš J, Ibrić S. Self-nanoemulsifying drug delivery systems (SNEDDS) and self-microemulsifying drug delivery systems (SMEDDS) as lipid nanocarriers for improving dissolution rate and bioavailability of poorly soluble drugs. In: Lipid nanocarriers for drug targeting. William Andrew Publishing; 2018. p. 473–508.

[128] Kumar S, Dilbaghi N, Rani R, Bhanjana G, Umar A. Novel approaches for enhancement of drug bioavailability. Rev Adv Sci Eng 2013;2(2):133–54.

[129] Talele SG, Gudsoorkar VR, Pharmacy MV. Novel approaches for solidification of SMEDDS. Int J Pharm Biosci 2016;15:90–101.

[130] Jaiswal P, Aggarwal G. Bioavailability enhancement of poorly soluble drugs by SMEDDS: a review. J Drug Deliv Ther 2013;3(1):98–109.

Role of inflammation, oxidative stress, and angiogenesis in polycystic ovary syndrome (PCOS): Current perspectives

Ankita Wal[a]*, Biswajit Dash*[b]*, Vaibhav Jaiswal*[c]*, Divyanshi Gupta*[a]*, and Arun Kumar Mishra*[d]

[a]PSIT-Pranveer Singh Institute of Technology (Pharmacy), Kanpur, Uttar Pradesh, India, [b]School of Health and Medical Sciences, Adamas University, Kolkata, West Bengal, India, [c]IES Institute of Pharmacy, IES University, Bhopal, Madhya Pradesh, India, [d]Pharmacy Academy, IFTM University, Moradabad, Uttar Pradesh, India

Key points

- Polycystic ovarian syndrome (PCOS) is a metabolic and endocrine condition marked by hirsutism, skin conditions such as acne, unpredictable menstrual periods, menstrual cramps, hyperandrogenism, oligo-anovulation, and polycystic ovaries.
- PCOS has a global prevalence ranging from 6% to 25%, primarily affecting women of reproductive age.
- PCOS is also associated with an increased risk of cardiovascular disease, infertility, diabetic complications, cancer, and menstrual issues.
- Statins, insulin sensitizers, antiandrogen medication, and laparoscopic ovarian drilling are often used to treat PCOS.

1 Introduction

Polycystic ovarian syndrome, also known as PCOS, is an endocrine condition that mostly affects women of reproductive age. Polycystic ovarian syndrome (PCOS) is a multifaceted illness characterized by hyperandrogenism and recurrent anovulation [1]. Polycystic ovarian syndrome has a wide range of symptoms, making it difficult to determine the severity of the problem [1]. According to De Leo et al.'s [2] study, this syndrome can be a risk factor in the advancement of various ailments, including high blood pressure, insulin resistance, congestive heart failure, and several metabolic disorders. PCOS is characterized by a combination of external (environmental) and internal (hormone and neural) variables that result in androgen and ovulatory dysfunction, which results in irregular menstruation [3,4], dysmenorrhea [5,6], infertility [7,8],

Targeting Angiogenesis, Inflammation and Oxidative Stress in Chronic Diseases. https://doi.org/10.1016/B978-0-443-13587-3.00018-7
Copyright © 2024 Elsevier Inc. All rights are reserved, including those for text and data mining, AI training, and similar technologies.

hypothyroidism, and chronic anovulation. The pituitary gland's malfunctioning feedback mechanism, which can cause LH hypersecretion, is also related to it.

Bharali et al. [6] undertook a meta-analysis of the prevalence of PCOS in India, resulting in a 10% PCOS prevalence, especially among women of reproductive age. The worldwide incidence was estimated to be 15%–25% [7–9]. According to the National Institute of Health Office of Disease Prevention, around 7% of women in India suffer from PCOS [10–12]. PCOS is triggered by a combination of variables including lifestyle, genetics, insulin resistance, oxidative stress, and low-grade inflammation [13]. If this illness is not addressed, it might lead to infertility [14]. PCOS is a neuroendocrine condition characterized by elevated hormones such as luteinizing hormone secretion frequency, luteinizing hormone serum concentration rise, amplitude rise, and LH/FSH ratio rise [15]. According to available PCOS research, this illness may be recognized by looking for cysts in the ovaries, no ovulation, and increased testosterone levels. Normal testosterone levels in women range between 15 and 70 ng/dL, whereas in PCOS, they rise to 80–90 ng/dL [16]. Several other features that induce similar symptoms in the condition, including hypothyroidism, hyperplasia, and elevated blood prolactin levels, can also be used to make the diagnosis [17]. To reduce the prevalence of the illness or risk factors, dietary and lifestyle changes such as increased consumption of fresh vegetables, fruits, and dairy products, and regular exercise are suggested [18]. PCOS is also linked to low-grade inflammation, angiogenesis, and oxidative stress. In this chapter, we examined in depth the PCOS and the pathologies associated with illness development, with a focus on insulin resistance, inflammation, angiogenesis, and oxidative stress, as well as all-possible PCOS therapies.

2   Polycystic ovarian syndrome

PCOS is a female sex hormone endocrine disease [19]. According to the National Institutes of Health, PCOS is characterized by high levels of androgen (hyperandrogenism) and oligo-ovulation [20]. The ovaries, together with the fallopian tube, uterus, and vagina, are components of the female reproductive system. The female reproductive system's ovaries are responsible for a lifelong supply of eggs, which are immature and kept in small fluid-filled structures called follicles [21]. Each month, the pituitary gland, situated at the base of the brain, secretes hormones into the bloodstream that guide the activity of the ovaries by secreting the hormone follicle-stimulating hormone (FSH) and luteinizing hormone [22]. These hormones reach the ovaries and cause several 100 immature eggs to grow, increasing the size of the follicles. As the eggs mature, these follicles emit estrogen (female sex hormone) at a predetermined level. Ovulation occurs when the pituitary gland delivers a rush of LH to the ovaries, forcing the most developed follicle to open and release its eggs. The free eggs proceed via the fallopian tube to be fertilized, after which all of the immature follicles and eggs are dispersed. If the eggs are not fertilized, the egg and uterine lining are shed during the subsequent menstrual cycle [23]. In PCOS, the pituitary gland may secrete an excessively large quantity of LH into the bloodstream, disrupting the regular menstrual cycle. As a result, follicles do not develop and ovulation does not occur, potentially resulting in infertility. Some immature follicles do not disintegrate or stay as fluid-filled sacs, but instead form cysts [24,25].

Furthermore, with PCOS, blood levels of insulin hormone generated by the pancreas are elevated. When too much insulin is paired with excessive amounts of LH, the ovaries produce an excess of testosterone, which prevents fertilization and leads to infertility in women [26,27]. PCOS is also linked to skin abnormalities such as excessive hair growth and acne. PCOS increases the risk of type 2 diabetes, heart disease, high blood pressure, blood cholesterol abnormalities, and endometrial cancer [28].

PCOS is generally explained with the help of different types of phenotypes like phenotype 1, a classic PCOS (combination of menstrual irregularities, hyperandrogenism, and oligo-ovulation), phenotype 2 (hyperandrogenism and oligo-ovulation), phenotype 3 (hyperandrogenism and absence of oligo-ovulation), and phenotype 4 (ovulation and nonhyperandrogenism) [29,30]. In the following book chapter, we look at the importance of mild inflammation, oxidative stress, and angiogenesis, as well as therapy and positive reproductive results suffering from polycystic ovary syndrome.

2.1 Etiology of PCOS

Endocrine and metabolic disorders are what are known as polycystic ovarian syndrome in women who are of reproductive age. In general, this illness is characterized by the development of an ovarian cyst, an increase in androgen, which interferes with menstruation, and several skin issues, including acne and hirsutism [31]. The cause of PCOS is largely unclear, but genetic, environmental, and lifestyle variables, including obesity and endocrine abnormalities, all seem to be important contributors. The various causes of polycystic ovarian syndrome in women are depicted in Fig. 1.

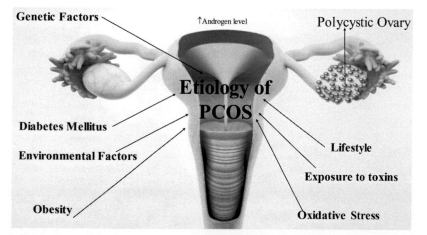

Fig. 1 Environmental factors, exposure to toxins, lifestyle comorbidities, and genetic factors influence the androgen level through the pituitary gland in the base of the brain which results in several symptoms like hirsutism, acne, and menstrual irregularities. All these factors give rise to PCOS.

Infertility, hirsutism, and acne-like clinical signs are caused by these variables, which are thought to be connected to insulin resistance, hyperandrogenism, and inflammation in PCOS [32]. As stated by Lim et al. [33], the same genetic and environmental variables that cause obesity also have a big impact on PCOS. These risk factors encourage the growth of insulin resistance and hyperinsulinemia as well as hyperandrogenism. Finally, many of the poor reproductive outcomes in PCOS may be caused by the interaction of inflammation, insulin resistance, and hyperandrogenism [34]. Obesity, oxidative stress, and inflammation indicators are also favorably connected with androgen levels in PCOS, according to González et al. [35], albeit the specific relationships between oxidative stress and inflammatory markers remain unclear.

3 Pathogenesis

Understanding the typical biochemical and molecular mechanisms behind steroidogenesis as well as the typical androgen physiology is essential to comprehending the pathophysiology of PCOS. Under typical circumstances, the synthesis of testosterone is roughly equally split between the adrenal glands and the ovaries. Negative feedback from the neuroendocrine system has very little impact on the regulation of androgen production. The adrenal glands and the ovaries both release androgens in response to tropic hormones, luteinizing hormones [36], and adrenocorticotropic hormones. The modulation of androgen production in response to stimulation by tropic hormones appears to be mostly dependent on intraglandular paracrine and autocrine pathways. Several studies also show the positive correlation between different mediators (inflammation or low-grade inflammation, oxidative stress, hyper androgens, insulin resistance, and angiogenesis) which contribute to PCOS [37,38]. Additionally, androgens have been found in studies to promote the rise of obesity by enhancing the process of differentiation of preadipocytes into adipocytes, particularly in the abdomen [39]. Chronic inflammation and high levels of cytokines, chemokines, and signs of oxidative stress characterized obesity as a metabolic illness [40]. It has been demonstrated that the activation of inflammatory pathways in adipocytes prevents the storage of triglycerides by increasing the release of free fatty acids, which may result in insulin resistance [41].

4 Androgen and PCOS

More than 60% of PCOS patients exhibited raised levels of several androgens, including increased testosterone, increased proandrogens androstenedione, and increased DHEAS. Hyperandrogenism is a distinguishing mark in PCOS patients [42].

The generation of androgens and all steroid hormones depends on the cleavage of cholesterol side chains by the enzymes cytochrome P450c17 and P450c455 [43]. Preovulatory follicles help to speed up the sluggish process of ovarian steroidogenesis. The 17-hydroxylase enzyme is created by cortisol, and its substrate, 17-ketosteroids, commonly known as androstenedione or dehydroepiandrosterone (DHEA), is the precursor of androgen [44]. The emergence of type 2 5-isomerase-3-hydroxysteroid dehydrogenase is facilitated by these hormones. Androgens are primarily converted to dihydrotestosterone rather than estradiol in small ovarian follicles before follicle selection because of the enhanced activity of the enzyme steroid 5′-reductase (5′RD) [45]. Type 1 and type 2 5RD isozymes are used by the theca, stroma, and granulosa cells to carry out this function, however, type 1 activity is more common in granulosa cells [34]. Minor pathways include the conversion of estradiol to estrone and the reconversion of testosterone to cells. These processes are complicated and include several stages, including homologous desensitization of LH and control of LH function, involving numerous hormones and growth factors. This dysregulation of androgen output was initially attributed to insulin excess, which is known to sensitize the ovary to LH by impeding the normal process of homologous desensitization to LH [46]. The excess androgen production and hyperactivity of the ovaries are, in brief, caused by intrinsic theca cell dysfunction, adrenocortical androgenic dysfunction, granulosa cell dysfunction, and disrupted folliculogenesis. As a result, ACTH hyperresponsiveness and DHEA hypersensitivity are additional characteristics of PCOS [47].

Androgen excess from the brain gives rise to various endocrine features (hyperandrogenism, LH hypersecretion), metabolic features (increased intraabdominal fat, decreased adiponectin, leptin resistance, dyslipidemia, nonalcoholic fatty liver), and reproductive features (irregular cycles, polycystic ovaries, and ovulatory dysfunction) through brain ovarian and brain adipocytes axis as shown in Fig. 1. Reduced levels of the sex hormone binding globulin lead to the production of free androgen and disruption of the metabolic profile due to insulin resistance and hyperinsulinemia. All of these factors contribute to hyperandrogenism which results in enlarged, multicystic ovaries [48]. All of these factors contribute to various reproductive, metabolic, and endocrine features like polycystic ovaries, ovulatory dysfunction, increased intraabdominal fat, decreased adiponectin, leptin resistance, dyslipidemia, nonalcoholic fatty liver, hyperandrogenism, and LH hypersecretion as shown in Fig. 2.

5   PCOS and low-grade inflammation

Patients with PCOS who have chronic inflammation run the risk of comorbid conditions and long-term cardiometabolic effects. PCOS and low-grade systemic inflammation commonly coexist. In terms of inflammation, circulating levels of several kinds of cytokines may play a part in the etiology of PCOS. Some studies have demonstrated the pathophysiological involvement that interleukin-1b, interleukin-1 receptor antagonist, interleukin-6, interleukin-17, and interleukin-18 play in the development of PCOS [49–51]. Increased levels of C-reactive protein, interleukin 18 [52], tumor necrosis factor, interleukin 6, and ferritin have all been linked to PCOS in women [53]. Additionally, MIP-1 (a chemokine) levels, which are necessary for leukocyte activation, are associated with PCOS. Inflammatory markers such as transforming factor 1, nuclear factor-kappa B, and omentin are also present.

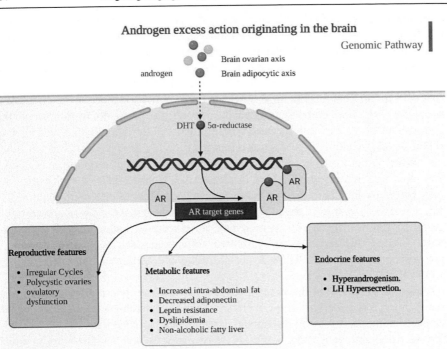

Fig. 2 Normal pathophysiology of polycystic ovarian syndrome due to the hyperresponse of the pituitary gland located in the base of the brain which results in increased secretion of androgen (testosterone and luteinizing hormone). The dysfunction of theca cell and granulosa cells at the androgen receptor site in the presence of 5α reductase leads to various reproductive, endocrine, and metabolic dysfunction.

5.1 C-reactive protein

In addition to many other inflammatory indicators, blood levels of C-reactive protein (CRP) are widely used to determine the presence of low-grade chronic inflammation. Interleukin-6 (IL-6) is released by activated immune cells like macrophages and adipocytes, and its release causes the formation of CRP, an acute-phase protein that is produced from the liver. The majority of research employs CRP as a biomarker to identify PCOS [54,55]. CRP is investigated as a trigger for inflammation that is typically formed in adipose tissue and the liver, followed by the activation of interleukin-6 and tumor necrosis factor [56].

5.2 Interleukins

According to Rostamtabar's study [57], interleukins are major proinflammatory mediators produced by mononuclear cells and adipose tissue cells. These interleukins are responsible for the synthesis of C-reactive protein which leads to inflammatory states in PCOS. Additionally connected to insulin resistance, cardiovascular disease, and obesity is the amount of IL-6 and IL-18. High body mass index

(BMI) and high insulin levels are linked to elevated interleukin levels, which are further linked to elevated androgen levels, according to Lin et al. [58]. From the recent literature survey, it was found that increased interleukin levels give rise to the follicle-stimulating hormone resulting in hyperandrogenism and oligo-ovulation in PCOS patients. As mentioned previously, obesity may be linked to an increase in proinflammatory markers, the penetration of macrophages in adipose tissue, and low-grade systemic inflammation. IL17 causes IL1, TNF, and IL6 to be released by adipose tissue macrophages, which aids in the development of the inflammatory state. IL-17A can affect inflammatory processes by changing adipogenesis and glucose metabolism [59,60].

5.3   Tumor necrosis factor (TNF-α)

TNF is a cytokine that is produced by many immune and nonimmune cells, including epithelial, endothelial, fibroblast, monocyte, and macrophage cells. The main TNF producers are the macrophages in adipose tissue. In response to the growth of visceral adipose tissue, adipokine synthesis, and TNF release are stimulated by adipose tissue, an endocrine organ. It was discovered that women with PCOS had greater blood levels of the cytokine TNF than those without the condition. Theca intra cells can grow in quantity and cause follicular hyperplasia when TNF levels are elevated [61]. TNF is produced as a result of the JNK pathway. Those with PCOS have higher follicular fluid (FF) TNF levels than those without PCOS. Yamamoto et al. claim that TNF reduces progesterone synthesis by suppressing the expression of genes essential in the hormone's synthesis. This decrease in progesterone production [62] results in the death of granulosa cells by apoptosis and autophagy which also plays important role in the progression of poly ovarian cyst syndrome in women's at reproductive age.

5.4   Peroxisome proliferator-activated receptor-γ and NF-κB

The proinflammatory disorder PCOS is associated with increased NF-κB transcription factor expression. Numerous factors promote the activation of NF-κB signaling, including oxidative stress, hyperglycemia, hyperinsulinemia, pharmacological agents, hormones, and inflammatory cytokines [63,64]. Furthermore, by boosting the M1 phenotype of macrophages, both NF-κB and AP1 promote the inflammatory state in adipose tissue. According to a recent study, PCOS patients had lower levels of PPAR expression than healthy controls, which suggests that a surge in LH and testosterone causes a decline in PPAR levels in ovarian granulosa cells [65]. In fact, the development of PCOS may be linked to an abnormal control of PPAR and LH levels [66].

5.5   Chemokines (MCP-1 and MIF) and other mediators

Monocyte, neutrophil, and lymphocyte migration to the site of infection are triggered by chemokines, proteins that develop in response to signals from inflammatory mediators [67]. MCP1 is a potent factor in monocyte attraction toward the site of the inflammatory process and is produced by a range of cell types. The National Center for Biotechnology Information claims that MCP1 may have a role in follicular dynamics, act as a potent source of IL1, and maybe regulate the ovulation process. PCOS and chronic inflammatory diseases are intimately related as a result of the elevation of these inflammatory markers [68]. Increased monocyte chemoattractant protein-1 expression and its levels in plasma are linked to hypoxia, which is increased in obese tissue. Hypoxia is believed to have an important function in the onset of inflammatory responses in obese PCOS patients. Furthermore, MCP1 overexpression is related to hyperglycemia and insulin resistance. In PCOS cases with insulin resistance, the elevation in glucose level is likely to trigger the inflammatory process more [69].

The development of illness also involves other inflammatory mediators such as adiponectin, vaspin resistin, omentin-1, and chemerin. Elevated chemerin levels are linked to PCOS, which causes ovarian dysfunction and follicular steroidogenesis [70].

6 PCOS and oxidative stress

According to the Gar D. research [45], advanced glycation end product is associated with hyperandrogenism owing to enzymes such as 17-hydroxylases, steroidogenic sudden regulating protein, and the cholesterol-related chain cleavage enzyme cytochrome. Reactive oxygen species are produced more often as a result of these enzymes.

Oxidative stress promotes DNA damage and methylation in the early stages of cancer by activating protooncogenes and inhibiting antioncogenes [71]. According to a literature review, oxidative stress promotes tumor and genetic instability and raises cancer risk in PCOS patients. It is also connected to several other disorders such as inflammation, insulin resistance, obesity, and so on. OS-generated ROS and proinflammatory chemicals can primarily induce IR via activation of connected pathways of signaling such as the nuclear factor-κB and JNK [72,73]. By stimulating cell proliferation signaling pathways, hyperinsulinemia, which is employed to offset IR, contributes to cancer etiology, and ultimately results in malignant transformation. Additionally, too much testosterone may contribute to obesity and cause OS, IR, and inflammation in vivo. OS is therefore thought to be a key player in the initiation of malignancies in PCOS [74]. Additional potential oxidative stress-mediated processes may contribute to the etiology of PCOS-related malignancies, although this is uncertain [75]. These unbound ROS cause a variety of expression, transcription factor, apoptotic, bacterial, and inflammatory effects to be dysregulated. Tyrosine kinase, Rho kinase, and transcription factors can all be activated more quickly by ROS whereas protein tyrosine phosphatase (PTP) can be inactivated. Since several OS markers are elevated in PCOS, it is thought that OS may contribute to the pathogenesis of PCOS [76]. To determine the likelihood of oxidative damage and associated diseases, as well as to help with the prevention and management of oxidative disorders, the measurement of oxidative stress and antioxidant biomarkers has been recommended [77,78]. Table 1 describes the impact of oxidative stress on PCOS, the variety of actions reactive oxygen species have on cell activity, and their connections to PCOS.

Table 1 Reactive oxygen species' diverse action on cell function and their relation with PCOS.

S. no.			Relation with PCOS	References
1.	Activation of redox-sensitive transcription factor	Regulate proinflammatory and cytokines expression, apoptosis, and cell differentiation	PCOS increases inflammation ($\uparrow$CRP, $\uparrow$IL-6, $\uparrow$IL-18)	[79]
2.	Activation of protein kinase	Increased extracellular stress leads to cell death (necrosis/apoptosis)	$\uparrow$ Action of tyrosine kinase, $\uparrow$ growth factor induces degradation of insulin receptor substance	[80,81]
3.	Opening of ion channel	$\uparrow$ ROS $\rightarrow$ $\uparrow$ Ca^{2+} Results in unstable homeostasis	PCOS showed dysregulation of calcium Results in irregular menses and reproductive dysfunction	[81,82]
4.	Protein oxidation	$\uparrow$ Formation of carbonyl product	$\uparrow$ AOPPs (advanced oxidation protein products) in PCOS	[83,84]
5.	Lipid peroxidation	$\uparrow$ Peroxyl radical and $\uparrow$ malondialdehyde	Biomarker of PCOS	[85]
6.	DNA oxidation	Free radicals damage DNA $\uparrow$ H$_2$O$_2$ $\rightarrow$ $\uparrow$ damage	H$_2$O$_2$ is associated with PCOS	[86]

7 PCOS and (IR) insulin resistance

IR in PCOS women is controlled by miRNA expression and DNA methylation. Nutrition, environment, and mood changes can all have a negative influence on insulin sensitivity [87]. Recent research suggests that in people with IR, a disruption of the intestinal microbes and abnormal levels of metabolites produced by bacteria may result in an inability of the immune system to properly signal, chronic low-grade inflammation, as well as rise in the synthesis of proinflammatory cytokines [88]. PCOS is now known to commonly coexist with significant insulin resistance and insulin secretion problems. In addition to obesity, these aberrations explain the substantially greater prevalence of glucose intolerance in PCOS. Interleukin-18, macrophage inflammatory protein-1 [89], and monocyte chemoattractant protein-1 levels have been observed to be elevated and linked with PCOS, as well as resistance to insulin and complications of metabolic syndrome [90]. Fig. 3 depicts the relationship between insulin resistance and PCOS.

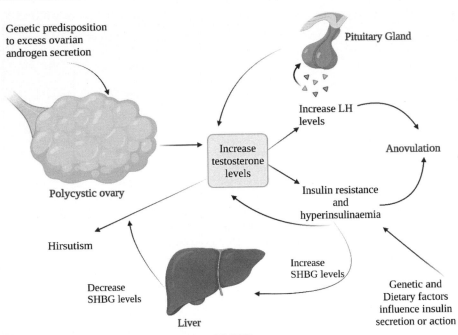

Fig. 3 Correlation of insulin resistance and PCOS.

8 PCOS and angiogenesis

Angiogenesis in the ovary is required for ovulation, follicular development, and the subsequent expansion and regression of the corpus luteum [91]. As illustrated in Fig. 4, all the components of the angiogenic factor family, including VEGF, angiopoietins, growth factor derived from platelets (PGF), transforming growth factor (TGF), and basic fibroblast growth factor (FGF), appear to be dysregulated in PCOS [92]. The elevated stromal vascularity found in PCOS is most likely due to an imbalance in angiogenic factors. Angiogenic factor dysregulation may have a role in the development and progression of PCOS, making it associated with ovulatory problems, inability to conceive, and a condition called ovarian hyperstimulation syndrome, which are all frequent in PCOS women [93,94]. VEGF is a glycoprotein that binds to heparin and functions as an endothelial cell (EC) mitogen, enhancing vascular permeability [91,95]. In research published in 2010, Peitsidis and Agrawal revealed the function of angiogenesis in the onset of PCOS [96]. Many studies show that these factors contribute to ovary stromal hypervascularity, ovulatory disorder, and decreased endometrial receptivity, all of which are distinguishing hallmarks of PCOS [97]. Changes in these angiogenic factors result in cyst production, increased primary follicles, and increased primordial follicles, all of which contribute to PCOS. Dysregulation of ovarian angiogenesis contributes to abnormal follicular development in PCOS patients [93]. Ovarian angiogenesis is a tightly regulated process that necessitates a precise balance of angiogenic factors, which is disrupted in PCOS patients with aberrant ovarian blood flow and angiogenesis. Such altered vasculature may

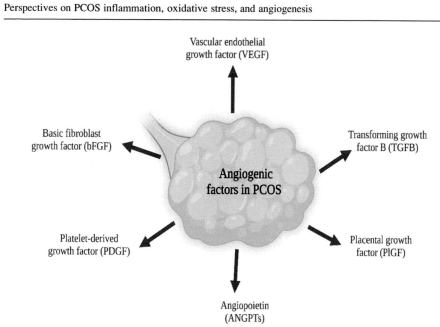

Fig. 4 Several angiogenic factors responsible for PCOS.

result in PCOS ovarian features such as abnormal follicular formation [94]. Anovulation and the formation of cysts are caused by development, an increase in the number of small follicles, and a failure to identify the dominant follicle. The equation summarizes the key angiogenic and ovarian changes observed in women with PCOS [98].

9   Current and novel treatment approaches

PCOS treatment in women is based on detecting and managing symptoms. Menstrual irregularities, androgen-related symptoms, and infertility due to anovulation are all possibilities. In addition to these phenotypes, recent research has shown that therapy should address additional characteristics such as higher antiMullerian hormone levels, plasma metabolomics, and gut microbiome composition, all of which are severe phenotypes in PCOS. Since there is no treatment for PCOS, only medications to manage symptoms are available, such as hormonal birth control, which prevents pregnancy, controls menstrual cycles, and suppresses testosterone synthesis in the ovaries. Antiandrogen medicines are used to treat excessive hair growth and acne. Metformin is a diabetes medicine that lowers insulin levels and regulates your menstrual cycle, and reproductive medications can promote ovulation. If fertility medication has not succeeded in restoring ovulation, then a laparoscopic ovarian drilling procedure is used.

9.1   Androgen receptor antagonists

Androgen receptor antagonists have been demonstrated to be beneficial in the treatment of PCOS symptoms, including irregular menstruation, hirsutism, acne, and androgenic alopecia. Paris et al. [99] studied flutamide as a competitive testosterone antagonist and reported that it reduced hirsutism, irregular menstruation or ovulation, and acne. This study also discussed obesity therapy (weight loss). Also beneficial for decreasing cholesterol (LDL and triglycerides) [99].

According to Rashid et al.'s study, finasteride and spironolactone are utilized to reduce androgen levels in PCOS patients [100]. These medications have also been shown to be useful in the treatment of hirsutism and in preventing the action of testosterone. Type 2 (5-reductase inhibitor) decreases hirsutism by inhibiting dihydrotestosterone synthesis [101,102].

9.2 Insulin sensitizers

Insulin-sensitizing drugs have been shown to lower androgen levels by enhancing insulin sensitivity. Metformin and thiazolidines have been shown to reduce glucose uptake, block hepatic glucose synthesis, and increase insulin sensitivity in peripheral tissue [103]. These medicines, according to the American College of Obstetricians and Gynecologists, are also beneficial in increasing ovulation, maintaining regular menstrual cycles, and lowering circulating testosterone levels [104].

According to the EHSRE and ASRM consensus statement, metformin should be confined to women with glucose intolerance alone, because metformin treatment does not boost ovulation and live-birth rates when compared to clomiphene citrate alone [105]. However, the Nestler [106] study's findings advocate metformin in the treatment of PCOS infertility, implying that its use should not be confined to patients with diabetes and should be used in conjunction with clomiphene citrate to increase ovulation rate [107].

New insulin sensitizers are also being introduced these days. Inositol, GLP-1 agonists, Liraglutide, Exenatide, DPP-4 inhibitors, and clomiphene citrate are all quite effective in the treatment of PCOS [108]. Table 2 discusses insulin-sensitizing agents and their role in PCOS patients.

10 Treatment for inflammation

Recently conducted research discovered that pentoxifylline has antiTNF properties that reduce inflammatory mediators and thereby protect ovarian cells. So far, it appears to be working by focusing on proinflammatory cytokines. Further study is needed, however, to assess the impact of antiinflammatory medicines on PCOS problems.

10.1 Revasterol treatment

It is found that Resveratrol had anti-inflammatory activities, achieved by reducing the levels of nuclear factor-κB and the products regulated by NF-κB. The administration of resveratrol was demonstrated to reduce the levels of interleukin-6, interleukin-1, cytokines such as interleukin-18, the nuclear factor-κB, and CRP in the blood [119,120]. Resveratrol is beneficial in the treatment of PCOS in women because it

Table 2 Various insulin-sensitizing agents used in PCOS along with their role.

Treatments	Agents	Role in PCOS	References
Insulin-senstizing agents used in PCOS patients	Metformin	Improve insulin sensitivity ↓ Androgen levels ↓ Testosterone ↓ Hirsutism score in PCOS patients Improve ovulation	[109–111]
	Inositol	Treat endocrine and metabolic disorders ↓ Circulating androgen ↓ LH, FSH, and testosterone	[112–114]
	Liraglutide, Exenatide (GLP-1 agonists)	↓ Fat ↓ Hyperglycemia Improved menstrual cycle	[115–117]
	Sitagliptin, Alogliptin, and Linagliptin (DPP-4 inhibitors)	↓ Blood glucose ↓ Fat Improve metabolic disorder	[118]

reduces testosterone, LH, and DHEAS levels. When paired with other medications, resveratrol is helpful for women with PCOS, notably for hyperlipidemia [121,122]. Revasterol treatment effectively suppresses inflammation in PCOS patients by inhibiting inflammatory mediators and decreasing levels of various hormones such as testosterone, follicle-stimulating hormone, luteinizing hormone, thyroid-stimulating hormone, dehydroepiandrosterone sulfate, and prolactin [123–126].

10.2 Statin treatment

Statins are useful in the treatment of PCOS because they reduce sex steroid synthesis, relieve dyslipidemia, and reduce ovarian androgen production by reducing androgen production by thecal cells [127,128]. Statins may be beneficial in the medical treatment of dyslipidemia, oxidative stress, and hyperandrogenemia in PCOS [129]. In a randomized controlled trial, statins were found to improve clinical and biochemical abnormalities associated with one ovulation dysfunction in women with PCOS [130]. On the other hand, statin therapy has been shown to reduce chronic inflammation and improve lipid profiles in PCOS [131,132].

10.3   Role of nutritional supplement in PCOS

Since ovarian follicle development and ovulation rates are largely regulated by nutrition-associated signaling pathways, nutritional deficiency is significantly linked to an etiology of PCOS [133,134]. All trans-retinol-treated theca interna cells generated higher dehydroepiandrosterone and mRNA accumulation of cytochrome P450 17-hydroxylase (CYP17), which is implicated in androgen production and retinol biosynthesis, according to research by Hahn et al. [135] and Tan et al. [136]. Through the activation of the RBP4 gene, obesity and impaired glucose metabolism may potentially contribute to the altered gonadal and adrenal steroid composition in obese PCOS women [136]. Vitamin B has also been related to an increase in homocysteine, which raises the risk of PCOS's reproductive system as well as heart disease. Kaya et al. reported that in PCOS women, insulin resistance, obesity, and higher Hcy levels were related to lower blood insulin B12 concentrations [137]. Folic acid treatment for 2–3 months offered effective advantages in lowering elevated serum Hcy levels, particularly in women who did not have insulin resistance.

Günalan et al. [138] found in their study that vitamin and mineral supplements can help with PCOS symptoms such as immature oocytes, hyperinsulinemia, hyperandrogenism, high BMI, cardiovascular illnesses, and mental and psychological disorders (as shown in Table 3).

10.4   Role of microparticulate and nanoparticulate drug delivery systems

Comparing particle drug delivery systems to traditional dosage methods reveals various benefits. These advantages include the capacity to use various routes of administration (e.g., oral, inhalation, parenteral), greater concentrations, less variation in GI transit time, minimal variation, low risk of dumping doses, reduced side effects, and hydrophilic and hydrophobic drug loading [149]. The effective use of metformin in PCOS therapy may depend on the development of microparticles and nanoparticle drug delivery systems, which may also help with metformin bioavailability, dosage frequency, gastrointestinal side effects, and toxicity [150]. Metformin nanoparticles decreased oxidative stress in the ovaries and liver by lowering the levels of malondialdehyde, androgenic hormone, and total nitrite while increasing diminished levels of the enzyme's superoxide dismutase and catalase, according to research by Butt et al. on the impact of nanoparticles of selenium on letrozole-induced PCOS in Wistar rats [151]. Wang et al. (2013) investigated the impact of nanoparticles of zinc oxide on PCOS by decreasing Ca ATPase production [152].

11   Treatment for acne and hirsutism

11.1   Cosmetic interventions in PCOS

Other treatments for PCOS (hirsutism) symptoms include the use of a specific tropical ointment and laser permanent hair removal. The most popular method for treating hirsutism is now using a light epilation laser [153]. Due to the high cost of this therapy,

Table 3 Several nutritional supplements used in PCOS.

S. no.	Nutraceuticals	Role in PCOS	References
1.	**Vitamin A** Retinols, Retinoids	↑ Antioxidant activity ↑ 17-α hydroxylase	[139]
2.	**Vitamin B** Folic acid, vitamin B6, vitamin B12	Regulates homocysteine in PCOS patients	[140]
3.	**Vitamin D**	Regulation of calcium in PCOS patients ↓ obesity ↑ Advanced glycosylated end receptors	[141]
4.	**Vitamin E**	↑ Antioxidant activity	[142]
5.	**Bioflavonoids**	↓ Inflammation Antidiabetic, treat metabolic syndrome in PCOS patients Improve ovarian and uterine morphologic appearance	[143]
6.	**Carnitine**	↑ Oocyte maturation ↓ Oxidative stress	[144,145]
7.	**Minerals** Calcium, chromium, magnesium, selenium, zinc	↓ Serum total cholesterol, LDL Improve lipid metabolism Role in insulin metabolism	[146]
8.	**Other supplements** Melatonin, N-acetyl-L-cysteine, omega-3 fatty acids, probiotics	Delayed onset of glucose intolerance, and hyperglycemia	[147,148]

some patients opted for tropical ointments like eflornithine hydrochloride, which permanently inhibits ornithine decarboxylase in the human epidermis, preventing cell division and synthetic processes, and slowing the growth of new hair [100,154].

11.2 Surgical treatment

11.2.1 Laparoscopic ovarian drilling

In PCOS individuals, the ovaries may acquire a thick outer layer, which might affect ovulation. Ovarian drilling boosts fertility by cutting through a tough outer layer [155]. Since ovarian drilling directly affects testosterone production, many women ovulate more often. Actually, ovarian drilling is a simple and minimally invasive procedure [156]. In this treatment, a tiny needle is inserted into the ovary, and an electric current is used to burn away small patches of ovarian tissue that over time create testosterone. This method reduces testosterone production and promotes the beginning of ovulation [157].

12　Future direction

Further investigation into the genetic causes and the development of PCOS is required to pinpoint preventive risk factors and efficient therapeutic modalities for this condition. These individuals' quality of life has increased because of the use of laser hair removal and other cosmetic procedures. The success and safety profiles of studies on conventional/folk medicine for the management of PCOS, however, are unclear and need to be further investigated. GLP-1 agonists, inhibitors of DPP-4, and SGLT2 antagonists are also used to treat PCOS, according to recent studies, however, there have been relatively few clinical trials for these medications. Therefore, further clinical trials are encouraged [158,159].

13　Conclusion

This review came to the conclusion that the disease PCOS results from luteinizing hormone dysregulation and follicle-stimulating hormone malfunction, both of which contribute to ovarian steroidogenesis. According to the aforementioned studies, 90% of PCOS patients reported having hyperandrogenic ovulation. Due to a dearth of research, the pathophysiologic and biochemical causes of the functioning of ovarian hyperandrogenism remain unknown. Furthermore, PCOS is a complicated and diverse endocrine and metabolic dysfunction condition brought on by the advancement of several variables including inflammation, glucose intolerance, oxidative stress, angiogenesis, and elevated androgen levels. The pathophysiology of polycystic ovarian syndrome has been linked to a few mediators, including tumor necrosis factor-alpha, white blood cells, cytokines, neutrophils, or interleukins, and proteins like VEGF, according to research. Additionally, it was shown that PCOS is linked to several risk factors, including irregular menstruation, cell proliferation, cell apoptosis, and several cardiovascular disorders. Inositol, statins, Letrozole, metformin, and a wide range of dietary supplements including vitamin D, calcium, and fatty acids such as omega-3 are all utilized to treat PCOS, according to the review. Clinical investigations utilizing GLP-1 agonists, inhibitors of DPP-4, and SGLT2 antagonists have produced favorable outcomes, but the results are still inconclusive, necessitating larger studies.

Acknowledgment

The pharmacy department at Pranveer Singh Institute of Technology in Kanpur, Uttar Pradesh, India, provided invaluable guidance and support in the writing of this comprehensive strategy review.

Conflicts of interest

The authors state no conflicts of interest exist.

Funding

None.

References

[1] Rosenfield RL, Ehrmann DA. The pathogenesis of polycystic ovary syndrome (PCOS): the hypothesis of PCOS as functional ovarian hyperandrogenism revisited. Endocr Rev 2016;37(5):467–520. https://doi.org/10.1210/er.2015-1104.

[2] De Leo V, Musacchio MC, Cappelli V, Massaro MG, Morgante G, Petraglia F. Genetic, hormonal and metabolic aspects of PCOS: an update. Reprod Biol Endocrinol 2016;14 (1). https://doi.org/10.1186/s12958-016-0173-x.

[3] Khan MJ, Ullah A, Basit S. Genetic basis of polycystic ovary syndrome (PCOS): current perspectives. Appl Clin Genet 2019;12(12):249–60. https://doi.org/10.2147/tacg. s200341.

[4] Witchel SF, Oberfield SE, Peña AS. Polycystic ovary syndrome: pathophysiology, presentation, and treatment with emphasis on adolescent girls. J Endocr Soc 2019;3 (8):1545–73. https://doi.org/10.1210/js.2019-00078.

[5] Deswal R, Narwal V, Dang A, Pundir CS. The prevalence of polycystic ovary syndrome: a brief systematic review. J Hum Reprod Sci 2020;13(4):261–71. https://doi.org/ 10.4103/jhrs.JHRS_95_18.

[6] Bharali MD, Rajendran R, Goswami J, Singal K, Rajendran V. Prevalence of polycystic ovarian syndrome in India: a systematic review and meta-analysis. Cureus 2022. https:// doi.org/10.7759/cureus.32351.

[7] Repaci A, Gambineri A, Pasquali R. The role of low-grade inflammation in the polycystic ovary syndrome. Mol Cell Endocrinol 2011;335(1):30–41. https://doi.org/10.1016/j. mce.2010.08.002.

[8] Rudnicka E, Suchta K, Grymowicz M, Calik-Ksepka A, Smolarczyk K, Duszewska AM, Smolarczyk R, Meczekalski B. Chronic low grade inflammation in pathogenesis of PCOS. Int J Mol Sci 2021;22(7). https://doi.org/10.3390/ijms22073789.

[9] Aboeldalyl S, James C, Seyam E, Ibrahim EM, Shawki HE-D, Amer S. The role of chronic inflammation in polycystic ovarian syndrome—a systematic review and meta-analysis. Int J Mol Sci 2021;22(5). https://doi.org/10.3390/ijms22052734.

[10] Ndefo UA, Eaton A, Green MR. Polycystic ovary syndrome: a review of treatment options with a focus on pharmacological approaches. PT 2013;38(6):336–55.

[11] Polycystic Ovary Syndrome (PCOS). National Health Portal of India. Nhp.gov.in; 2015. https://www.nhp.gov.in/disease/endocrinal/ovaries/polycystic-ovary-syndrome-pcos. [accessed 22.05.23].

[12] Sidra S, Tariq MH, Farrukh MJ, Mohsin M. Evaluation of clinical manifestations, health risks, and quality of life among women with polycystic ovary syndrome. PLoS One 2019;14(10), e0223329. https://doi.org/10.1371/journal.pone.0223329.

[13] Jitendra Patel A, Pravin Thakor A. Prospective use of Tephrosia Purpurea in remedial treatment of PCOS: study in Wistar rat. Int Res J Biol Sci 2012;1(3):1–6.

[14] Pereira SS, Alvarez-Leite JI. Low-grade inflammation, obesity, and diabetes. Curr Obes Rep 2014;3(4):422–31. https://doi.org/10.1007/s13679-014-0124-9.

[15] Maiorino MI, Bellastella G, Giugliano D, Esposito K. From inflammation to sexual dysfunctions: a journey through diabetes, obesity, and metabolic syndrome. J Endocrinol Invest 2018;41(11):1249–58. https://doi.org/10.1007/s40618-018-0872-6.

[16] Xiong Y, Liang X, Yang X, Li Y, Wei LN. Low-grade chronic inflammation in the peripheral blood and ovaries of women with polycystic ovarian syndrome. Eur J Obstet Gynecol Reprod Biol 2011;159(1):148–50. https://doi.org/10.1016/j.ejogrb.2011.07.012.

[17] Boulman N, Levy Y, Leiba R, Shachar S, Linn R, Zinder O, Blumenfeld Z. Increased C-reactive protein levels in the polycystic ovary syndrome: a marker of cardiovascular disease. J Clin Endocrinol Metab 2004;89(5):2160–5. https://doi.org/10.1210/jc.2003-031096.

[18] Rudnicka E, Kunicki M, Suchta K, Machura P, Grymowicz M, Smolarczyk R. Inflammatory markers in women with polycystic ovary syndrome. Biomed Res Int 2020;2020:1–10. https://doi.org/10.1155/2020/4092470.

[19] Monteiro R, Azevedo I. Chronic inflammation in obesity and the metabolic syndrome. Mediators Inflamm 2010;2010:1–10. https://doi.org/10.1155/2010/289645.

[20] Mažibrada I, Djukić T, Perović S, Plješa-Ercegovac M, Plavšić L, Bojanin D, Bjekić-Macut J, Simić PD, Simić T, Savić-Radojević A, Mastorakos G, Macut D. The association of hs-CRP and fibrinogen with anthropometric and lipid parameters in non-obese adolescent girls with polycystic ovary syndrome. J Pediatr Endocrinol Metab 2018. https://doi.org/10.1515/jpem-2017-0511.

[21] Marciniak A, Nawrocka Rutkowska J, Brodowska A, Wiśniewska B, Starczewski A. Cardiovascular system diseases in patients with polycystic ovary syndrome—the role of inflammation process in this pathology and possibility of early diagnosis and prevention. Ann Agric Environ Med 2016;23(4):537–41. https://doi.org/10.5604/12321966.1226842.

[22] Duleba AJ, Dokras A. Is PCOS an inflammatory process? Fertil Steril 2012;97(1). https://doi.org/10.1016/j.fertnstert.2011.11.023.

[23] Goodarzi MO, Dumesic DA, Chazenbalk G, Azziz R. Polycystic ovary syndrome: etiology, pathogenesis and diagnosis. Nat Rev Endocrinol 2011;7(4):219–31. https://doi.org/10.1038/nrendo.2010.217.

[24] Huang A, Brennan K, Azziz R. Prevalence of hyperandrogenemia in the polycystic ovary syndrome diagnosed by the National Institutes of Health 1990 criteria. Fertil Steril 2010;93(6):1938–41. https://doi.org/10.1016/j.fertnstert.2008.12.138.

[25] Diamanti-Kandarakis E, Kouli CR, Bergiele AT, Filandra FA, Tsianateli TC, Spina GG, Zapanti ED, Bartzis MI. A survey of the polycystic ovary syndrome in the Greek Island of Lesbos: hormonal and metabolic profile. J Clin Endocrinol Metab 1999;84(11):4006–11. https://doi.org/10.1210/jcem.84.11.6148.

[26] Rababa'h AM, Matani BR, Yehya A. An update of polycystic ovary syndrome: causes and therapeutics options. Heliyon 2022;8(10), e11010. https://doi.org/10.1016/j.heliyon.2022.e11010.

[27] Yu O, Christ JP, Schulze-Rath R, Covey J, Kelley A, Grafton J, Cronkite D, Holden E, Hilpert J, Sacher F, Micks E, Reed SD. Incidence, prevalence, and trends in polycystic ovary syndrome diagnosis: a United States population-based study from 2006 to 2019. Am J Obstet Gynecol 2023. https://doi.org/10.1016/j.ajog.2023.04.010. S0002-9378(23)002417.

[28] Azziz R. Polycystic ovary syndrome. Obstet Gynecol 2018;132(2):321–36. https://doi.org/10.1097/aog.0000000000002698.

[29] Singh A, Vijaya K, Sai Laxmi K. Prevalence of polycystic ovarian syndrome among adolescent girls: a prospective study. Int J Reprod Contracept Obstet Gynecol 2018;7(11):4375. https://doi.org/10.18203/2320-1770.ijrcog20184230.

[30] Abdel-Aziz A, El-Sokkary A, El-Refaeey A, El-Sokkary M, Osman H, El-Saeed R. Association between follicle stimulating hormone receptor (FSHR) polymorphism and polycystic ovary syndrome among Egyptian women. Int J Biochem Res Rev 2015; 5(3):198–206. https://doi.org/10.9734/ijbcrr/2015/13896.

[31] Davis SR, Wahlin-Jacobsen S. Testosterone in women—the clinical significance. Lancet Diabetes Endocrinol 2015;3(12):980–92. https://doi.org/10.1016/S2213-8587(15)00284-3.

[32] Rojas J, Chávez M, Olivar L, Rojas M, Morillo J, Mejías J, Calvo M, Bermúdez V. Polycystic ovary syndrome, insulin resistance, and obesity: navigating the pathophysiologic labyrinth. Int J Reprod Med 2014;2014:1–17. https://doi.org/10.1155/2014/719050.

[33] Lim SS, Norman RJ, Davies MJ, Moran LJ. The effect of obesity on polycystic ovary syndrome: a systematic review and meta-analysis. Obes Rev 2012;14(2):95–109. https://doi.org/10.1111/j.1467-789x.2012.01053.x.

[34] Anagnostis P, Tarlatzis BC, Kauffman RP. Polycystic ovarian syndrome (PCOS): long-term metabolic consequences. Metabolism 2018;86(2018):33–43. https://doi.org/10.1016/j.metabol.2017.09.016.

[35] González F, Rote NS, Minium J, Kirwan JP. Reactive oxygen species-induced oxidative stress in the development of insulin resistance and hyperandrogenism in polycystic ovary syndrome. J Clin Endocrinol Metab 2006;91(1):336–40. https://doi.org/10.1210/jc.2005-1696.

[36] Amer SA, Alzanati NG, Warren A, Tarbox R, Khan R. Excess androgen production in subcutaneous adipose tissue of women with polycystic ovarian syndrome is not related to insulin or LH. J Endocrinol 2019;241(1):99–109. https://doi.org/10.1530/JOE-18-0674.

[37] Tilg H, Moschen AR. Inflammatory mechanisms in the regulation of insulin resistance. Mol Med 2008;14(3–4):222–31. https://doi.org/10.2119/2007-00119.tilg.

[38] Guilherme A, Virbasius JV, Puri V, Czech MP. Adipocyte dysfunctions linking obesity to insulin resistance and type 2 diabetes. Nat Rev Mol Cell Biol 2008;9(5):367–77. https://doi.org/10.1038/nrm2391.

[39] da Silva ACR, Ferro JA, Reinach FC, Farah CS, Furlan LR, Quaggio RB, Monteiro-Vitorello CB, Sluys MAV, Almeida NF, Alves LMC, do Amaral AM, Bertolini MC, Camargo LEA, Camarotte G, Cannavan F, Cardozo J, Chambergo F, Ciapina LP, Cicarelli RMB, Coutinho LL. Comparison of the genomes of two Xanthomonas pathogens with differing host specificities. Nature 2002;417(6887):459–63. https://doi.org/10.1038/417459a.

[40] Orio F, Palomba S, Spinelli L, Cascella T, Tauchmanovà L, Zullo F, Lombardi G, Colao A. The cardiovascular risk of young women with polycystic ovary syndrome: an observational, analytical, prospective case-control study. J Clin Endocrinol Metab 2004;89(8):3696–701. https://doi.org/10.1210/jc.2003-032049.

[41] Deligeoroglou E, Vrachnis N, Athanasopoulos N, Iliodromiti Z, Sifakis S, Iliodromiti S, Siristatidis C, Creatsas G. Mediators of chronic inflammation in polycystic ovarian syndrome. Gynecol Endocrinol 2012;28(12):974–8. https://doi.org/10.3109/09513590.2012.683082.

[42] Kaya C, Pabuccu R, Berker B, Satıroglu H. Plasma interleukin-18 levels are increased in the polycystic ovary syndrome: relationship of carotid intima-media wall thickness and cardiovascular risk factors. Fertil Steril 2010;93(4):1200–7. https://doi.org/10.1016/j.fertnstert.2008.10.070.

[43] Escobar-Morreale HF, Botella-Carretero JI, Villuendas G, Sancho J, San-Millán JL. Serum Interleukin-18 concentrations are increased in the polycystic ovary syndrome:

relationship to insulin resistance and to obesity. J Clin Endocrinol Metab 2004;89 (2):806–11. https://doi.org/10.1210/jc.2003-031365.

[44] Glintborg D, Andersen M, Richelsen B, Bruun JM. Plasma monocyte chemoattractant protein-1 (MCP-1) and macrophage inflammatory protein-1α are increased in patients with polycystic ovary syndrome (PCOS) and associated with adiposity, but unaffected by pioglitazone treatment. Clin Endocrinol (Oxf) 2009;71(5):652–8. https://doi.org/10.1111/j.1365-2265.2009.03523.x.

[45] Garg D, Merhi Z. Relationship between advanced glycation end products and steroidogenesis in PCOS. Reprod Biol Endocrinol 2016;14(1). https://doi.org/10.1186/s12958-016-0205-6.

[46] Chang RJ, Cook-Andersen H. Disordered follicle development. Mol Cell Endocrinol 2013;373:51–60. https://doi.org/10.1016/j.mce.2012.07.011.

[47] Erickson GF, Magoffin DA, Lee Jones K. Theca function in polycystic ovaries of a patient with virilizing congenital adrenal hyperplasia. Fertil Steril 1989;51(1):173–6. https://doi.org/10.1016/s0015-0282(16)60450-8.

[48] Erickson G, Yen S. New data on follicle cells in polycystic ovaries: a proposed mechanism for the genesis of cystic follicles. Semin Reprod Med 1984;2(3):231–43. https://doi.org/10.1055/s-2008-1068381.

[49] Toker S, Shirom A, Shapira I, Berliner S, Melamed S. The association between burnout, depression, anxiety, and inflammation biomarkers: C-reactive protein and fibrinogen in men and women. J Occup Health Psychol 2005;10(4):344–62. https://doi.org/10.1037/1076-8998.10.4.344.

[50] Zhao Y, Zhang C, Huang Y, Yu Y, Li R, Li M, Liu N, Liu P, Qiao J. Up-regulated expression of WNT5a increases inflammation and oxidative stress via PI3K/AKT/NF-κB signaling in the granulosa cells of PCOS patients. J Clin Endocrinol Metab 2015;100 (1):201–11. https://doi.org/10.1210/jc.2014-2419.

[51] Regidor P-A, Mueller A, Sailer M, Gonzalez Santos F, Rizo JM, Moreno Egea F. Chronic inflammation in PCOS: the potential benefits of specialized pro-resolving lipid mediators (SPMs) in the improvement of the resolutive response. Int J Mol Sci 2020; 22(1):384. https://doi.org/10.3390/ijms22010384.

[52] Ojeda-Ojeda M, Murri M, Insenser M, Escobar-Morreale H. Mediators of low-grade chronic inflammation in polycystic ovary syndrome (PCOS). Curr Pharm Des 2013;19(32):5775–91. https://doi.org/10.2174/1381612811319320012.

[53] Alanbay I, Ercan CM, Sakinci M, Coksuer H, Ozturk M, Tapan S. A macrophage activation marker chitotriosidase in women with PCOS: does low-grade chronic inflammation in PCOS relate to PCOS itself or obesity? Arch Gynecol Obstet 2012;286(4): 1065–71. https://doi.org/10.1007/s00404-012-2425-0.

[54] dos Santos ACS, Soares NP, Costa EC, de Sá JCF, Azevedo GD, Lemos TMAM. The impact of body mass on inflammatory markers and insulin resistance in polycystic ovary syndrome. Gynecol Endocrinol 2015;31(3):225–8. https://doi.org/10.3109/09513590.2014.976546.

[55] Sproston NR, Ashworth JJ. Role of C-reactive protein at sites of inflammation and infection. Front Immunol 2018;9(754). https://doi.org/10.3389/fimmu.2018.00754.

[56] Stanimirovic J, Radovanovic J, Banjac K, Obradovic M, Essack M, Zafirovic S, Gluvic Z, Gojobori T, Isenovic ER. Role of C-reactive protein in diabetic inflammation. Mediators Inflamm 2022;2022:1–15. https://doi.org/10.1155/2022/3706508.

[57] Rostamtabar M, Esmaeilzadeh S, Tourani M, Rahmani A, Baee M, Shirafkan F, Saleki K, Mirzababayi SS, Ebrahimpour S, Nouri HR. Pathophysiological roles of chronic

low-grade inflammation mediators in polycystic ovary syndrome. J Cell Physiol 2020; 236(2):824–38. https://doi.org/10.1002/jcp.29912.

[58] Lin Y-S, Tsai S-J, Lin M-W, Yang C-T, Huang M-F, Wu M-H. Interleukin-6 as an early chronic inflammatory marker in polycystic ovary syndrome with insulin receptor substrate-2 polymorphism. Am J Reprod Immunol 2011;66(6):527–33. https://doi.org/ 10.1111/j.1600-0897.2011.01059.x.

[59] Khanna D, Khanna S, Khanna P, Kahar P, Patel BM. Obesity: a chronic low-grade inflammation and its markers. Cureus 2022;14(2). https://doi.org/10.7759/cureus.22711.

[60] Ansari S, Haboubi H, Haboubi N. Adult obesity complications: challenges and clinical impact. Ther Adv Endocrinol Metab 2020;11, 204201882093495. https://doi.org/ 10.1177/2042018820934955.

[61] Gonzalez F, Thusu K, Abdel-Rahman E, Prabhala A, Tomani M, Dandona P. Elevated serum levels of tumor necrosis factor alpha in normal-weight women with polycystic ovary syndrome. Metabolism 1999;48(4):437–41. https://doi.org/10.1016/s0026-0495 (99)90100-2.

[62] Yamamoto Y, Kuwahara A, Taniguchi Y, Yamasaki M, Tanaka Y, Mukai Y, Yamashita M, Matsuzaki T, Yasui T, Irahara M. Tumor necrosis factor alpha inhibits ovulation and induces granulosa cell death in rat ovaries. Reprod Med Biol 2014;14(3):107–15. https:// doi.org/10.1007/s12522-014-0201-5.

[63] Decara J, Rivera P, López-Gambero AJ, Serrano A, Pavón FJ, Baixeras E, Rodríguez de Fonseca F, Suárez J. Peroxisome proliferator-activated receptors: experimental targeting for the treatment of inflammatory bowel diseases. Front Pharmacol 2020;11. https://doi. org/10.3389/fphar.2020.00730.

[64] Zhang J, Zhang Y, Xiao F, Liu Y, Wang J, Gao H, Rong S, Yao Y, Li J, Xu G. The peroxisome proliferator-activated receptor γ agonist pioglitazone prevents NF-κB activation in cisplatin nephrotoxicity through the reduction of p65 acetylation via the AMPK-SIRT1/p300 pathway. Biochem Pharmacol 2016;101:100–11. https://doi.org/ 10.1016/j.bcp.2015.11.027.

[65] Wang Y, Li N, Wang Y, Zheng G, An J, Liu C, Wang Y, Liu Q. NF-κB/p65 competes with peroxisome proliferator-activated receptor gamma for transient receptor potential channel 6 in hypoxia-induced human pulmonary arterial smooth muscle cells. Front Cell Dev Biol 2021;9. https://doi.org/10.3389/fcell.2021.656625.

[66] Wu Z, Fang L, Li Y, Yan Y, Thakur A, Cheng J-C, Sun Y-P. Association of circulating monocyte chemoattractant protein-1 levels with polycystic ovary syndrome: a meta-analysis. Am J Reprod Immunol 2021;86(2), e13407. https://doi.org/10.1111/aji.13407.

[67] Ciaraldi TP, Aroda V, Mudaliar SR, Henry RR. Inflammatory cytokines and chemokines, skeletal muscle and polycystic ovary syndrome: effects of pioglitazone and metformin treatment. Metabolism 2013;62(11):1587–96. https://doi.org/10.1016/j.metabol.2013. 07.004.

[68] Wang L, Qi H, Baker PN, Zhen Q, Zeng Q, Shi R, Tong C, Ge Q. Altered circulating inflammatory cytokines are associated with anovulatory polycystic ovary syndrome (PCOS) women resistant to clomiphene citrate treatment. Med Sci Monit 2017;23: 1083–9. https://doi.org/10.12659/msm.901194.

[69] Tilg H, Moschen AR. Adipocytokines: mediators linking adipose tissue, inflammation and immunity. Nat Rev Immunol 2006;6(10):772–83. https://doi.org/10.1038/nri1937.

[70] Kukla M, Menżyk T, Dembiński M, Winiarski M, Garlicki A, Bociąga-Jasik M, Skonieczna M, Hudy D, Maziarz B, Kusnierz-Cabala B, Skladany L, Grgurevic I, Wójcik-Bugajska M, Grodzicki T, Stygar D, Rogula T. Anti-inflammatory adipokines:

chemerin, vaspin, omentin concentrations and SARS-CoV-2 outcomes. Sci Rep 2021;11(1), 21514. https://doi.org/10.1038/s41598-021-00928-w.

[71] Zuo T, Zhu M, Xu W. Roles of oxidative stress in polycystic ovary syndrome and cancers. Oxid Med Cell Longev 2016;2016:1–14. https://doi.org/10.1155/2016/8589318.

[72] Hilali N, Vural M, Camuzcuoglu H, Camuzcuoglu A, Aksoy N. Increased prolidase activity and oxidative stress in PCOS. Clin Endocrinol (Oxf) 2013;79(1):105–10. https://doi.org/10.1111/cen.12110.

[73] Agarwal A, Aponte-Mellado A, Premkumar BJ, Shaman A, Gupta S. The effects of oxidative stress on female reproduction: a review. Reprod Biol Endocrinol 2012;10(1):49. https://doi.org/10.1186/1477-7827-10-49.

[74] Victor VM, Rovira-Llopis S, Bañuls C, Diaz-Morales N, Martinez de Marañon A, Rios-Navarro C, Alvarez A, Gomez M, Rocha M, Hernández-Mijares A. Insulin resistance in PCOS patients enhances oxidative stress and leukocyte adhesion: role of myeloperoxidase. PloS One 2016;11(3), e0151960. https://doi.org/10.1371/journal.pone.0151960.

[75] Yang P, Feng J, Peng Q, Liu X, Fan Z. Advanced glycation end products: potential mechanism and therapeutic target in cardiovascular complications under diabetes. Oxid Med Cell Longev 2019;2019:1–12. https://doi.org/10.1155/2019/9570616.

[76] Deepika MLN, Nalini S, Maruthi G, Ramchander V, Ranjith K, Latha KP, Rani VU, Jahan P. Analysis of oxidative stress status through MN test and serum MDA levels in PCOS women. Pak J Biol Sci 2014;17(4):574–7. https://doi.org/10.3923/pjbs.2014.574.577.

[77] Murri M, Luque-Ramírez M, Insenser M, Ojeda-Ojeda M, Escobar-Morreale HF. Circulating markers of oxidative stress and polycystic ovary syndrome (PCOS): a systematic review and meta-analysis. Hum Reprod Update 2013;19(3):268–88. https://doi.org/10.1093/humupd/dms059.

[78] Fenkci V, Fenkci S, Yilmazer M, Serteser M. Decreased total antioxidant status and increased oxidative stress in women with polycystic ovary syndrome may contribute to the risk of cardiovascular disease. Fertil Steril 2003;80(1):123–7. https://doi.org/10.1016/s0015-0282(03)00571-5.

[79] Wang X, Martindale JL, Liu Y, Holbrook NJ. The cellular response to oxidative stress: influences of mitogen-activated protein kinase signalling pathways on cell survival. Biochem J 1998;333:291–300.

[80] Pollak M. The insulin and insulin-like growth factor receptor family in neoplasia: an update. Nat Rev Cancer 2012;12:159–69.

[81] Akbarali HI. Oxidative stress and ion channels. In: Laher I, editor. Systems biology of free radicals and antioxidants. Berlin, Heidelberg: Springer Berlin Heidelberg; 2014. p. 355–73.

[82] Dalle-Donne I, Aldini G, Carini M, Colombo R, Rossi R, Milzani A. Protein carbonylation, cellular dysfunction, and disease progression. J Cell Mol Med 2006;10:389–406.

[83] Kaya C, Erkan AF, Cengiz SD, Dünder I, Demirel ÖE, Bilgihan A. Advanced oxidation protein products are increased in women with polycystic ovary syndrome: relationship with traditional and nontraditional cardiovascular risk factors in patients with polycystic ovary syndrome. Fertil Steril 2009;92:1372–7.

[84] Abuja PM, Albertini R. Methods for monitoring oxidative stress, lipid peroxidation and oxidation resistance of lipoproteins. Clin Chim Acta 2001;306:1–17.

[85] Torun AN, Vural M, Cece H, Camuzcuoglu H, Toy H, Aksoy N. Paraoxonase-1 is not affected in polycystic ovary syndrome without metabolic syndrome and insulin resistance, but oxidative stress is altered. Gynecol Endocrinol 2011;27:988–92.

[86] Dinger Y, Akcay T, Erdem T, Ilker Saygili E, Gundogdu S. DNA damage, DNA susceptibility to oxidation and glutathione level in women with polycystic ovary syndrome. Scand J Clin Lab Invest 2005;65:721–8.

[87] Corbould A, Kim Y-B, Youngren JF, Pender C, Kahn BB, Lee A, Dunaif A. Insulin resistance in the skeletal muscle of women with PCOS involves intrinsic and acquired defects in insulin signaling. Am J Physiol Endocrinol Metab 2005;288(5):E1047–54. https://doi.org/10.1152/ajpendo.00361.2004.

[88] Dunaif A, Wu X, Lee A, Diamanti-Kandarakis E. Defects in insulin receptor signaling in vivo in the polycystic ovary syndrome (PCOS). Am J Physiol Endocrinol Metab 2001;281(2):E392–9. https://doi.org/10.1152/ajpendo.2001.281.2.e392.

[89] Diamanti-Kandarakis E. Insulin resistance in PCOS. Endocrine 2006;30(1):13–8. https://doi.org/10.1385/endo:30:1:13.

[90] Yilmaz O, Calan M, Kume T. The relationship between serum lipocalin-2 levels and insulin resistance in patients with PCOS. Endocr Abstr 2014. https://doi.org/10.1530/endoabs.35.p664.

[91] Peitsidis P, Agrawal R. Role of vascular endothelial growth factor in women with PCO and PCOS: a systematic review. Reprod Biomed Online 2010;20(4):444–52. https://doi.org/10.1016/j.rbmo.2010.01.007.

[92] Tal R, Seifer DB, Arici A. The emerging role of angiogenic factor dysregulation in the pathogenesis of polycystic ovarian syndrome. Semin Reprod Med 2015;33(3):195–207 [Thieme Medical Publishers].

[93] Chen AY, Seifer DB, Tal R. The role of angiogenic factor dysregulation in the pathogenesis of polycystic ovarian syndrome. In: Polycystic ovary syndrome: current and emerging concepts. Springer; 2022. p. 449–87.

[94] Sharma PS, Sharma R, Tyagi T. VEGF/VEGFR pathway inhibitors as anti-angiogenic agents: present and future. Curr Cancer Drug Targets 2011;11:624–53. https://doi.org/10.2174/156800911795655985.

[95] Di Pietro M, Pascuali N, Parborell F, Abramovich D. Ovarian angiogenesis in polycystic ovary syndrome. Reproduction 2018;155(5):R199–209.

[96] Patil K, Joseph S, Shah J, Mukherjee S. An integrated in silico analysis highlighted angiogenesis regulating miRNA-mRNA network in PCOS pathophysiology. J Assist Reprod Genet 2022;39(2):427–40.

[97] Ma T, Cui P, Tong X, Hu W, Shao LR, Zhang F, Li X, Feng Y. Endogenous ovarian angiogenesis in polycystic ovary syndrome-like rats induced by low-frequency electro-acupuncture: the CLARITY three-dimensional approach. Int J Mol Sci 2018;19(11):3500.

[98] Shim WS, Ho IA, Wong PE. Angiopoietin: a TIE(d) balance in tumor angiogenesis. Mol Cancer Res 2007;5:655–65. https://doi.org/10.1158/1541-7786.MCR-07-0072.

[99] Rodriguez Paris V, Bertoldo MJ. The Mechanism of Androgen Actions in PCOS Etiology. Med Sci (Basel) 2019;7(9):89. https://doi.org/10.3390/medsci7090089.

[100] Rashid R, Mir SA, Kareem O, Ali T, Ara R, Malik A, Amin F, Bader GN. Polycystic ovarian syndrome-current pharmacotherapy and clinical implications. Taiwan J Obstet Gynecol 2022;61(1):40–50. https://doi.org/10.1016/j.tjog.2021.11.009.

[101] Ryan GE, Malik S, Mellon PL. Antiandrogen treatment ameliorates reproductive and metabolic phenotypes in the letrozole-induced mouse model of PCOS. Endocrinology 2018;159(4):1734–47.

[102] Lumachi F, Rondinone R. Use of cyproterone acetate, finasteride, and spironolactone to treat idiopathic hirsutism. Fertil Steril 2003;79(4):942–6.

[103] Grundy SM. Obesity, metabolic syndrome, and coronary atherosclerosis. Circulation 2002;105(23):2696–8.

[104] American College of Obstetricians and Gynecologists. ACOG practice bulletin no. 41: polycystic ovary syndrome. Obstet Gynecol 2002;100:1389–402.

[105] Eshre TT, ASRM-Sponsored PCOS Consensus Workshop Group. Consensus on infertility treatment related to polycystic ovary syndrome. Fertil Steril 2008;89(3):505–22.

[106] Nestler JE. Metformin in the treatment of infertility in polycystic ovarian syndrome: an alternative perspective. Fertil Steril 2008;90(1):14–6.

[107] Palomba S, Orio Jr F, Falbo A, Manguso F, Russo T, Cascella T, Tolino A, Carmina E, Colao A, Zullo F. Prospective parallel randomized, double-blind, double-dummy controlled clinical trial comparing clomiphene citrate and metformin as the first-line treatment for ovulation induction in nonobese anovulatory women with polycystic ovary syndrome. J Clin Endocrinol Metab 2005;90(7):4068–74.

[108] Nestler JE. Metformin for the treatment of the polycystic ovary syndrome. N Engl J Med 2008;358(1):47–54.

[109] Aboulghar MA, Mansour RT. Ovarian hyperstimulation syndrome: classifications and critical analysis of preventive measures. Hum Reprod Update 2003;9(3):275–89.

[110] Mathur R, Alexander CJ, Yano J, Trivax B, Azziz R. Use of metformin in polycystic ovary syndrome. Am J Obstet Gynecol 2008;199(6):596–609.

[111] Harborne L, Fleming R, Lyall H, Norman J, Sattar N. Descriptive review of the evidence for the use of metformin in polycystic ovary syndrome. Lancet 2003;361 (9372):1894–901.

[112] Genazzani AD. Inositol as putative integrative treatment for PCOS. Reprod Biomed Online 2016;33(6):770–80.

[113] Gateva A, Unfer V, Kamenov Z. The use of inositol(s) isomers in the management of polycystic ovary syndrome: a comprehensive review. Gynecol Endocrinol 2018; 34(7):545–50.

[114] Gambioli R, Forte G, Aragona C, Bevilacqua A, Bizzarri M, Unfer V. The use of D-chiro-inositol in clinical practice. Eur Rev Med Pharmacol Sci 2021;25(1):438–46.

[115] Siamashvili M, Davis SN. Update on the effects of GLP-1 receptor agonists for the treatment of polycystic ovary syndrome. Expert Rev Clin Pharmacol 2021;14(9):1081–9.

[116] Rasmussen CB, Lindenberg S. The effect of liraglutide on weight loss in women with polycystic ovary syndrome: an observational study. Front Endocrinol 2014;5:140.

[117] Kahal H, Aburima A, Ungvari T, Rigby AS, Coady AM, Vince RV, Ajjan RA, Kilpatrick ES, Naseem KM, Atkin SL. The effects of treatment with liraglutide on atherothrombotic risk in obese young women with polycystic ovary syndrome and controls. BMC Endocr Disord 2015;15(1):1–9.

[118] Anam AK, Inzucchi SE. Newer glucose-lowering medications and potential role in metabolic management of PCOS. In: Polycystic ovary syndrome: current and emerging concepts. Cham: Springer International Publishing; 2022. p. 527–53.

[119] Helvaci N, Yildiz BO. Current and emerging drug treatment strategies for polycystic ovary syndrome. Expert Opin Pharmacother 2023;24(1):105–20.

[120] Hassan S, Shah M, Malik MO, Ehtesham E, Habib SH, Rauf B. Treatment with combined resveratrol and myoinositol ameliorates endocrine, metabolic alterations and perceived stress response in women with PCOS: a double-blind randomized clinical trial. Endocrine 2023;79(1):208–20.

[121] Brenjian S, Moini A, Yamini N, Kashani L, Faridmojtahedi M, Bahramrezaie M, Khodarahmian M, Amidi F. Resveratrol treatment in patients with polycystic ovary syndrome decreased pro-inflammatory and endoplasmic reticulum stress markers. Am J Reprod Immunol 2019;83(1), e13186. https://doi.org/10.1111/aji.13186.

[122] Fadlalmola HA, Elhusein AM, Al-Sayaghi KM, Albadrani MS, Swamy DV, Mamanao DM, El-Amin EI, Ibrahim SE, Abbas SM. Efficacy of resveratrol in women with polycystic ovary syndrome: a systematic review and meta-analysis of randomized clinical trials. Pan Afr Med J 2023;44–134.

[123] Huo P, Li M, Le J, Zhu C, Yao J, Zhang S. Resveratrol improves follicular development of PCOS rats via regulating glycolysis pathway and targeting SIRT1. Syst Biol Reprod Med 2023;69(2):153–65.

[124] Liang Y, Xu ML, Gao X, Wang Y, Zhang LN, Li YC, Guo Q. Resveratrol improves ovarian state by inhibiting apoptosis of granulosa cells. Gynecol Endocrinol 2023;39(1):2181652.

[125] Zanjirband M, Baharlooie M, Safaeinejad Z, Nasr-Esfahani MH. Transcriptomic screening to identify hub genes and drug signatures for PCOS based on RNA-Seq data in granulosa cells. Comput Biol Med 2023;154, 106601.

[126] Lang X, Liu W, Chen Q, Yang X, Chen D, Cheng W. Resveratrol inhibits multiple organ injury in preeclampsia rat model. Acta Biochim Pol 2023;70(1):131–5.

[127] Chiti H, Parsamanesh N, Reiner Ž, Jamialahmadi T, Sahebkar A. Statin therapy and sex hormones. In: Principles of gender-specific medicine. Academic Press; 2023. p. 551–71.

[128] Kodaman PH, Duleba AJ. Statins in the treatment of polycystic ovary syndrome. Semin Reprod Med 2008;26(1):127–38 [©Thieme Medical Publishers].

[129] Sokalska A, Piotrowski PC, Rzepczynska IJ, Cress A, Duleba AJ. Statins inhibit growth of human theca-interstitial cells in PCOS and non-PCOS tissues independently of cholesterol availability. J Clin Endocrinol Metab 2010;95(12):5390–4.

[130] Banaszewska B, Spaczyński R, Pawelczyk L. Statins in the treatment of polycystic ovary syndrome. Ginekol Pol 2010;81(8):618–21.

[131] Puurunen J, Piltonen T, Puukka K, Ruokonen A, Savolainen MJ, Bloigu R, Morin-Papunen L, Tapanainen JS. Statin therapy worsens insulin sensitivity in women with polycystic ovary syndrome (PCOS): a prospective, randomized, double-blind, placebo-controlled study. J Clin Endocrinol Metab 2013;98(12):4798–807.

[132] Sun J, Yuan Y, Cai R, Sun H, Zhou Y, Wang P, Huang R, Xia W, Wang S. An investigation into the therapeutic effects of statins with metformin on polycystic ovary syndrome: a meta-analysis of randomised controlled trials. BMJ Open 2015;5(3), e007280.

[133] Yu J, Yaba A, Kasiman C, Thomson T, Johnson J. mTOR controls ovarian follicle growth by regulating granulosa cell proliferation. PLoS One 2011;6:21415.

[134] Wickenheisser JK, Nelson-DeGrave VL, Hendricks KL, Legro RS, Strauss JF, McAllister JM. Retinoids and retinol differentially regulate steroid biosynthesis in ovarian theca cells isolated from normal cycling women and women with polycystic ovary syndrome. J Clin Endocrinol Metab 2005;90:4858–65.

[135] Hahn S, Backhaus M, Broecker-Preuss M, Tan S, Dietz T, Kimmig R, et al. Retinol-binding protein 4 levels are elevated in polycystic ovary syndrome women with obesity and impaired glucose metabolism. Eur J Endocrinol 2007;157:201–7.

[136] Tan BK, Chen J, Lehnert H, Kennedy R, Randeva HS. Raised serum, adipocyte, and adipose tissue retinol-binding protein 4 in overweight women with polycystic ovary syndrome: effects of gonadal and adrenal steroids. J Clin Endocrinol Metab 2007;92: 2764–72.

[137] Kaya C, Cengiz SD, Satiroglu H. Obesity and insulin resistance associated with lower plasma vitamin B12 in PCOS. Reprod Biomed Online 2009;19:721–6.

[138] Günalan E, Yaba A, Yılmaz B. The effect of nutrient supplementation in the management of polycystic ovary syndrome-associated metabolic dysfunctions: a critical review. J Turk Ger Gynecol Assoc 2018;19(4):220–32. https://doi.org/10.4274/jtgga.2018.0077.

[139] Wood JR, Nelson VL, Ho C, Jansen E, Wang CY, Urbanek M, et al. The molecular phenotype of polycystic ovary syndrome (PCOS) theca cells and new candidate PCOS genes defined by microarray analysis. J Biol Chem 2003;278:26380–90.

[140] Loverro G, Lorusso F, Mei L, Depalo R, Cormio G, Selvaggi L. The plasma homocysteine levels are increased in polycystic ovary syndrome. Gynecol Obstet Invest 2002;53:157–62.

[141] Krul-Poel YH, Snackey C, Louwers Y, Lips P, Lambalk CB, Laven JS, et al. The role of vitamin D in metabolic disturbances in polycystic ovary syndrome: a systematic review. Eur J Endocrinol 2013;169:853–65.

[142] Izadi A, Ebrahimi S, Shirzai S, Taghizadeh S, Parized M, Farzadi L, et al. Hormonal and metabolic effects of coenzyme q10 and/or vitamin E in patients with polycystic ovary syndrome. J Clin Endocrinol Metab 2018;104:319–27.

[143] Romualdi D, Costantini B, Campagna G, Lanzone A, Guido M. Is there a role for soy isoflavones in the therapeutic approach to polycystic ovary syndrome? Results from a pilot study. Fertil Steril 2008;90:1826–33.

[144] Shah KN, Patel SS. Phosphatidylinositide 3-kinase inhibition: a new potential target for the treatment of polycystic ovarian syndrome. Pharm Biol 2016;54:975–83.

[145] Dumesic DA, Abbott DH. Implications of polycystic ovary syndrome on oocyte development. Semin Reprod Med 2008;26:53–61.

[146] Mazloomi S, Sharifi F, Hajihosseini R, Kalantari S, Mazloomzadeh S. Association between hypoadiponectinemia and low serum concentrations of calcium and vitamin D in women with polycystic ovary syndrome. ISRN Endocrinol 2012;2012, 949427.

[147] Tamura H, Nakamura Y, Terron MP, Flores LJ, Manchester LC, Tan DX, et al. Melatonin and pregnancy in the human. Reprod Toxicol 2008;25:291–303.

[148] Guo Y, Qi Y, Yang X, Zhao L, Wen S, Liu Y, et al. Association between polycystic ovary syndrome and gut microbiota. PloS One 2016;11, e0153196.

[149] Duncan WC, Nio-Kobayashi J. Targeting angiogenesis in the pathological ovary. Reprod Fertil Dev 2013;25(2):362. https://doi.org/10.1071/rd12112.

[150] Cetin M, Sahin S. Microparticulate and nanoparticulate drug delivery systems for metformin hydrochloride. Drug Deliv 2015;23(8):2796–805. https://doi.org/10.3109/10717544.2015.1089957.

[151] Butt MA, Shafique HM, Mustafa M, Moghul NB, Munir A, Shamas U, Tabassum S, Kiyani MM. Therapeutic potential of selenium nanoparticles on letrozole-induced polycystic ovarian syndrome in female Wistar rats. Biol Trace Elem Res 2023. https://doi.org/10.1007/s12011-023-03579-2.

[152] Wang D, Guo D, Bi H, Wu Q, Tian Q, Du Y. Zinc oxide nanoparticles inhibit Ca2+-ATPase expression in human lens epithelial cells under UVB irradiation. Toxicol In Vitro 2013;27(8):2117–26. https://doi.org/10.1016/j.tiv.2013.09.015.

[153] DeUgarte CM, Woods KS, Bartolucci AA, Azziz R. Degree of facial and body terminal hair growth in unselected black and white women: toward a populational definition of hirsutism. J Clin Endocrinol Metab 2006;91(4):1345–50. https://doi.org/10.1210/jc.2004-2301.

[154] Hamzavi I, Tan E, Shapiro J, Lui H. A randomized bilateral vehicle-controlled study of eflornithine cream combined with laser treatment versus laser treatment alone for facial hirsutism in women. J Am Acad Dermatol 2007;57(1):54–9.

[155] Galan N. Laparoscopic ovarian drilling to treat PCOS infertility. Verywell Health; 2022. https://www.verywellhealth.com/ovarian-drilling-to-treat-pcos-2616330. [accessed 22.05.23].

[156] Li TC, Saravelos H, Chow MS, Chisabingo R, Cooke ID. Factors affecting the outcome of laparoscopic ovarian drilling for polycystic ovarian syndrome in women with anovulatory infertility. BJOG 1998;105(3):338–44. https://doi.org/10.1111/j.1471-0528.1998.tb10097.x.

[157] Amer SAK. Ovulation induction using laparoscopic ovarian drilling in women with polycystic ovarian syndrome: predictors of success. Hum Reprod 2004;19(8):1719–24. https://doi.org/10.1093/humrep/deh343.

[158] El Mouhayyar C, Riachy R, Khalil AB, Eid A, Azar S. SGLT2 inhibitors, GLP-1 agonists, and DPP-4 inhibitors in diabetes and microvascular complications: a review. Int J Endocrinol 2020;2020, 1762164. https://doi.org/10.1155/2020/1762164.

[159] Gilbert MP, Pratley RE. GLP-1 analogs and DPP-4 inhibitors in type 2 diabetes therapy: review of head-to-head clinical trials. Front Endocrinol 2020;11. https://doi.org/10.3389/fendo.2020.00178.

Index

Note: Page numbers followed by *f* indicate figures and *t* indicate tables.

Printed in the United States
by Baker & Taylor Publisher Services